BALKAN BIODIVERSITY

PATTERN AND PROCESS IN THE EUROPEAN HOTSPOT

Balkan Biodiversity

Pattern and Process in the European Hotspot

Edited by

Huw I. Griffiths†
Department of Geography,
University of Hull, U.K.

Boris Kryštufek
Science and Research Centre Koper,
University of Primorska, Slovenia
Slovenian Museum of Natural History,
Ljubljana, Slovenia

and

Jane M. Reed
Department of Geography,
University of Hull, U.K.

KLUWER ACADEMIC PUBLISHERS

DORDRECHT / BOSTON / LONDON

A C.I.P. Catalogue record for this book is available from the Library of Congress.

ISBN 1-4020-2853-9 (HB)
ISBN 1-4020-2854-7 (e-book)

Published by Kluwer Academic Publishers,
P.O. Box 17, 3300 AA Dordrecht, The Netherlands.

Sold and distributed in North, Central and South America
by Kluwer Academic Publishers,
101 Philip Drive, Norwell, MA 02061, U.S.A.

In all other countries, sold and distributed
by Kluwer Academic Publishers,
P.O. Box 322, 3300 AH Dordrecht, The Netherlands.

Printed on acid-free paper

Cover photos:
Top left: *Dinaromys bogdanovi*, photographed by Alenka Kryštufek
Top right: *Carlina uczkae*, photographed by Boštjan Surina
Middle left: *Proteus anguinus*, photographed by Ciril Mlinar
Middle right: *Monolistra spinosa*, photographed by Boris Sket
Bottom left: *Leptodirus hochenwartii*, photographed by Slavko Polak
Bottom right: *Gyraulus janinensis*, photographed by Mick Frogley

Printed in the Netherlands.

IN MEMORIAM

Huw Idwal Griffiths
(1958 – 2002)

Huw at the ESF Balkan Biodiversity Workshop, Koper, September 2001

This volume has become a memorial to our ebullient and gifted first editor, Dr. Huw I Griffiths, who died at the age of 44 in 2002. In his own words, he regarded the meeting which gave rise to this collection of papers as a 'testimony to his life's work'. The diversity and productivity of his short career was enormous, and we know he will not be forgotten.

David Watts and Alexei Korniushin

Sadly, we would also like to pay tribute to two other contributors who have passed away since the ESF Workshop. The respected biogeographer, Dr. David Watts, was a good friend and colleague of Huw in Hull and died at the age of 68 in 2003. Dr Alexei Korniushin, a young molluscan expert with a growing international reputation, passed away suddenly early in 2004.

CONTENTS

List of Contributors ix

1 Pattern and process in Balkan Biodiversity – an overview 1
Boris Kryštufek and Jane M. Reed

2 The physical geography of the Balkans and nomenclature of place names 9
Jane M. Reed, Boris Kryštufek and Warren J. Eastwood

Part 1: Terrestrial **23**

3 East Mediterranean vegetation and climate change 25
Warren J. Eastwood

4 The Balkans as prime glacial refugial territory of European temperate trees 49
P. Chronis Tzedakis

5 Quaternary biotic interactions in Slovenia and adjacent regions: the vegetation 69
David Watts †

6 A quantitative assessment of Balkan mammal diversity 79
Boris Kryštufek

7 Karyotypic variation in mammals of the Balkan Peninsula 109
Jan Zima

8 Late Pleistocene rodent dispersal in the Balkans 135
Gerhard Storch

9 Early hominids in the Balkans 147
Nena Galanidou

10 The biodiversity of amphibians and reptiles in the Balkan Peninsula 167
Georg Džukić and Miloš L.Kalezić

11 A zoogeographical review of the spiders (Araneae) of the Balkan Peninsula 193
Christo Deltshev

Part 2: Aquatic **201**

12 Distribution pattern of the aquatic fauna of the Balkan Peninsula 203
Petru M. Bănărescu

13 The bivalve mollusc fauna of ancient lakes in the context of the historical biogeography of the Balkan region 219
Alexei V. Korniushin †

14 A faunistic review of the modern and fossil molluscan fauna from Lake Pamvotis, Ioannina, an ancient lake in NW Greece: implications for endemism in the Balkans — 243
Michael R. Frogley and Richard C. Preece

15 Fossil ostracods, faunistics and the evolution of regional biodiversity — 261
Huw I. Griffiths † and Michael R. Frogley

16 The potential of diatoms as biodiversity indicators in the Balkans — 273
Jane M. Reed

17 The current status of Adriatic fish biodiversity — 291
Lovrenc Lipej and Jakov Dulčić

Part 3: Subterranean — **307**

18 A census of the obligate subterranean fauna of the Balkan Peninsula — 309
Boris Sket, Kaloust Paragamian and Peter Trontelj

19 Diversity of Copepoda (Crustacea) in the unsaturated zone of Karstic caves of Slovenia — 323
Anton Brancelj and Tanja Pipan

Subject Index — 333

LIST OF CONTRIBUTORS

Names of corresponding authors are marked with an asterisk in Chapter headings

Petru M. Bănărescu: Institute of Biology, Spl. Independentei 296, Bucharest, Romania. Email: pbana@ibio.ro.

Anton Brancelj: National Institute of Biology, Večna pot 111, 1000 Ljubljana, Slovenia. Email: anton.brancelj@uni-lj.si.

Christo Deltshev: Institute of Zoology, Bulgarian Academy of Science, 1 Tsar Osvodboditel Blvd., Sofia 1000, Bulgaria. Email: zoology@bgcict.acad.bg.

Jakov Dulčić: Institute of Oceanography and Fisheries, 21000 Split, Croatia. Email: dulcic@izor.hr.

Georg Džukić: Institute for Biological Research "Siniša Stanković", 29 Novembra 142, 11060 Belgrade, FR Yugoslavia.

Warren J. Eastwood: School of Geography, Earth and Environmental Sciences, University of Birmingham, Edgbaston, Birmingham B15 2TT. Email: w.j.eastwood@bham.ac.uk..

Michael R. Frogley: Centre for Environmental Research, University of Sussex, Brighton, BN1 9QJ, UK. Email: m.r.frogley@sussex.ac.uk.

Nena Galanidou: Department of History and Archaeology, University of Crete, 74100 Rethymno, Greece. Email: ngalanidou@phl.uoc.gr.

Huw I. Griffiths (deceased): formerly of the Department of Geography, University of Hull, Cottingham Road, Hull, HU6 7RX, UK. Address for correspondence: Jane M Reed (below)

Miloš L. Kalezić: Institute for Biological Research "Siniša Stanković", 29 Novembra 142, 11000 Belgrade, FR Yugoslavia, and Institute of Zoology, Faculty of Biology, Studentski trg 16, 11000 Belgrade, FR Yugoslavia. Email: nkalezic@eunet.yu.

Alexei V. Kornuishin (deceased): formerly of the Institute of Zoology, The National Academy of Sciences, B. Khmelnitsky str. 15, 01601 Kiev, Ukraine. Address for correspondence: Jane M Reed (below)

Boris Kryštufek: Science and Research Centre, University of Primorska, Garibaldijeva 18, 6000 Koper, and Slovenian Museum of Natural History, PO Box 290, 1001 Ljubljana. Email: boris.krystufek@zrs-kp.si.

Lovrenc Lipej: Marine Biology Station, National Institute of Biology, Fornače 41, 6330 Piran, Slovenia. Email: lipej@mbss.org.

Kaloust Paragamian: Natural History Museum of Crete, PO Box 2208, Irakleion, Crete, 71409, Greece.

Tanja Pipan: Karst research Institute ZRC SAZU, Titov trg 2, Postojna, Slovenia. E-mail: pipan@zrc-sazu.si.

Richard C. Preece: University Museum of Zoology, Downing Street, Cambridge, CB2 3EJ, UK. Email: rcp1001@cus.cam.ac.uk.

Jane M. Reed: Department of Geography, University of Hull, Cottingham Road, Hull, HU6 7RX, UK. Email: j.m.reed@hull.ac.uk.

Boris Sket: Biology Department, Biotechnical Faculty, University of Ljubljana, Večna pot 111, PO Box 2995, 1001 Ljubljana, Slovenia. Email: b.sket@uni-lj.si.

Gerhard Storch: Forschungsinstitut Senckenberg, Abteilung Terrestrische Zoologie, Senckenberganlage 25, 60325 Frankfurt am Main, Germany. Email: gstorch@sng.uni-frankfurt.de.

Peter Trontelj: Biology Department, Biotechnical Faculty, University of Ljubljana, Vecna pot 111, PO Box 2995, 1001 Ljubljana, Slovenia.

P. Chronis Tzedakis: School of Geography, University of Leeds, Leeds LS2 9JT, UK. Email: polychrt@geography.leeds.ac.uk.

David Watts (deceased): formerly of the Department of Geography, University of Hull, Cottingham Road, Hull, HU6 7RX, UK. Address for correspondence: Jane M Reed (above)

Jan Zima: Institute of Vertebrate Zoology, Academy of Sciences of the Czech Republic, Květná 8, 60365 Brno, Czech Republic. Email: jzima@brno.cas.cz.

1. PATTERN AND PROCESS IN BALKAN BIODIVERSITY - AN OVERVIEW

BORIS KRYŠTUFEK* AND JANE M. REED

> "On that morning the vizier was tired after a sleepless night, but calm and composed. His eyelids were heavy, his face as if lined with ice in the morning chill ... he was thinking about Bosnia, his distant, hilly and gloomy homeland (there had always been something gloomy about having Bosnia on his mind!) that even the rays of Islam could only partially enlighten, and in which life with no courteousness or homeliness is indigent, deaf and coarse. And how many more such countries must there be in this godly world? How many wild rivers with no bridges or ferries? How many places lacking potable water and mosques without any adornment or beauty?"
> (Andrić, 1951; our translation)

The morose manner in which the grand vizier, Jusuf İbrahim, is reported by the celebrated writer, the Bosnian Serb, Ivo Andrić, to have recalled his Bosnian homeland probably reflects equally the opinions of the two men. In the 19th century, İbrahim had been transferred from Bosnia in his early childhood to İstanbul, the most splendid city of its time, as part of a 'harač', or tax in blood, which was paid in the form of children by Christian families for them to be converted to Islam and trained as yanitsarees. Jusuf İbrahim, the poor Bosnian child, went on to attain the highest position in the Ottoman's court. Both men made successful careers in the outside world, but both apparently harboured deep-seated resentment about their early childhoods in Bosnia.

Not long after Andrić recounted the fate of the grand vizier, the American politician, E. Stillman (1966), described the Balkans in a similar fashion, "The Balkans begin in Istria, the peninsula jutting into the Adriatic Sea on the border of Italy. There the northern Italian landscape of cypress trees, pink and tawny stucco farmhouses and green fields passes shockingly into a savage caricature of itself … All that is easy-going and prettily charming ends in Italy." Stillman evidently did not include sunny Greece in his perception of the Balkan Peninsula, and was referring specifically to his gloomy vision of the Balkan communist countries of Yugoslavia, Albania, Bulgaria, and Romania.

It was not simply the influence of communism which generated feelings of unease amongst foreign visitors to the mountainous Balkans. Earlier on, the environment had already been described as an "aching wilderness, the bare bones of a half-created world" (Durham, 2000) by the adventurous English lady, Edith Durham, who explored Montenegro, Serbia and Albania at the very start of the 20th century. The words of Stillman (1966) have the same resonance, "Even the natural landscape has the aspect of a lunar purgatory: great white, lonely peaks, barren of soil or any green, slag and boulders strewn about like the bones of a monumental antediluvian beast … Balkan in Turkish means mountain, and the Balkans are the ultimate mountain world - dark, passionate, replete with violence. Much of the Balkans is barren (and) the essential character of the region is difficult to define. The common feature is not linguistic … The unity is certainly not political … The region is a kind of minuscule universe."

H.I. Griffiths et al. (eds.), Balkan Biodiversity, 1–8.

Yet, this minuscule universe of dark and passionate landscape, replete with violence, is Europe's most splendid centre, or "hotspot" of biodiversity. As will become evident in this volume, in addition to the outstanding diversity and endemism of its cave and ancient lake environments, its high terrestrial and riverine biodiversity reflects at least in part the importance of the peninsula as a glacial refuge for plant and animal species, and as a crossroads for faunal and floral exchange with Central Europe on the one hand, and Asia Minor on the other.

From a human perspective, however, the region has also been a point of intersection and conflict between a number of very different cultures, ideologies and religions, and as a result has been somewhat retarded in its political development and national maturity. Some parts of the Balkans still have a reputation for being unfriendly, if not downright hostile, to visitors, including natural scientists. This is most regrettable, since the natural sciences initially received a tremendous boost during the 19th century, when the study of natural history formed a vital focus for national identity and was at the heart of the renaissance of the Balkan nation states following their independence from the Ottoman Empire. Naturalists of this period are still considered as outstanding personalities of their time, and their work still forms an essential part of national identity. During the 20th century, however, the entire region entered a phase of endless ethnic and political tension, bloody wars, the repeated intervention of great powers and economic decline, which may be summarised conveniently in the term, balkanisation. This process is sadly evident in the sciences; over the last decade Balkan scientists have worked more prominently in the international arena, but there is often still virtually no contact within the Balkan states between students of different linguistic and political backgrounds.

Within biodiversity research as a whole there is a tendency for specialists in different fields of floral and faunal research to work independently, using techniques which are not necessarily comparable in spite of having common goals. The problem is obviously more acute in the Balkans, where its history is such that no comprehensive study of biodiversity has ever been undertaken beyond the national level. Biodiversity patterns remain poorly documented and the underlying processes are far from understood. Even in comprehensive reviews such as the recent ecological history of the European Mediterranean (Grove & Rackham, 2001), the entire peninsula, with the exception of Greece, is often left blank in spite of its extensive Mediterranean coastal zone.

As a first step towards overcoming the problems caused by a history of lack of communication between Balkan scientists, and to foster an international and multidisciplinary work ethic, the University of Hull, together with the Science and Research Centre of the Republic of Slovenia, Koper, organised an exploratory workshop entitled "Pattern and Process in Balkan Biodiversity". The workshop was generously supported by the European Science Foundation and took place from September 25-28, 2001 in Koper, Slovenia. The town of Koper is conveniently situated in the north-western corner of Istria, on Stillman's peninsula which juts out into the Adriatic Sea, exactly where in his words the Balkans begin.

The workshop brought together 21 delegates from the UK, Germany, Slovenia, the Czech Republic, Croatia, the Federal Republic of Yugoslavia (now known separately as Serbia and Montenegro), Bulgaria, Romania, the Ukraine and Greece, linking in through research collaboration with a much larger group. As its title suggests, its primary aim was to develop a more complete and integrated appreciation of pattern and process in Balkan biodiversity, from both evolutionary and phylogeographic perspectives, to highlight research areas which need addressing, and to formulate plans for future collaborative research. The programme and abstracts were published in the Slovenian journal, *Annales* (*Annals for Istrian and Mediterranean Studies, Seria historia naturalis*, 11, 1[23], 2001), and the series of review papers produced as a result of the meeting are now offered to the international audience in this volume, published by courtesy of Kluwer Academic Publishers.

The volume draws attention to the importance of the Balkans as a centre of biodiversity and, thereby, to its high conservation value. Due to the high species richness of the tropical latitudes, the Balkans have never really figured in global assessments of biodiversity. The classic study of plant diversity by Myers (1988, 1990), for example, confirmed the general latitudinal gradient in biodiversity, in highlighting only four centres of diversity outside the tropics (out of a total of 18), all of which were in non-European Mediterranean climate zones. Along with some neighbouring regions such as Spain and Turkey, the Balkans made the grade in a subsequent study based instead on levels of endemism, wherein the Balkan Mountains, central and southern Greece and Crete were included in a list of 240 centres of plant diversity (International Union for the Conservation of Nature, cited in World Conservation Monitoring Centre, 1992). Since Europe as a whole is low in bird diversity (International Council for Bird Preservation, cited in World Conservation Monitoring Centre, 1992) it could be assumed that the same is true of other faunal groups, but this has not been demonstrated rigorously. Many studies use one or two groups as a proxy for overall biodiversity based on the assumption that there is significant correlation between, for example, endemism and total biodiversity, or bird diversity and mammal/reptile/amphibian diversity, but assessment should ideally be based on the synthesis of detailed data for a range of taxonomic groups (World Conservation Monitoring Centre, 1992). The lack of synthesised data is especially acute for microscopic organisms, which are notoriously difficult to enumerate, or even to identify with confidence to species level (Jarvis, 2000). Previously, the most important review paper to focus on Europe is that of Gaston & David (1994), wherein the significance of the Balkans as a European hotspot was indeed emphasised on the basis of a number of plant and animal groups.

Thus, it was the specific intention of the workshop to include as wide a range of elements as possible of the Balkan flora and fauna. Faunal papers range from the level of the microsopic to mammals and even humans, and floral papers range from algae to vegetation communities. They span both modern and palaeoecological research, and address issues of biodiversity at the level of the gene through to the ecosystem in terrestrial, aquatic and cave (hypogean) ecosystems.

In demonstrating regional patterns of biodiversity, many papers confirm the notion that the Balkan Peninsula, and its mountainous western region in particular, is indeed a hotspot of European biodiversity. This applies not only to terrestrial and riverine groups but also to those of aquatic and hypogean environments. For vegetation, the high plant diversity and endemism of the Mediterranean region as a whole is well attested, but, unlike most of Western Europe, these characteristics extend into the continental interior in the case of the Balkans, where 6,530 species of native plants are currently known (Polunin, cited in Eastwood). For mammals, the group as whole is most species rich in the mountainous regions of the southern Dinarides, the Šara-Pindhos Mts., the Rila-Rhodopes, and the Balkan Mts

(Kryštufek). A similar pattern is demonstrated for the herpetofauna, with the mountains of the southern Dinarides being by far the most species rich (Džukić & Kalezić). In addition to species-level diversity, the Peninsula appears to exhibit a greater degree of distinct intraspecific and interspecific variation than the rest of Europe. The herpetofauna displays marked interspecific plasticity in several important biological phenomena (Džukić & Kalezić), while 25 mammal species out of a total of 123 considered are polymorphic in their karyotypes and 24 species are polytypic (Zima), for example.

The extremely high level of endemism in the Balkans compared to other parts of Europe is perhaps the most striking element of comparative patterns of biodiversity. The volume does not deal in great depth with patterns of vegetation across the Balkans; in the case of animals, however, the patterns of variation again show clear peaks in the western Balkans. 379 spiders out of a total of 1,409 (= 26.9 %) are reported to be endemic to the peninsula (Deltshev), for example, and 28% of amphibians and 21% of reptiles (Džukić & Kalezić). In river environments, the rivers of the Adriatic watershed (in the western Balkans) are the most important centre of freshwater endemism (Bănărescu). More famously, and by something of a geological coincidence, the outstanding endemism of the deep and geologically ancient Balkan lakes (Ohrid and Prespa) (Frogley & Preece; Korniushin; Reed) and the hypogean environment of caves in the Dinarides in particular (Brancelj & Pipan; Sket *et al.*) also points to maximum biodiversity in the western Balkans.

The processes underlying these patterns of biodiversity are more complex to define. High environmental stability, topographic and climatic diversity, and the presence of land bridges are key themes. It is likely that high terrestrial diversity is explained in part by the role of the Balkan Peninsula as a glacial refugium for forest communities and their associated fauna over the *ca.* 1.8 million years of the Quaternary period (Eastwood, Tzedakis, Watts). This hypothesis is supported by evidence from a pollen sequence spanning the last 430,000 years from Ioannina, western Greece, for example, which showed the continuous presence of tree populations in the region during glacial and interglacial phases (Tzedakis). It is also supported by the argument that many of the Balkan endemic spiders have evolved in situ (Deltshev) and, based on karyotypic data for certain mammals, that the Mediterranean region was a centre of endemism rather than simply a source for the postglacial colonisation of Europe (Zima). For fish and hydrobioid snails, Bănărescu identifies seven main biogeographic regions (and two additional lacustrine zones) which have varying affinities to Holarctic *vs.* Palaearctic faunas and differing degrees of endemism, indicating the complexity of their phylogeography and the contribution to biodiversity of the Balkans' connection to different biogeographic realms. On the other hand, Storch presents mammalian fossil evidence which suggests a rather gradual and continuous process of evolutionary change during the Late Quaternary, arguing that direct faunal exchange between Asia Minor and the Balkans did not play a major role in modern rodent biodiversity. This probably reflects the fact that, of the three large southern European peninsulas (Iberian, Italian and Balkan), the Balkan Peninsula is only one which is fairly open to the rest of Europe. The Pyrenees and Alps effectively isolate the other peninsulas and function as a biogeographic filter. Such a view is supported by modern patterns of distribution, wherein both mammals (Kryštufek) and non-endemic spiders (Deltshev) show strong affinities with the European (Palaearctic) faunas.

The diversity of taxa in ancient lake environments has also been linked to long-term environmental stability (Frogley & Preece, Griffiths & Frogley, Korniushin). Levels of endemism for Ohrid are comparable to other ancient lakes such as in the East African Rift Valley, or Lake Baikal, and involve many different taxonomic groups. The lakes of Ohrid and Prespa are unusual in the high number of endemic bivalve species in the genera Pisidium and Dreissena (Korniushin), for example, while Ohrid also exhibits high endemism amongst gastropods (Frogley & Preece). Adaptive radiation in a stable environment is suggested in both cases as an important underlying mechanism for speciation. In addition, Frogley & Preece highlight the potential importance of basin morphometry, arguing that the dominance of eurytopic taxa in shallower lakes is a function of the lack of adaptive radition in the wider and less stable littoral zone, while arguing that the level of endemism has been maintained over time in Ohrid, but has decreased in shallower lakes. In biogeographic terms, Korniushin points to a Paratethian origin for the bivalve fauna. At the microscopic level, the possible complexity of biogeographic origins is stressed in Griffiths & Frogley's review of modern and fossil ostracod data; in a similar vein to Lake Baikal, the authors suggest that taxa of Aegean lakes, and Ohrid in particular, are probably of different ages, the modern fauna being a mixture of ancient relicts and other species which have evolved locally more recently, during the Neogene. Possibly in the same way, the diatom flora of Balkan lakes, and ancient lakes in particular, is reported to be exceptional in the number of endemic taxa; the current level of taxonomic uncertainty (and lack of research) is such, however, that it is impossible as yet to compare diatom biodiversity with confidence between different lakes or regions (Reed).

The biodiversity of the obligate subterranean fauna is perhaps even more spectacular (see Brancelj & Pipan for a review of copepods, and Sket *et al.* for an overall review), with *ca.* 1,000 terrestrial and 680 aquatic species. These environments are famous; Slovenia is well known as the cradle of the biospeleology, and subterranean interstitial aquatic invertebrates were discovered for the first time in the Balkans by S. Karaman in the 1930s. The diversity of the caves is such that they contain several unique phylogenetic relicts which do not occur in caves outside the Balkan Peninsula, such as freshwater sponges, cnidarians, clams, tubeworms, and the only non-American stygobiont amphibian (Sket *et al.*). While the carbonate geology and varied topography of these parts of the Balkans, which led to the formation of large karstic caverns, is an obvious contributory factor in the high hypogean biodiversity, both studies highlight the need for additional data, particularly from caves outside the western Balkans, in order to identify the underlying processes, and also to understand the environmental preferences and range limits, of different taxa.

In terms of global biodiversity, one may argue convincingly that the ancient lake and cave ecosystems are of global significance. In contrast, while the biodiversity of the terrestrial flora and fauna are certainly high compared to other parts of Europe, it has been demonstrated that in the simplest terms (which are those upon which assessments are usually based) Balkan terrestrial biodiversity is indeed relatively modest on a global scale. Kryštufek demonstrates, for example, that mammalian species richness tends to be considerably higher in the neighbouring regions of the Eastern Mediterranean: Anatolia, the Caucasus, Iran, and the Levant, and that Balkan mammal biodiversity (including that of bats) is on a par with other temperate regions of the world.

In the case of the Balkans, however, simple figures of relative species richness do not bring out the real importance of the region in terms of its conservation value. The additional significance of the Balkans as an important region of environmental stability for evolution and the maintenance of biodiversity during Quaternary glacial fluctuations has already been stressed. As in most regions, conservation issues are further complicated by the effects of human impact on the environment. The first significant evidence for human impact in the Balkans may be discerned as early as the Palaeolithic (Galanidou), and the degradation of Mediterranean forest ecosystems in particular has a long history (Grove & Rackham, 2001). In the rest of the peninsula, however, human impact appears to have been fairly minor until well into the Holocene; top predators such as the lion survived in Bulgaria until as late as the Iron Age (Ninov, 1999), for example. More recently, there has been marked acceleration in environmental degradation due in a large part to agricultural intensification, but the Balkans are again outstanding in the degree to which forest ecosystems have survived in many regions, even if in a modified state. Most notably, some of the largest remaining European populations of large carnivores (wolf, lynx and bear) survive in the mountains, which is a direct reflection of the quality (composition and area remaining intact) of the forest communities. The Balkans thus offer great potential at the European scale for conserving the "last untouched" wilderness on the continent. Furthermore, the ability of species in the Balkans to survive dramatic climatic fluctuation has taken on an additional significance with issues surrounding global warming. On the other hand, poor planning of resource exploitation (as was the case with the karstic cave of Crnulja, in Popovo polje, Herzegovina, where extensive colonies of the tubeworm Marifugia were destroyed; *cf.* Sket *et al.*) poses an immediate threat to the future existence of numerous small-range endemics. The recognition of hotspots in species richness should thus be a high priority issue.

The final aim of the workshop was to highlight potential areas for future research. In spite of the foregoing, it was also evident throughout the meeting that, in many respects, we are far from achieving the goal of understanding pattern and process in Balkan biodiversity. This is partly due to differences in the amount of research in different regions, which is itself in part a consequence of recent Balkan political history, resulting in the fragmentation of research effort and lack of financial support. At the most simple level of α-diversity, the need for further taxonomic research to generate additional morphological and genetic data is clear. The problem is particularly evident amongst smaller organisms, where taxonomic groups are constantly being split and new species are being described. The geographical and environmental limits of species ranges are also poorly understood at present, all of which makes regional assessment of patterns of biodiversity difficult to define and prevents us from understanding the underlying processes more fully.

Another obvious conclusion to be drawn was that there is an urgent need for further palaeoenvironmental research in the Balkans. With the exception of Greece, palaeolimnological research is largely in its infancy compared to Western Europe, Africa or Turkey, for example. While the potential may be limited in karstic regions in particular by the lack of suitable sites, the generation of such data would facilitate the understanding of the temporal dynamics and underlying processes of the evolution of biodiversity in the region, both in terms of aquatic ecosystems themselves, and, from the pollen record preserved therein, of the regional terrestrial vegetation. In the context of conservation, the preserved fossil remains in lake or bog sediment records can also provide an unrivalled record of environmental change which, apart from palynological research, has been largely untapped; perhaps the most valuable contribution of diatoms, for example (Reed) is as proxies for ecosystem change, such that they may be used to assess the degree to which aquatic environments are being degraded, or even to identify the loss of entire aquatic ecosystems over time.

The Exploratory Workshop was a resounding success in laying the foundations for future collaborative research. However, less than a year later, on 12th June, 2002, Huw Griffiths, the principal organiser and the inspiration behind the workshop, died of cancer at the age of only 44 years. With him we lost not only a most devoted friend and husband, but also a genuine scientist with an extraordinarily broad range of interests and the enthusiasm to pull together people of widely differing research backgrounds. This tragic loss has affected us all deeply and it is our challenge, and that of other scientists, to continue some of this multidisciplinary work without him. This volume provides only the first glimpse at the complex issues surrounding Balkan biodiversity.

We are most grateful to the European Science Foundation, without whose generous support there would have been no Exploratory Workshop and no book. Particular thanks are due to Prof. Gabor Vida, our delegate from the European Science Foundation, for his genuine interest in the topic and for his active support. The project was administered by the University of Hull, and the meeting was organised by the Science and Research Centre of the Republic of Slovenia, Koper. Additional sponsorship was provided by the Ministry of Science and Technology of the Republic of Slovenia. We thank the anonymous referees for commenting on earlier versions of the manuscripts. Jane Bunting and Mick Frogley are thanked for providing initial assistance with English editing. John Garner (Hull) was of valuable assistance in production of graphics, and Paul McSherry provided initial help with formatting. Mia Asta (Ljubljana) is acknowledged for her efficient help in preparing texts in a camera ready format, supported by funding from by the Science and Research Centre of the Republic of Slovenia Koper. Our thanks are also extended to Dr. Darko Darovec, the director of the Centre, for his interest and continuous support in the organisation of the workshop and production of this book. Finally, our warm thanks are due to all those at Kluwer who provided rapid and helpful advice at every stage in the production process: namely, our Publisher, Henny van Hoogervorst, and her team members Gloria Verhey, Amber Tanghe and Esther Verdries.

EDITORS' NOTE: CONVENTIONS ON CITING DATES AND AGES

For papers which discuss the geological and Pleistocene chronology of events related to Balkan biodiversity, we have attempted to provide consistency by adopting a single convention to cite dates and ages. For specific radiocarbon chronologies, k yr. BP refers to uncalibrated radiocarbon dates and cal. k yr. BP refers to calibrated radiocarbon ages. Where authors have referred to events in more general terms of 'years ago', the convention is ka and Ma for thousands and millions of years ago, respectively.

REFERENCES

Andrić, I. 1951. [*Pripovedka o vezirjevem slonu in druge zgodbe*]. Slovenski knjižni zavod v Ljubljani, Ljubljana. (in Slovene).
Gaston, K.J. & David, R. 1994. Hotspots across Europe. *Biodiversity Letters* 2, 108-116.
Durham, E. 2000. *High Albania*. Phoenix Press, London.
Grove, A.T. & Rackham, O. 2001. *The nature of Mediterranean Europe. An ecological history*. Yale University Press, London.
Jarvis, P.J. 2000. *Ecological Principles and Environmental Issues*. Prentice & Hall, Harlow, Essex.
Myers, N. 1988. Threatened biotas: 'hot spots' in tropical forest. *Environmentalist* 8, 187-208.
Myers, N. 1990. The biodiversity challenge: expanded hot-spots analysis. *Environmentalist* 10, 243-256.
Ninov, L.K. 1999. Vergleichende Untersuchungen zur Jagd und zum Jagdwild während des Neolithikums und Äneolithikums in Bulgarien. In: *The Holocene history of the European vertebrate fauna. Modern aspects of research* (N. Benecke, ed.), pp.323-338. Archäologie in Eurasien, Bd. 6, Verlag Marie Leidorf, Rahden/Westf.
Stillman, E. 1966. *The Balkans*. Life World Library. Time-Life International, Nederland, N.V.
World Conservation Monitoring Centre, 1992. *Global Biodiversity. Status of the Earth's Living Resources*. Chapman & Hall, London.

2. THE PHYSICAL GEOGRAPHY OF THE BALKANS AND NOMENCLATURE OF PLACE NAMES

JANE M. REED*, BORIS KRYŠTUFEK AND WARREN J. EASTWOOD

INTRODUCTION

The complexity of Balkan physical geography, together with its location in an important transition zone of faunal and floral influences, will have influenced greatly the evolution of Balkan biodiversity. This chapter gives a broad description of the geology, topography and climate of the Balkan Peninsula which serves as a reference point for subsequent specialist chapters on the biodiversity of the Balkan flora and fauna. Since some of the geographic boundaries of the Balkans are not well demarcated the chapter begins with a discussion of how the territory is defined. For the international audience, it includes a brief summary of the nomenclature used for geographic features whose names vary locally.

DEFINITION OF THE STUDY REGION

The Balkan Peninsula forms an irregular, inverted triangle of land which extends from Central Europe in the north to the Eastern Mediterranean in the south, and is bounded on most sides by the Adriatic, Ionian, Aegean and Black Seas (Figure 1). While the term *Balkan* is clearly associated with the modern political states of former Yugoslavia, Bulgaria, Albania and Greece, the exact delineation of its boundaries is something of a matter for debate.

The eastern boundary is often defined at the Bosphorus Strait which runs through Istanbul, so as to include the western, 'European' region of Turkey. The boundary then follows the Black Sea coast north; some would limit its extension in the north to the Bulgarian border, whereas others include the territory of the delta of the River Danube (*ie.* the Dobrugea region) in southeastern Romania. Dobrugea forms part of the larger region of Dobrudja (or Dobrudzha) which extends further south into Bulgaria and is located at an interesting intersection of floral and faunal influences from the east, north and south (Storch, this volume). The entire territory used to be part of Romania, but its southern portion was given to Bulgaria following changes in political boundaries after the Second World War (2WW) (Stillman, 1966).

The northern boundaries are not clearly demarcated by mountain chains, so are not clearly defined. The River Danube is usually taken as the main divide. In the northeast, this definition excludes Romania; others prefer to include either the whole of Romania, or the province of Dobrugea (see above) and westwards along the southern foothills of the Carpathians, to encompass the entire Danube delta and valley. The northwestern Balkans is demarcated in part by the Adriatic coast and offshore islands of Croatia, but there is debate over the location of the western boundary on the Adriatic coast, and how best to define the boundary between the foothills of the Alps (Julian Alps) and the Dinarides (Dinaric Alps; see Džukić &

H.I. Griffiths et al. (eds.), Balkan Biodiversity, 9–22.

Kalezić, this volume, for discussion). It depends largely on whether the River Drava (running along the Slovenian border) or River Sava (central Slovenia; rivers not named on Figure 1) is taken as the dividing line; many tend simply to follow the modern political state boundary of Slovenia, together with the peninsula of Istria to the south of it.

Figure 1. Map showing the major political boundaries, topography and rivers of the Balkan Peninsula

As several authors remark, the history of scientific research has often been confined within national boundaries, many of which follow the course of rivers or coastline, so a logical and simple approach is to use political boundaries to define the Balkans. Since the bulk of research has taken place post-2WW, this is valid in spite of dramatic changes to political boundaries in the earlier 20th century. Thus, a common definition of the Balkan Peninsula used by authors here comprises the modern political states of Slovenia, Croatia, Bosnia-Herzegovina, Serbia-Montenegro, Albania, Macedonia, Greece and Bulgaria, often including European Turkey, and often including all or part of Romania. Serbia-Montenegro is equivalent to the territory known as Yugoslavia following the break up of former Yugoslavia in

1992; in February 2003 Montenegro gained virtual independence from Serbia as a federal state only loosely affiliated with the latter. Macedonia is the Former Yugoslav Republic of Macedonia (FYROM), as opposed to the Greek territory of Macedonia.

NOMENCLATURE OF PLACE NAMES

An understanding of the physical geography of the Balkans may be confounded by the bewildering array of terms in different languages for geographical features such as mountains, lakes or national regions. In this volume some non-English names are in common usage so are adopted in preference to the anglicised name; thus, we have the Dinarides (Dinaric Alps), Thraki (Thrace - northeastern Greece and European Turkey, also known in Turkey as Trakya) and the southern Greek peninsula of the Peloponnisos (Peloponnese). In other cases the local nomenclature varies according to different Balkan languages and no single term is well recognised internationally, so the anglicised version has been adopted. Lakes are the most obvious examples; Lake Skutari (also known in English as Scutari), for example, lies on the border of Albania and Montenegro and has local names of Shkodra and Skadar (or Skadarsko jezero), respectively, while Lake Dojran on the border of Macedonia and Greece is similarly known as Dojransko jezero and Doïrani, respectively. Dalmatia (or Dalmacija, Dalmatien, in Croatia), the Balkan Mountains (English, Balkanides; Bulgarian, Stara Planina) and the Pindus (or English, Pindos; Greek, Pindhos). Mountains are other examples. The southern Carpathian Mountains are also known as the Transylvanian Alps in English, and many smaller mountain ranges within the large chains have their own names (*e.g.* the Serbo-Macedonian Massif to the west of the Rhodopes, or the Istranca Mountains in the east of the Rhodopes).

GEOLOGY

Overview

On a geological timescale, changes in land connections and climate associated with long-term latitudinal shifts in continental land masses will have influenced the evolution of the Balkan fauna and flora. Within the Eastern Mediterranean as a whole, key geological issues centre on continental plate movement and the evolution and extent of the Tethys Sea (the precursor of the Mediterranean), which are debated in detail in Dixon & Robertson (1984).

In the more recent past, the current topographic configuration of the Balkans is largely a function of Alpine orogenic activity which culminated in the mid Tertiary. In the simplest possible terms, the Balkans may be classified physiographically as being part of The Alpine Mountains, one of four major physiographic regions of Europe which extends from the Betic Cordillera of southern Spain through to the Caucasus Mountains in the south of the former USSR (Figure 2; Ager, 1980). Alpine uplift began in the Triassic (Early Mesozoic; *ca.* 200 Ma) and climaxed in the mid-Cretaceous (Late Mesozoic; *ca.* 100 Ma) and mid-Tertiary (*ca.* 30 Ma). Late Tertiary (Miocene and Oligocene; *ca.* 25-2 Ma) sediments underwent local deformation thereafter, with thrust often causing overlap over Neogene (Miocene/Oligocene) sediments. The region of Dobrugea is an exception, with an older geology composed of flat-lying Mesozoic (*ca.* 240-65 Ma) and Tertiary (*ca.*

65-2 Ma) sediment resting on older (Carboniferous to Triassic) folded rocks, which is part of the same tectonic division as northwestern Europe and much of Spain (Ager, 1980). The rugged landscape of western Crete is also composed of ancient metamorphic rock, which has been reworked from the sea floor and overlies more recent limestones (Grove & Rackham, 2001).

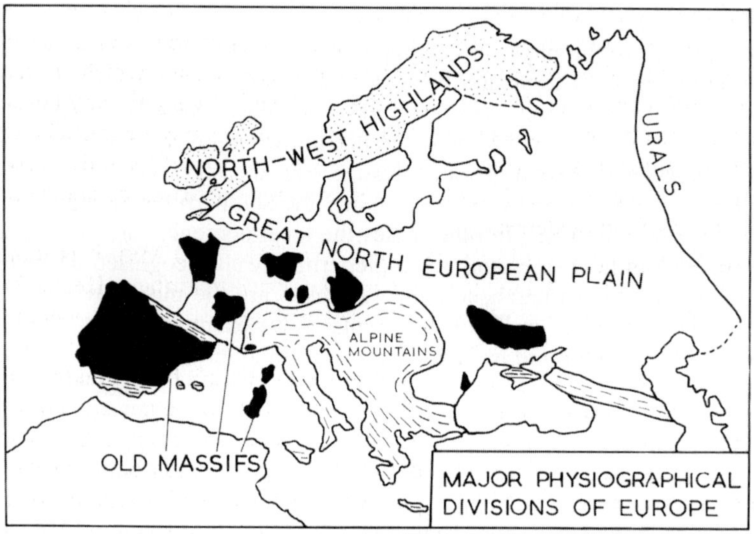

Figure 2. Map showing the major physiological divisions of Europe (reproduced with permission, from Ager, 1980).

Many regions of the Balkans are still tectonically active and major uplift - particularly isostatic - has continued throughout the Quaternary (Ager, 1980). The convergent boundary between the African and Anatolian continental plates runs through southern Greece into Turkey, while a transform fault between the Anatolian and Eurasian continental plates runs in an east-west direction through western Turkey and central Greece (*e.g.* Press & Siever, 1986). The Balkans (and Turkey) are renowned for severe earthquake events (*e.g.* Greece: Karakostas, 2003; Montenegro: Boore *et al.*, 1981; Slovenia: Bajc *et al.* 2001), and landslides are relatively common. Ager (1980) notes ironically that a Romanian earthquake in 1977 destroyed the Institute of Bucharest in which were stored much of the current geological data.

Plate tectonics and the Tethys Sea

The Tethys Sea is thought to have originated during the Permo-Triassic periods of the Late Palaeozoic (upper chronostratigraphic boundary *ca.* 240 Ma) or Early Mesozoic. During its earlier evolution it probably formed a westward-narrowing, wedge-shaped gulf between the Eurasian and African land masses which made up the single supercontinent of Pangaea, in isolation from the super-ocean of Panthalassa (Robertson & Dixon, 1984). It was subsequently destroyed by subduction as the giant super-continent, Pangaea, broke up. The possible configuration of continental plates over time presents an 'abundance of controversies and unsolved problems' (Robertson & Dixon, 1984, p. 56). The situation is even more problematic by the Miocene (Late Tertiary, *ca.* 25 Ma), when the additional complexity of local tectonic movement, uplift and subsidence associated with Alpine orogeny is such that plate interaction effects can no longer be distinguished (Robertson & Dixon, 1984).

Robertson & Dixon (1984) distinguish two major phases in its evolution. The *Palaeotethys* was extant in the Palaeozoic; following ocean closure in the Early Mesozoic (Triassic-Jurassic), with subduction of much of the Tethys oceanic crust, there followed the birth and growth of Mesozoic ocean basins which are termed the *Neotethys*. Estimates as to the width and shape of the Tethys gulf vary considerably, and even the distinction between an old and new Tethys may be artificial (Smith & Spray, 1984), since it is not clear how long the destruction of the older Tethys took; remnants may have persisted until the Tertiary but processes of subduction have removed much of the evidence.

During the Late Mesozoic, Cretaceous, period (*ca.* 130 Ma), it is thought that Pangaea finally separated into the Eurasian *Laurasia* (Northern Hemisphere) and African *Gondwanaland* (Southern Hemisphere), following the opening of the Atlantic Ocean at around 173 Ma. At around the same time, a series of 'microplates' including the Adriatic plate and the Greek block, broke off from Gondwanaland and started to move towards Eurasia (Mršić, 1997). While the details are still uncertain, in the context of Balkan land bridges the most pertinent point is that it is almost universally accepted that there was definitely no connection between the African and Eurasian land masses until the Upper Cretaceous (Late Mezozoic, *ca.* 120 Ma), when two projecting blocks (the Apulian platform and the Rhodope-Moesia) came into glancing contact (Robertson & Dixon, 1984).

During the Tertiary, from *ca.* 70-65 Ma, complex patterns of progressive convergence and collision followed as the African and Eurasian land masses approached each other and there were complex patterns of microplate motion. The leading edge of the African plate was subducted under the European plate, causing major deformation of the floor of the ancient Tethys and, in simple terms, splitting the edge of the European plate into smaller islands and peninsulas which then moved independently (Grove & Rackham, 2001). This culminated at some stage during the Tertiary in the major impingement of the Adriatic-Apulian promintory with Eurasia which caused the latter to begin a phase of anti-clockwise rotation (Robertson & Dixon, 1984), and represents the climax of Alpine orogeny around 20 Ma.

As recently as *ca.* 5.7 - 5.4 Ma, the loss through subsidence of the inflow from the Atlantic into the Mediterranean via the Gibraltar Straits is thought to have turned the Mediterranean Sea into a series of relatively shallow, hypersaline basins (Grove & Rackham, 2001). This event, known as the Messinian Salinity Crisis, will have had huge impact on Balkan ecology, being accompanied by major erosion of river valleys and caves in addition to causing a complete change in the ecology of the marine waters.

TOPOGRAPHY AND LOCAL GEOLOGICAL VARIATION

Mountains and lowlands

The name *Balkan* itself comes from a Turkish word for a forested mountain (Hupchick & Cox, 2001). Almost 70% of the Balkan Peninsula is high relief, comprising mountain chains rising from narrow strips along the coasts of the Adriatic, Ionian and Aegean Seas of the northern Mediterranean Basin and, to the east, the Black Sea which adjoins the Mediterranean via the Bosphorus Strait and Sea of Marmara in Turkey (Figure 1). Apart from Croatia (1,830 m a.s.l.) and Bosnia-Herzegovina (2,386 m a.s.l.), the modern Balkan states all contain highest mountain peaks of 2,500 - 3,000 m a.s.l. The highest peak is the Rila (2,926 m a.s.l.) in the north of the Rhodopes chain (Bulgaria). More generally, there is a contrast in relief between the eastern and western Balkans. In the east, the slopes of the Balkan Mountains are very gentle, such that they do not appear to be real mountains when viewed from the Danube (Furlan, 1977), although they do give way to high ranges to the south. In the west, the Dinarides and Pindus tend to rise very steeply from the coastal strip, and the boundary between the northern Dinarides and the Pannonian Plain is clearly defined.

The north of the Peninsula contains most of the lowland territory of the Balkans. In the west this is the southernmost extension of the Pannonian Plain of southern Hungary, which includes the river valleys of the Danube tributaries. With the inclusion of Romania, the most extensive area of lowlands is formed by the Danube valley in the east, including the wide plains of the Wallachia region of Romania, and running between the Carpathians and the Balkan Mountains. Other than this, the most extensive lowlands are in central Bulgaria (the Thracian Plain of the River Marissa, north of the Rhodopes) and European Turkey, together with a number of smaller river valleys.

Each of the different mountain chains has a different tectonic and sedimentary history; they are all characterised by complex folding and faulting as a consequence of Alpine orogenesis, but to differing degrees. Outcrops of more ancient, metamorphic, magmatic and sedimentary rocks are common throughout (*e.g.* marble, blueschists, granite, gneiss or sandstone) and vary considerably in their local character. Over 50 different types of rock are known in Slovenia alone (Mršić, 1997). The lowlands are often overlain by thick layers of Late Tertiary or Quaternary sediment.

In the western Balkans, the Dinarides are singled out by the dominance of carbonate rocks; they extend in a southeasterly direction ultimately to continue into the Greek Pindus Mountains and the Greek islands, running parallel and close to the coastline. Although these chains show elements of continuity with each other, a major distinction may be made in geological terms across the transition known as the Scutari-Peć Line which runs through northern Albania (Ager, 1980), south of which the geology is dominated instead by volcanic or metamorphic rocks. The Dinarides, named after Mt. Dinara, Croatia, are one of the most famous karstic regions of the world. The name *karst* itself originates from the Slovenian region of Kras (Italian, Carso) (Kranjc, 1997). Structurally, the mountains are almost a mirror image of the Apennines (Ager, 1980), but the latter do not exhibit the same degree of carbonate development. Structures in the Dinarides all run in a NW-SE direction, with thrusts and folds cut by faults which run mainly on northwest and northeast trending lines. Being composed of soft rock, prone to dissolution, much of the hydrology in the Dinarides is dominated by groundwater flow and many parts, such as in Montenegro and Dalmatia, are characterised by poor soil and a lack of surface water (Ager, 1980). On the other hand, Kranjc (1997) acknowledges this reputation for a harsh landscape, but notes that many regions have sufficient soil formation to support dense forests, while features such as *dolines* (solution subsidence structures) provide patches of highly fertile soil and contribute to regional biodiversity. The *poljes* of the Dinarides and other regions of carbonate geology such as the Peloponnese, below, provide characteristic flat plains within mountainous regions, being formed by subsidence between faults; Grove & Rackham (2001) note that the mountain plains of Crete, which were formed in this manner and lie at an altitude of 680 m a.s.l., grow the best wine on the island.

Moving further south, the Pindus Mountains are dominated by volcanic and metamorphic rocks. These mountains exhibit extremely complex folding or *décollement* of upper rocks over basal rocks, in a similar fashion to the Triassic evaporites of the Jura Mountains in the European Alps, but on a larger scale. The zone from Shkoder, Albania to near the Greek border, for example, displays 'seemingly endless sections in fantastically contorted flysch, radiolarites and thin-bedded pelagic limestones' (Ager, 1980, p. 511). The political boundaries of Greece also include a series of island arcs formed through tectonic activity; the outer series is composed of resistant Triassic and Mesozoic carbonate rocks, running from Albania south through the Peloponnesos and Ionian islands, then east through Crete and Rhodes into southern Anatolia (Grove & Rackham, 2001). To the north of this lie further arcs composed of volcanic islands and, to the north of these, more islands which are dominated by marble, gneiss and other metamorphic rocks.

When the Balkans are defined to include Romania, the eastern Balkans encompasses the southerly *Carpathian Bend* of the Carpathians as it curves around in Romania to cross the Danube into Serbia at the narrow crossing point of the Iron Gates; the Balkan Mountains in Bulgaria are a continuation of this chain. The Carpathians are of particularly complex geology and the origins of the Bend are still something of a mystery (Ager, 1980). Early Palaeozoic green schists, intruded by granites, are overlain by marine and non-marine Carboniferous and Permian clastics and volcanics, followed by Mesozoic shales and sandstones, with limestone restricted to the upper stratigraphic levels. The geology of the Balkan Mountains is more simple as a result of their position relative to the motion of continental microplates, but they still exhibit overthrust lithology and large folds, with outcrops of older rocks such as granite.

Between the Balkan Mountains and Carpathians the region of Dobrugea is characterised by older geology. The Măcin Mountains of northern Dobrugea, for example, are a metamorphic complex dominated by Precambrian schists and amphiboles, flanked by metamorphosed sandstones and shales and overlain by fossiliferous Silurian limestones and pyretic shales. Alpine-type seas flooded the massif in the Early to Mid-Triassic leaving conglomerates, sandstones and red limestones similar to Austria (Ager, 1980), followed by submarine lavas.

To the south of the Balkan Mountains, the Rhodopes run in a similar direction but are again of rather contrasting geological origins; in the west of these mountains, the Serbo-Macedonian Massif (not named on Figure 1) is considered by some to be a separate chain on the grounds that it was much more affected by Alpine orogeny, and has a character more similar to the mountains of the west which it almost meets, in Macedonia. While the Rhodopes may be comparatively stable (Ager, 1980), even they show evidence for being affected by orogenic activity, including metamorphism towards the end of the Palaeozoic. In contrast again to the geology of other chains, they comprise a lower series of crystalline gneiss and migmatites and an upper crystalline series of metamorphosed sedimentary rocks such as the marbles of western Thraki, which are cut by granites.

Finally, in the extreme southeastern Balkans, the Thracian depression of European Turkey provides yet another example of extreme local variability, being the smallest of the European Tertiary basins. Shaly and calcareous Silurian and Devonian rocks are overlain by 3000m of pre-Pliocene calcareous Tertiary sediment. Marine transgression in the Eocene resulted in the formation of conglomerates and fossiliferous limestones, overlain by fresh and brackish Miocene flysch.

In addition to providing high local variability in soil types and microhabitats, the implications of all this for geologically recent Quaternary floral and faunal colonisation routes are clear. The western and northern Balkans are open to western and Central Europe, and the Balkans are connected in the east, via the Bosphorus and island chains, with Asia Minor (the Anatolian region of central Turkey) and thereby to Asia. To the south, the Mediterranean Sea separates the Balkans from northern Africa. On the other hand, the presence of mountain chains, whose orientation varies, may have acted as an important barrier to migration within (*e.g.* see discussion in Tzedakis, this volume).

Caves

The Balkans are rich in caves, which are distributed across the region from Slovenia to Bulgaria. Cave systems are even known below the present Mediterranean Sea surface, and the origins of most are linked to major erosion during the Messinian Salinity Crisis *ca.* 5 Ma (Grove & Rackham, 2001). Those of the Dinarides are probably the most famous. The main karst regions of the Dinarides comprise the high limestone plateaux (or High Karst Zone; Ager, 1980) from Postojna, Slovenia in the north (incorporating the Kras region and extending into Italian territory in the Bay of Trieste), to Lake Skutari in the south, on the northern Albanian border. A second important zone, with the same geomorphological characteristics, is the Dalmatian zone which runs along the Croatian coast. The Dalmatian zone lacks the characteristic deep, eroded valleys of the High Karst, although its elongated offshore islands are a similar form of eroded anticline and syncline. Cave origins are associated with the dissolution and erosion of soft bedrock, and it is in the Kras limestone region where the most spectacular deep valleys, caves and underground streams such as Postojnska jama and the World Heritage site of Škocjanske jame have formed. The limestones are all shallow water deposits, ranging in age from Mid Triassic to Late Cretaceous, and overlie Triassic red beds and a Palaeozoic basement which outcrops in places, such as west of Sarajevo (Bosnia-Herzegovina). Their deposition is related to the presence of a shallow Mesozoic sea; limestones are intercalated in places with flysch (a clastic rock often rich in the reworked fossils of extinct species, produced by the erosion of uprising fold structures during Alpine orogeny), laid down during Eocene marine transgression (Kranjc, 1997), and the alternation between freshwater, brackish and marine limestones reflects the complexity of tectonic activity throughout the Mesozoic and Alpine orogenic phases.

Lakes

Geological processes will have had a major influence on aquatic biodiversity. During the Tertiary, Griffiths & Frogley (this volume), for example, describe how the Tethys was repeatedly isolated from and connected to the *Paratethys*, a series of shallow tectonic basins existing in the foreland of uplifting mountain chains in what is now the northern Balkans (extending from the present-day Pannonian Plain to the Black and Caspian Sea regions). The origins of most Balkan lakes are in fact linked to tectonism. This includes the ancient lakes of Ohrid (Albania-Macedonia), Prespa (Macedonia-Greece-Albania) and Ioannina (northwest Greece), all of which are thought to have formed in the Late Tertiary (2-3 Ma or more). Lake basins of this age are rare globally and their occurrence in the Balkans again singles out the region as a unique region of Europe.

Balkan lakes are of varied origins. The largest lake, Ohrid, is outstanding, being a steep-sided graben formed through rifting in a similar fashion to Lakes Tanganyika and Malawi in the East African Rift system, or Lake Baikal. Most other lakes, including Prespa and Ioannina, are a product of karstic processes and, thus, are a type of lake particularly well represented in the Balkans compared to many other parts of Europe. Prespa and Ioannina are both dissolution basins which have undergone continuous subsidence since their formation (*e.g.* see Frogley & Preece, this volume; Reed, this volume). Examples of more recent subsidence basins include Lake Dojran (Macedonia-Greece) and smaller lakes such as Kastoriás (Greece). While strictly speaking it is outside the confines of the Balkan region, Frogley & Preece (this volume) also discuss the important karstic palaeolake of Pannon in the Pannonian Plain of southern Hungary. As another phenomenon peculiar to carbonate environments, the World Heritage Plitvice National Park system of lakes and waterfalls in Croatia is a gorge-filled tufa-dammed river system. The well-known Balkan lake, Lake Skutari, is an exception, its origins being linked to the closure by dune building of an inlet of the Adriatic Sea; the west side of the lake is a high limestone wall formed by thrusting within the Dinarides (Ager, 1980).

Finally, in contrast to the high relative diversity of karstic lakes, the Balkans lack the numerous glacial lakes of neighbouring temperate Europe and the volcanic crater lakes of Italy and Turkey. Glacial lakes are restricted to small, remote mountain lakes (*e.g.* in Slovenia and Macedonia), which are cirques formed by glacial scouring.

CLIMATE

Overview

The mountains of the Balkans and those of the Alps and Pyrenees further west effectively separate Mediterranean climate zones from those of Europe (Barry & Chorley, 1998), acting as a barrier to the penetration of warm air masses further north. The geographic location of the Balkans (*ca.* 35° - 48° N) falls mainly within the temperate zone (Furlan, 1977). Only the southern reaches fall within the latitudinal range of a typical Mediterranean climate (*ca.* 30° - 35° N) where there is a transition between the subtropical high pressure dry belt to the south, and the belt of prevailing westerly winds to the north. Thus, the regional climate is dominated by the movement of polar air masses but tropical air is also drawn in, in connection with depressions on the polar front.

The mountainous character of the Balkans imparts extreme variability on the climate of the Peninsula (Figure 3), with abrupt transitions and associated variability in vegetation types; see Eastwood (this volume). Climatic diversity is a feature of the Mediterranean as a whole, but few other regions of the world exhibit such strong temperature contrasts as in the Balkans. They are particularly marked in the western Balkans, where mountains rise very steeply from the coast (Furlan, 1977). Furlan (1977) suggests three main Balkan climate zones. The coastal regions and islands enjoy a Mediterranean climate, with mild, wet winters and hot, dry summers; precipitation occurs when westerlies move south in winter, and summer drought occurs when air from the dry belt moves north. Secondly, the climate of the mountainous regions is Alpine, with strong altitudinal shifts in precipitation and temperature. The third zone comprises the northern lowlands which are cut off from the influence of the Mediterranean but completely open to Central and Eastern Europe; these have a continental climate with cold, dry winters, warm summers and high precipitation which peaks in May-June. In addition to this, the northeastern Balkans are also more exposed to northern or Arctic influence from the Black Sea and beyond. On a smaller scale, the aspect of mountains can impart very high local variability due to rain shadow effects, often with a humid climate on one side and an extreme arid climate on the other (Grove & Rackham, 2001).

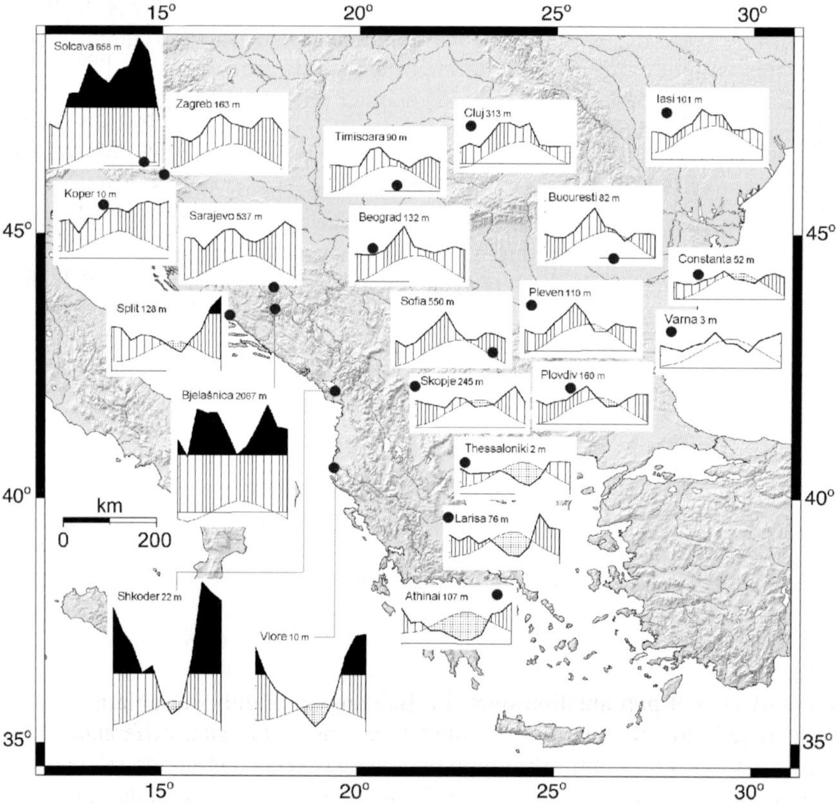

Figure 3. Climogram showing annual patterns of variation in mean monthly temperature (thin line) and precipitation (thick line) from January to December. Data taken from Furlan (1977) and Rodić (1970).

Temperature

Temperature variation across the Balkans is dictated mainly by latitude and height (Furlan, 1977). Temperature contrasts are by far the most marked in winter, when the Mediterranean coastal strips are protected from the invasion of polar and arctic air by the Dinarides, Pindus and Rhodopes mountains. Moving northwest along the Ionian and Adriatic coastline, mean January temperature decreases steadily from 12ºC in northern Crete to around 5ºC in the Gulf of Trieste, for example (e.g. Koper, Slovenia). To the northeast, it tends to decrease by around 0.8ºC per degree of latitude towards the northwest Aegean (e.g. Thessaloniki 5.5°C), and by around 1.3ºC towards the Danube delta, north of Constanta, Romania (-0.8°C). The cold climate of the Danube delta is due both to the lack of shelter afforded to the Black Sea coast, and to differences in the nature of marine currents; there is a warm current from the south, but the Black Sea coast is cooled further by a cold current coming from the north. Low Januarry temperatures are common inland from the Black Sea coast (e.g. Timişoara, Romania, 90 m a.s.l., -1.6°C), until the mountains of the Carpathians and Balkan Mountains provide some resistance to cold air movement. Mean January temperatures in the Pannonian Plain of southern Hungary is around 2°C higher as a result. Southeast of the Carpathians, the coldest region of Europe - Bukovina, Romania - has January temperatures of around -4.1°C for nearby Iasi, at an altitude of 110m a.s.l.

In the mountains, temperature tends to fall with altitude, but the Dinarides exhibit temperature inversions which are typical of the Alps. Mean January temperature for the peaks of the Julian Alps are only around 2.5°C lower than those of the Carpathians, indicating that temperature is distributed fairly evenly inland in the Balkans.

In the summer, the strong contrasts in temperature between Mediterranean and inland zones disappear. In general, summer temperatures are high throughout, and, if anything, temperatures inland tend to be higher than those on the coast (*ca.* 23ºC in northern Crete, 26ºC on the coast of the Gulf of Trieste, 22° - 23°C on the Black Sea coast, 28ºC on the Peloponnese plain [200 m a.s.l.] and 21° in Bukova [100 m a.s.l.], for example).

In the context of Balkan biodiversity, it is often extremes of temperature or the length of the growing season, for example, which control the growth of vegetation and thereby, indirectly, the distribution of fauna. In general, the topographic diversity of the Balkans, and its combination of Mediterranean and continental influences, ensures great diversity in these climatic elements.

Precipitation

Patterns of annual precipitation over the Balkans are highly variable and are again dictated largely by the presence of mountain chains, with major differences between western and eastern coasts, and between lowlands and mountains (Furlan, 1977). The peaks of the southern Dinarides exhibit the highest mean annual precipitation values for Europe, the highest being Crkvice, Montenegro, located at 1,050 m a.s.l., with mean annual precipitation of >4,600 mm $yr.^{-1}$; not marked on Figure 3). This is due to a combination of the orientation of topographic barriers, the steepness of the slopes rising from the coast, and the influence of southwesterly air flow which is very high in humidity. Precipitation decreases from here in all directions; mean values for the Adriatic coast are often <1000 mm $yr.^{-1}$ and for the Pannonian Plain 600 - 700 mm $yr.^{-1}$, for example. The coast of Dobrugea (*ca.* 350 - 400 mm $yr.^{-1}$) is particularly arid.

In terms of seasonal variation, the central and Eastern Mediterranean tend to exhibit winter maxima, but they tend to fall in autumn and/or spring in the northern Balkans (Figure 3), in the transition zone between the Mediterranean summer-dry climate and the continental interior. Much of the 'true' continental interior is under the influence of anticyclones during winter, so spring or summer maxima are common (Barry & Chorley, 1998). In contrast to temperature, it is in summer rather than winter that regional contrasts are most strong. The main areas of summer precipitation are in the north and northeast, on the northern slopes of the Carpathian and Balkan Mountains (e.g. Cluj, Romania or Pleven, Bulgaria), in line with central European patterns of air flow with an intermittent polar influence. The rest of the Balkans are very dry in summer, decreasing southwards to *ca.* 30 mm in Shkoder (Shkodra), to zero July-August precipitation in the Aegean islands. The northern Adriatic, in the lee of the Alps, and the seas west of Crete, are two of the four main Mediterranean regions of cyclogenesis which generate internally most of the depressions in the Mediterranean during winter and spring; from the northern Adriatic, storms move into the mountains of Central Europe. The colder, continental interior exhibits the greatest number of days with snowfall (Furlan, 1977).

Prevailing winds

Patterns of wind direction are again dictated largely by the presence of mountain barriers, and show considerable variability across the Peninsula. Northerly winds should prevail across the Balkans throughout the year, according to the pressure gradient, but this is prevented by topographic effects, such that the northern Balkans is dominated instead by easterlies and westerlies (Furlan, 1977). There are also a number of strong, local (and often specifically seasonal) winds, of variable prevailing direction, which are well enough defined to have earned themselves individual names. The cool, dry *bora*, for example, is the sister of the *mistral* further west (Grove & Rackham, 2001); it originates in Central Europe and is most pronounced near the Gulf of Trieste as it blows at up to 130 km hr^{-1} through a narrow pass separating the Julian Alps from the Dinarides and causes snow drifts inland and storminess on the coast. Incursions of polar air similarly cause the *föhn* to the lee of the Dinarides and Julian Alps, along the Sava River to the Pannonian Plain, affecting Zaghreb and Sarajevo for *ca.* 70 and 100 days, respectively, and causing raised temperatures in winter and, to a lesser extent, summer. The southeastern Pannonian Plain and northern Carpathians are also affected by the *košava* (October to April) which may last from a couple of days up to several weeks; it causes raised temperatures but is dry, so is regarded as a cold wind. In Bulgaria and European Turkey, the cold *poyraz* wind often blows strongly from the Black Sea in March. As a final example, the *burja* of Slovenia is a result of the contact between mild Mediterranean and cold continental climate systems along the Soča and Vipava.

CONCLUSIONS

Following brief discussion of how the Peninsula territory may be delimited, and the nomenclature we have used in this volume for geographic features, this chapter has desribed something of the complexity of Balkan geology, topography and climate, with a focus on the character of its mountains and lowlands, caves and lacustrine environments. The roots of Balkan biodiversity are firmly planted in the geological origins of the Peninsula, being a product of continental plate tectonics and the evolution of the Tethys Sea, followed by a major Alpine phase of mountain building which climaxed around 20 Ma. The region has remained highly active, with the origins of many landforms linked intimately to processes of uplift and subsidence. The topographic and climatic diversity of the Balkans is extreme. This 'wondrous variety' (Grove & Rackham, 2001, p. 37) is a characteristic of the Mediterranean region as a whole. From the descriptions above, however, it is clear that the Balkans contain some of the most extreme environments, and can boast the additional diversity of their cave and lake environments which is unsurpassed in other parts. Certainly, compared to other temperate regions of Europe, the Balkans are quite outstanding. Coupled with the importance of the geographic location in a transitional zone open to a variety of floral and faunal influences, it is perhaps not surprising that the papers which follow in this volume tend to lend support to the notion that the Balkans are indeed the centre of European biodiversity.

ACKNOWLEDGEMENTS

With thanks to Martyn Pedley for his careful criticism of the geological sections of the chapter.

REFERENCES

Ager, D.V. 1980. *The Geology of Europe.* McGraw-Hill, London.

Bajc, J., Aoudia, A., Sarao, A., Suhadolc, P. 2001. The 1998 Bovec-Krn mountain (Slovenia) earthquake sequence. *Geophysical Research Letters* 28, 1839-1842.

Barry, R.G. & Chorley, R.J. 1998. *Atmosphere, Weather and Climate.* 7th edition. Routledge, London.

Boore, D.M., Sims, J.D., Kanamori, H. and Harding, S. 1981. The Montenegro, Yugoslavia, Earthquake of April 15, 1979 - source orientation and strength. *Physics of the Earth and Planetary Interiors* 27, 133-142.

Dixon, J.E. & Robertson, A.H.F. (eds.) 1984. *The Geological Evolution of the Eastern Mediterranean.* Geological Society Special Publication. The Geological Society, London.

Furlan, D. 1977. The Climate of Southeast Europe. In *Climates of Central and Southern Europe. World Survey of Climatology Volume 6.* (C.C. Wallén, ed.), pp.185-235. Elsevier, Amsterdam.

Grove,A.T. & Rackham, O. 2001. *The Nature of Mediterranean Europe. An Ecological History.* Yale University Press, New Haven.

Hupchick, D.P. & Cox, H.E. 2001. *The Palgrave Concise Historical Atlas of The Balkans*. Palgrave, New York.

Karakostas, V.G., Papadimitriou, E.E., Karakaisis, G.F., Papazachos, C.B., Scordilis, E.M., Vargemezis, G. & Aidona, E. 2003. The 2001 Skyros, Northern Aegean, Greece, earthquake sequence: off-fault aftershocks, tectonic implications, and seismicity triggering. *Geophysical Research Letters* 30, article no. 1012.

Kranjc, A. 1997. *Slovene Classical Karst. Kras.* Knanstvenoraziskovalni center SAZU, Založba ZRC: Inštitut za raziskovanje krasa ZRC SAZU, Ljubljana, Slovenia.

Mršić, N. 1997. *Biotic Diversity in Slovenia.* Academy of the Sciences and Arts of Slovenia, Ljubljana, Slovenia.

Press, F. & Siever, R. 1986. *Earth.* 4th edition. W.H. Freeman, New York.

Robertson, A.H.F. & Dixon, J.E. 1984. Introduction: aspects of the geological evolution of the Eastern Mediterranean. In *The Geological Evolution of the Eastern Mediterranean* (J.E. Dixon & A.H.F. Robertson, eds.), Geological Society Special Publication, The Geological Society, London, pp. 1-74.

Rodić, D. 1970. *Geografija Jugoslavije.* Naučna knjiga, Beograd.

Smith, A.G. & Spray, J.G. 1984. A half-ridge transform model for the Hellenic-Dinaric ophiolites. In *The Geological Evolution of the Eastern Mediterranean* (J.E. Dixon & A.H.F. Robertson, eds.), Geological Society Special Publication, The Geological Society, London, pp. 629-644.

Stillman, E. 1966. *The Balkans*. Life World library. Time-Life International, Nederland, N.V.

PART 1: TERRESTRIAL

3. EAST MEDITERRANEAN VEGETATION AND CLIMATE CHANGE

WARREN J. EASTWOOD*

INTRODUCTION

The countries of the Balkans and the eastern Mediterranean, located at the junction of three continental regions, are located in a climatically diverse region that has had a profound effect on the development of the fauna and flora. Zolitschka *et al.* (2000) highlight the importance of palaeoecological research in this region across three broad fronts; these being the potential of obtaining very long records of environmental change from basins that have not been over-ridden by extensive glaciations (unlike northern Europe); the fact that the region is a 'frontier zone' where the tropical (monsoonal) climatic system of northern Africa meets and interacts with the North Atlantic climatic system, and the long history of human occupation and civilisation in this region.

The aim of this paper is to provide an overview of vegetation change in southeast Europe and southwest Asia with a view to examining the linkages and highlighting the gaps in our present understanding between these two vast areas. The palaeoecological review will focus mostly on Holocene vegetation change with a brief overview of pre-Holocene vegetation changes; the latter are dealt more fully by Tzedakis (this volume) and Watts (this volume). Future palaeoecological investigations should adopt a circum-Mediterranean perspective in order to understand climate change and environmental change more fully and to highlight the linkages necessary in interdisciplinary research programmes in this region.

Modern Vegetation: Balkans

The vegetation of southeast Europe and the Balkans is strongly influenced by climate, altitude, soils and human disturbance (Polunin, 1980; Willis, 1994). On a regional scale the vegetation may be divided into two vegetation communities: a Mediterranean vegetation community in the south (comprising elements of warm deciduous forest) and a Central European (continental) plant community in the north (comprising elements of cold deciduous forests, Figure 1); the line on Figure 1 broadly shows the boundary between these two major vegetation communities. The Mediterranean vegetation community comprises evergreen forest (*Pinus halepensis*, evergreen *Quercus* and *Juniperus*), with Mediterranean deciduous forest (deciduous *Quercus*, *Carpinus orientalis*, *Fraxinus ornus*, *Ostrya carpinifolia*, *Pistacia terebinthus* and *Acer*) in inland and cooler areas. This unit is found from sea level to ~700 m a.s.l. (Figure 1), while from 700-1700 m a.s.l. coniferous forest comprising *Abies* and *Juniperus* is found, with sub-alpine and alpine vegetation from 1700-3000 m a.s.l. (units 4, 5 and 6 on Figure 1). Central European (continental) plant community comprising mixed oak-hornbeam forest (*Quercus*, *Carpinus* with *Fagus*, *Ulmus*, *Tilia*, *Fraxinus* and *Acer*) is found from 0-700 m a.s.l. (unit 3 on Figure 1),

H.I. Griffiths et al. (eds.), Balkan Biodiversity, 25–48.

with montane forest comprising mostly *Fagus* and above this coniferous forest comprising *Abies*, *Picea* and *Pinus* from 700-1700 m a.s.l. (unit 4 on Figure 1). Sub-alpine and alpine vegetation together with a scrub layer comprising *Pinus*, *Juniperus* and *Alnus* is found from 1700 to 3000 m a.s.l. (units 7, 8 and 9 on Figure 1).

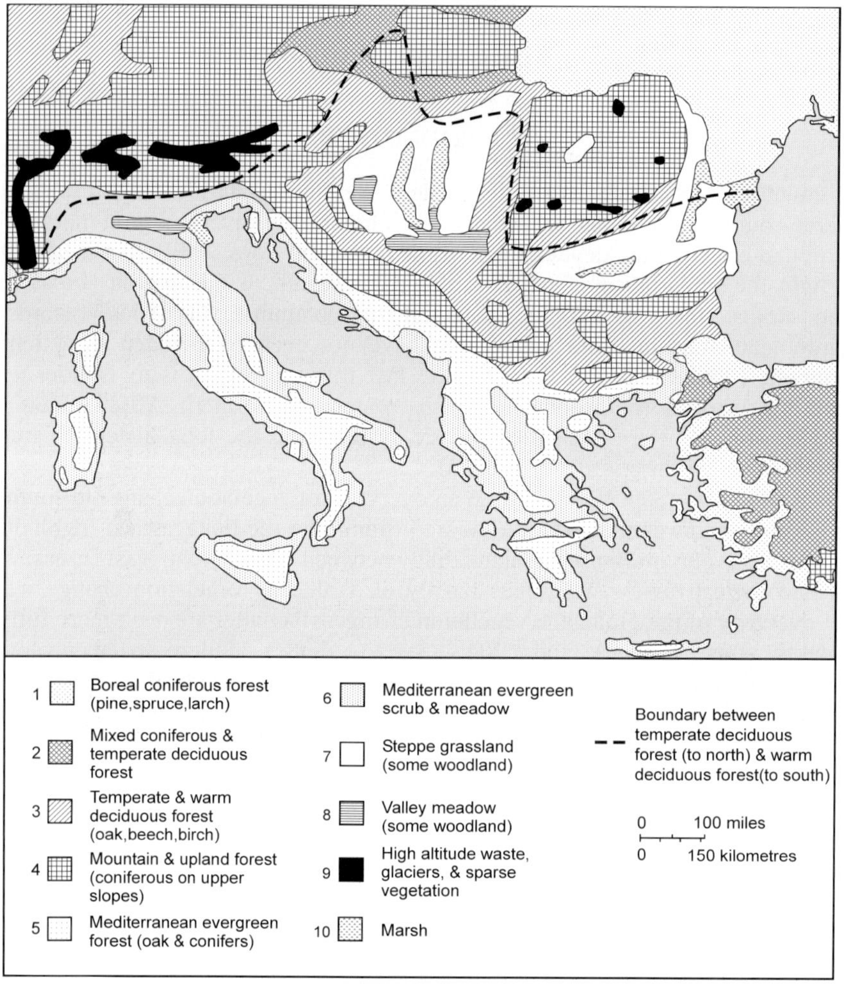

Figure 1. Simplified vegetation map of southeastern Europe (redrawn from Norfalise, 1987).

Modern Vegetation: Southwest Asia

The vegetation of southwest Asia (Asia minor (Anatolia), the 'Near' and 'Middle East') is also strongly influenced by climate, altitude, soils and human disturbance and six major vegetation types can be identified (Figure 2; Kürschner, 1984; van Zeist & Bottema, 1991; Zohary, 1973). Low density xerophytic shrubs, herbs and sedges are found in arid and semi-arid desert regions with precipitation less than 300 mm $yr.^{-1}$ (units 9, 10 and 11 on Figure 2), while steppic areas are

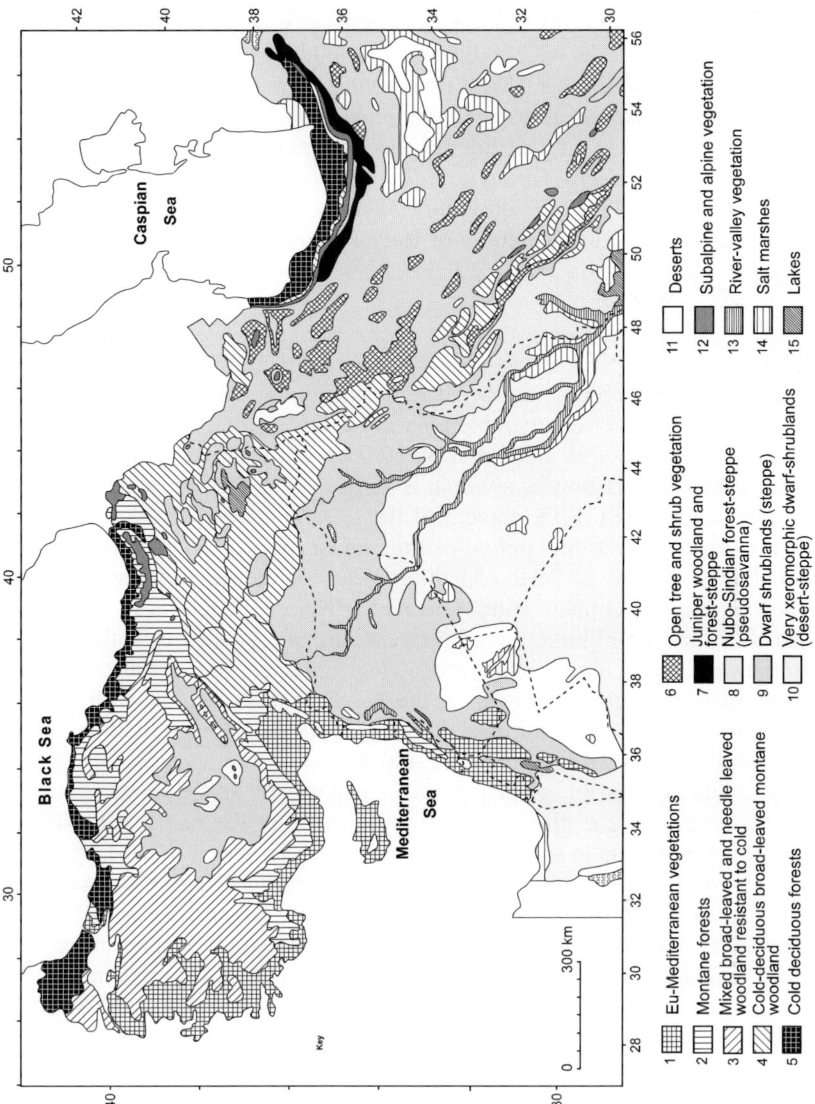

Figure 2. Simplified vegetation map of southwest Asia (redrawn from Norfalise, 1987).

treeless unless along watercourses (*e.g.*, the Tigris and Euphrates rivers, unit 13 on Figure 2). Dwarf shrubs as well as grasses are important in these regions (*Artemisia,* Poaceae/Gramineae*,* Asteraceae/Compositae*, Plantago*). There is also a steppe-forest transitional zone ranging from xeric woodland to steppe with scattered trees and shrubs comprising *Pinus*, deciduous *Quercus*, *Pistacia*, *Juniperus*, *Amygdalus* and Poaceae/Gramineae (units 6, 7 and 8 on Figure 2). Eu-Mediterranean woodland (including a xeric variety) comprising *Quercus ilex*, *Q. coccifera*, *Pinus halepensis*, *Pistacia lentiscus*, *Olea europea* and *Ceratonia* occurs from sea level to ~400 m a.s.l. with precipitation in the range 300-1000 mm $yr.^{-1}$ (unit 1 on Figure 2). Oro or montane Mediterranean forest varies in composition across the region. Deciduous oak is dominant in the Zagros Mountains of Iran (as it is in Greece) (unit 4 on Figure 2), while pine and cedar is widely dispersed in southwest Turkey and the Lebanon at elevations of 400-2000 m a.s.l. with precipitation around ~600 mm $yr.^{-1}$ (unit 2 on Figure 2). Mesic-Euxinian forest with summer-green vegetation comprising *Fagus*, *Carpinus* and *Castanea* with deciduous oak and pine with rhododendron is found near the coasts in northern Turkey and the southern coast of the Caspian Sea (van Zeist & Bottema, 1991) (unit 5 on Figure 2). This Mesic-Euxinian forest in northern Turkey is an extension into Anatolia of Balkan elements of vegetation (van Zeist & Bottema, 1991).

Biodiversity

Mediterranean Europe including the Balkans is a region with a very rich biodiversity, far richer than any other comparable area in Europe, a total of 25,000 plant species for the Mediterranean area alone and 6,530 species of native plants in the Balkans (Polunin, 1980). The region as a whole is surpassed in species richness only by the tropics (Mooney, 1988). Fifty percent of the flora is endemic to the region and the flora has evolved over a long period of time under highly variable climatic regimes. It is possible to identify a 'Palaeo-Mediterranean' contingent that has evolved under pre-Pliocene tropical climatic conditions and a 'Neo- Mediterranean' group that has originated from immigration and speciation since the establishment of a Mediterranean-type climate (Scarascia-Mugnozza *et al.*, 2000). Forest trees represent an important component in the Mediterranean region with 100 species compared to 30 for Central Europe; the genus *Quercus* alone has 20 species in the Mediterranean region. Many Holarctic and Eurasian tree species had their ranges severely contracted during glacial periods in northern and Central Europe. Their continued survival to the present day is attributable to the fact that these tree taxa survived glacial periods in suitable refugial areas in southeast Europe, spreading out once again into central and northern Europe once interglacial climatic conditions and other factors controlling migration were suitable (Bennett *et al.*, 1991; see also Tzedakis, this volume). The Mediterranean floral region is particularly rich in endemics; for example, in Greece one plant in five is an endemic (the small Athos Peninsula has 16 and Mt Olympus has 19 endemics). According to Polunin (1980), the Balkan Peninsula has 700 Mediterranean plant species with one-in-four plants for the Balkan region being endemic. Turkey in particular has convergence from three distinct floristic zones, and has a highly diverse variety of forest which today is limited to *ca.* 26% of the total land area of the country (Scarascia-Mugnozza *et al.*, 2000). Furthermore, modern day Turkey and bordering regions contain the wild relatives of many domesticated food crop plants and are one of the eight major gene centres on earth (Harlan, 1995; Simmonds, 1986). The wild progenitors of such cultivated crop plants as lentil, chickpea, wheat, peach, almond and pistachio are native to Turkey (Kaya & Raynal, 2001). Vertebrate density for the Mediterranean region is also very high, but mammal species have declined due to extinction events and human impact from Neolithic times onwards. Today, there are ~200 species of terrestrial mammals, of which 25% are endemic. Over 340 species of avifauna breed in the Mediterranean region and the region has several RAMSAR wetland sites that are especially important for over-wintering and migratory birds (Figure 3). Lastly, but not least, the Mediterranean and Black Seas as well as the interior lakes of southeast Europe and southwest Asia are important for global fish stocks, local fishing industries and their intrinsic habitat conservation value.

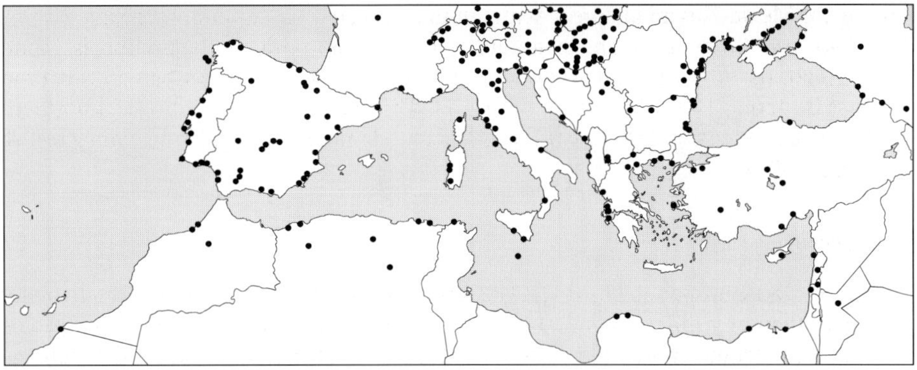

Figure 3. Map of the Mediterranean basin showing RAMSAR sites.

PALAEOECOLOGY OF BALKANS & SOUTHWEST ASIA

Archives as a source for palaeoenvironmental reconstructions

Past environmental changes can be reconstructed at a variety of temporal and spatial scales by a range of 'palaeo' or proxy indicator methods, each of which has certain inbuilt spatial and temporal restrictions and limitations (Flenley, 2003). Pollen analysis (palynology) can be undertaken on a range of deposits including archaeological, cave, peat, lacustrine and marine cores and lends itself in particular to a multi-indicator approach incorporating other allied palaeo indicators such as plant macrofossil, charcoal, invertebrate remains and stable isotope analyses. Palynology, in particular, is only a suitable methodology where sediments are preserved under anaerobic or waterlogged conditions; therefore, vast areas of southwest Asia do not readily yield sediments suitable for pollen analysis. The 'off-site' component of the technique also complements environmental archaeology where the reconstruction of former vegetation patterns associated with archaeological sites can provide a much fuller picture of landscape-human interactions (*e.g.* Atherden *et al.*, 1993; Atherden, 2000; Atherden & Hall, 1999; Eastwood *et al.*, in press, b). Pollen analysis has also been attempted from many 'on-site' archaeological deposits, but the technique in this context is fraught with difficulties and limitations (*cf.* Bottema, 1975). The analysis of strand lines and fossil beach terraces is useful for palaeolimnology and the former extent of basins can be informative for lake level reconstructions and past changes in climate (Roberts *et al.*, 1999; Karabiyikoğlu *et al.*, 1999; Roberts and Wright, 1993). Other proxy indicator methods such as Packrat/Hyrax midden analysis are particularly informative for the dryer parts of the region, but this indicator suffers from patchy preservation, limited geographical distribution and discontinuous records (Fall, 1990). Tree-ring and speleothems have the potential to produce continuous records of environmental change at high resolution, but may be spatially and temporally limited (Kunihlom *et al.*, 1996; Bar-Matthews *et al.*, 1997; Manning *et al.*, 2001). Archaeological and historical records of vegetation and landscape change

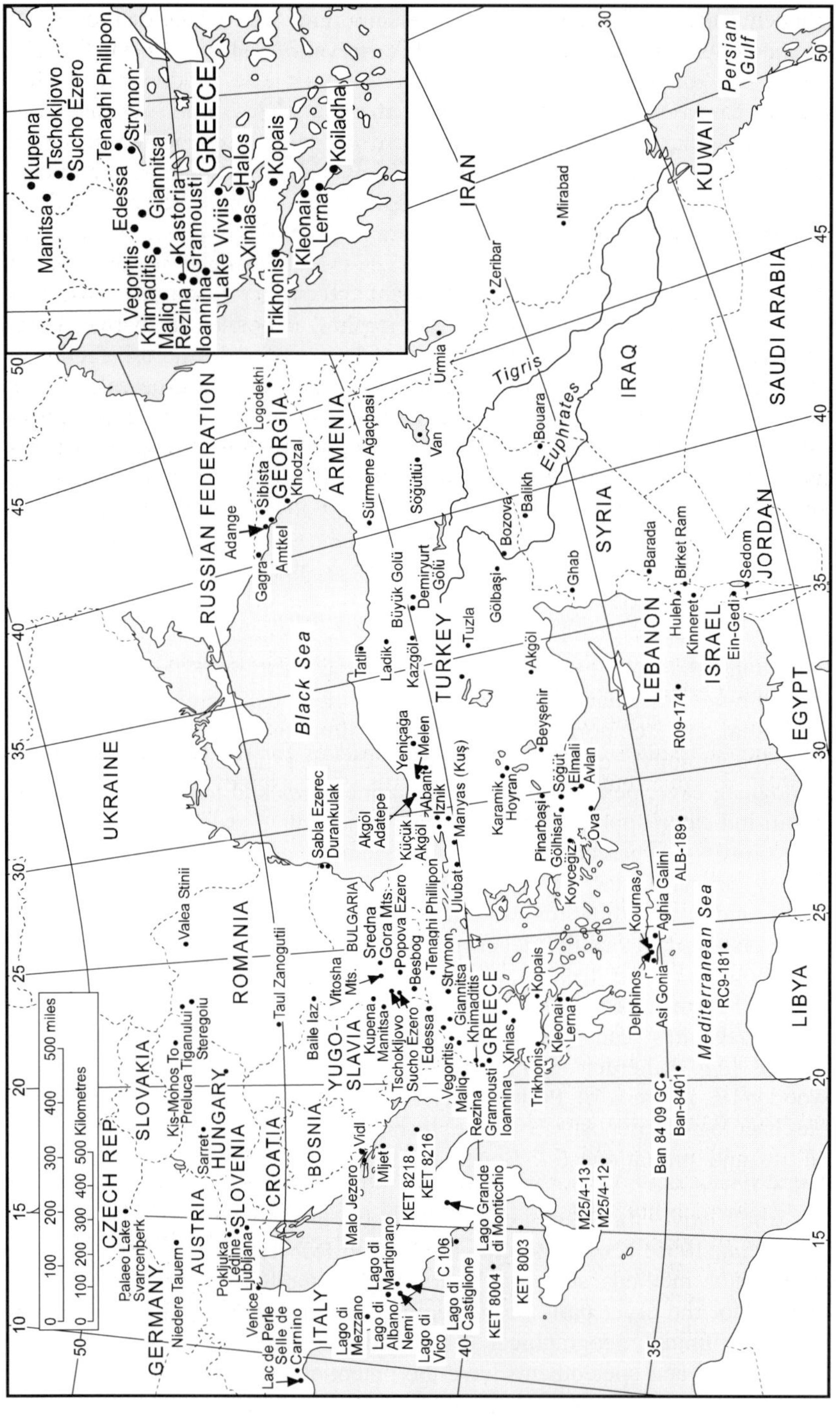

Figure 4. Map of the eastern Mediterranean showing palaeoecological sites.

complement many of these proxy or palaeo methods (Atherden *et al.*, 1993; Atherden, 2000; Atherden and Hall, 1999; Eastwood *et al.*, 1998). Clearly, the greater number of palaeoecological sites for a given area produces a much fuller picture of vegetation change and increases the spatial resolution. Dating events in the palaeoecological record can be a problem for some sites located in the eastern Mediterranean. Very long sequences that are older than ~40 kyr. BP are outside the range of radiocarbon dating and other dating techniques must be employed such as correlation with the oxygen isotope stratigraphy (Tzedakis *et al.*, 1997; Magri & Tzedakis, 2000; Tzedakis *et al.*, 2001). The Mediterranean region has several volcanic provinces comprising active, dormant and extinct volcanoes, most of which have contributed to layers of volcanic ash (tephra) deposited in marine, lacustrine and terrestrial peat deposits. The presence of layers of volcanic ash facilitates the application of tephrochronology that provides an independent timescale, which can pre-date the radiocarbon timescale, and which can, in some instances provide a 'check' on radiocarbon ages, particularly for eruptions whose age is known by other means (Eastwood *et al.*, 1999a; Narcisi & Vezzoli, 1999; Wulf *et al.*, 2001). Long palaeoecological records originating from deep volcanic crater lakes (*e.g.* crater lakes of Italy and the Cappadocian Volcanic Province of Turkey) and tectonic subsidence basins (*e.g.* Ioannina) may contain sequences of laminated sediments which provide an alternative and robust dating framework and yield high resolution environmental change data (Roberts *et al.*, 2001; Frogley *et al.*, 2001). However, volcanic outgassing of old CO_2 may have a deleterious effect on ^{14}C ages for younger time periods (Keenan, 2002).

Figure 4 shows some of the sites that have yielded published pollen information and provides a visual spatial picture of pollen sites for the eastern Mediterranean. Noticeably, the distribution of sites is spatially restricted; the Balkan Peninsula, southwest and northwest Turkey and Italy are relatively well represented, while interior areas of Turkey, the countries of the former Yugoslavia and inland areas of Southwest Asia have yielded relatively fewer sites. However, this inventory is continuously being added to and sites that have been investigated in Albania, Romania and countries of the former Yugoslavia have provided new palaeoecological data (*e.g.* Denèfle *et al.*, 2000; Björkman *et al.*, 2002, 2003). This present inventory, does not, however give an indication of the numerous palaeoecological studies that have been published in 'national' outlets as well as the temporal longevity and temporal quality of the palaeoecological data; many sediment cores that originate from the Balkans, Italy, Turkey and southwest Asia extend back only to the last glacial period; others, however extend much further. This is the single most important aspect of palaeoecological research in this region, compared to northern and Central Europe. The absence of extensive ice sheets has meant the availability of long and continuous terrestrial records some of which record several glacial-interglacial cycles (Zolitschka *et al.*, 2000; van Zeist & Bottema, 1977; Watts *et al.*, 1996a, b; Tzedakis, 1994; Tzedakis *et al.*, 1997; Frogley *et al.*, 1999; Magri, 1999; Magri & Tzedakis, 2000; Tzedakis *et al.*, 2001; Roberts *et al.*, 2001).

PRELUDE TO THE HOLOCENE

Pollen diagrams from southeast Europe and southwest Asia show predominantly an open, discontinuous 'glacial' vegetation comprising *Artemisia*-Chenopodiaceae steppe, which suggests a drier climate, and in this respect the southern Europe and southwest Asia regions has its parallel with much of unglaciated northern Europe. However, unlike much of unglaciated northern Europe, most pollen diagrams from southeast Europe and Turkey show a continuous presence of both coniferous and deciduous tree taxa, the presence and abundance of which varies from site to site (Tzedakis, 1993). For example, at Ljubljana in Slovenia (Willis, 1994; Šercelj, 1966), the last glacial period fluctuated between periods of a predominantly open steppe-dominated landscape with *Pinus* and *Picea* and periods dominated by mixed woodland comprising *Abies*, *Betula*, *Alnus*, *Corylus*, *Quercus*, *Tilia*, *Ulmus*, *Acer*, *Salix* and *Juglans*. Cores from Ioannina in northern Greece show a continuous, yet fluctuating presence of *Quercus*, *Abies*, *Ulmus*, *Pinus*, *Corylus*, *Fagus*, *Juniperus* (Figure 5; Bottema, 1974; Tzedakis, 1993, 1994). In addition to these, sites from the volcanic region of central Italy (Follieri *et al.*, 1998), for example Lago di Monticchio (Figure 6; Watts *et al.*, 1996a, b), record the continuous presence of *Pinus*, *Abies*, *Betula*, *Carpinus, Quercus*, *Alnus*, *Fagus*, *Salix* and *Juniperus*. It has been suggested that pollen present in sediments of glacial age is the product of long distance transport with refugial areas in the Near East as the source. However, there is very little palynological evidence in pollen diagrams from southwest Asia to support this hypothesis, while some of the pollen types present in pollen diagrams from southeast Europe have poor pollen production and dispersal capabilities (*e.g. Tilia*, *Ulmus* and *Fraxinius*) suggesting local/extra local deposition. The more or less continuous presence of pollen of these tree taxa in pre-Holocene sediments suggests that these areas were important glacial refugia for most if not all of these tree taxa. Some pollen diagrams contain 40% AP (a large component of which is coniferous pollen), which suggests that trees were a significant component of the glacial landscape of southeast Europe. Further evidence comes from Willis *et al.* (2000), who studied macrofossil charcoal and pollen from various sequences in Hungary. When these charcoal fragments were identified down to species level, the results confirmed the presence of trees during this period. Combined high resolution pollen and macrofossil analyses at Preluca Tiganului (730 m a.m.s.l.) and pollen analysis at Steregoiu (790 m a.m.s.l.) in north-west Romania show persistently high levels of *Pinus* (>40%) together with macrofossil evidence for *Pinus*, *Salix* and *Betula* (Wohlfarth *et al.*, 2001; Björkman *et al.*, 2002). Thus, the evidence appears to suggest that there was a low but persistent presence of forest trees in southern Europe during the last glacial period, most probably located in small but shifting refugial areas with suitable microclimates at a range of altitudes (Denèfle *et al*, 2000). Indeed, recent palaeoecological data (Willis *et al.*, 2000; Wohlfarth *et al.*, 2001; Björkman *et al.*, 2002; 2003) indicate the presence of woodland in more northerly locations than hitherto supposed and suggest the occurrence of 'cryptic refugia' (*sensu* Stewart & Lister, 2001) or secondary refugia (Brewer *et al.*, 2002; Willis and Whittaker, 2000). See Tzedakis (this volume) for a more detailed discussion of this topic.

Overall, the pollen results for the glacial period in the eastern Balkans (Romania and Bulgaria) suggest that *Picea* was the dominant tree, whereas in the western and southern Balkans *Pinus* was the dominant tree. Although no modern analogue exists for these full glacial assemblages, Willis (1994) suggests that a close approximation might be the southern edge of the present Boreal forest.

As in southeast, Central and northern Europe, pollen diagrams from Turkey and southwest Asia show that the late Pleistocene vegetation comprised *Artemisia*-

Chenopodiaceae steppe throughout the region. In particular, pollen diagrams from the Zagros Mountains region of Iran (*e.g.* Zeribar; Figure 7) suggest that this lowland steppe graded upslope to alpine vegetation, without intervening woodland. This indicates a very arid environment and winter precipitation was most

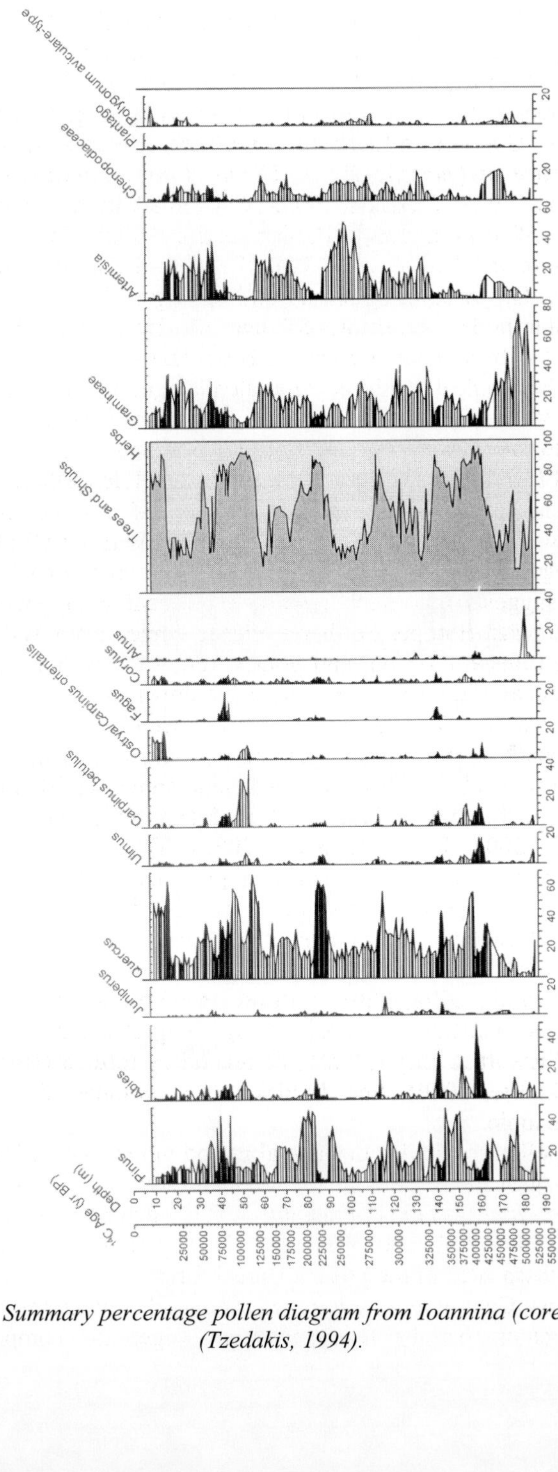

Figure 5. Summary percentage pollen diagram from Ioannina (core 249), Greece (Tzedakis, 1994).

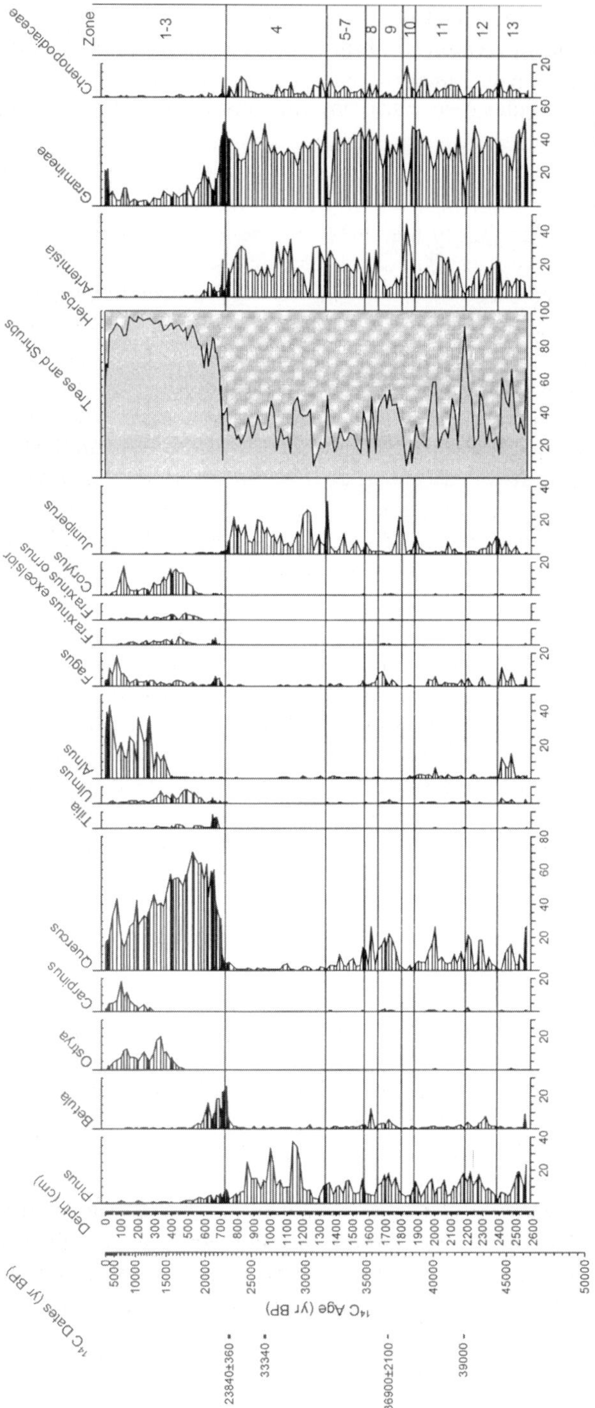

Figure 6. Summary percentage pollen diagram from Lago di Monticchio, Italy (Watts et al.*, 1996a).*

probably very much reduced as a result of increased strength and influence of Siberian winter anti-cyclonic activity (Roberts & Wright, 1993). Additionally, summers may have been perhaps less severely dry due to the absence or near absence of pollen of *Olea*, *Pistacia* and evergreen *Quercus* (indicative of a Mediterranean climate with summer drought) in pollen diagrams at this time period from this region (although these pollen types may have been affected by cold winters). Moreover, at one pollen site (Mirabad), the presence of a herb pollen type of the Chenopodiaceae family, which today occurs in the Pamir region that receives summer as well as winter rain, suggests that the glacial climate may have been characterised by increased cloudiness and by greater effective moisture as a result of decreased temperature and evaporation. Evidence of high lake levels on the Iranian and Anatolian plateaux during the period 20-18 k yr. BP appears to support this inference, but more secure dating is needed to substantiate this conclusion. In the Zagros Mountains region, steppe or steppe-tundra must be the assumed vegetation because of the positive evidence of trees at higher elevations in the Greek Mountains.

The identification of a Younger Dryas-type climatic reversal during the lateglacial period in pollen diagrams from the eastern Mediterranean has met with limited success (Bottema, 1995). At Xinias (Bottema, 1978) and Ioannina (Tzedakis, 1993, 1994; Bottema, 1974) there is an increase in the pollen of coniferous trees and a decrease in Gramineae/Poaceae, *Artemisia* and Chenopodiaceae prior to the Pleistocene-Holocene boundary that is suggestive of a Younger Dryas-type event (Bottema, 1978; Willis, 1994). The Younger Dryas only appears to be registered at higher elevation sites in Italy, yet appears to be absent from the Maliq pollen diagram, Albania at 800 m elevation (Denèfle *et al.*, 2000). In most pollen diagrams from Turkey and other parts of southwest Asia, the Younger Dryas is not readily discernible as a biozone and the period itself, identified in pollen diagrams on the basis of radiocarbon dating, does not show a characteristic palynological pattern. At Hula (Huleh) in northern Israel, new pollen data appear to show a period of climatic aridity and a reversion to almost full glacial conditions coinciding largely with the Younger Dryas (Baruch & Bottema, 1999). At Akgöl (Adabağ) in the Ereğli marshes at the eastern end of the Konya basin in south-central Turkey, Bottema & Woldring (1984) identified a Younger Dryas-type climatic reversal (zone 1d on Figure 8).

New multi-proxy data (pollen, stable isotopes, diatoms, mineralogy) on a sediment core with sequences of laminated sediments during the late Pleistocene and early Holocene from Eski Acıgöl in central Anatolia (Roberts *et al.*, 2001) provide a good time control for examining environmental changes for this important time period. These data show an abrupt change at the onset of the Holocene with an abrupt increase in Gramineae/Poaceae and a marked decrease in the steppic indicators of *Artemisia* and Chenopodiaceae, suggesting an increase in moisture during the growing season. In addition to pollen analysis, the Eski Acıgöl core and new core sequences from Akgöl Adabağ have been analysed using other techniques such as stable isotope and diatom analyses (Leng *et al.*, 1999; Roberts *et al.*, 2001). The isotopic record from authigenic carbonates for the new Akgöl Adabağ core shows that marked hydrological changes are recorded for the new core as well as the older (1977) core during the Late-Glacial and early Holocene. The large variations in $\delta^{18}O$ values suggest that evaporative effects will have been a more important control than temperature (Roberts, *in press*).

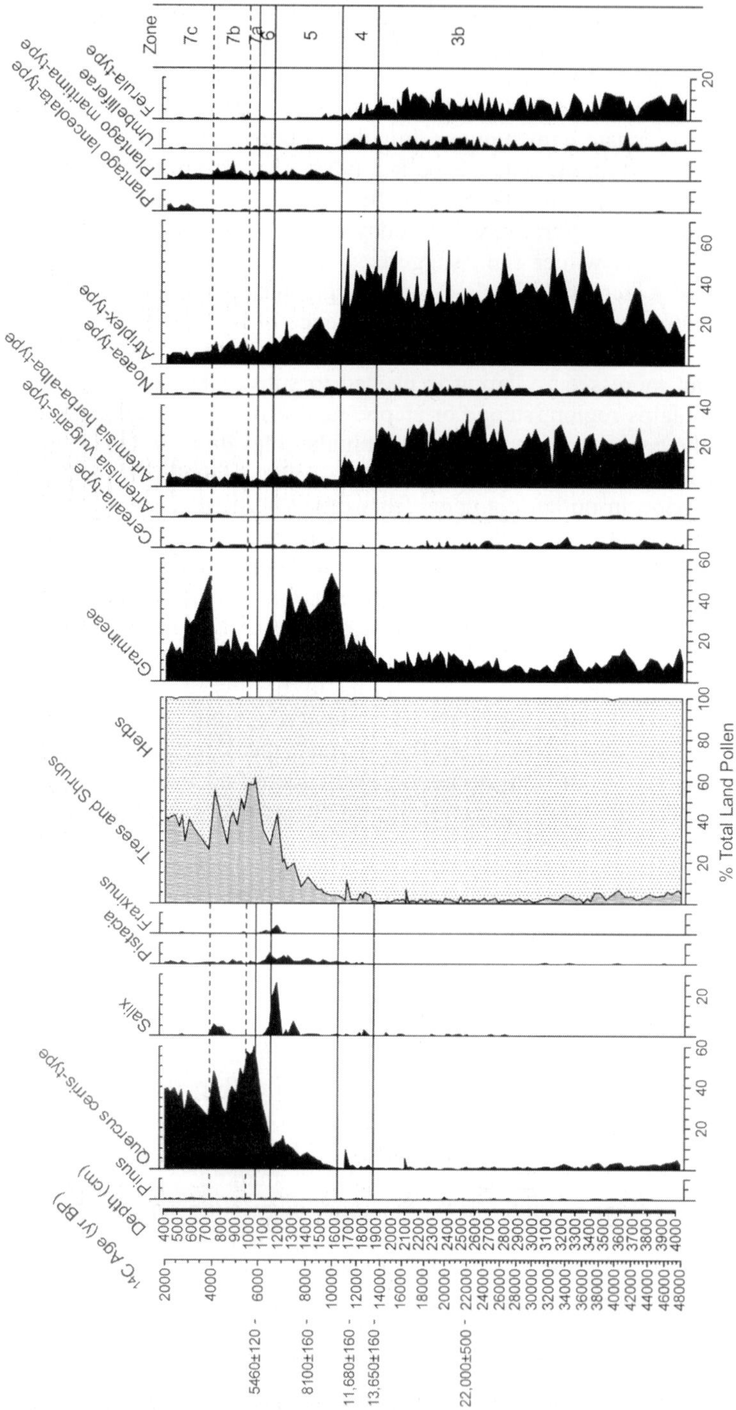

Figure 7. Summary percentage pollen diagram from Zeribar, Iran (van Zeist & Bottema, 1977).

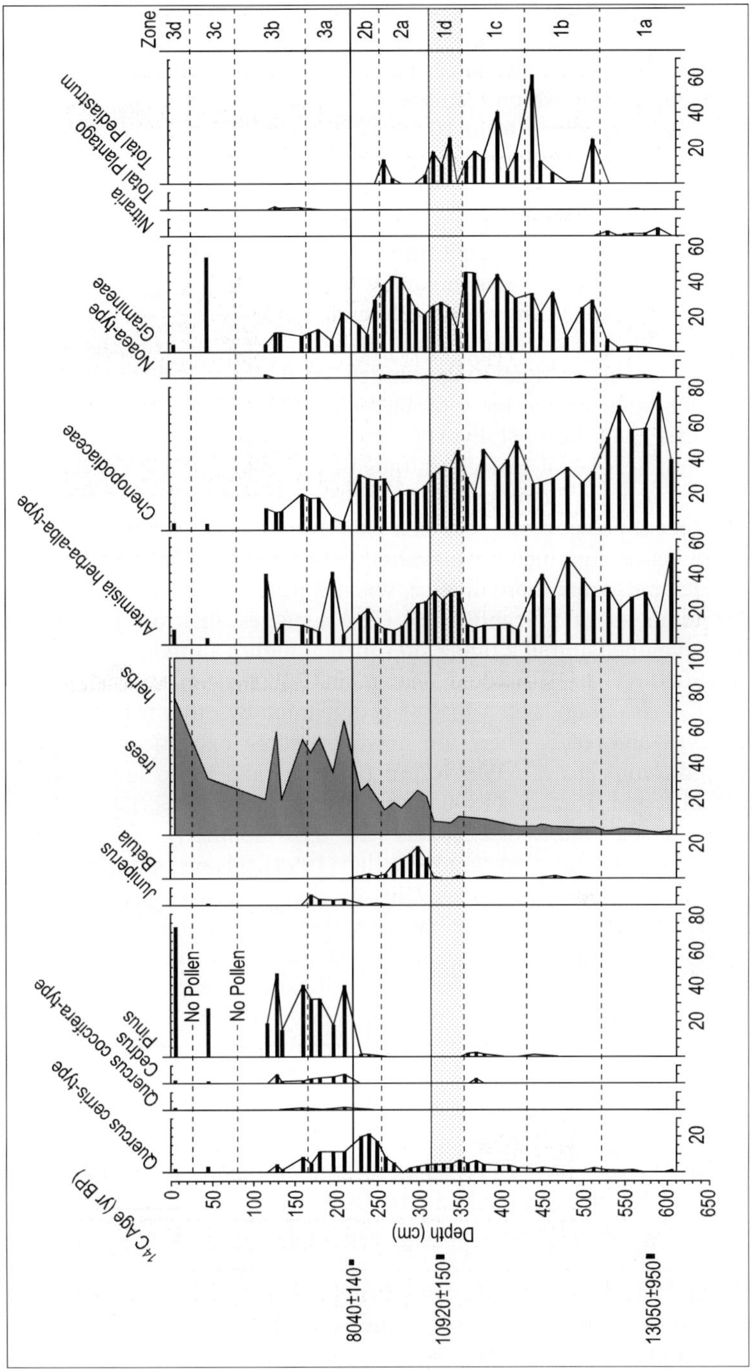

Figure 8. Summary percentage pollen diagram from Akgöl (Adabağ), Ereğli marshes, Konya, Turkey (Bottema & Woldring, 1984).

Holocene: Southeast Europe

In southeastern Europe, the most notable effect of climatic warming at the lateglacial-Holocene transition resulted in the more or less immediate expansion of mixed deciduous woodland from 10,500 and 9,500 yr. BP. However, the dominant taxon in the early postglacial woodland varied greatly from site to site. In Greece, the pollen evidence appears to suggest that mixed woodland comprising *Quercus* was prevalent, while in Bulgaria mixed woodland comprising *Quercus, Ulmus* and *Tilia* was widespread. In Romania, mixed woodland comprising *Pinus* and *Picea* was evident, while in Slovenia mixed woodland comprising *Ulmus*, *Betula* and *Picea* was widespread (Willis, 1994). Pollen data also suggest that *Ulmus*, *Tilia* and *Corylus* had widespread refugia throughout the region, whereas *Carpinus* and *Pistacia* had a smaller refugial range limited to Greece and the Dalmatian coastline. These refugial ranges were most probably also altitudinally-based; Rezina marsh (1800 m a.m.s.l.) had the most diverse forest assemblage with 12 temperate tree taxa present at the late glacial-Holocene transition (Willis, 1992; 1994), whereas low altitude sites such as Gramousti Lake (285 m a.m.s.l.) had much less diverse woodland. Recent palaeoecological data from Maliq Lake, Albania (800 m a.m.s.l.), a site occupying an intermediate elevation between Rezina and Gramousti, suggest that forest elements were more diverse, with 22 tree taxa compared to 12 for Rezina and 9 for Gramousti, and Denèfle *et al.* (2000) suggest that forest refugia may have been more developed at mid altitude sites than at higher altitude sites.

From ~8000 yr. BP a marked change in Balkans vegetation occurs with an increase in *Corylus* abundance usually accompanied by other taxa including *Ulmus*, *Tilia*, *Fraxinus* and *Acer*. There are however, some exceptions: an expansion of *Phillyrea* and *Juniperus* at Malo Jezero (Croatia) and *Abies* and *Pinus* at Edessa (120 m a.s.l.) and Rezina (1800 m a.s.l.) are from sites that are located in mountainous areas, whereas at Kopais (the most southerly site; 100 m a.s.l.) no vegetational change was registered. Willis (1994) advanced three hypotheses to account for these perceived changes. Climate change was reasoned to be essentially unidirectional and not significant. Gradual soil maturation may have allowed certain taxa to gain a competitive edge, while the establishment time (the time taken from when a population is first present in the woodland until expansion) of *Corylus* may have taken approximately 1000 years to become firmly established and out compete the pioneer population. Willis (1994) concludes that more than one variable may have been at work.

Willis (1994) also reviews the apparent delayed increase of *Carpinus orient./Ostrya*, *Abies*, and *Fagus* between 7.5-5 k yr. BP; an apparent establishment time of ~3500 years. Willis (1994) suggests similar hypotheses as outlined above for *Corylus*; including, anthropogenic disturbance, climatic change, establishment time and soil deterioration. Roberts (2002) has suggested that low impact Neolithic activities in southwest Asia may have had a profound effect on woodland readvance during the early Holocene by way of 'managing' the landscape with fire and thus hindering woodland advance. This hypothesis needs to be tested for southwest Asia via micro and macro charcoal analysis from lake and other sedimentary sequences. In southeast Europe, there is only palaeoecological evidence for Neolithic agriculture from about 6000 yr. BP onwards, although archaeological data record that Neolithic agriculture commenced between 8000 and 6000 yr. BP. It is hypothesised that Neolithic cultures undertook selective removal of other tree types thereby favouring the expansion of *Carpinus orientalis/Ostrya*, *Carpinus betulus* and *Fagus*. Willis (1994) reasons that the change occurred at almost all sites in the Balkans and was not confined to known areas of Neolithic activity. Climate change and perhaps changes in the seasonal distribution of temperature and moisture may have been a causal factor although more work on the autecology of these tree species needs to be carried out. Longer establishment times of these tree taxa, which are independent of climatic change and anthropogenic activities, is a plausible hypothesis because the same effect is seen in other long pollen diagrams for earlier/previous interglacial periods. Finally, Willis (1994) advanced the theory that soil deterioration could be an important causal mechanism as the tree taxa are able to tolerate lighter, poorer, often disturbed soils and this could have been a function of natural soil deterioration.

Pollen diagrams from the Balkans also show the expansion of *Pistacia* between 9-8 k yr. BP together with a notable decline of this pollen type at 8 k yr. BP, and then again between 4-2 k yr. BP. Willis (1994) considers various hypotheses to account for this palynological conundrum. Migration could be invoked to explain its decline, but *Pistacia* was already present in Greece during full glacial times. Climate change was also ruled out as a possible mechanism because no significant climate change occurred between 9-8 k yr. BP; the first millennia of the Holocene are characterised by a warmer and drier climate. The response time of *Pistacia* to climatic warming is feasible as it may have been slower to become established taking 1000 years to become detectable in pollen diagrams. Although this hypothesis may explain the rise of *Pistacia*, its decline may have been linked to increasing human impact and the opening-up of the canopy as the ecology of *Pistacia* is dry open woodlands (Willis, 1994).

From around 4500-2000 yr. BP palynological evidence records the decrease in the density and diversity of the woodland and an increase in the abundance of open ground herbaceous pollen types (Gramineae/Poaceae, Cerealia-type, *Plantago*) leading to an anthropogenic landscape. Moreover, the appearance of 'new' forest types such as *Juglans*, *Castanea*, *Olea*, *Platanus* and *Fagus* seems to indicate increasing anthropogenic intervention especially trees characteristic of fructiculture or arboriculture (*Juglans*, *Castanea*, *Olea*, *Platanus*) (Bottema, 1982). *Juglans* and *Castanea* are very low pollen producers and their pollen dispersal is not particularly good; their appearance in pre-Holocene pollen diagrams suggest that they existed in refugia in the Balkans (Slovenia and Greece) and their pollen is evident in low amounts at other sites (Willis, 1994; Bottema, 1982).

The palynological record of the Balkans highlights some very important environmental changes; notably that it contained important refugial areas for temperate deciduous trees. Most pollen records do not detect a Younger Dryas climatic reversal; indeed there is, more or less, an immediate expansion of woodland with no detectable time lag at the onset of the Holocene as evidenced in pollen diagrams from northern Europe and some pollen diagrams from southwest Asia (see below). Although woodland expansion was immediate, the dominant taxon varied across the region with *Quercus* in Greece and parts of Bulgaria (Brewer *et al*., 2002), *Betula* woodland in Bulgaria (Tonkov, 2003), *Ulmus-Betula-Picea* woodland in Slovenia and *Picea* woodland in Romania (Ravazzi, 2002); with *Tilia*, *Ulmus* and *Corylus* present at most sites and *Pistacia* and *Carpinus orientalis/Ostrya* restricted spatially and temporally (Willis, 1994). After the increase of woodland taxa during the early Holocene, the palynological record shows the expansion and decline of *Pistacia* at *ca*. 9-8 k yr. BP, a change in forest dominants at *ca*. 8000 yr. BP with the spread of *Carpinus orientalis/Ostrya*, *Abies*, *Carpinus betulus* and *Fagus* at *ca.* 7.5-5 k yr. BP. The development and establishment of the present-day landscape with new taxa comprising *Juglans*, *Castanea*, *Olea*, *Platanus* occurred at *ca.* 4.5-2 k yr. BP (Willis, 1994).*Holocene: Southwest Asia*

In southwest Asia, the Holocene is characterised by expansion of trees into areas previously dominated by steppe although a transitional phase is evident reflected by adjustments of individual taxa to the changed climatic conditions. At the intramontane valley site of Lake Zeribar (1300 m a.s.l.) the late glacial is characterised by *Artemisia*-Chenopodiaceae and Umbelliferae/Apiaceae (Figure 7). After ~10.5 k yr. BP *Quercus* and *Pistacia* become continuous together with Gramineae/Poaceae and *Plantago* indicating warmth. By 9 k yr. BP the pollen evidence suggests the presence of grass steppe with some trees, and after 9 k yr. BP *Quercus-Pistacia* increase steadily. It is only after 5 k yr. BP that *Quercus* begins to dominate AP (van Zeist & Bottema, 1977). The pollen diagram from the site of Lake Mirabad (300 km to SW; 800 m a.s.l.), which is 500 m lower than Zeribar, and located at the base of the Zagros Mountains, shows that *Quercus* was not depressed to lower elevations but was eliminated entirely from the region most probably because of aridity (Roberts & Wright, 1993). The palynological record from Lake Van (1650 m a.s.l.) resembles Zeribar and Mirabad, with AP at <10% and comprises the same steppic elements together with *Ephedra*, suggesting an environment even drier than Zeribar (van Zeist & Bottema, 1991). The pollen data for the late Pleistocene and early Holocene at Van then record a gradual change as *Quercus* pollen increases to 50% from 6.4-3.4 k yr. BP. The pollen record from Lake Urmia (Bottema, 1986) shows the same sequence as the other pollen records for interior parts of southwest Asia, and despite this part of Southwest Asia having relatively few sites, the pollen records provide a degree of regional coherence (Roberts & Wright, 1993).

The Ghab depression in the Levant (Dead Sea-Jordan rift) records a steppic environment with herbaceous pollen types during the glacial period although there is deciduous *Quercus* at some levels (van Zeist & Bottema, 1991). Between 11 k yr. BP and 10 k yr. BP the forest expanded rapidly and by 9,000 yr. BP the area was well wooded with Mediterranean plant types comprising evergreen *Quercus*, *Pistacia* and *Olea* indicating a Mediterranean climate regime. By 6000 yr. BP deciduous *Quercus* and other pollen types had decreased, possibly reflecting a decrease in precipitation or an increase in summer drought. This is different to, and opposite to the climatic interpretation of the pollen data for the Zeribar region. At the Hula (Huleh) Marsh (Dead Sea-Jordan rift) there are significant amounts of *Quercus* pollen, but the overall vegetation is dominated by steppe vegetation between 25-14 k yr. BP. *Quercus* expanded markedly between 14-10 k yr. BP and reduced between 9-6 k yr. BP. The new core from Hula (Huleh) records a similar palynological picture with a phase of severe aridity and steppe vegetation from 16 to 10.5 k yr. BP, to a humid period from 14.5-11 k yr. BP. A return to arid conditions from 10.5-9.5 k yr. BP coincides with the Younger Dryas chronozone and the early Holocene was marked by a gradual return of trees suggesting increased humid conditions until *ca.* 7000 yr. BP when anthropogenic impacts are discernible (Baruch & Bottema, 1999). It is possible to compare the new record from Hula with the new core from Ghab (van Zeist & Bottema, 1991); both cores show a high degree of commensurability and provide evidence that late Pleistocene and early Holocene climatic changes in the Levant were broadly synchronous.

Important multi-proxy data from laminated sediments from the crater-lake site of Eski Acıgöl in central Anatolia show that there was a marked increase in Gramineae/Poaceae during the early Holocene suggesting an increase in moisture supply during the growing season. Maximum AP values comprising *Quercus* and *Corylus* at Eski Acıgöl were achieved around 8 k Cal. yr. BP (~6.9 k yr. BP), and although a slight decrease in mesic arboreal taxa occurs around 6.5 k Cal. yr. BP (~5.5 k yr. BP), the expansion of *Quercus* woodland continues until about 4.5-4 k Cal. yr. BP (~4 k yr. BP), whereupon *Pinus* becomes the dominant arboreal pollen type together with the appearance of steppic pollen types (Roberts *et al.*, 2001). Stable isotope data for this site record a shift of $\delta^{18}O$ to more negative values at precisely the same time as the shift from herb to grass steppe, thereby providing independent evidence supporting a more favourable water balance. However, the response of arboreal pollen to the increase in effective moisture availability was much slower, taking *ca.* 3000 years to reach maximum AP values. Roberts *et al.* (2001, p. 734) note that the Eski Acıgöl data provide evidence that the response of vegetation to climatic change was taxon specific, and that specific indices such as the AP/NAP ratio mask important floristic changes for the early Holocene period.

In southwest Turkey, the pollen record from Karamık Batalığı, today located in the dry transition zone between forested mountains and dry steppe, shows herb-steppe conditions even at 9000 yr. BP with an expansion of AP at around 8000 yr. BP. A similar situation is seen at Söğüt (Figure 8), Beyşehir and Gölhisar (Figure 9) with steppe vegetation predominant until the early Holocene and then the expansion of *Pinus*, *Quercus* and *Cedrus*. Maximum AP at Söğüt (van Zeist *et al.*, 1975) was achieved at ~9000 yr. BP, while at Beyşehir (van Zeist *et al.*, 1975) it is achieved at 7500 yr. BP, these ages being based on percentage pollen data. At Gölhisar, although pollen percentage data show high values of *Pinus* for the early Holocene (40-60%), implying a landscape comprising pine-dominated woodland, total land pollen concentration values are extremely low and it is not until ~8000 yr. BP, that these begin to increase, signalling an increase in vegetation density. Furthermore, high percentage values of a fern spore indicative of open ground conditions is found during this time period (Eastwood *et al.*, 1999b). That *Pinus* pollen can be a nuisance in the comparison of sites and eventual vegetation reconstructions is highlighted by Tzedakis *et al.* (1997) who considered that percentage values of *Pinus* may represent noise arising from long distance transport particularly during phases of open vegetation. This is entirely applicable for vegetation reconstructions for the Pleistocene-Holocene time period. The comparison of pollen percentage (relative) data and pollen concentration (absolute) data assists in evaluating noise versus a genuine vegetation signal (Tzedakis *et al.*, 1997). The Gölhisar data are important for highlighting a possible conflict between vegetation reconstructions based on pollen percentage data and those based on pollen concentration or absolute data (Eastwood *et al.*, 1999b). After 6000 yr. BP at Beyşehir *Pinus* began to dominate except for a period from about 3.5-1.5 k yr. BP when a period of human impact based on fructiculture or arboriculture (the Beyşehir Occupation phase) is detected at these sites (Bottema *et al.*, 1986; Eastwood *et al.*, 1998). Thus, intercomparison of sites across southwest Asia indicates important regional and local variations in forest composition during the early to mid Holocene. The late glacial-Holocene transition witnessed the advance of forest over large areas of the eastern Mediterranean region and a diachronous pattern of vegetation change is discernible: pollen sites located close to the sea record that this transition was more or less completed by 9000 yr. BP. In southwest Turkey this advance was not complete until after 9000 yr. BP, while at some sites (*e.g.* Gölhisar) pollen concentration data suggest that woodland readvance was not complete until 8000 yr. BP. At other sites, for example, Zeribar, Mirabad and to a lesser extent Van, maximum AP values were not achieved until mid-Holocene times. In short, the more inland the site the greater the time for maximum AP values to be achieved and the more delayed start to forest re-advance. This can be contrasted with southeast Europe where the pollen evidence suggests a more or less immediate response of woodland expansion soon after the late glacial-Holocene transition. One of the most important questions surrounding palaeoecological investigations in southwest Asia, and which is in need of explanation, is the delay in the readvance of woodland at interior sites in southwest Asia and Anatolia (Roberts, 1982). Clearly, the delay may be apparent and attributable to inaccurate and/or imprecise radiocarbon ages. This is an important issue, as many sites in southwest Asia yield sediment cores with high clay content and very low organic matter values. Furthermore, many of these sites are located in limestone or marl catchments and ^{14}C ages may have been susceptible to a hard water error.

Conversely, the delay in woodland may be real and may relate to a host of factors that can be invoked to account for the delay in woodland as shown by pollen data for southwest Asia. These range from inherent establishment times and rates of dispersal of taxa that are dependent upon dispersal mechanisms and growth rates. The 'starting positions' of taxa are also important if refugia were located some distance away, while 'suitable' micro-scale environmental conditions such as aspect, soil moisture and edaphic factors may have been important. Competition and physiographic barriers together with human disturbance are important factors; particularly the latter because of the timing of Neolithic impacts in this part of the world from *ca.* 10 k Cal. yr. BP. Finally, climatic factors including humidity and seasonality may have been important. This last factor, in particular has received attention due to the interpretations placed on pollen data originating from southwest Asia. The presence of strongly steppic pollen types in pollen diagrams for the late Pleistocene and early Holocene has led to the inference that climatic aridity may have been the controlling factor in causing a delay in forest and woodland expansion during the early Holocene. However, new stable isotope data from Eski Acıgöl (Roberts *et al.*, 2001), Akgöl Adabağ (Leng *et al.*, 1999), Lake Zeribar (Stevens *et al.*, 2001), Gölhisar (Eastwood *et al.*, in press-a) and Lake Van (Wick *et al.*, 2003, in press) show a shift in $\delta^{18}O$ to more negative values at the same time as changes in the species composition of non-arboreal pollen types suggesting greater moisture availability. Palynological data show that there appears to be significant time lags with the onset of increasing moisture availability and maximum AP values.

CONCLUSIONS

This review outlines past research in southeast Europe and southwest Asia, as well as outlining our current understanding. Some clear conclusions have emerged. The spatial distribution of palaeoecological sites could be improved, particularly in southwest Asia, but this is problematic and dependent upon the location of suitable wetland sites (lakes, marshes, peat deposits). In order to unravel the complexities of

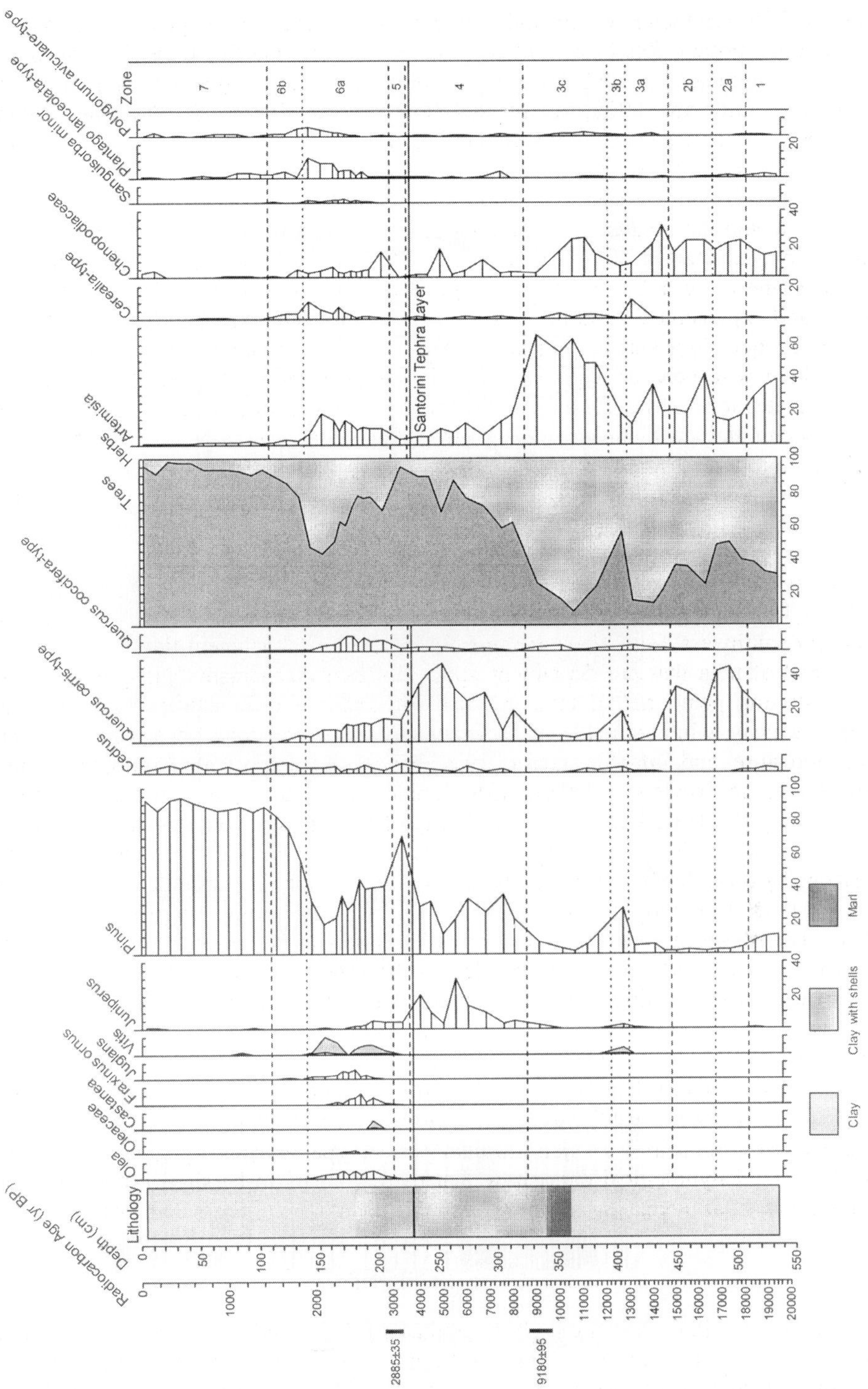

Figure 9. Summary percentage pollen diagram from Söğüt, southwest Turkey (van Zeist et al., 1975).

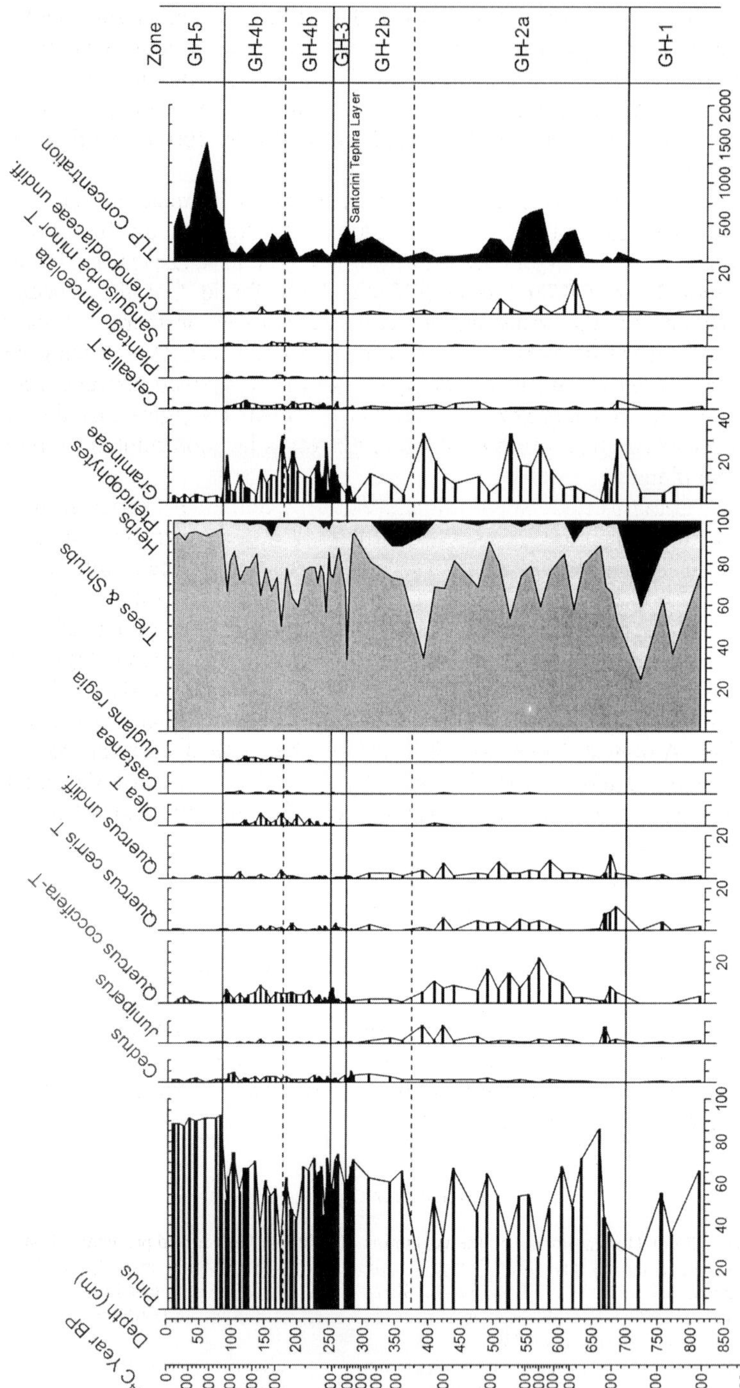

Figure 10. Summary percentage pollen diagram from Gölhisar, southwest Turkey (Eastwood et al., *1999b).*

site specific factors and conflicting lines of data, future investigations need to adopt multi-proxy/indicator approaches. Coupled proxy investigations need to consider incorporating stable isotope research in any future (re)investigation of sites. The use of pollen analysis together with other proxy methods such as using plant macrofossil analysis, charcoal (Björkman *et al.*, 2002; Birks & Birks, 2000) would appear to be one way forward.

Feedbacks are important and are inherent in the climate system. Reale & Dirmeyer (2000) and Reale & Shukla (2000) reconstructed the likely vegetation based on pollen data for North Africa during the Roman Classical Period and then applied a global circulation model (GCM). They found that the albedo effects significantly altered the atmospheric circulation over northern Africa with a northward shift of the Inter-Tropical Convergence Zone (ITCZ) producing increased precipitation. Model results were compared with pollen, historical and geographic data together with other proxy sources, and the results suggest that deforestation around the Mediterranean during the last 2000 years has contributed to the dryness of the current climate.

If we are going to understand climate and environmental change in the eastern Mediterranean region more completely, we need to appraise new sites for their palaeoecological potential. Existing sites that have not been published in the mainstream literature also need to be re-appraised for possible future palaeoecological study. Many of these sites which have been published in national and 'in-house' journals, are undated and report only pollen (percentage) analytical data and have the potential to yield important palaeoecological data using a multi-indicator/proxy approach. Spatially, there are important regions that are in need of investigation. Amongst these are sites located between southwest Asia and the Balkans; namely, Turkey (Anatolia and Thrace) and countries of Eastern Europe; future investigations may yield important results for teleconnections between these two diverse regions.

ACKNOWLEDGEMENTS

The European Science Foundation are thanked for a travel grant to attend the Slovenia conference. Kevin Burkhill and Anne Ankcorn are also thanked for producing most of the graphics. Margaret Atherden is thanked for her helpful comments and improvement of a draft manuscript. The (http://www.ngdc.noaa.gov/paleo/epd.html) European Pollen Database is acknowledged for providing pollen data.

REFERENCES

Atherden, M.A. 2000. Human impact on the vegetation of southern Greece and problems of palynological interpretation: a case study from Crete. In *Landscape and land use in postglacial Greece* (P. Halstead and C. Frederick, eds.), pp. 62-78. Sheffield Academic Press, Sheffield.

Atherden, M.A. & Hall, J.A. 1999. Human Impact on Vegetation in the White Mountains of Crete since AD 500. *The Holocene* 9, 183-193.

Atherden, M.A., Hall, J. & Wright, J.C. 1993. A pollen diagram from the northeast Peloponnese, Greece: implications for vegetation history and archaeology. *The Holocene* 3, 351-356.

Bar-Matthews, M., Ayalon, A., & Kaufman, A., 1997. Late Quaternary Paleoclimate in the Eastern Mediterranean Region from Stable Isotope Analysis of Speleothems at Soreq Cave, Israel. *Quaternary Research* 47, 155-168.

Baruch, U. & Bottema, S. 1999. A new pollen diagram from Lake Hula: Vegetation, climatic and anthropogenic implications. In *Ancient Lakes: Their Cultural and Biological Diversity* (H. Kawanabe, G.W. Coulter & A.C. Roosevelt, eds.), pp. 75-86. Kenobi Publications, Belgium.

Bennett, K.D., Tzedakis, P.C. & Willis, K.J. 1991. Quaternary refugia of north European trees. *Journal of Biogeography* 18, 103-115.

Birks, H.H. & Birks, H.J.B. 2000. Future uses of pollen analysis must include plant macrofossils, *Journal of Biogeography* 27, 31-35.

Björkman, L., Feurdean, A., Cinthio, K. Wohlfarth, B. & Possnert, G. 2002. Lateglacial and early Holocene vegetation development in the Gutaiului Mountains, northwestern Romania. *Quaternary Science Reviews* 21, 1039-1059.

Bjorkman, L., Feurdean, A. & Wohlfarth, B. 2003. Late-Glacial and Holocene forest dynamics at Steregoiu in the Gutaiului Mountains, Northwest Romania. *Review of Palaeobotany and Palynology* 124, 79-111.

Bottema, S. 1974. *Late Quaternary vegetational history of northwestern Greece*. Unpublished PhD Thesis, University of Groningen, The Netherlands.

Bottema, S. 1975. The interpretation of pollen spectra from prehistoric settlement (with special attention to Liguliflorae). *Palaeohistoria* 17, 17-35.

Bottema, S. 1978. The lateglacial in the eastern Mediterranean and the Near East. In *The Environmental History of the Near and Middle East Since the Last Ice Age* (W.C. Brice, ed.), pp. 15-28. Academic Press, London.

Bottema, S. 1982. Palynological investigations in Greece with special reference to pollen as an indicator of human activity. *Palaeohistoria* 24, 257-289.

Bottema, S. 1986. Late Quaternary pollen diagram from Lake Urmia (northwestern Iran). *Review of Palaeobotany and Palynology* 47, 241-261.

Bottema, S. 1995. The Younger Dryas in the Eastern Mediterranean. *Quaternary Science reviews* 14, 883-891.

Bottema, S. & Woldring, H. 1984. Late Quaternary vegetation and climate of Southwest Turkey II. *Palaeohistoria* 26, 123-149.

Bottema, S., Woldring, H. & Aytuğ, B. 1986. *Palynological investigations on the relations between prehistoric man and vegetation in Turkey: The Beyşehir Occupation Phase*. Proceedings of the 5th Optima Congress, September, 1986. Istanbul. pp. 315-328.

Brewer, S., Cheddadi, R., de Beaulieu, J. L. & Reille, M. 2002. The spread of deciduous *Quercus* throughout Europe since the last glacial period. *Forest Ecology and Management* 156, 27-48.

Denèfle, M., Lézine, A-M., Fouache, E. & Dufaure, J-J. 2000. A 12,000-year pollen record from Lake Maliq, Albania. *Quaternary Research* 54, 423-432.

Eastwood, W.J., Roberts, N. & Lamb, H.F. 1998. Palaeoecological and archaeological evidence for human occupance in southwest Turkey: The Beysehir Occupation Phase. *Anatolian Studies* 48, 69-86.

Eastwood, W.J., Pearce, N.J.G., Westgate, J.A., Perkins, W.T., Lamb, H.F. & Roberts, N. 1999a. Geochemistry of Santorini tephra in lake sediments from Southwest Turkey. *Global and Planetary Change* 21, 17-29.

Eastwood, W. J., Roberts, N., Lamb, H. F. & Tibby, J. C. 1999b. Holocene environmental change in southwest Turkey: a palaeoecological record of lake and catchment-related changes. *Quaternary Science Reviews* 18, 671-696.

Eastwood, W.J., Leng, M.J., Roberts, N. & Davis, B. 2003, in press-a. Holocene climate change in the eastern Mediterranean region: a comparison of stable isotope and pollen data from a lake record in southwest Turkey. (*Journal of Quaternary Science*).

Eastwood, W.J., Roberts, N. & Boyer, P. 2003, in press-b. Pollen analysis at Çatalhöyük. In *Excavations at Çatalhöyük* (I. Hodder *et al*., eds.). BIAA, McDonald Institute of Archaeology monograph series.

Fall, P.L. 1990. Deforestation in Southern Jordan: evidence from fossil Hyrax middens. In *Man's role in the shaping of the Eastern Mediterranean landscape* (S. Bottema, G. Entjes-Nieborg & W. van Zeist, eds.), pp. 271-281. A. A. Balkema, Rotterdam.

Flenley, J. 2003. Some prospects for lake sediment analysis in the 21st century. *Quaternary International* 105, 77-80.

Follieri, M., Giardini, M., Magri, D. & Sadori, L. 1998. Palynostratigraphy of the last glacial period in the volcanic region of Central Italy. *Quaternary International* 47-48, 3-20.

Frogley, M.R., Tzedakis, P.C. & Heaton, T.H.E. 1999. Climate variability in northwest Greece during the last interglacial. *Science* 285, 1886-1889.

Frogley, M.R., Griffiths, H.I. & Heaton, T.H.E. 2001. Historical biogeography and Late Quaternary environmental change of Lake Pamvotis, Ioannina (north-western Greece): evidence from ostracods. *Journal of Biogeography* 28, 745-756.

Harlan, J. R. 1995. *The Living Fields: Our Agricultural Heritage*. Cambridge University Press, Cambridge.

Karabiyikoğlu, M., Kuzucuoğlu, C., Fontugne, B., Kaiser, B. & Mouralis, D. 1999. Facies and depositional sequences of the late Pleistocene Göçü shoreline, Konya Basin, Central Anatolia: Implications for reconstructing lake-level changes. *Quaternary Science Reviews* 18 593-610.

Kaya, Z. & Raynal, D.J. 2001. Biodiversity and conservation of Turkish forests. *Biological Conservation* 97, 131-141.

Keenan, D.J. 2002. Why early-historical radiocarbon dates downwind from the Mediterranean are too early. *Radiocarbon* 44, 1, 225-237.

Kuniholm, P.I., Kromer, B., Manning, S.W., Newton, M., Latini, C.E. & Bruce, M.J. 1996. Anatolian tree-rings and the absolute chronology of the east Mediterranean 2220-718BC. *Nature* 381, 780-783.

Kürschner, H. 1984. Der östliche Orta Toroslar (Mittlerer Taurus) und angrenzende Gebiete. Ein formationskundliche Darstellung der Vegetation Südost-Anatoliens, Beih. *Tübinger Atlas des Vorderen Orients*, Reihe A (Naturwiss.), 15, Wiesbaden.

Leng, M.J., Roberts, N., Reed, J.M. & Sloane, H.J. 1999. Late Quaternary palaeohydrology of the Konya Basin, Turkey based on isotope studies of modern hydrology and lacustrine carbonates. *Journal of Paleolimnology* 22, 187-204.

Magri, D. 1999. Late Quaternary vegetation history at Lagaccione near Lago di Bolsena (central Italy). *Review of Palaeobotany and Palynology* 106, 171-208.

Magri, D. & Tzedakis, P.C. 2000. Orbital signatures and long-term vegetation patterns in the Mediterranean. *Quaternary International* 73-74, 69-78.

Manning, S.W., Kromer, B, Kuniholm, P.I. & Newton, M.W. 2001. Anatolian tree-rings and a new chronology for the east Mediterranean Bronze-Iron Ages. Science 294, 2532-2535.

Mooney, A.H. 1988. Lessons from Mediterranean-Climate regions. In *Biodiversity* (E.O. Wilson, ed.), pp.156-165. National Academy Press, Washington.

Narcisi, B. & Vezzoli, L. 1999. Quaternary stratigraphy of distal tephra layers in the Mediterranean-an overview. *Global and Planetary Change* 21, 31-50.

Norfalise, A. 1987. *Map of the Natural Vegetation of the Member Countries of the European Community and the Council of Europe*. 1:3,000,000, second edition, Office for the Official Publications of the European Communities, Luxembourg. 80 pp.

Polunin, O. 1980. *Flowers of Greece and the Balkans-A Field Guide*. Oxford University Press, Oxford.

Ravazzi, C. 2002. Late Quaternary history of spruce in southern Europe. *Review of Palaeobotany and Palynology* 120, 131-177.

Reale, O. & Dirmeyer, P. 2000. Modeling the effects of vegetation on Mediterranean climate during the Roman Classical Period: Part I: Climate history and model sensitivity. *Global and Planetary Change* 25, 163-184.

Reale, O. & Shukla, J. 2000. Modeling the effects of vegetation on Mediterranean climate during the Roman Classical Period: Part II. Model simulation. *Global and Planetary Change* 25, 185-214.

Roberts, N. 1982. Forest re-advance and the Anatolian Neolithic. In *Archaeological aspects of woodland economy* (M. Bell & S. Limbrey, eds.), pp. 231-246. British Archaeological Reports - International Series, 146.

Roberts, N. 2002. Did prehistoric landscape management retard the post-glacial spread of woodland in Southwest Asia? *Antiquity* 76, 1002-10.

Roberts, N. *in press*. Introduction to Palaeoenvironments of the Konya Plain. In *Excavations at Çatalhöyük* (I. Hodder *et al*. eds.). BIAA, McDonald Institute of Archaeology monograph series.

Roberts, C.N. & Wright, H.E. 1993. Vegetational, lake level, and climatic history of the Near East and Southwest Asia. In *Global climates since the last glacial maximum* (H.E. Wright, Jr., J.E. Kutzbach, T. Webb III, W.F. Ruddiman, F.A. Street-Perrott, & P.J. Bartlein, eds.), pp. 194-220. University of Minnesota Press, Minneapolis [COHMAP volume].

Roberts, N., Black, S., Boyer, P., Eastwood, W.J., Griffiths, H.I., Lamb, H.F., Leng, M.J., Parish, R., Reed, J.M., Twigg, D. & Yigitbasioğlu, H. 1999. Chronology and stratigraphy of Late Quaternary sediments in the Konya Basin, Turkey: Results from the KOPAL Project. *Quaternary Science Reviews* 18, 611-630.

Roberts, N., Reed, J.M., Leng, M.J., Kuzucuoglu, C., Fontugne, M., Bertaux, J., Woldring, H., Bottema, S., Black, S., Hunt, E. & Karabiyikoglu, M. 2001. The tempo of Holocene climatic change in the eastern Mediterranean region: new high-resolution crater-lake sediment data from central Turkey. *Holocene* 11, 721-736.

Scarascia-Mugnozza, G., Oswald, H., Piussi, P. & Radoglou, K. 2000. Forests of the Mediterranean region: gaps in knowledge and research needs. *Forest Ecology and Management* 132, 97-109.

Šercelj, A. 1966. Pelodne analize Pleistocenskih in Holocenskih sedimentov Ljubljanskega Barje. *Slovenska Akademiza Zananosti in Umentnosta Classis IV Dissertationes* 9, 431-472 (in Slovene).

Simmonds, N.W. 1986. *Evolution of Crop Plants*. Longman Scientific and Technical, Harlow.
Stevens, L.R., Wright, H.E. & Ito, E. 2001. Proposed changes in seasonality of climate during the Lateglacial and Holocene at Lake Zeribar, Iran. *The Holocene* 11, 747-755.
Stewart, J.R. & Lister, A.M. 2001. Cryptic northern refugia and the origins of the modern biota. *Trends in Ecology & Evolution* 16, 608-613.
Tonkov, S. 2003. Holocene palaeovegetation of the Northwestern Pirin Mountains (Bulgaria) as reconstructed from pollen analysis. *Review of Palaeobotany and Palynology* 124, 51-61.
Tzedakis, P.C. 1993. Long-term tree populations in northwest Greece through multiple Quaternary climatic cycles. *Nature* 364, 437-440.
Tzedakis, P.C. 1994. Vegetation change through glacial-interglacial cycles: a long pollen sequence perspective. *Philosophical Transactions of the Royal Society of London* B 345, 403-432.
Tzedakis, P.C., Andrieu, V., de Beaulieu, J.-L., Crowhurst, S., Follieri, M., Hooghiemstra, H., Magri, D., Reille, M., Sadori, L., Shackleton, N.J. & Wijmstra, T.A. 1997. Comparison of terrestrial and marine records of changing climate of the last 500,000 years. *Earth and Planetary Science Letters* 150, 171-176.
Tzedakis, P.C., Andrieu, V., de Beaulieu, J.-L., Birks, H.J.B., Crowhurst, S., Follieri, M., Hooghiemstra, H., Magri, D., Reille, M. & Sadori, L. 2001. Establishing a terrestrial chronological framework as a basis for biostratigraphical comparisons. *Quaternary Science Reviews* 20, 1583-1592.
van Zeist, W. & Bottema, S. 1977. Palynological investigations in western Iran. *Palaeohistoria* 24, 19-85.
van Zeist, W. & Bottema, S. 1991. *Late Quaternary vegetation of the Near East*, in Beihefte zum Tubinger Atlas des vorderen orients. Reihe A (Naturwissenscaften) Nr. 18, Wiesbaden: Dr. Ludwig Reichert Verlag, 156 pp.
van Zeist, W., Woldring, H. & Stapert, D. 1975. Late Quaternary vegetation and climate of southwestern Turkey, *Palaeohistoria* 17, 55-143.
Watts, W. A., Allen, J.R.M. & Huntley, B. 1996a. Vegetation history and palaeoclimate of the last glacial period of Lago Grande di Monticchio, southern Italy. *Quaternary Science Reviews* 15, 133-153.
Watts, W.A., Allen, J.R.M., Huntley, B. & Fritz, S.C. 1996b Vegetation history and climate of the last 15,000 years at Laghi Di Monticchio, southern Italy. *Quaternary Science Reviews* 15, 113-132.
Wick, L., Lemcke, G. & Sturm, M. 2003, in press. Evidence of Late-glacial and Holocene climatic change and human impact in eastern Anatolia: high resolution pollen, charcoal, isotopic, and geochemical records from the laminated sediments of Lake Van, Turkey. *The Holocene*.
Willis, K.J. 1992. The Late Quaternary vegetational history of northwest Greece. II. Rezina Marsh. *New Phytologist* 121, 119-138.
Willis, K.J. 1994. The vegetational history of the Balkans. *Quaternary Science Reviews* 13, 769-788.
Willis, K.J., Rudner, E. & Sümegi, P. 2000. The full glacial forests of central and southeast Europe. *Quaternary Research* 53, 203-213.
Willis, K.J. & Whittaker, R.J. 2000. The refugial debate. *Science* 287, 1406-1407.
Wohlfarth, B., Hannon, G., Feurdean, A., Ghergari, L., Onac, B.P. & Possnert, G. 2001. Reconstruction of climatic and environmental changes in NW Romania during the early part of the last deglaciation (~15,000-13,600 cal yr BP). *Quaternary Science Reviews* 20, 1897-1914.
Wulf, S., Negendank, J.F.W. Brauer, A., Mingram, J. & Zolitschka, B. 2001. Application to tephrastratigraphy of lacustrine sediment records in the Mediterranean – a case study from Lago Grandi di Monticchio (southern Italy). *Terra Nostra*. 6th ELDP Workshop, Potsdam.
Zohary, D. 1973. *Geobotanical Foundations of the Middle East*. (2 volumes), Stuttgart-Amsterdam.
Zolitschka, B., Wulf, S. & Negendank, J.F.W. 2000. Circum-Mediterranean lake records as archives of climatic and human history. *Quaternary International* 73-74, 1-5.

4. THE BALKANS AS PRIME GLACIAL REFUGIAL TERRITORY OF EUROPEAN TEMPERATE TREES

P. C. TZEDAKIS*

INTRODUCTION

Every subdiscipline investigating the fossil record has its own holy grail, be they soft body tissues, or ancient DNA, or missing links in human ancestry, the level of its desirability determined not only by the importance in understanding particular aspects of the history of life, but also by the degree of its elusiveness. In the case of European Quaternary palaeobotany the search for cold stage refugia of temperate trees would probably qualify under this category. Ever since the first full-glacial pollen diagrams from southern Europe (*e.g.* Wijmstra, 1969; Florschütz *et al.*, 1971) showed steppe-dominated landscapes, implying that forest biomes had not simply shifted *en mass* southwards as ice sheets expanded, the whereabouts of temperate elements of the flora has been a topic of continuous discussion. The prevailing hypothesis has been that remnant tree populations found refuge in the southern peninsulas of Europe where they survived in suitable microhabitats in mid-altitude zones and in locally moist sites in lowland and coastal areas (*e.g.* Beug, 1968, 1975; Frenzel, 1968, 1979; Lang, 1970; van der Hammen *et al.*, 1971; Bennett *et al.*, 1991). The presumed small size of such populations and the relative lack of full-glacial evidence means that direct palaeobotanical detection has been difficult, and it is precisely this aspect of the problem that provides the element of fascination in the search for glacial refugia. Indeed, when reviewing the large number of publications on this issue, one is often struck by a tendency (if not a desire) to infer the presence of refugial populations on the basis of sometimes indirect and even tenuous evidence. It seems that the designation of a site as refugial territory confers an air of respectability, which sets it apart from other less fortunate areas. However, in addition to being a palaeobotanical trophy, the identification of refugial areas has important implications for conservation strategies. If an area's particular environmental setting leads to the long-term survival of populations through multiple climatic oscillations, then a case can be made that the area represents a conservation priority and be designated as protected. In view of this, it is important that palaeobotanists exercise considerable restraint and evaluate the evidence critically, when attempting to identify the geographical location of refugial areas.

West (1980) suggested that information on the glacial locations of tree populations could be obtained through examination of their postglacial migration routes. Huntley & Birks (1983) followed this suggestion by making the first European compilation of pollen records 0-13,000 yr. BP and, on the basis of the locations of first expansion of tree populations, proposed 17 refugial areas in the closing parts of the last glacial stage. From these areas, trees were able to colonise Central and northern Europe during interglacials, while at the onset of the ensuing glacial contracting northern populations were thought to follow the reverse southward route towards refugial areas. During the southward journey under rapidly

H.I. Griffiths et al. (eds.), Balkan Biodiversity, 49–68.

deteriorating climate conditions, the Pyrenees, Alps and Balkan mountains would delay spread, with some species inevitably failing to reach the refugial havens. Thus, the gradual extirpation of many Tertiary tree species from Europe during the course of the Pleistocene was seen by Reid (1935) as a function of the E-W orientation of mountain chains. By comparison, the absence of such barriers in eastern North America and Asia allowed a less restricted access to lower latitudes, contributing to the survival of Tertiary species.

However, after examining the available late interglacial and early glacial pollen diagrams from Central Europe, Bennett *et al.* (1991) found no evidence for a southward migration of tree populations. This, in conjunction with the observation that at present most of the north European tree species are also found in southern European mountains, led Bennett *et al.* (1991) to suggest that during the population contraction phase at the onset of a stadial or glacial there is no reverse movement towards southern refugial areas, but rather northern populations degrade *in situ*. The northern population crashes were caused by abrupt changes in climate regimes associated with oceanic and atmospheric circulation reorganisations, which precluded the possibility of a gradual southward migration. In southern Europe local tree populations would also contract, but a subset of them would be able to remain in suitable areas for survival. From these glacial stations trees were able to expand during intervals of favourable climate conditions, but a part of the population would always remain in the south, providing the long-term continuity. This hypothesis was corroborated by evidence from a long pollen sequence spanning the last 430,000 years from Ioannina, western Greece, which showed the continuous presence of tree populations in the region during both glacial and interglacials (Tzedakis, 1993). Thus, the maintenance of tree species in Europe becomes inexorably linked to the long-term persistence of part of the populations in the south. Northern populations of most species are irrelevant to their long-term survival because they disappear at the end of interglacials; it is the failure of a species' southern populations to survive during either an interglacial or a glacial interval that ultimately leads to its extinction from Europe. A corollary of this is that the E-W orientation of mountains in Europe is unlikely to have led to the disappearance of many Tertiary species by acting as a barrier to their spread (Bennett *et al.*, 1991). Thus, the enhanced rate of extinction in Europe relative to eastern North America and Asia is seen as a function of the much reduced area available for survival of refugial populations south of the Alps (Bennett *et al.*, 1991; Huntley, 1993). This has important implications for conservation strategies, because the pressures exerted on southern populations by anthropogenic practices during the course of the present interglacial may lead to a significant increase of extinction rates by the end of the next climatic cycle (Tzedakis, 1993).

Of the different refugial areas, the Balkan Peninsula consistently emerges as the most taxon-rich (Huntley & Birks, 1983; Birks & Line, 1993). Bennett *et al.* (1991) suggest that topographical variability allowed populations to migrate altitudinally and provided a range of habitats with particular microclimatic conditions. In addition, especially for the western Balkans, the proximity to the Adriatic and Ionian Seas is important as it provided a source of moisture. However, in 1993 despite the publication of long pollen records like that from Ioannina, documenting the continuous presence of tree pollen during glacial periods, our overall understanding of the geographical distribution of glacial tree populations in the Balkans remained extremely limited, owing to the lack of pollen sequences extending into the last glacial maximum (LGM) (Figure 1). For clarification, it should be noted that the

LGM, as defined by the EPILOG project, is the interval between 19,000 and 23,000 cal. yr. BP (*ie.* 16,100-19,500 yr. BP), centred on 21,000 cal. yr. BP (18,000 yr. BP) (Mix *et al.*, 2001). This interval (Figure 1) is coeval with lowest sea level stand (maximum ice volume) and is characterised by low termperatures and absence of extensive millennial-scale variability (Mix *et al.*, 2001)). Within the last ten years a number of developments have take place which go some way to alleviating this state of relative ignorance: (i) the publication of a number of new pollen sequences as well as plant macrofossil studies; (ii) a rapid increase of studies of molecular markers revealing the genetic structure of modern populations and providing important insights into species' Quaternary histories and migration routes; and (iii) a new appreciation of the complexity of climatic conditions during glacial intervals.

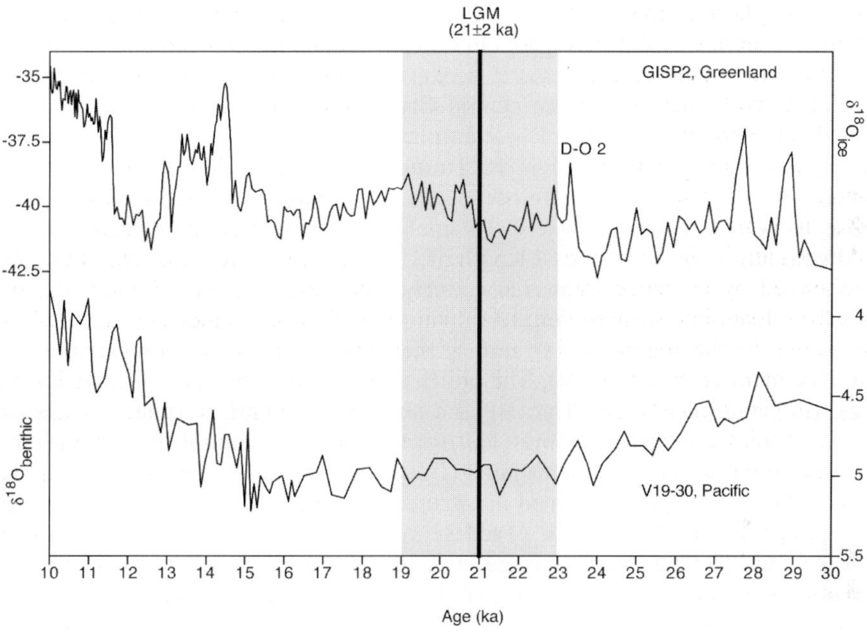

Figure 1. Definition of the Last Glacial Maximum (LGM) according to the EPILOG Project (Mix et al., *2001). 1a. Oxygen isotope composition of ice in the GISP2 record from Greenland, indicating variations in air temperature (Grootes & Stuiver, 1997; Meese* et al., *1997). 1b. Oxygen isotope composition of benthic foraminifera in equatorial Pacific core V19-30, as a broad signal of global ice volume (Shackleton* et al., *1983).*

The aim of this paper is to review these developments with particular reference to the Balkans and attempt to provide some provisional answers to the following issues:

(1) Can refugial species distribution within the Balkans be inferred in more detail?
(2) What was the northernmost extent of tree populations during the glacial maximum?
(3) What was the impact of extreme and rapid climate oscillations in the North Atlantic on refugial populations?

(4) What are the evolutionary implications of persistence of tree populations in refugial sites?

PALAEOBOTANICAL EVIDENCE

In 1993 there were only five terrestrial pollen sequences from the Balkans that extended into the LGM (see Figure 2 for location of sites). Arguably the most famous of Balkan sequences (and indeed European), is Tenaghi Philippon, northeast Greece, containing a continuous record of the last 1 Ma (Wijmstra, 1969; Wijmstra & Smit, 1976; van der Wiel & Wijmstra, 1987a,b); its LGM section showed low presence of *Pinus*, *Juniperus*, *Betula* and *Alnus* pollen. In central Greece, Bottema (1979) produced a pollen record from Lake Xinias, extending well into the last glacial with a basal date of 46,900 yr. BP; the diagram shows low but continuous presence of *Pinus*, deciduous *Quercus*, *Betula*, *Juniperus* and *Abies* pollen along with traces of evergreen *Quercus*, *Pistacia*, *Ulmus*, *Carpinus orientalis/Ostrya* and *Corylus* in the LGM. Also from central Greece, a series of diagrams is available from the Kopais basin, on the Boeotian plain. The earliest palynological work on Kopais was undertaken by Greig & Turner (1974) and Turner & Greig (1975) covering the Lateglacial and Holocene and was followed by Allen (1986, 1990) whose cores (pollen analysis by J.C. Ritchie) extended into the full glacial. Core KA of Allen (1986, 1990) had a total length of 19m and two radiocarbon dates of 9,900 and 12,520 yr. BP at 7 and 9 m, respectively. The lower part of the pollen diagram shows minimal arboreal pollen (AP) values with low presence of *Pinus* and *Juniperus*. A fourth site from Greece is the Ioannina basin, western Greece, first investigated by Bottema (1974). The pollen record, supported by three radiocarbon dates, extended to *ca.* 45,000 yr. BP and showed the LGM presence of *Quercus*, *Abies*, *Pinus*, *Betula*, *Fagus*, *Ulmus*, *Carpinus betulus*, *Acer* and *Alnus*. Two decades later, a longer sequence from Ioannina (I-249) became available, spanning the last 480 thousand years, as determined through correlations with Tenaghi Philippon and the marine isotope stratigraphy (Tzedakis, 1993, 1994). I-249 showed the same LGM patterns as Bottema's sequence, in terms of continuous presence of pollen of temperate trees throughout the last glacial, and indeed during earlier glacial cycles. Finally the fifth record was from Slovenia, in Ljubljana Moor (Šercelj, 1966), a 135 m sequence whose chronological framework remains uncertain. Two radiocarbon dates at 40.0 m and 57.8 m have provided infinite ages of >48,000 yr. BP and >51,000 yr. BP and inorganic non-polleniferous sediments are encountered at ten separate intervals along the core. Despite the chronological uncertainty, the Late Pleniglacial appears to be between 18 and 35 m and shows the presence of *Pinus* along with some *Picea* and *Betula*; temperate tree pollen is absent during this interval, but present during the intervals before and after. In addition to the terrestrial sequences, two pollen records are available from marine cores in the Adriatic Sea (Grüger, 1975). Cores Adr. 309 and Adr. 353 share very similar pollen stratigraphies and extend from the Holocene back into the Pleniglacial. Given the prevailing westerly circulation, the aerial pollen transport may be dominated by southern Italian sources. However, pollen may also enter into the sediments through river transport from the western Balkan coast, which during the last glacial sea-level regression would have been located nearer to the coring sites. No radiocarbon dates or oxygen isotope stratigraphy are available, but both diagrams show continuous presence of deciduous *Quercus* and *Betula*, along with intermittent presence of

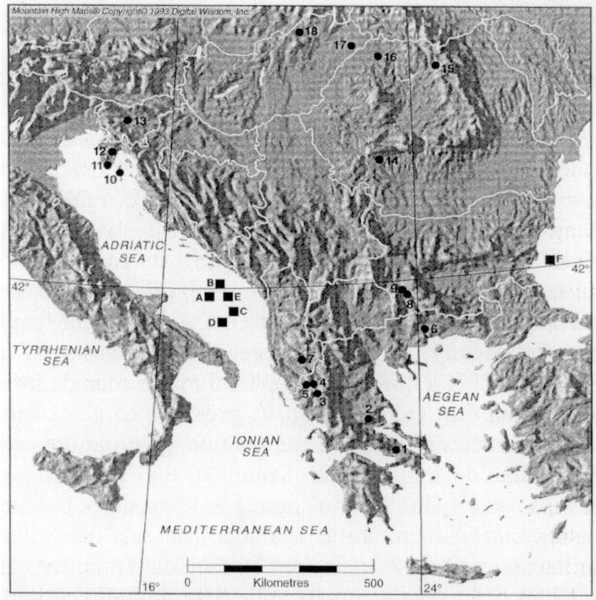

Figure 2. Location of sites discussed in text.

Terrestrial sites
1. Kopais, Greece (38º26'N, 23º03'E; 95m a.s.l.)
2. Xinias, Greece (39°3'N; 22°16'N; 480m a.s.l.)
3. Ioannina, Greece (39º45'N, 20º51'E; 470m a.s.l.)
4. Lake Gramousti (39°78'N, 20°35'E; 285m a.s.l.)
5. Lake Tseravinas, Greece (c. 39°78'N, 20°30'E; 450m a.s.l.)
6. Tenaghi Philippon, Greece (41º10'N; 24º20'E; 40m a.s.l.)
7. Lake Maliq, Albania (40°21'N, 20°25'E; 818m a.s.l.)
8. Sucho Ezero, Bulgaria (42°04'N; 23°35'E; 2095m a.s.l)
9. Sedmo Rilsko, Bulgaria (c. 42°N; 23°E; 1900m a.s.l.)
10. Lake Vrana, Island of Cres, Croatia (c. 44°50'N, 14°30'E; 10-15m a.s.l.)
11. Pula, Istria, Croatia (c. 44°52'N; 14°E)
12. Vela Draga sites, Istria, Croatia (45°18'-45°21'N; 14°9'-14°13'E; c. 200m a.s.l.)
13. Ljubljana Moor, Slovenia (45°59'N; 14°25'E; 300m a.s.l.)
14. Taul Zanogutii, Romania (45°19'N, 22°48'E; 1840m a.s.l.)
15. Iezerul Calimani, Romania (47°19'N, 25°16'E; 1650m a.s.l.)
16. Preluca Tiganului, Romania (47°48'83"N, 23°31'91"E; 730m a.s.l.)
17. Bátorliget marsh, Hungary (c. 47°45'N, 22°28'E; 130m a.s.l.)
18. Kis-Mohos Tó peat bog, Hungary (48°24'40"N, 20°24'30"E; 310m a.s.l.)

Marine sites
A. Adr. 309, Adriatic Sea (41º49.1'N, 17º11.4E; 929m water depth)
B. Adr. 353 Adriatic Sea (42°07.2'N, 17°38'E; 1207 water depth)
C. KET 8216 Adriatic Sea (41°31'N, 17°59'E; 1166m water depth)
D. IN9 Adriatic Sea (41°20'N, 17°40'E; 1234m water depth)
E. MD90-917 Adriatic Sea (41°50'N, 17°50'E)
F. Black Sea South (42°06'N; 28°30'E; 94m water depth)

Picea, *Abies*, *Juniperus*, *Corylus*, *Ulmus* and *Alnus* during the Late Pleniglacial. Because of its overrepresentation in marine sediments, *Pinus* is outside the main pollen sum, so its relative contribution is more difficult to assess, but its presence in the surrounding landmasses is not disputed. A more recent pollen sequence from marine core KET 8216 in the south Adriatic (Rossignol-Strick *et al.*, 1992) is supported by both an oxygen isotope stratigraphy and ten AMS radiocarbon determinations, with a basal date of 16,700 yr. BP. The lower part of the pollen diagram shows presence of deciduous *Quercus* (up to 10%) *Picea*, *Abies* and *Alnus* (up to 5%) and low occurrence of evergreen *Quercus*, *Corylus*, *Ulmus* and *Fagus*. Again, *Pinus* dominates the glacial pollen spectra.

Although the last decade has not witnessed a significant increase in the number of new Balkan sequences extending into the LGM, a prominent feature has been an improvement in the quality of the new records in terms of resolution and chronological control. This allows the application of more rigorous criteria in assessing the palaeobotanical evidence for the presence of glacial tree populations. Thus, here I consider records fulfilling the following minimum requirements: (1) pollen sequences extending into the LGM (and ideally beyond) as determined by radiometric dates from the LGM interval or at least two dates bracketing the LGM, or failing that, using independent, non-palaeobotanical data (such as oxygen isotope stratigraphies in marine pollen sequences). (2) charcoal fragments from palaeosols or archaeological sites of LGM age as determined through direct radiocarbon dating. This represents a departure from earlier practice, where presence of arboreal pollen in early post-LGM sediments has been used to infer presence of refugial populations. However, given that initial warming and associated ice melting started as early as *ca.* 19,000 cal. yr. BP (16,100 yr. BP), with a sea-level rise of 15 m in 500 years (Lambeck *et al.*, 2002), such arboreal appearances may represent rapid immigration rather than *in situ* refugial populations. In addition, charcoal fragments of trees which have been assigned a LGM age on the basis of their relative stratigraphical position (rather than direct dating) are not considered as they may represent reworked material into a sequence. As regards palynological data, the determination of regional presence is a well-known problem given species differences in pollen productivity and dispersibility. Here, continuous percentage curves through the LGM are taken to reflect at least minimum abundance within the hydrological catchment. The following adjustments to this are applied: for palynologically over-represented taxa *Pinus* and *Quercus* minimum values of 10% and 5% are required to infer local presence; for under-represented taxa such as *Abies*, *Pistacia* even discontinuous appearance constitutes evidence for presence (see Tzedakis, 1993 for more detailed discussion).

On this basis, the following sequences are discussed: (1) Xinias: a new investigation from Lake Xinias (Digerfeldt *et al.*, 2000), aiming to reconstruct lake-level changes on the basis of sediment and pollen analyses from a series of cores with several radiocarbon dates, provided additional pollen records that extended back to *ca.* 45,000 cal. yr. BP. The results allow a more precise determination of the LGM interval and reproduce many of the features of Bottema's (1979) earlier diagram, showing the presence of *Pinus*, *Quercus*, *Betula*, *Alnus*, *Juniperus* and *Abies* populations in the area. (2) Kopais: two new pollen records have emerged from Kopais in the last decade. A published pollen diagram from core K93, a 60 m sediment sequence from the centre of the basin (Tzedakis, 1999) has recently been supplemented by six new AMS radiocarbon measurements spanning the interval

10,310-31,780 yr. BP (Griffiths *et al.*, 2002; Tzedakis *et al.*, 2002). The K93 record suggests the LGM presence of deciduous *Quercus*, *Pinus*, *Juniperus* and *Abies* as well as small populations of evergreen *Quercus* and *Pistacia* in the area. The second record was derived from a 120 m core from the basin margins (Okuda *et al.*, 2001). The Lateglacial and Holocene section of the core was supported by six radiocarbon dates (some of them showing chronological reversals), with the oldest being 12,360±85 yr. BP. The LGM interval showed similar presence of arboreal taxa as K93. (3) Ioannina: A new high-resolution record (mean sampling interval 225 yrs) with improved chronological control from Ioannina (I-284) spanning the last 133 ka, has recently become available (Tzedakis *et al.*, 2002). The upper part of the sequence is supported by 19 AMS dates (Frogley, 1997; Lawson, 2001). The LGM is bracketed by dates of 15,330±140 and 20,760±230 yr. BP and the pollen record corroborates earlier results by Bottema (1974) and Tzedakis (1993). Mean AP values during the LGM are 35% (min. 24%; max. 54%) and the record suggests the presence of deciduous *Quercus*, *Abies*, *Pinus*, *Betula*, *Fagus*, *Ulmus*, *Carpinus Betulus*, *Ostrya/Carpinus orientalis*, and *Alnus* within the Ioannina catchment. (4) Marine pollen records: results from two marine cores extending into the LGM in the Adriatic and supported by radiocarbon dating and isotope stratigraphies have been published. Core IN9 from the central Adriatic shows LGM presence of *Pinus*, *Quercus*, *Abies*, *Corylus*, *Ostrya*, *Betula* and *Ulmus* pollen (Zonneveld, 1996). A higher resolution record (MD90-917) from the same area of the central Adriatic shows LGM presence of *Pinus*, deciduous *Quercus*, *Abies*, *Picea*, *Carpinus*, *Fagus*, *Alnus* and *Ulmus* (Combouriet-Nebout *et al.*, 1998). Both sequences show similarities with the earlier diagrams from marine cores in the area (Grüger, 1975; Rossignol-Strick *et al.*, 1992), integrating the pollen signal from adjacent Italy and the western Balkans and pointing to the presence of temperate tree populations in these broad areas. (5) Finally, anthracological evidence has recently emerged from a palaeolithic site in the karst cave Sandalja II, in Istria, Croatia where charcoal samples were collected from hearths (Culiberg & Šercelj, 1995). Although not strictly within the LGM interval, of interest is the identification of three *Fagus sylvatica* pieces, dated at 21,740±45 yr. BP. Lower in the sequence another two fragments of *Fagus* are dated at 25,340±170 yr. BP.

In addition to the above sequences, a number of other records extending into the post-LGM have recently become available from the Balkans. Although they cannot be used as evidence for LGM presence of tree populations, they may provide some idea of relative proximity to refugia and complement what is decidedly a very low spatial coverage of palaeoecological sites in the Balkans. In western Greece, a pollen diagram is available from Lake Tseravinas (450 m a.s.l.), supported by five AMS radiocarbon dates (Turner & Sánchez Goñi, 1997). The base of the diagram has a date of 15,300±160 yr. BP and shows low but continuous presence of *Quercus*, *Pinus*, *Betula* and *Abies* pollen. Similar results are also found in nearby pollen records from Lake Gramousti, *ca.* 285 m a.s.l. (Willis, 1992) and Lake Maliq, *ca.* 818 m a.s.l., Albania (Denèfle *et al.*, 2000), which, however, extend only to *ca.* 12,000-13,000 yr. BP. In fact Gramousti shows the lowest Lateglacial arboreal values, which may be a reflection of the relatively reduced presence of tree populations at lower altitudes. In Bulgaria, there are no sites extending beyond the Lateglacial (*i.e.* 13,000 yr. BP). Those available from the Rila Mountains, SW Bulgaria show the presence of *Pinus*, *Betula*, *Abies*, *Quercus* and *Corylus* at *ca.* 12,000 yr. BP, followed by their near-complete disappearance during the Younger

Dryas (Bozilova *et al.*, 1996; Bozilova & Tonkov, 2000). Although of lower resolution, additional evidence is available from a marine sequence from the southwestern Black Sea, extending to *ca.* 15,000 yr. BP and showing presence of *Quercus*, *Betula*, *Alnus*, *Corylus* and *Fagus* pollen in the lower part of the diagram (Shopov *et al.*, 1992). In Croatia, a new pollen diagram is available from Lake Vrana, on the Island of Cres, on the Dalmatian coast (Schmidt *et al.*, 2000). Chronological control is provided by four AMS dates on bulk samples, with a basal date of 14,445±45 yr. BP. The base of the diagram shows high values of *Pinus* (up to 50%) and low but continuous percentages of *Betula*, *Picea*, *Abies*, deciduous and evergreen *Quercus*, *Ulmus*, *Carpinus betulus*, *Acer* and *Fagus*. However, the authors consider this assemblage to be redeposited by erosion and slumping on the basis of echography. They also suggest that a hiatus is present above that zone, with the Alleröd interval missing (Schmidt *et al.*, 2000). An alternative interpretation is that, given the carbonate nature of the sediments, the basal date is prone to 'hard water' effects and a more realistic age could be as much as two thousand years younger (*i.e.* 12,500 yr. BP) and therefore belonging to the Alleröd. Just to the north, in the Istrian peninsula, a new charcoal study from karstic caves provides additional information on the Lateglacial presence of trees (Fletcher, in press). Two of the caves examined, Pupicina and Vesanska have levels dating to *ca.* 11,500 yr. BP. The charcoal assemblages from these levels are dominated by *Pinus*, but of interest is the occurrence of fragments of deciduous *Quercus*, *Acer*, *Fraxinus* and *Fagus* in the latter part of the Lateglacial interstadial and the Younger Dryas chronozones. In Slovenia, a more recent diagram from Zamedvejca, Ljubljana Moor (Culiberg & Šercelj, 1996), extends according to the authors back to 14,000 yr. BP, but this an extrapolated age, with the oldest available radiocarbon date being 10,800±200 yr. BP. The base of the diagram suggests presence of *Pinus*, with some *Picea* and *Betula*; *Quercus*, *Tilia* and *Ulmus* pollen appear later in the Lateglacial. In Romania, although a country with a long and distinguished palynological tradition under E. Pop, none of the early sequences investigated was radiocarbon dated. The situation is beginning to be rectified, with two new records supported by several radiocarbon dates recently published by Farcas *et al.* (1999). One site, Iezerul Calimani in the northeast branch of the Carpathians has a conventional radiocarbon date from the basal pollen zone of 14,800±1100 yr. BP. The pollen diagram from this basal zone shows high values of *Pinus* (*ca.* 60%) and low but continuous (<5%) curves of *Picea* and *Betula*. Noteworthy in the same zone is the appearance of pollen grains of *Quercus*, *Ulmus*, *Corylus* and *Alnus*, but their occurrence is too sporadic and low to suggest the presence of local populations. More importantly, the basal radiocarbon date is the only one from the Lateglacial interval (the next date up the sequence is at 9760±600 yr. BP) and should therefore be treated with caution. The second site is Taul Zaogutii in the western part of southern Carpathians. The basal zone shows relatively low presence of *Pinus* (20%) with some *Betula* and *Alnus*, interspersed with non-polleniferous samples. However, its age is unclear, although the ensuing zone has a date of 11,140±75 yr. BP. A third site from Romania, Preluca Tiganului, a former crater lake in NW Romania, has also been investigated and results are beginning to emerge in a series of instalments (Wohlfarth *et al.*, 2001). Although the sequence extends only back to *ca.* 14,700 cal. yr. BP (*ca.* 12,500 yr. BP), the level of detail, combining pollen, plant macrofossil, charcoal and sediment analyses on a multiple cores supported by AMS radiocarbon dates, is unparalleled in Balkan studies. The lower parts of the sequence show high (40-50%) values of *Pinus* along with low but continuous *Betula* percentages. Plant macrofossils of both *Pinus* and *Betula* are encountered slightly later (*ca.* 14,500 cal. yr. BP), followed by macrofossils of *Populus*, *Picea* and *Larix* (*ca.* 14,100 cal. yr. BP).

Finally, although geopolitically outside the Balkans, recent palynological and charcoal evidence from Hungary provides relevant information on the Pleniglacial and Lateglacial distribution of trees. A pollen record from Kis-Mohos Tó in northeastern Hungary shows the continuous presence of *Pinus* (up to 40%), *Picea*, *Larix*, *Betula* and *Juniperus* along with sporadic occurrences of *Quercus*, *Corylus* and *Ulmus* grains during the Lateglacial (Willis *et al.*, 1997). Chronological control is provided by nine bulk and four AMS dates, but there is only one pre-Holocene date, taken from the base of the sequence (12,495±95 yr. BP). Further to the east, near the Hungarian-Romanian border, palynological results from Bátorliget marsh are also available from the Lateglacial, showing similar patterns to Kis-Mohos Tó: continuous presence of *Pinus* (up to 40%), *Picea*, *Betula*, *Juniperus* and *Salix* along with sporadic occurrences of *Quercus*, *Ulmus*, *Corylus*, *Abies*, *Alnus* and, notably, *Ostrya*/*Carpinus orientalis* grains (Willis *et al.*, 1995). Chronological control is based on bulk conventional dates, from the Holocene section of the core, with the oldest date being 9339±130 yr. BP at 85.5 cm depth. The pre-Holocene sediments extend from 103 to 179 cm depth and are characterised by allogenic inorganic material, representing erosion from the surrounding slopes and the authors refrain from assigning a basal age to the sequence (Willis *et al.*, 1995). However, in a separate publication Willis *et al.* (2000), extrapolating from the early Holocene dates, present the Bátorliget data on a timescale extending back to 17,200 cal. yr. BP (*ca.* 15,000 yr. BP). A more conservative approach based on the identification of the Younger Dryas chronozone between 100-125 cm and pollen-stratigraphical similarities with the Kis-Mohos Tó record, would suggest that the basal age of the Bátorliget sequence is probably nearer 13,000 yr. BP and that the occurrence of pollen of temperate trees is of Lateglacial age. Direct evidence on the Pleniglacial distribution of trees has been provided by an important macroscopic charcoal study from palaeosols buried in loess and archaeological sites in Hungary (Willis *et al.*, 2000). Data from thirty sites are presented supported by radiocarbon dates on charcoal material and show the presence of *Pinus*, *Picea*, *Larix*, *Betula* and *Salix* with dates ranging from 32,500 to 16,750 yr. BP along with one site where *Carpinus betulus* is identified and dated to 26,962±657 yr. BP. These results establish the presence of boreal trees in that area during the LGM and represent an important advance in our understanding. However, the authors also invoke the existence of oases of thermophilous trees in favourable microhabitats. This is based on (i) the occurrence of temperate tree pollen from the Bátorliget and Kis-Mohos Tó, which, however, is more likely to be Lateglacial in age; and (ii) the occurrence of one charcoal sample (out of 30) of *Carpinus betulus*, which in fact is Marine Isotope Stage 3 in age and could easily represent immigration from the south during an interstadial phase. An alternative interpretation of the evidence would suggest presence of boreal tree populations in Hungary, with temperate tree immigration during interstadials.

DNA EVIDENCE

The advent of modern molecular techniques for studying the genetic diversity of modern populations has recently provided significant insights into species lineages, migration routes and refugial locations (see reviews by Hewitt, 1996, 1999, 2000). A number of comprehensive studies on plant and animal species sampled across their range are now beginning to emerge from Europe (*e.g.* Taberlet *et al.*, 1998; Hewitt, 1999; Petit *et al.*, 2002) and these show that the southern European peninsulas were the source of most of the postglacial colonisation of Europe. The Balkans emerge as the most important of the three refugial peninsulas, being the source for all species in the east and for many species in the west of northern Europe (*e.g.* Hewitt, 1999, 2000).

Hewitt (1996, 1999, 2000) envisaged a 'leading edge' expansion from populations at the northern limits of the refugial areas, with long-distance dispersers establishing colonies far ahead of the main population and rapidly expanding to fill the area before others arrived. Over time, multiple founding events would lead to loss of alleles and to increased homozygosity. This tendency to homozygosity may be increased by climatic reversals, like the Younger Dryas, which would cause population bottlenecks. One implication of the leading edge expansion is that refugial populations further to the south may have been blocked and therefore may have had a much reduced contribution to the postglacial recolonisation of northern Europe (Hewitt, 2000). In the case of the Balkans, this predicts that populations from the southern part of the peninsula would have distinct genetic signatures not found in northern Europe. This can indeed be seen in the genomic structure of *Alnus glutinosa* (King & Ferris, 1998) with unique haplotypes in Greece and Bulgaria, while the rest of northern Europe shares haplotypes from the northern Balkans. Hewitt (1999) suggests a similar situation for deciduous *Quercus* and *Fagus*, with the northern Balkans contributing to the northward colonisation. However, it must be borne in mind that the sampling coverage in these studies did not extend to the southern Balkans and therefore, this hypothesis could not be properly tested. A similar lacuna in areal coverage unfortunately persists even in more recent compilations (*e.g.* Petit *et al.*, 2002).

The genetically-driven notion of the northern Balkans as the source of the postglacial colonisation of Europe by a number of temperate trees may at first appear to be in conflict with the palaeobotanical record. With the exception of *Corylus* with sporadic Lateglacial occurrences throughout Central Europe and *Alnus* whose Lateglacial expansion appears to originate in the northern Balkans (Huntley & Birks, 1983), temperate tree species have not left any palynological indication of LGM stations near this northern edge. However, before the inadequacies of the fossil record are prematurely emphasised, it is important to bear in mind the history of species migration and understand the nature of the information provided by molecular genetics. Lateglacial population expansion may include advances and extirpations due to climatic oscillations (*e.g.* the Younger Dryas) and thus northern remnants of an earlier advance may form what Brewer *et al.* (2002) termed "secondary refugia" from which subsequent waves of expansion originate. Survival in these secondary refugia during the Younger Dryas millennium may have been accompanied by genetic modifications of the original signature of primary refugia. It is possible, therefore, that the molecular evidence is picking up the "secondary" signal from the northern boundary, which has overprinted the original signature. Separating the complicating effects of these series of events and resolving the location of the northern edge during glacial maxima, therefore, requires spatially detailed genetic and fossil evidence which is still largely unavailable (Tzedakis *et al.*, 2003).

EMERGENT PICTURE

On the basis of the fossil and genetic evidence outlined above, the following generalisations on the distribution of refugial populations in the Balkans can be made. It now appears that populations of *Pinus*, *Picea*, *Larix*, *Betula*, *Salix* and *Juniperus* were present in the northern Balkans (and indeed further north) during the LGM. However, with the exception of *Alnus* and also *Corylus*, the LGM stations of temperate trees were located further south, in mid-altitude zones and coastal areas. Although there are still considerable gaps in our knowledge, especially from the central Balkans, it must be remembered that with moisture availability being a limiting bioclimatic parameter for temperate tree growth, proximity to moisture sources (*i.e.* S. Adriatic, Ionian, Aegean and Black Seas) was critically important. The identification of refugial populations in western Greece and eastern Bulgaria is consistent with this view. In this respect, the find of macroscopic charcoal of *Fagus* with a date of 21,740±45 yr. BP in the Istrian cave (Culiberg & Šercelj, 1995) is puzzling. The estimated sea-level drop for the LGM would have led to a significant change in the extent of the Adriatic Sea, with the northern half subaerially exposed. This means that areas such as Istria would have been starved of a moisture supply and that the survival of species with distinct moisture requirements such as *Fagus*, would have been difficult. Perhaps the progradation of the Po River into the northern Adriatic may have had an influence on local climatic conditions. Alternatively, the age of the *Fagus* charcoal, which in fact is lying just outside the LGM interval, may belong to interstadial D-O 2 (Figure 1) and therefore represent immigration during a short interval of climatic amelioration. What emerges from the review of the evidence is that the northern edge of temperate tree populations would, in any case, have been south of 45°N, probably even further to south within the Balkan hinterland. Further to the south still, populations of Mediterranean species (*Olea*, *Pistacia*, *Phillyrea* and evergreen *Quercus*), whose distributions are today controlled more by minimum temperatures than by moisture availability, would have been in low altitude areas near the coast in southern and western Greece and possibly the Dalmatian coast.

This view of the LGM distribution of temperate refugial populations has been recently challenged by Stewart & Lister (2001) and Stewart (2003) who invoke the existence of "northern cryptic refugia" of temperate trees with seriously disjunct distributions north of the Iberian, Italian and Balkan peninsulas. This claim is based on three lines of evidence:

(1) The occurrence of fossils of temperate mammals, which today would normally be associated with deciduous woodland, in glacial deposits in northern Europe. However, this association is questionable, as closer examination of the present day distribution of every mammalian species referred to by Stewart & Lister (2001) reveals that their occurrence is not restricted to deciduous woodland, but includes open, montane and coniferous habitats (*e.g.* Mitchell-Jones *et al.*, 1999).

(2) Genetic evidence from cold water fish, land snails, sedges, rockferns and small mammals, indicating northern glacial refugial locations, in addition to the southern peninsulas. However, in view of species' individualistic responses, one fails to see how this evidence could be used to make any inferences on the locations of refugia of thermophilous trees. With the exception of *Corylus avellana* whose genomic structure suggests that postglacial recolonization of Europe originated from outside the classic refugial peninsulas (Palmé &

Vendramin, 2002), no genetic evidence exists for disjunct refugia of temperate trees far north of the southern European peninsulas. Only evidence from boreal tree species appears to suggest more northerly refugial distributions: a recent examination of cpDNA variation of *Betula pendula* reveals that the highest levels of variation are found near the Urals, Central Europe and S. Sweden and suggests that postglacial recolonisation originated from intermediate latitudes (Palmé *et al*., 2003). By contrast, molecular markers of *Fagus* (Demesure *et al*., 1996), *Quercus* (Petit *et al.,* 2002) and *Carpinus* (Grivet & Petit, 2003) show that the highest levels of variation are found in southern refugial peninsulas. This is in agreement with the palaeobotanical evidence discussed earlier, which shows a clear latitudinal distinction between boreal and temperate tree species in terms of their glacial distributions.

(3) Plant macrofossils: (a) *Carpinus betulus* charcoal in loess deposits in Hungary (Willis *et al*., 2000) (b) *Taxus* charcoal of LGM age from an archaeological setting in Slovakia (Litynska-Zajac, 1995; and (c) macrofossils of *Quercus*, *Corylus*, *Ulmus* and *Alnus* during the early postglacial (8500-8000 yr. BP) in the Scandes Mountains, as indicative of proximity of glacial refugia (Kullman, 1998). However, (a) as discussed earlier, is not LGM in age, but belongs to MIS 3; the charcoal pieces from the archaeological site in (b) are not dated directly and their chronological framework is not clear; and finally the evidence from (c) is not necessarily in conflict with the leptokurtic model of arrival of long-distance dispersers ahead of the main population.

As regards the nature of palynological evidence, the suggestion has been made that pollen productivity may have been drastically reduced during adverse climatic conditions, and thus trees may have been present in the landscape but were palynologically silent (Willis *et al*., 2000). This is certainly possible for individual trees, but it is difficult to see how tree populations could be sustained for four thousand years or more with severely curtailed sex lives. If temperate trees had somehow managed to survive undetected during the LGM north of the Alps, then during the Lateglacial warming their pollen productivity and reproductive capacity should have increased and this should have been detected palynologically in the form of northern nuclei from which dispersal originated. No such expansions are observed (*e.g.* Huntley & Birks, 1983). Instead isopoll maps show the steady advance of the expansion wave originating from southern refugial areas. In addition, no temperate tree expansions are detected during interstadial phases of MIS 3 in sites north of the Alps, which show increases in *Pinus*, *Picea*, *Betula*, *Larix*, *Juniperus* and *Salix* with occasional traces of *Corylus* (*e.g.* de Beaulieu & Reille, 1992; Müller *et al*., 2003). This represents a significant problem for the "northern cryptic refugia" concept, for even if the occasional temperate tree macrofossil of glacial age from this area comes to light, the absence of population expansion during interstadial periods of climatic amelioration needs to be addressed. A possible explanation may be that small size and reduced viability of populations and/or reduced competitive ability because of cold-stage adaptation, prevented any significant expansion. Hence, whether or not small northern refugia of temperate trees existed, their contribution to European recolonisation and their importance to the long-term survival of species appears to be quite limited.

In conclusion, the most likely scenario for the postglacial migration histories is a model similar to that proposed by Brewer *et al.* (2002) for deciduous *Quercus*, which may be more widely applicable to most temperate trees (excluding *Alnus* and

Corylus). This envisages the presence of primary refugia in the southern Balkans, south of *ca.* 45°C, which formed the initial nuclei for range expansion during the Lateglacial. The northward recolonisation was arrested during the Younger Dryas climatic reversal, with a subset of the expanding populations managing to survive in the northern Balkans, forming secondary refugia. From there range expansion resumed at the onset of the Holocene.

A LONGER-TERM VIEW

Perhaps the most important development in our understanding of Quaternary environments during the last decade has been the realisation of the pervasive and extreme nature of millennial-scale climate variability, especially during intervals of increased ice volume (*e.g.* Dansgaard *et al.*, 1993; Bond *et al.*, 1993) (Figures 3a and 3b). This has dispelled previous notions of glacial monotony and replaced them with a view of dynamic and unstable climate regimes. The discovery of dramatic millennial-scale air temperature fluctuations (Dansgaard-Oeschger [D-O] events) coeval with iceberg discharges and sea surface temperature variations (the most extreme of which are known as Heinrich events) in the North Atlantic throughout the last 110 ka has raised important questions about the downstream impact on terrestrial ecosystems. High-frequency oscillations in pollen values during the last glacial period have been known from sequences in southern Europe (*e.g.* Follieri *et al.*, 1998), but it is only recently that improved age estimates for these changes have indicated a link to North Atlantic variability (*e.g.* Allen *et al.*, 1999). Moreover, the phase relationship between North Atlantic climate oscillations and vegetation response has now been established by comparing pollen and proxy indicators of the ocean environment within the same deep-sea sequences off Portugal (*e.g.* Sánchez Goñi *et al.*, 2000; Roucoux *et al.*, 2001) and the western Mediterranean (Sánchez Goñi *et al.*, 2002). These sequences have provided the first unequivocal evidence of the immediate response of vegetation to North Atlantic millennial-scale variability in terms of contractions and expansions of tree populations. Rapid transmission of this variability through reduced moisture content of eastward-moving low-pressure systems and advection of polar air would have led to abrupt changes in terrestrial ecosystems across southern Europe, with the largest tree population crashes associated with Heinrich events and intermediate contractions corresponding to D-O stadials (Sánchez Goñi *et al.*, 2002).

Examination of palaeoclimate timeseries reveals that the amplitude of Heinrich events surpasses LGM levels in terms of temperature and precipitation declines, with Heinrich Event 4 (*ca.* 39,000 cal. yr. BP) being the most extreme of all. Records from long pollen sequences show that absolute minima in arboreal pollen values are coeval with Heinrich events and in most sites point to an almost complete disappearance of tree populations from the landscape (*e.g.* Follieri *et al.*, 1998). This requires a re-evaluation of our criteria used to infer the survival of refugial populations as the LGM is no longer viewed as coincident with the most extreme bioclimatic conditions in the Mediterranean. Instead, it is becoming apparent that what is now needed to establish the long-term presence of refugial populations during the last glacial is continuous records extending at least back to the last interglacial complex. In addition, any notion of long-term persistence of tree populations established under sampling resolution of *ca.* 1000 years or coarser, must now be re-assessed through more detailed records that are able to resolve any higher frequency variability.

In the Balkans very few records fulfil such criteria. They are all in Greece and are characterised by distinct local environmental settings, thus providing some initial insight into regional variability. The Ioannina basin is an intramontane plateau on the western flank of the Pindus mountain range and has a sub-Mediterranean climate (mean January temperature [T_{jan}] 4.6°C; mean July temperature [T_{jul}] 24.9°C; annual precipitation [P_{ann}] 1200 mm). Precipitation values are the highest recorded for an inland station in Greece and are a function of orographic uplift of air charged with moisture from the nearby Ionian Sea. Kopais, on the other hand is located on the Boeotian plain in the rain-shadow of the Pindus and has a eu-Mediterranean climate with little summer rainfall (mean T_{jan} 9°C; mean T_{jul} 27°C; P_{ann} 470 mm). Finally Tenaghi Philippon, located on the landlocked Drama plain, has a distinctly more continental climate with colder winters and incursions of cold polar air (mean T_{jan} 3.4°C; mean T_{jul} 23.9°C; P_{ann} 600 mm). Figure 3c shows the total temperate tree pollen percentages (i.e. arboreal pollen minus [*Pinus* and *Juniperus*]) from Ioannina I-284, Kopais K93 and Tenaghi Philippon over the interval 11,000-52,000 cal. yr. BP (see Supplementary Online Information in Tzedakis *et al.* [2002] for a full description of age models). Tenaghi Philippon (Wijmstra, 1969) shows complete absence of temperate tree populations during Heinrich events, interspersed with low interstadial increases (mainly *Quercus*). Kopais (Tzedakis, 1999; Tzedakis *et al.*, 2002) also shows significant population crashes during Heinrich events; temperate tree pollen percentages (mostly deciduous *Quercus* with some presence of evergreen *Quercus* and *Abies*) are higher than Tenaghi Philippon but are often too low to infer local presence during stadials. However, large and abrupt interstadial expansions (50-75%) suggest that refugial populations were never too far away. Ioannina, western Greece (Tzedakis *et al.*, 2002), on the other hand, shows a distinctly different pattern. Minima in temperate tree pollen percentages are usually above 20%, except during Heinrich Event 2 when they drop to 10%. The level of representation and continuity of pollen curves suggests the persistence of local populations of deciduous *Quercus*, *Abies*, *Ulmus*, *Corylus* and *Carpinus* throughout the climatic oscillations of the last 52 ka and indeed the last 133 ka. These results show that even within Greece there are distinct regional patterns in terms of the distribution of specific refugial centres.

During the last glacial, tree growth in Greece would have been limited by: (i) increased aridity; (ii) lower atmospheric CO_2 content, which in turn leads to reduced water-use efficiency of plants, exacerbating water stress; and (iii) minimum winter temperatures. Although all three sites experienced climatic oscillations throughout the last glacial, their impact on vegetation communities was determined by the extent to which these changes crossed ecological thresholds controlled by local intrinsic properties (Tzedakis *et al.*, 2002). The persistence of tree populations at Ioannina during the last climate cycle is seen as a function of continued moisture availability from the nearby Ionian Sea, counteracting the effects of increased aridity and reduced CO_2 concentrations. An additional factor is the degree of topographical variability, which determines the extent to which populations can shift altitudinally in response to climate change. Given that both temperature and CO_2 concentration increase with decreasing altitude, vertical migration allows populations, at least partly, to evade extirpation, provided that there is sufficient topographical variability to supply a range of microclimates suitable for survival. Of the three sites considered

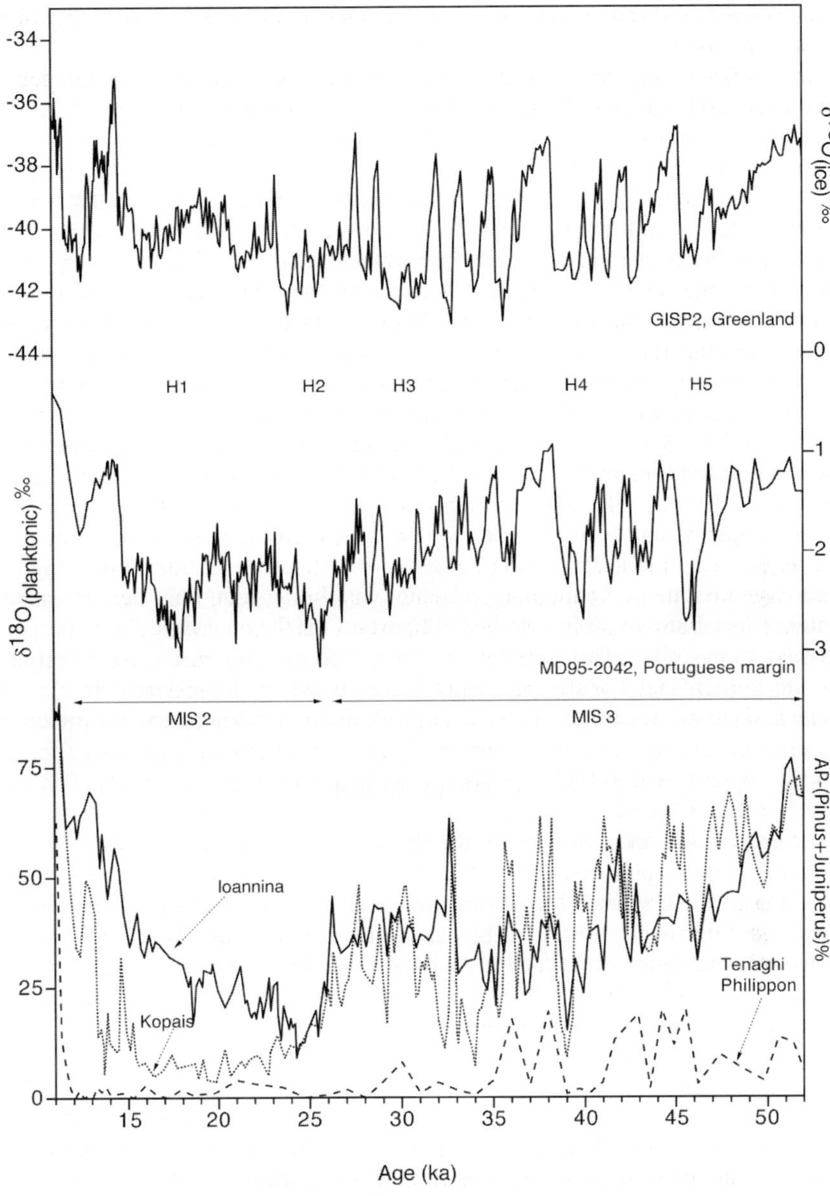

Figure 3. North Atlantic millenial-scale variability during the last glacial and response of vegetation in Greece over the interval 11-52 ka. 3a. Oxygen isotope composition of ice in the GISP2 record from Greenland, indicating variations in air temperature (Grootes & Stuiver, 1997; Meese et al., *1997). 3b. Oxygen isotope composition of planktonic foraminifera in marine core MD95-2042, off Portugal (Shackleton* et al., *2000), indicating variations in sea surface temperature. 3c. Changes in temperate tree population size as indicated by (Arboreal Pollen-[Juniperus+Pinus]) frequencies at three sites: Ioannina I-284, continuous line (Tzedakis* et al., *2002); Kopais K93, dotted line (Tzedakis, 1999) and Tenaghi Philippon (dashed line) (Wijmstra, 1969). Timing of Heinrich Events 1 to 5 (H1-H5) and Marine Isotope Stages (MIS) 2 and 3 also shown.*

here, Ioannina is the one located within the topographically most diverse landscape. Thus the synergy of these local factors at Ioannina appears to have buffered the most extreme effects of Quaternary climate variability, creating an environment of relative ecological stability. When combined with the longer I-249 record (Tzedakis, 1993), what emerges is that populations of many temperate tree species have persisted in the general area over several hundred millennia, at varying glacial and interglacial abundances. This long-term persistence coupled with genetic isolation may have evolutionary implications.

Although the prevailing view has been that Quaternary climate instability has inhibited speciation, molecular genetic data reveal that species continued to form in places where ecological stability allowed accumulation of genetic divergence. Thus while in lowland tropical forests most species formed before the Quaternary, clusters of recently diverged lineages are found in tropical mountains where moisture availability and varied topography are thought to have provided relatively stable habitats (*e.g.* Fjeldså & Lovett, 1997; Hewitt, 2000). By extension, Ioannina may provide a middle latitude analogue to the tropical mountain speciation centres. The western Balkans and the Pindus mountains consistently emerge as hotspots of endemism. This is partly a reflection of the geographical position and geological history of the region (*e.g.* Blondel & Aronson, 1999). However, local buffering from the extreme effects of Quaternary climate variability, not only led to reduced extinction rates, but may have also been important for the emergence of new species (Tzedakis *et al.*, 2002). This may provide an added urgency to the conservation of such long-term refugial areas and a combined palaeoecological-genetic approach could be used to pinpoint those populations that should be designated as protected.

CONCLUSIONS

The review of the evidence from the Balkans outlined above has adopted a conservative approach in designating the distribution of refugial sites. In doing so, it is entirely possible that areas have been excluded either because the criteria were too stringent, or published information has been unwittingly overlooked. However, the proposed scheme represents a falsifiable hypothesis which can be tested and refined continually as new evidence emerges. One hopes that ten years from now we will have a much more complete and detailed picture of the position of the northern edge, of the differences in species composition between various refugia and of the genetic character of populations.

As regards conservation issues, it is becoming increasingly clear that the protection of present populations in refugia is important for the long-term survival of species (and possibly for the emergence of new ones). This is a conservation priority not simply for the remote posterity (assuming that the next significant glaciation will be upon us in *ca.* 50,000 years from now (Berger & Loutre, 2002), but also in more proximal timescales. The evidence of the impact of millennial and centennial-scale variability in the North Atlantic on tree populations in southern Europe through aridity crises suggests that changes associated with future disruption of the North Atlantic ocean circulation, predicted by most global climate change scenarios (Stocker *et al.*, 2001) may place significant stress on present populations. Therefore, identifying areas of relative ecological stability where populations are more likely to survive the effects of such climate changes and ensuring their protection from anthropogenic pressures has now taken a new more urgent dimension. It is thus important that palaeobotanists proceed carefully by proposing protected area certification only when sufficient evidence, pointing to long-term local persistence of populations through multiple climatic oscillations, has accumulated.

ACKNOWLEDGEMENTS

I am indebted to the late Huw Griffiths for inviting me to attend the ESF Exploratory Workshop on Pattern and Process in Balkan Biodiversity, at Koper, Slovenia. This was the first and also the last time I ever met him, but the memory of his inexhaustible energy, joy of scientific enquiry and deep humanity remains. I thank Godfrey Hewitt, Ian Lawson, Katy Roucoux and Richard Preece for constructive comments on the manuscript and Alison Manson for assistance with graphics.

REFERENCES

Allen, H.D. 1986. *Late Quaternary of the Kopais Basin, Greece: Sedimentary and Environmental History.* PhD Thesis, University of Cambridge, Cambridge.

Allen, H.D. 1990. A postglacial record from the Kopais Basin, Greece. In: *Man's Role in the Shaping of the Eastern Mediterranean* (S. Bottema, G. Entjes-Nieborg & W. van Zeist, eds), pp. 173-182 Balkema, Rotterdam.

Allen, J.R.M., Brandt, U., Brauer, A., Hubberten, H.-W., Huntley, B., Keller, J., Kraml, M., Mackensen, A., Mingram, J., Negendank, J.F.W., Nowaczyk, N.R., Oberhänsli, H., Watts, W.A., Wulf, S. & Zolitschka, B. 1999. Rapid environmental changes in southern Europe during the last glacial period. *Nature* 400, 740-743.

Bennett, K.D., Tzedakis, P.C. & Willis, K.J. 1991. Cold stage refugia of north European trees. *Journal of Biogeography* 18, 103-115.

Berger, A. & Loutre, M.F. 2002. An exceptionally long interglacial ahead? *Science* 297, 1287-1288.

Beug, H.-J. 1968. Probleme der Vegetationgeschichte in Südeuropa. *Berichte der Deutschen Botanischen Gesellschaft* 80, 682-689.

Beug, H.-J. 1975. Changes of climate and vegetation belts in the mountains of Mediterranean Europe during the Holocene. *Biuletyn Geologiczny Warszawa* 19, 101-110.

Birks, H.J.B. & Line, J.M. 1993. Glacial refugia of European trees - a matter of chance? *Dissertationes Botanicae* 196, 283-291.

Blondel, J. & Aronson, J. 1999. *Biology and Wildlife of the Mediterranean Region.* Oxford University Press, Oxford.

Bond, G., Broecker, W., Johnsen, S., McManus, J., Labeyrie, L., Jouzel, J. & Bonani, G. 1993. Correlations between climate records from North Atlantic sediments and Greenland ice. *Nature* 365, 143-147.

Bottema, S. 1974. *Late Quaternary Vegetation History of Northwestern Greece.* PhD Thesis, Rijksuniversiteit te Groningen, Groningen.

Bottema, S. 1979. Pollen analytical investigations in Thessaly (Greece). *Palaeohistoria* 21, 19-40.

Bozilova, E.D. , Filipova, M., Filipovich, L. & Tonkov, S. 1996. Bulgaria. Type region BG-e, Vitosha, Rila, Pirin and Rhodopes Mountains. In: *Palaeoecological events during the last 15000 years: Regional Syntheses of Palaeoecological Studies from Lakes and Mires in Europe* (B. Berglund, H.J.B. Birks, M. Ralska-Jasiewiczova & H.E. Wright, eds), pp. 719-721. John Wiley & Sons, Chichester.

Bozilova, E.D. & Tonkov, S.B. 2000. Pollen from Lake Sedmo Rilsko reveals southeast European postglacial vegetation in the highest mountain area of the Balkans. *New Phytologist* 148, 315-325.

Brewer, S., Cheddadi, R., de Beaulieu, J.L., Reille, M. *et al.*, 2002. The spread of deciduous Quercus throughout Europe since the last glacial period. *Forest Ecology and Management* 156, 27-48.

Combouriet-Nebout, N., Paterne, M., Turon, J.-L. & Siani, G. 1998. A high-resolution record of the last deglaciation in the central Mediterranean Sea: palaeovegetation and palaeohydrological evolution. *Quaternary Science Reviews* 17, 303-317.

Culiberg, M. & Šercelj, A. 1995. Anthracotomical and palynological research in the palaeolithic site Sandalja II (Istria, Croatia). *Razprave IX, Slovenska Akademija znanosti in umetnosti* 36, 49-47.

Culiberg, M. & Šercelj, A. 1996. Slovenia. In: *Palaeoecological events during the last 15000 years : Regional Syntheses of Palaeoecological Studies from Lakes and Mires in Europe* (B. Berglund, H.J.B. Birks, M. Ralska-Jasiewiczova & H.E. Wright, eds), pp. 687-700. John Wiley & Sons, Chichester.

Dansgaard, W., Johnsen, S.J., Clausen, H.B., Dahl-Jensen, D., Gundestrup, N.S., Hammer, C.U., Hvidber, C.S., Steffensen, J.P., Sveinbjörnsdottir, A.E., Jouzel, J. & Bond, G. 1993. Evidence for general instability of past climate from a 250-kyr ice-core record. *Nature* 364, 218-220.

de Beaulieu, J.-L. & Reille, M. 1992. The last climatic cycle at Grande Pile. *Quaternary Science Reviews* 11, 431-438.

Demesure, B., Comps, B. & Petit, R.J. 1996. Chloroplast DNA phylogeography of the common beech (Fagus sylvatica L.) in Europe. *Evolution* 50, 2515-2520.

Denèfle, M., Lézine, A.-M., Fouache, E. & Dufaure, J.-J. 2000. A 12,000-year record from Lake Maliq, Albania. *Quaternary Research* 54, 423-432.

Digerfeldt, G., Olson, S. & Sandgren, P. 2000. Reconstruction of lake-level changes in lake Xinias, central Greece, during the last 40,000 years. *Palaeogeography, Palaeoclimatology, Palaeoecology* 158, 65-82.

Farcas, S., de Beaulieu, J.-L., Reille, M., Coldea, G., Diaconeasa, B., Goeury, C., Goslar, T. & Jull, T. 1999. First ^{14}C datings of Late Glacial and Holocene pollen sequences from Romanian Carpathes. *Comptes Rendues de l'Académie des Sciences de Paris, Sciences de la vie* 322, 798-807.

Fjeldså, J. & Lovett, J.C. 1997. Geographical patterns of old and young species in African forest biota: the significance of specific montane areas as evolutionary centres. *Biodiversity and Conservation* 6, 325-346.

Fletcher, W.J. in press. Late-glacial and Holocene vegetation and prehistoric wood-use in the Istrian Karst, Croatia. *BAR International Series*.

Florschütz, F., Menéndez Amor, J. & Wijmstra, T.A. 1971. Palynology of a thick Quaternary successions in southern Spain. *Palaeogeography, Palaeoclimatology, Palaeoecology* 10, 233-264.

Follieri,M., Giardini, M., Magri, D. & Sadori, L. 1998. Palynostratigraphy of the last glacial period in the volcanic region of central Italy. *Quaternary International* 47/48, 3-20.

Frenzel, B. 1968. The Pleistocene vegetation of northern Eurasia. *Science* 161, 637-649.

Frenzel, B. 1979. Europe without forest. *Geographical Magazine* August 1979, 756-761.

Frogley, M.R. 1997. *The biostratigraphy, palaeoecology and geochemistry of a long lacustrine sequence from NW Greece*. PhD Thesis, University of Cambridge, Cambridge.

Greig, J.R.A. & Turner, J. 1974. Some pollen diagrams from Greece and their archaeological significance. *Journal of Archaeological Science* 1,177-194.

Griffiths, S.J., Street-Perrot, A., Holmes, J.A., Leng, M.J. & Tzedakis, P.C. 2002. Chemical and isotopic composition of modern water bodies in the Lake Kopais Basin, central Greece. *Sedimentary Geology* 148, 79-103.

Grivet, D. & Petit, R.J. 2003. Chloroplast DNA phylogeography of the hornbeam in Europe: Evidence for a bottleneck at the outset of postglacial colonization. *Conservation Genetics* 4, 47-56.

Grootes, P. M. & Stuiver, M. 1997. $^{18}O/^{16}O$ variability in Greenland snow and ice with 10^3 to 10^5 year time resolution. *Journal of Geophysical Research* 102 (C12), 26,455-26,470.

Grüger, E. 1975. Pollenanalyse spätzpleistonzäner und holozäner Sedimente aus der Adria. *Geologische Jahrbuch* A29, 3-32.

Hewitt, G.M. 1996. Some genetic consequences of ice ages, and their role in divergence and speciation. *Biological Journal of the Linnean Society* 58, 247-276.

Hewitt, G.M. 1999. Post-glacial re-colonization of European biota. *Biological Journal of the Linnean Society* 68, 87-112.

Hewitt, G.M. 2000. The genetic legacy of the Quaternary ice ages. *Nature* 405, 907-913.

Huntley, B. 1993. Species richness in north-temperate zone forests. *Journal of Biogeography* 20, 163-180.

Huntley, B. & Birks, H.J.B. 1983. *An Atlas of Past and Present Pollen Maps for Europe: 0-13000 years ago*. Cambridge University Press, Cambridge.

King, R.A. & Ferris, C. 1998. Chloroplast DNA phylogeography of *Alnus glutinosa* (L.) Gaertn. *Molecular Ecology* 7, 1151-1161.

Kullman, L. 1998. Non-analogous tree flora in the Scandes Mountains, Sweden, during the early Holocene - macrofossil evidence of rapid geographic spread and response to palaeoclimate. *Boreas* 27, 153-161.

Lambeck, K., Yokoyama, Y. & Purcell, T. 2002. Into and out of the Last Glacial Maximum: sea-level change during Oxygen Isotope Stages 3 and 2. *Quaternary Science Reviews* 21, 343-360.

Lawson, I.T. 2001. *The Late Glacial and Holocene environmental history of Greece*. PhD Thesis, University of Cambridge, Cambridge.

Lang, G. 1970. Florengeschichte und mediterran-mitteleuropäische Florenbeziehungen. *Feddes Repertorium* 81, 315-355.

Litynska-Zajac, M. 1995. Anthracological analysis. In: *Complex of Upper Palaeolithic sites near Moravany, Western Slovakia* (J. Hromada & J. Kozlowski, eds), pp. 74-79. Jagellonian University Press, Krakow.

Meese, D.A., Gow, A.J., Alley, R.B., Zielinski, G.A. Grootes, P.M., P.A., Ram, Taylor, K.C., Mayewski, M. & Bolzan, JF. 1997. The Greenland Ice Sheet Project 2 depth-age scale: Methods and results. *Journal of Geophysical Research* 102(C12), 26,411-26,424.

Mitchell-Jones, A.J., Amori, G., Bogdanowicz, W., Kryštufek, B., Reijnders, P.J.H., Spitzenberger, F., Stubbe, M., Thissen, J.B.M., Vohralik, V. & Zima, J. 1999. *The atlas of European mammals*. T. & A.D. Poyser, London.

Mix, A., Bard, E. & Schneider, R. 2001. Environmental processes of the ice age: land, oceans, glaciers (EPILOG). *Quaternary Science Reviews* 20, 627-657.

Müller, U.C., Pross, J. & Bibus, E. 2003. Vegetation response to rapid climate change in Central Europe during the past 140,000 yr based on evidence from Füramoos pollen record. *Quaternary Research* 59, 235-245.

Okuda, M., Yasuda, Y. & Setoguchi, T. 2001. Middle to Late Pleistocene vegetation history and climate changes at Lake Kopais, Southeast Greece. *Boreas* 30, 73-82.

Palmé, A.E. & Vendramin, G.G. 2002. Chloroplast DNA variation, postglacial recolonization and hybridization in hazel, Corylus avellana. *Molecular Ecology* 11, 1769-1779.

Palmé, A.E., Su, Q., Rautenberg, A., Manni, F. & Lascoux, M. 2003. Postglacial recolonization and cpDNA variation of silver birch, *Betula pendula*. *Molecular Ecology* 12, 201-212.

Petit, R., Brewer, S., Bordács, S., Burg, K., Cheddadi, R., Coart, E., Cottrell, J., Csaikl, U.M., van Dam, B.C., Deans, J.D., Espinel, S., Fineschi, S., Finkeldey, R., Glaz, I., Goicoechea, P.G., Jensen, J.S., König, A.O., Lowe, A.J., Mandsen, S.F., Mátyás, G., Munro, R.C., Popescu, F., Slade, D., Tabbener, H., de Vries, S.M.G., Ziegenhagen, B., de Beaulieu, J.-L. & Kremer, A. 2002. Identification of refugia and post-glacial colonisation routes of European white oaks based on chloroplast DNA and fossil evidence. *Forest Ecology and Management* 156, 49-74.

Reid, E.M. 1935. British floras antecedent to the Great Ice Age. Discussion on the origin and relationship of the British Flora. *Proceedings of the Royal Society of London* B118, 197.

Rossignol-Strick, M., Planchais, N., Paterne, M. & Duzer, D. 1992. Vegetation dynamics and climate during the Deglaciation in the South Adriatic Basin from a marine record. *Quaternary Science Reviews* 11, 415-423.

Roucoux, K.H., Shackleton, N.J., de Abreu, L., Schönfeld, J. & Tzedakis, P.C. 2001. Combined marine proxy and pollen analyses reveal rapid vegetation response to North Atlantic millennial-scale climate oscillations. *Quaternary Research* 56, 128-132.

Sánchez Goñi M.F., Turon, J.-L., Eynaud, F. & Gendreau, S. 2000. European climatic response to millenial-scale changes in the atmosphere-ocean system during the Last Glacial period. *Quaternary Research* 54, 394-403.

Sánchez Goñi, M.F, Cacho, I., Turon, J.-L., Guiot, J., Sierro, F.J., Peypouquet, J.-P., Grimalt, J. & Shackleton, N.J. 2002. Synchroneity between marine and terrestrial responses to millennial scale climatic variability during the last glacial period in the Mediterranean region. *Climate Dynamics* 19, 95-105.

Schmidt, R., Müller, J., Drescher-Schneider, R., Krisai, R., Szeroczynska, K. & Baric, A. 2000. Changes in lake levels and trophy at Lake Varna, a large karstic lake on the Island of Cres (Croatia), with respect to palaeoclimate and anthropogenic impacts during the last approx. 16,000 years. *Journal of Limnology* 59, 113-130.

Šercelj, A. 1966. Pelodne analize Pleistocenskih in Holocenskih sedimentov Ljubljanskega barja. *Razprave IX, Slovenska Akademija znanosti in umetnosti* 9, 431-472.

Shackleton, N.J., Imbrie, J. & Hall, M.A. 1983. Oxygen and carbon isotope record of East Pacific core V19-30: implications for the formation of deep water in the late Pleistocene North Atlantic. *Earth and Planetary Science Letters* 65, 233-244.

Shackleton, N.J., Hall, M. A. & Vincent, E. 2000. Phase relationships between millennial scale events 64,000-24,000 years ago. *Paleoceanography* 15, 565-569.

Shopov, V.L., Bozilova, E.D. & Atanassova, J.R. 1992. Biostratigraphy and radiocarbon data of upper Quaternary sediments from the western part of the Black Sea. *Geologica Balcanica* 22, 59-70.

Stewart, J.R. 2003. Comment on "Buffered vegetation changes in a Quaternary refugium: evolutionary implications". *Science* 299, 825a.

Stewart, J.R. & Lister, A.M. 2001. Cryptic northern refugia and the origins of the modern biota. *Trends in Ecology and Evolution* 16, 608-613.

Stocker, T.F., Knutti, R. & Plattner, G.-K. 2001. The future of the thermohaline circulation - a perspective. In: *The Oceans and Rapid Climate Changes: Past, Present, and Future* (D. Seidov, B.J. Haupt & M.

Maslin, eds), pp. 277-293. Geophysical Monograph 126, American Geophysical Union, Washington D.C.

Taberlet, P., Fumagalli, L., Wust-Saucy, A.-G. & Cosson, J.-F. 1998. Comparative phylogeography and postglacial colonization routes in Europe. *Molecular Ecology* 7, 453-464.

Turner, C. & Sánchez Goñi M.F. 1997. Late Glacial landscape and vegetation in Epirus. In: *Klithi: Palaeolithic settlement and Quaternary landscapes in northwest Greece. Volume 2 Klithi in its local and regional setting* (G. Bailey, ed.), pp. 559-585. McDonald Institute Monographs, Cambridge.

Turner, J. & Greig, R.A. 1975. Some Holocene pollen diagrams from Greece. *Review of Palaeobotany and Palynology* 20, 171-204.

Tzedakis, P.C. 1993. Long-term tree populations in northwest Greece through multiple Quaternary climatic cycles. *Nature* 364, 437-440.

Tzedakis, P.C. 1994. Vegetation change through glacial-interglacial cycles: a long pollen sequence perspective. *Philosophical Transactions of the Royal Society of London* B345, 403-432.

Tzedakis, P.C. 1999. The last climatic cycle at Kopais, central Greece. *Journal of the Geological Society, London* 155, 425-434.

Tzedakis, P.C., Lawson, I.T., Frogley, M.R., Hewitt, G.M. & Preece, R.C. 2002. Buffered vegetation changes in a Quaternary refugium: evolutionary implications. *Science* 297, 2044-2047.

Tzedakis, P.C., Lawson, I.T., Frogley, M.R., Hewitt, G.M. & Preece, R.C. 2003. Reply to comment on "Buffered vegetation changes in a Quaternary refugium: evolutionary implications". *Science* 299, 825b.

van der Hammen, T., Wijmstra, T.A. & Zagwijn, H. 1971. The floral record of the Late cenozoic of Europe. In: *The Late Cenozoic Glacial Ages* (K.K. Turekian, ed.), pp. 391-424. Yale University Press, New Haven.

van der Wiel, A.M. & Wijmstra, T.A. 1987a. Palynology of the lower part (78-120) of the core Tenaghi Philippon II, Middle Pleistocene of Macedonia, Greece. *Review of Palaeobotany and Palynology* 52, 73-88.

van der Wiel, A.M. & Wijmstra, T.A., 1987b. Palynology of 112.8-197.8 m interval of the core Tenaghi Philippon III, Middle Pleistocene of Macedonia. *Review of Palaeobotany and Palynology* 52, 89-117.

West, R.G., 1980. Pleistocene forest history in East Anglia. *New Phytologist* 85, 571-622.

Willis, K.J. 1992. The late Quaternary vegetational history of northwest Greece. I. Lake Gramousti. *New Phytologist* 121, 101-117.

Willis, K.J., Sümegi, P., Braun, M. & Tóth, A. 1995. The late Quaternary environmental history of Bátorliget, N.E. Hungary. *Palaeogeography Palaeoclimatology Palaeoecology* 118, 25-47.

Willis, K.J., Braun, M., Sümegi, P. & Tóth, A. 1997. Does soil change cause vegetation change or vice versa? A temporal perspective fom Hungary. *Ecology* 78, 740-750.

Willis, K.J., Rudner, E. & Sümegi, P. 2000. The Full-Glacial forests of central and southeastern Europe. *Quaternary Research* 53, 203-213.

Wijmstra, T.A. 1969. Palynology of the first 30 metres of a 120 m deep section in northern Greece. *Acta Botanica Neerlandica* 18, 511-527.

Wijmstra, T.A. & Smit, A. 1976. Palynology of the middle part (30-78 metres) of the 120 m deep section in northern Greece (Macedonia). *Acta Botanica Neerlandica* 25, 297-312.

Wohlfarth, B., Hannon, G., Feurdean, A., Ghergari, L., Onac, B.P. & Possnert, G. 2001. Reconstruction of climatic and environmental changes in NW Romania during the early part of the last deglaciation (~15,000-13,600 cal yr BP). *Quaternary Science Reviews* 20, 1897-1914.

Zonneveld, K.A.F. 1996. Paleoclimatic reconstruction of the last deglaciation (18-8 ka B.P.) in the Adriatic Sea region; a land-sea correlation based on palynological evidence. *Palaeogeography Palaeoclimatology Palaeoecology* 122, 89-106.

5. QUATERNARY BIOTIC INTERACTIONS IN SLOVENIA AND ADJACENT REGIONS: THE VEGETATION

DAVID WATTS*†

INTRODUCTION

Slovenian vegetation is one of the most diverse in the world outside of the tropics, in terms of species numbers, yet surprisingly little is known of its vegetation history, the existence of past refugia, and the Quaternary/Holocene spread of the major vegetation communities. This paper considers this latter sequencing, the possible location of early Holocene refugia and their species dominants, and some anthropogenic and genetic influences.

GENERAL PATTERNS

Slovenia, lying at *ca.* 46ºN, 14ºE at the meeting place of the Alps and the Central European, Mediterranean, Dinaric and Pannonia regions, is an extraordinarily diverse region physically and biotically, with a relief consisting of plains, highlands, rocky mountains, karst valleys, ponds *etc*, 40% of which is underlain by carbonate rocks, mainly well karstified and dolomatised. (Cimerman & Gunde-Cimerman, 1996). In the karstic southwest of the country, a centre of biodiversity with one of the richest floras in the extra-tropical world (and certainly Europe), there are 3,100 species of plants in *ca.* 20,000 km^2, with over 80 endemics in this serendipitous mixture of mountain and sea coast regions. In Slovenia as a whole, there are 900 registered soil types, with 71 different earthworm species and 1000 different vegetation communities. It may be considered broadly to be a complex ecotone region, between which Alpine, Moderate Mediterranean, Submediterranean areas, Dinaric and Pannonic elements are widely spread and most plant species are of Central European, North European, Holarctic and Mediterranean derivations, along with the many endemics, and some cosmopolitan elements.

Overview

Despite the burgeoning amount of knowledge concerning the present-day flora (Figure 1), still relatively little is known in detail about the vegetation history of the Quaternary and the Holocene. In part, this is due to the paucity of suitable sites (lakes and mires) for pollen analysis in karstic areas, a feature which is well known to those working in limestone areas both here and elsewhere in the world. A second biotic feature is that the diversity of landscapes seems to be reflected in an equal diversity of pollen diagrams, where they do exist. The consequence is that, in Slovenia, no two pollen diagrams are entirely alike: and nowhere do such diagrams point to where refugia of vegetation might have been located, except in the most general sense. Elsewhere in the broad region, pollen and molluscan data from

H.I. Griffiths et al. (eds.), Balkan Biodiversity, 69–78.

Hungary (Willis, 1994) strongly indicate that trees grew as far north as north Hungary, growing in microenvironmental, particularly favourable sites, in the last full-glacial (32,500 to 16,500 yr. BP). Trees present included at least (one site only) *Pinus sylvestris*, *Picea* sp., *Betula* sp., *Juniperus* sp., *Salix* sp., *Larix* and (Willis *et al.*, 2000). A third major biotic feature is that, at the edge of Slovenia itself, the sea level of the Adriatic dropped by 100 m during the last glaciation, withdrawing southwards to a line between Ancona and Split (Šercelj, 1996): later, at the beginning of the Holocene, it rose again to –25 m and was still 2 m below present day levels in Roman times (Figure 2).

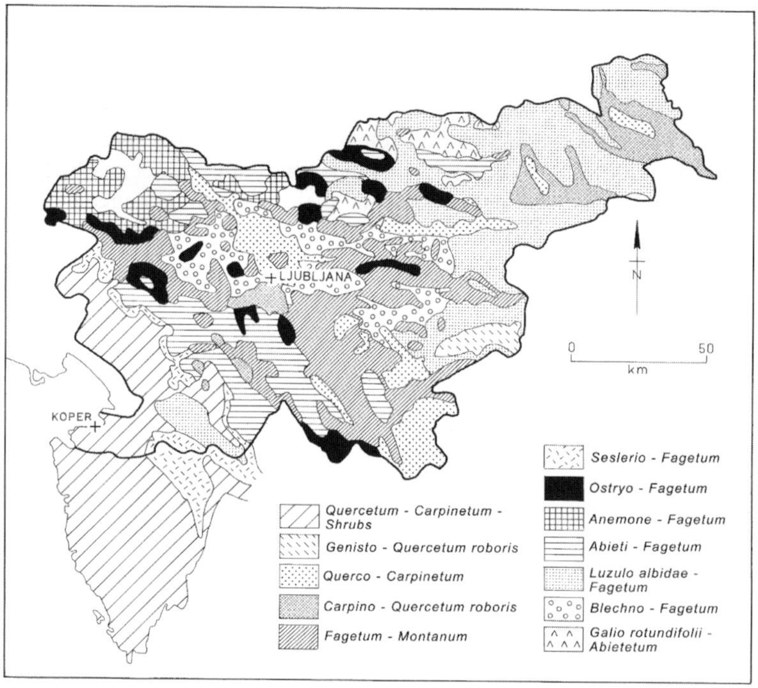

Figure 1. The major vegetation groups of Slovenia: a generalised map. (Redrawn from Šercelj, 1996, by permission of the Slovenian Academy for the Sciences and Arts.)

Lake deposits in Pula maar in central Hungary reveal that in the late Pliocene and Pleistocene there were significant and major fluctuations in the vegetation, with both population increases and decreases. Principal components analysis (PCA) indicates that two axes only, accounting for *ca.* 60% of the variation, characterise the data (Figure 3). The principle directions of variance lie between pollen from species which are now found in the northern boreal forest, *e.g. Pinus*, *Tsuga*, *Larix*, *Alnus* and *Betula*; and those found in subtropical/temperate forests, *e.g. Sequoia*, *Nyssa*, *Eucommia*, *Pterocarya*, *Sciadopitys*, *Taxodium*, *Fagus*, *Quercus*, *Ulmus* and *Corylus* (Willis & Whitaker, 2000). Some of these latter groups survived into the Holocene, but many did not, as the overall temperature cooled. A major threshold of change occurred at between ~ 2.72 to 2.67 Ma, *i.e.* just prior to the Pleistocene, when there was a disappearance of subtropical taxa, including *Sequoia*, *Nyssa*, *Eucommia*, *Pterocarya* and *Sciadopitys*, a period which represents the final appearance of many of these species in northern Europe, and which corresponds with the first extreme (glacial) excursion in the marine oxygen-isotope record, and the global development of large terrestrial ice-sheets in the northern hemisphere. Willis *et al.* (1989) suggest that the subtropical taxa are responding to increased isolation, in refuges which were becoming thus ever smaller; this is linked in with climatic change associated with Milankovich orbital periodicities, in particular precession (19-23 ka) and obliquity (41 ka), when the boreal and tropical pollen percentages are exactly out of phase with each other. In addition, internally driven non-linear responses of the climatic system over a 124 ka variance were as important (if not more important) in the relationships between vegetation, ice volume and isolation, relating in particular to the presence of increased amounts of dust in terrestrial systems especially at 95-124 ka.

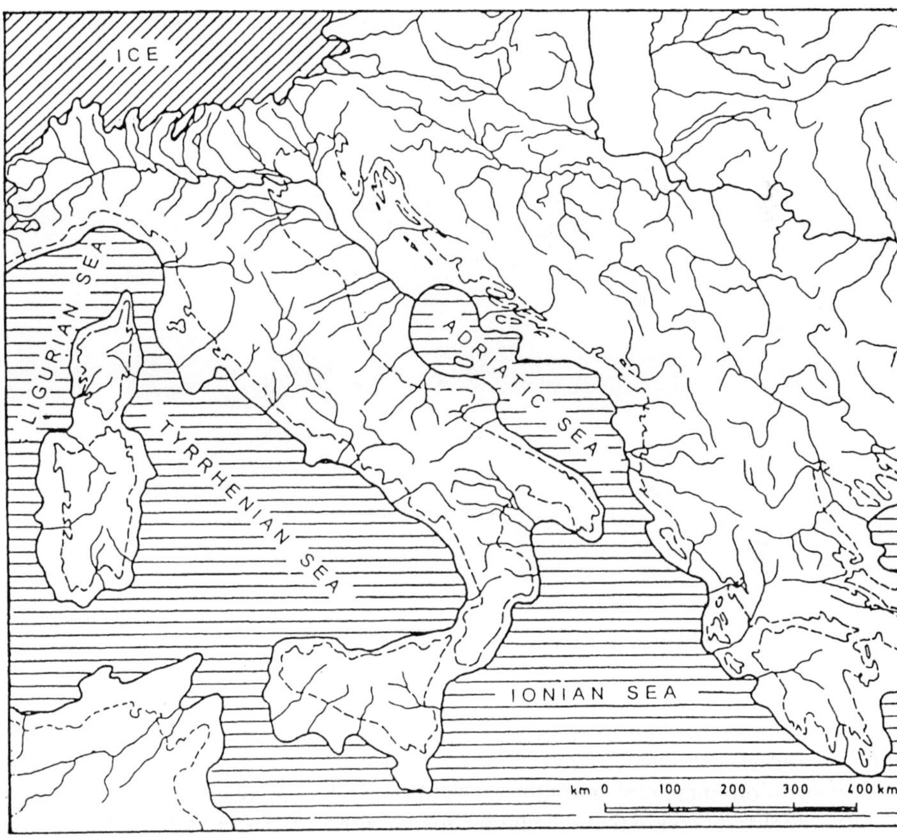

Figure 2. The presence of ice, and lowering of sea level, at the maximum of the Last Glaciation (ca. 20,000 yr. BP) The broken line represents present-day sea level. (Redrawn from Šercelj, 1996, by permission of the Slovenian Academy for the Sciences and Arts.)

Small ice sheets first formed in the northern hemisphere at ~ 10 Ma, with progressive intensification from ~ 2.75 Ma, and 41 ka cycles as a further feature. The Holocene as such commenced at *ca.* ~ 10 ka. North of the Alps, during the Late Glacial-Holocene transition, conifers dominated with broad-leaved deciduous species coming in after 10,000 yr. BP. In Hungary, under circumstances which were no doubt exceptional, in micro-environmental sites which were favourable, eight different tree types were present between 32,500 to 16,500 yr. BP, *viz. Pinus sylvestris, Pinus cembra, Betula* sp., *Picea* sp., *Juniperus* sp., *Larix* sp., *Carpinus betulus* and *Salix* sp. (Bennett *et al.*, 1991). To the south, the Mediterranean littoral was covered by a herb-dominated steppe, in which *Artemisia* sp. and some chenopods were most common, with possibly a few trees (Huntley & Birks, 1985). Over the last 430,000 years, at Ioannina, western Greece, Tzedakis (this volume) has shown that trees including *Quercus, Abies, Pinus, Betula, Fagus, Ulmus, Carpinus betulus* and *Acer* were continuously present throughout the last glacial. In between, in the Balkans, including Slovenia, there appears to have been a mixed system in which a cold steppe was pre-eminent, especially in the east, with increasing numbers of small-scale refugia in the west, along karst lakes and rivers, beside karst springs and in amphitheatre-like valleys that faced south – although no major refugium has yet been discovered.

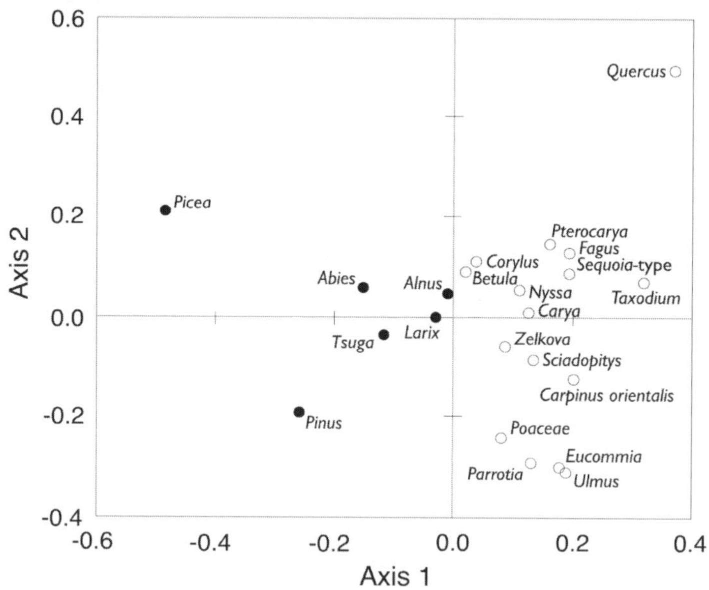

Figure 3. PCA covariance scatter plot of pollen base, Pula maar, central Hungary, late Pliocene. Only the first two axes are significant, accounting for ca. *60% of the total variance (after Willis* et al., *1999, reprinted by permission from* Nature *(397, 685-688, 1999) copyright 2004. Macmillan Publishers Ltd. www.nature.com/nature). Key: • types found in present northern Boreal forest; ∘ types found in subtropical/temperate forest.*

REFUGIA

Indirectly, the presence of a relatively large number of small or very small refugia, which may not have left traces in the pollen record, may perhaps be inferred by the relatively rapid colonisation of trees in Slovenia in the early Holocene, over a period of only 3,000 years (10,000 – 7,000 yr. BP) with, apparently, no tree immigration from outside the Balkan region. In any event, where tree migration was recorded, it was very slow (estimated at 25 m yr.$^{-1}$ in the case of manna ash – Huntley & Birks, 1985). But in the vast majority of cases in the Balkans, tree immigration from outside may be largely discounted, and it is clear that successional stages from pioneer to climax took place only within the framework of what was available for colonisation within the small refugia. Furthermore, within those refugia, no great readjustment to massive temperature changes would have been required for range expansion: mean winter temperatures reveal a cold and arid climate (a little like Korea, in East Asia), and mean summer temperatures at 16-18°C, were only 5°C lower than those of today. Moreover, there was no major body of ice to remove within the Balkans (Figure 2), as also in Korea (Kong & Watts, 1993), before recolonisation could begin; it simply was not present, despite the fact that 60% of land lies at over 3000 m a.s.l. Also, the fact that, where they exist, no two pollen diagrams are entirely similar, argues for mainly local developments.

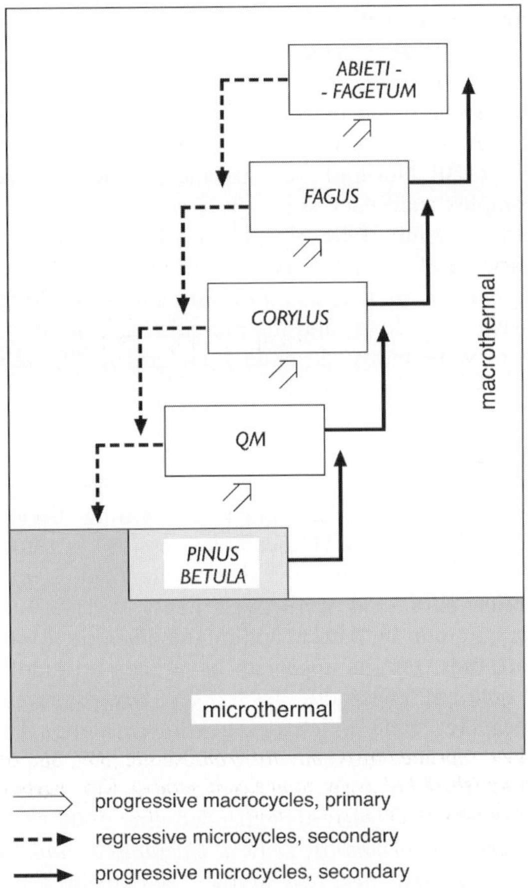

Figure 4. Suggested vegetational successional changes in Slovenia during the Holocene. (Redrawn from Šercelj, 1996, by permission of the Slovenian Academy for the Sciences and Arts.)

Although there are many variations, the major general succession in Slovenia during the Holocene suggested by Šercelj (1996) is shown in Figure 4. For Slovenia, in the *Quercetum-mixtum* phase, *Ulmus* had a leading role, but there are also massive representations too of *Tilia*, *Quercus* and *Acer* along with the already-present *Fagus*, *Corylus* and *Abies*. In the *Abieti-Fagetum* phase, *Picea* is sometimes very important, with 20-40% of pollen.

In parts of the northern Balkans, there are a few distinctive deviations away from this general pattern (Šercelj, 1996). Thus in the vicinity of Lake Bled (NW Slovenia), there is a notable presence of walnut, signifying a quite highly settled area. In the beginning of the pollen sequence of the extreme northeast of Slovenia, *Tilia* was extremely important; and on the higher moors generally (*e.g.* Ribnica, at *ca.* 1500 m a.s.l.), the mountain pine (*Pinus mugo*) commonly came into the sequence. Refuges of broad-leaved trees appear to have formed the basis of succession in central Slovenia, including genera such as *Rhamnus*, *Fagus*, *Salix*, *Acer*, *Ostrya*, *Fraxinus*, *Tilia* and *Ulmus*. Among submediterranean elements in the vicinity of Koper, on the southeast Slovenian coastline, *Carpinus orientalis*, *Quercus pubescens*, *Fraxinus ornus*, *Ostyra carpinifolia*, *Cornus mas*, *Cotinus*, *Coggyria* and *Buxus sempervirens* were present. It should be noted that most of the genera are typical of the broad Euro-Siberian-North American zone. In any event, no more primary succession occurred after 7000 yr. BP. North of the Alps, the situation was very different, with a *Corylus-Pinus* complex succeeded solely by *Corylus* in the Boreal, since *Corylus* had no real competitors. South of the Alps, already by the *Quercetum-mixtum* phase, a considerable variety of species recolonisation was present, with invasive representations of *Ulmus*, *Tilia*, *Quercus*, *Fraxinus* and *Acer*, along with the already-present *Fagus*, *Corylus* and *Abies*, and the pre-existing conifers (*Pinus* and *Picea*). All early complexes were partly dependent on grazing by reindeer, mammoth, European bison, moose, cave bear, alpine marmot, hare and their predators.

Species dominance in refugia

Some discussion has recently taken place concerning the nature of species dominants (and, accordingly, early Holocene succession) in refugia in the Balkans as a whole. Willis (1994) has pointed out that, while *Tilia*, *Ulmus* and *Corylus* are found on most pollen sites, and were most likely widespread, mixed *Quercus* woodland was most common in Greece, and mixed *Quercus*, *Ulmus*/*Betula*/*Picea* in Slovenia. In general, most refugia appear to have been at middle altitudes, higher altitudes being too cold and lower altitudes too dry. Several sucessional features do not, however, appear to relate to changing climate alone. Thus, *Pistacia* was common in southern Greece between 7000-8000 yr. BP. Its expansion may be caused by the presence of an open tree landscape in the early Holocene; and it became less common as trees (*Quercus*) expanded after 5000 yr. BP. The relatively late expansion of *Carpinius orientalis*, *Ostyra*, *Carpinus betulus*, *Fagus* and (in the Dalmatian coastlands) *Quercus ilex*, may perhaps be explained by a possibly longer establishment time, and (in respect of their tolerances of poor or lighter soils which were often degraded) soil erosion, which is thought to have commenced in the Balkans at various times between 7500 to 5000 yr. BP.

Although there were no major climatic changes in the Balkans during the early Holocene, relatively recent simulations of the COHMAP data of southeast Europe suggest that, although precipitation levels were lower, or similar to those today, there was a significant increase in available moisture, between 9000 and 6000 yr. BP, in particular as winters became cooler and summers became warmer (Huntley & Prentice, 1993; Wright, 1993). The seasonality of the climate was thus becoming more marked. The effects of these changes on vegetation are not clear, partly because little is known in detail of the cold tolerances of the several species involved. Species such as *Carpinus orientalis/ostrys*, *Carpinus betulus* and *Fagus* do not seem able to tolerate relatively high precipitation levels, and are not seen in such localities today (Ellenburg, 1988). In contrast *Quercus*, *Fraxinus* and *Acer* thrive in wet soil, but are in decline today (Polunin & Walters, 1985). One partial explanation may relate to the lower January temperatures. Accordingly, Huntley & Prentice (1993) suggest that, although *Carpinus*, *Abies*, *Fagus*, *Corylus* and *Quercus ilex* are all suited to roughly the same precipitation band (700 to 1200 mm yr^{-1}) the first three of these may be more tolerant of colder winter (January) temperatures than the others – however, this does not account for all types concerned.

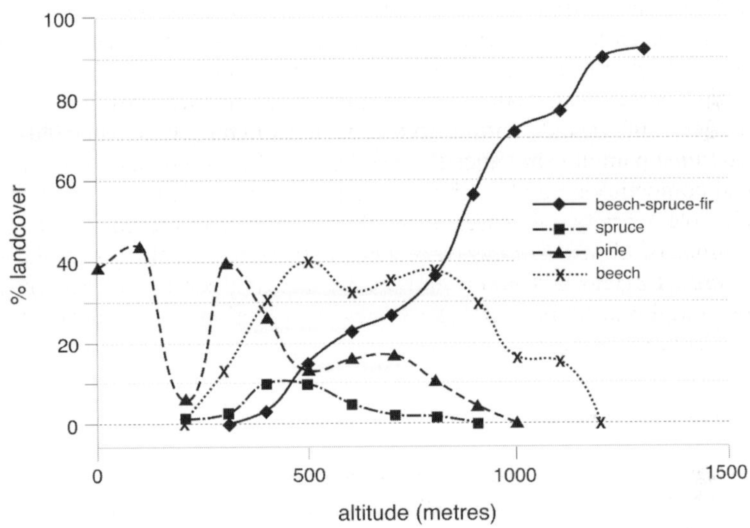

Figure 5. Present variation in forest type with altitude, expressed as a percentage of total land cover taken from modern satellite surveys (Ferrier, unpublished data, used with permission).

ANTHROPOGENIC INFLUENCES

From *ca.* 4,500 yr. BP, pollen data from sediment cores display a widespread anthropogenic influence, which ultimately created the fine details of the modern landscape. Prior to the beginning of the Holocene, it is likely that Palaeolithic populations, with the use of fire, had very little overall impact on the environment as a whole. More burning and clearing occurred in the Mesolithic phase, between 10,000 to 7,500 yr. BP, when much of the region was densely settled. As the Neolithic developed, from 7,000 to 5,500 yr. BP, and especially from 4,500 yr. BP, a decline in the amount and diversity of species in woodlands was heralded so as to make way for agriculture. At the same time, grass-like weeds and open-ground herbaceous types, such as *Cerealia* – type and *Plantago lanceolata* became much more common than formerly. In cleared areas and weed areas, many soils became depleted of nutrients, acidified and, in many places, podsolised, leading to the establishment of Ericacea, which of course are highly tolerant of acid soils; and probably too, a second burgeoning expansion, after 3,000 yr. BP, of *Fagus* and *Abies*, which are also well known for their ability to colonise acid and/or podzolic soils. Additionally, new tree types appeared, notably *Juglans*, *Castanea*, *Platanus* and *Olea*. Although present in the Balkans in the last glacial (they have poor pollen production and dispersal, so may have gone unremarked in pollen cores), most of these appear to have become established as landscapes opened up, and as a result of economic demand by ancient Greeks/Romans or their predecessors.

Today, as anthropogenic influence has continued to intensify, several major plant communities may be delimited, in which several of the major tree species noted above are still present as conspicuous features, but some (*e.g. Corylus*) are not. The first group of these form a series of Mediterranean-climate communities, in which may be distinguished (a) a Mediterranean evergreen forest, including *Pinus halepensis*, *Quercus coccifera*, *Q. ilex* and *Juniperus oxycedrus*; and, (b) in cooler territories, *Ostrya carpinifolia*, *Pistacea terebinthus*, *Fraxinus ornus* and *Acer monspesulanum*, forming a mixed Mediterranean forest. From 700-1700 m a.s.l. coniferous forests display a mixture of *Abies* and *Juniperus* species, with *Pinus mugo* above 1700 m a.s.l. Many of these divisions (Šercelj, 1996) may still be seen in modern satellite images, either according to altitude, or in patch differentials, reflecting recent planting strategies (Figure 5).

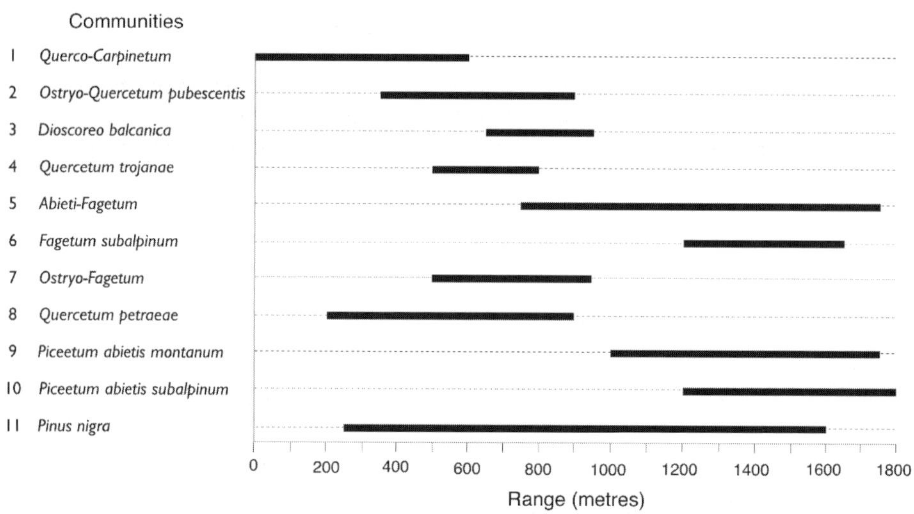

Figure 6. Present altitudinal range of selected Slovenian vegetation groups, derived by the author from modern satellite survey data (Ferrier, unpublished data, used with permission).

GENETIC INFLUENCE ON VEGETATION

One other major consequence attends the biotic changes of the Holocene over the last three million years or so, and this relates to the genetic composition of the various species involved. Within the climatic oscillations, it was necessary for species to be able to move, to adapt, or to go extinct, and examples of all these features are present in the record of the Balkans. I have noted that subtropical plant species went extinct at ~ 2.72 – 2.67 Ma, towards the beginning of the Pleistocene, in the duration of the first cold spell. Yet since then, no known species appear to have gone extinct and their percent ranges of tolerance are, on the whole, rather large (Figure 6), totally unlike, for example those of some species present in similar environmental circumstances on the Korean peninsula (Kong & Watts, 1993). Accordingly, during climatic warming, plant species here would expand their ranges, meet new neighbours and expand their environments – yet very few have produced new species. It used to be thought that the Pleistocene cold and warm phases increased the rate of change of species yet, for plants, there is very little evidence of it here; Pleistocene cold phases appear to have inhibited speciation. On the other hand, genetic divergence for birds, mammals, frogs, reptiles and fishes seems to have taken place fairly evenly during the Pliocene and Pleistocene and seems to have proceeded in a fairly unhindered way during the latter (Hewitt, 2000; Avise *et al.*, 1998). A further genetic consideration is that the expansion of *Corylus* and other species north of the Alps appears to have been by a 'leading-edge' model, in which those individuals nearest to fresh land would have, through opportunity, colonised first, with a limited representation of genes, making for a generically largely uniform population, and presumably blocking access by individuals with a more varied gene structure from more southerly populations. Later, with declining conditions, such groups of species would die off *in situ*, displaying no tendency to move south.

ACKNOWLEDGEMENTS

The author would like to thank Graham Ferrier (Hull) for supplying the modern satellite data used in this study. Grateful thanks are also due to the late Dr Huw Griffiths, who introduced me to the Balkans and was of immense help in the preparation of this paper.

REFERENCES

Avise, J.C., Walker, D. & Johns, G.C. 1998. Speciation durations and Pleistocene effects on vertebrate phytogeography. *Proceedings of the Royal Society of London* B 263, 1707-1712.

Bennett, K.D., Tzedakis, P.C. & Willis, K.J. 1991. Quaternary refugia of north European trees. *Journal of Biogeography* 18, 103-116

Cimerman, A. & Gunde-Cimerman, N. 1996. *Biodiversity in Slovenia*. Ljubljana

Ellenburg, H. 1988. *Vegetation Ecology of Central Europe*. 4th edition. Cambridge University Press, Cambridge.

Hewitt, G. 2000. The genetic legacy of the Quaternary ice ages. *Nature* 405, 907-915

Huntley, B. & Birks, H.J.B. 1983. *An atlas of past and present pollen maps for Europe, 0 – 13,000 years ago*. Cambridge University Press, Cambridge.

Huntley, B. & Prentice, I.C. 1993. In *Global climates since the last Glacial Maximum* (H.E. Wright Jr., J.E. Kutzbach, T. Webb III, W.F. Ruddiman, F.A. Street-Perrot, & P.J Bartlett eds.), pp. 136-169. University of Minnesota Press, Minneapolis.

Kong, W.S. & Watts, D. 1993. *The plant geography of Korea, with an emphasis on the alpine zones*. Kluwer, Dordrecht.

Polunin, O. & Walters, M. 1985. *A guide to the vegetation of Britain & Europe*. Oxford University Press, Oxford.

Šercelj, A. 1996. The origins and development of forests in Slovenia. *Slovenian Academy of Sciences and Arts* IV. 35: 5-142.

Willis, K.J. 1994. The vegetational history of the Balkans. *Quaternary Science Reviews* 13, 769-788.

Willis, K.J. & Whitaker, R.J. 2000. The refugial debate. *Science* 287, 1406-1407.

Willis, K.J., Kleczkowski, A. & Crowhurst, S.J. 1999. 124,000 – year periodicity in vegetation change during the late Pliocene epoch. *Nature* 397, 685-688.

Willis, K.J., Rudner, E. & Sumegi, P. 2000. The full glacial forests of central and southeastern Europe. *Quaternary Research* 53, 203-213.

Wright, H.E. 1993. Environmental determinism in Near Eastern prehistory. *Current Anthropology* 34, 458-469.

6. A QUANTITATIVE ASSESSMENT OF BALKAN MAMMAL DIVERSITY

BORIS KRYŠTUFEK*

INTRODUCTION

Although naturalists have long recognised the importance of faunal and floristic diversity in the Balkan Peninsula, extensive field research did not really begin in earnest until after the retreat of the Ottoman Empire from this part of Europe (*cf.* Miller, 1912; Doflein, 1921; Dautbegović, 1988, *etc.*). The continuous state of political instability in the Balkans over the last two centuries, resulting from regional nationalism and the interference of great powers (Glenny, 2000), discouraged collaboration among mammalogists and other natural scientists and inhibited the compilation of faunal data for the peninsula as a whole. Thus, the process of "balkanisation" manifests itself even in the natural sciences.

Detailed mammal studies with corresponding dot maps are available for parts of the Balkans, particularly in the case of insectivores and rodents (*e.g.* Markov, 1957; Ondrias, 1966; Vohralík, 1985; Niethammer, 1986; Vohralík & Sofianidou, 1987, 1991, 1992; Kryštufek, 1991; Petrov, 1992; Kryštufek & Petkovski, 2003). No attempt has been made to prepare a synthesis of the mammalian fauna of the Balkan Peninsula, however, an issue covered so far only at the European scale (Niethammer & Krapp, 1978, 1982, 1983, 1986; Stubbe & Krapp, 1993a,b; Krapp, 2001; Mitchell-Jones *et al.*, 1999). Not surprisingly, zoogeographical assessments are also few and fragmentary. Approaches have tended to be descriptive at best, with no clear methodology specified (*e.g.* Petrov, 1979). Most common are comparisons of mammal faunas between zoogeographical regions which have been defined on an *ad hoc* basis (*e.g.* Petrov, 1985), between geographic areas (*e.g.* Tvrtković *et al.*, 1985) or between vegetational entities (*e.g.* Gerasimov & Minkova, 1999). Hosey (1982) provides a rare example of a thorough study of biogeographic processes operating in the Balkans and affecting its present mammal composition.

Gaston & David (1994) demonstrate that the distribution of hot spots across Europe is not random but follows a pattern. Highest scores for species richness are in the latitudinal belt across the southern part of the continent and values achieved in the Balkans are particularly high. The faunal heterogeneity of this part of Europe is also evident from the superimposed European distributional limits of insectivores and rodents (Kryštufek & Griffiths, 1999; Figure 1). Contrary to this, a synthetic map of Europe's mammal species richness suggests hot spots are located in Central Europe (Mitchell-Jones *et al.*, 1999). The mammal maps in Mitchell-Jones *et al.* (1999) are derived from a presence/absence matrix of the 194 European mammals in individual 50 x 50 km squares of the Universal Transfer Mercator grid. In this approach, the lack of records for a species within a square was treated as equal to its actual absence, causing a source of bias in the evaluation of species density. A brief look at distribution maps of widespread and common mammals persuades one that sampling efforts were much less intense in the Balkan countries than in Central and

H.I. Griffiths et al. (eds.), Balkan Biodiversity, 79–108.

northern Europe, thus probably resulting in the underestimation of mammal richness for the Balkan Peninsula.

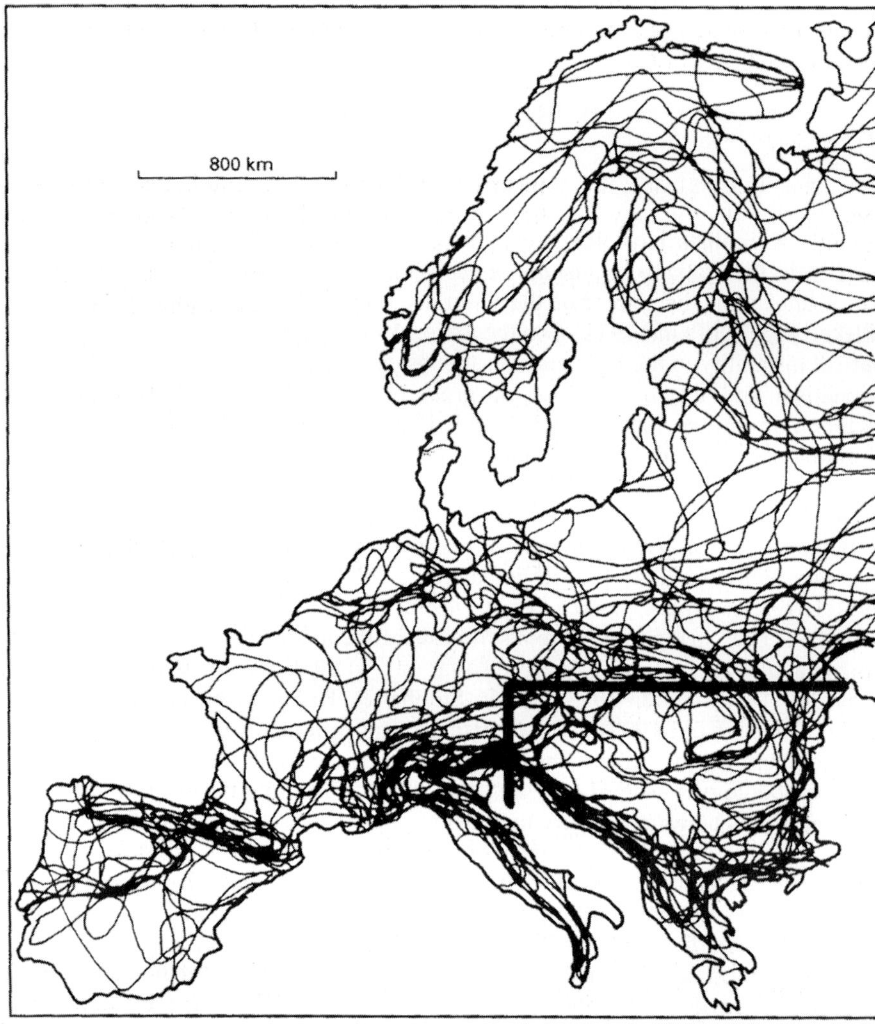

Figure 1. *Map showing the superimposed European distributional limits of 76 insectivore and rodent species, indicating the faunal heterogeneity of the Balkan Peninsula. The definition of the peninsula used in this study, is framed. (Modified from Kryštufek & Griffiths, 1999, by permission of the publishers of the Journal of Biogeography: Blackwell Publishing, Oxford.)*

This paper aims to (1) describe patterns in mammal richness of the Balkan Peninsula, (2) compare its mammalian diversity with other regions, and (3) assess the level of endemism. The Appendix provides additional details on taxonomic scope, definition of the region and a number of unavoidable methodological considerations.

SPECIES RICHNESS

The study encompassed 86 species of insectivores, bats, rodents and mustelids (see Appendix). Evident introductions (*Rattus* spp., commensal *Mus* spp.) were excluded. Rodents represented the highest proportion of species (41.9% of the total) and mustelids the least (11.6%).

Large scale patterns

As is evident at a glance, the number of species per 100 x 100 km square was not evenly distributed in any of the groups. Individual groups, however, showed different spatial patterns (Figure 2). Although frequency distributions of species square^{-1} values were right skewed in all the taxonomic groups studied, none of them deviated significantly from the normal distribution. The application of parametric statistics was therefore valid (Table 1).

Table 1. Descriptive statistics for the number of species per 100 x 100 km square. S – total number of species (species pool); S.D. – standard deviation; Min–max – range; C.V. – coefficient of variation; K-S d – Kolmogorov-Smirnov d (none of the tests deviated significantly from a normal distribution at P< 0.05).

Taxon	S	Mean	S.D.	Min - max	C.V.	Skewness	K-S d
Insectivora	13	7.4	1.673	4 – 12	22.6	-0.205	0.120
Chiroptera	27	20.1	2.912	13 – 25	14.5	-0.807	0.086
Rodentia	35	15.4	3.141	5 – 21	20.4	-1.157	0.106
Mustelidae	10	7.2	1.677	4 – 10	23.4	-0.410	0.110
Total	85	50.2	5.316	35 – 60	10.6	-0.711	0.061

From four to twelve insectivores occurred in individual squares, *i.e.* between one third and 92.3% of the total insectivorous species pool. Approximately one half of squares contained seven or eight species (= 54–62% of the total) and the arithmetic mean was slightly over a half of the insectivore species pool (= 57%). Hot spots in species richness (defined as the upper quartile squares; see Kryštufek & Griffiths, 2001) were of sporadic occurrence, with the highest concentration in the southern Dinarides (to the south-east of the Neretva River) and on the northern Šara-Pindhos Mts.. The mountains along the Adriatic coast in Montenegro, Albania and western Macedonia (Former Yugoslav Republic of Macedonia; FYROM) were particularly rich in insectivores (= 10 or 11 species square^{-1}). Another three hot spot regions coincided with eastern Thraki (or Thrace; particularly the Istranca Mts.), the southern Carpathians and with the contact zone of the Alps and the Dinarides. The Peloponnisos and the Panonnian Plain were poor in insectivores (Figure 2). The lowest numbers of species square^{-1} were invariably in the coastal zones of the southernmost parts of the peninsula.

Number of bats square^{-1} varied from 13 to 25 (= 48–93% of all bats analysed); *ca.* one half of squares contained 19–22 species (= 70–81% of all bats). The relative mean was high (= 74% of the bats), being surpassed only by the mustelids. Bats attained the highest values of species richness along the Adriatic coast and in the southern part of the peninsula, *i.e.* in the belt between 42° N and the Peloponnisos.

Species densities were low in the northern half of the peninsula, particularly in the north-east (Figure 2).

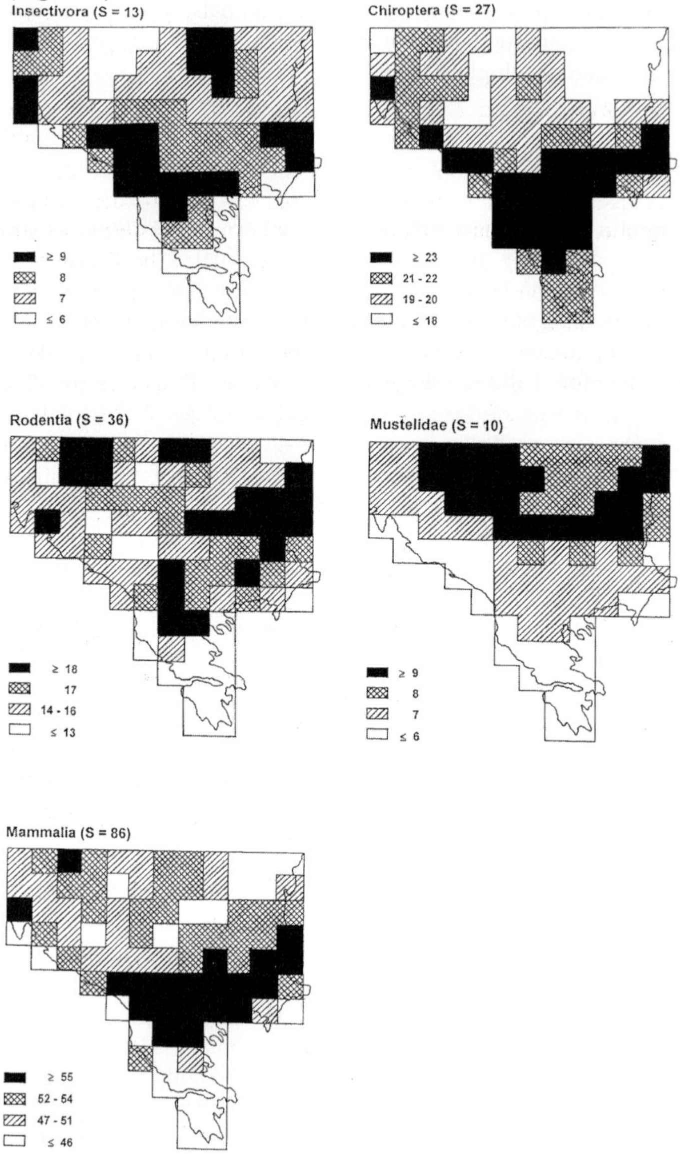

Figure 2. Diagram showing patterns of geographic variation of species richness within 100 x 100 km squares separately for four mammalian groups and for the combined mammal species pool (see Appendix for the taxonomic scope). Values were grouped according to the quartiles which are given with corresponding S-values. S – species richness.

Between five and 21 rodents were counted in individual squares (= 14–58% of all rodents) and approximately one half of squares had from 14 to 17 species (= 39–47%); the relative mean was lower (= 43% of all rodents) than in any other group. Rodent hot spot squares did not form a single contiguous block, but were rather dispersed. The majority were in the eastern half of the peninsula, mainly in Bulgaria, but also in the northern Šara-Pindhos Mts., and in the contact zones of the Panonnian Plain with the Carpathians and with the eastern Alps, respectively. The southern Pindhos Mts. and the Peloponnisos were particularly poor in rodents (Figure 2).

Mustelids numbered from four to ten species (= 40–100% of the group's total) and showed the highest relative mean (= 72%). Approximately one half of squares had seven or eight species out of ten. All the hot spots were within a single block which was focused in the lowlands of the northern part of the peninsula, mainly along the Danube. All ten mustelids occurred only in the easternmost squares of the hot spots' block, *i.e.* along the Danube but to the east of the Djerdap Canyon. The density decreased gradually going southwards and squares in southern Greece (Peloponnisos included) were particularly impoverished (Figure 2).

Pooling all the above groups gave 35 to 60 species square^{-1} (= 41–70% of the total). Approximately one half of squares had 47 to 54 species (= 55–63% of the total) and the mean was slightly over one half (= 58%) of the total. Hot spots in species richness formed a single contiguous block in the mountains between approximately 40° and 43° N, thus extending from the southern Adriatic to the Black Sea coast. This belt coincided with the southern Dinarides, the Šara-Pindhos Mts., the Rila-Rhodopes and the Balkan Mts. The southern Pindhos Mts., the Peloponnisos peninsula, and the northeastern corner of the study area exhibited low species richness (Figure 2).

Pearson correlation coefficients for pairwise comparisons of species number square^{-1} were from –0.58 (bats *vs.* rodents) to +0.85 (rodents *vs.* all mammals). Three Pearson's correlation coefficients (out of a total of ten) did not differ significantly from zero ($P>0.05$): insectivores *vs.* bats and mustelids, respectively, and rodents *vs.* bats. The correlation matrix is summarised as a tree diagram (Figure 3). Rodents mimicked most closely the pattern of the total mammalian fauna, which, however, is not surprising; the two data-sets were autocorrelated and rodents made up nearly 50% of the total number of mammals.

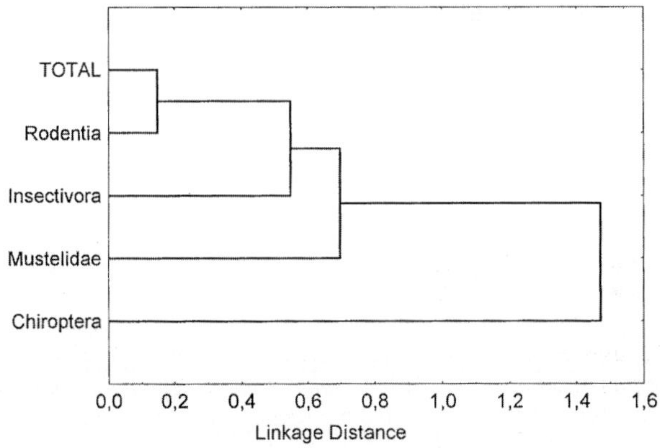

Figure 3. Ward's tree diagram for five mammalian groups based on a correlation matrix of species richness within 100 x 100 km squares. Correlations were converted to distances computed as 1 – r. Cophenetic correlation r = 0.67.

Bats exhibited the most distinct distribution suggesting a unique spatial pattern in species density. As is evident from the scatter of bats against the total number of mammals, this volant group was particularly species rich in several squares with a low number of mammals (Figure 4). The percentage of bats correlated negatively with the total number of mammals, and the Pearson correlation coefficient was highly significant (r= -0.45, P<0.00001). Thus the higher the number of mammals per square, the lower the proportion of bats tended to be (Figure 4). Since the number of species square^{-1} was more constant in bats than in non-volant mammals, species-rich squares reflected the accumulation of the latter. More precisely, squares exhibiting impoverished mammalian richness tend to lack non-volant mammals. Kryštufek & Griffiths (1999) suggest that the low number of small terrestrial mammals (insectivores and rodents) in the Mediterranean belt at the foothills of the Dinarides might result from local extirpations following clearance, burning and grazing, with consequent vegetational changes and ecosystem degradation. Volant mammals with advanced dispersal abilities presumably cope better with local restrictions in resource availability and are thus better pre-adapted for survival in the habitat mosaic of the Mediterranean landscape. This might be further exacerbated by the general tendency of bats to increase in species density with increasing average temperature.

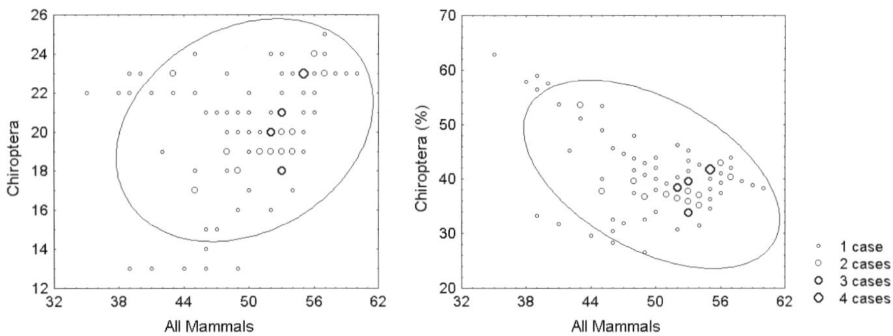

Figure 4. Bivariate frequency scatter-plots of bat species per square (left) and percentage of bats within total mammalian species (right) against the mammalian total within 100 x 100 km squares. The 95% range ellipse is also shown.

The percentage of bats ranged from 26.5–62.9% of the mammals' total within a square (mean=40.4, *S.D.*=6.979). In ten squares their proportion was more than half that of all the mammals and in another ten it was less than one third of the faunal total. Without exception, bat-dominated squares were located in the coastal regions of the southernmost peninsula, while the squares with a prevalence of non-volant mammals (>66.7% of all mammals) clustered in the northeastern corner of the study area (Figure 5).

The coincidence of hot spot squares is evident from Figure 6. Twenty-eight squares (=31% of the total area) were not a hot spot for any of the four groups. Forty squares were hot spots for a single group, 21 squares for two groups while only two squares were hot spots for three groups. Hot spot squares for ≥2 groups showed a recognisable geographic pattern, forming a semicircular band from the southern Dinarides across the northern Šara-Pindhos Mts., the Rila-Rhodope Mts., and the eastern Balkan Mts. towards the Carpathians. The two squares with three hot spots each were from the northern Šara-Pindhos Mts. and were contiguous. The number of hot spots correlated positively with the total number of species square^{-1} (r=0.62, P<0.00001).

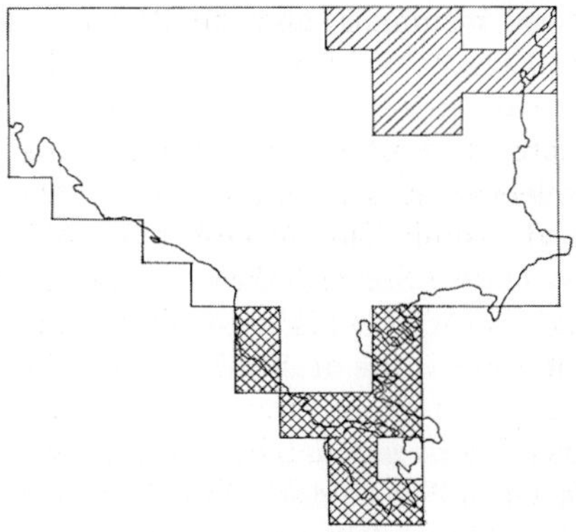

Figure 5. Map showing regions of the Balkan Peninsula with a high proportion of bats (> 50% of all mammals; densely shaded) and high predominance of non-volant mammals (bats < 33% of all mammals; lightly shaded).

Associations among groups in hot spots were tested by two-by-two contingency tables (see Appendix). Of the six possible associations, three were significant at P<0.05: two were negative (insectivores *vs.* mustelids χ^2=8.13; bats *vs.* mustelids χ^2=7.76) and one was positive (rodents *vs.* mustelids χ^2=7.74). These results do not follow closely those suggested by the inter-correlations among the groups which were based on species number square^{-1} (*cf.* Figure 3).

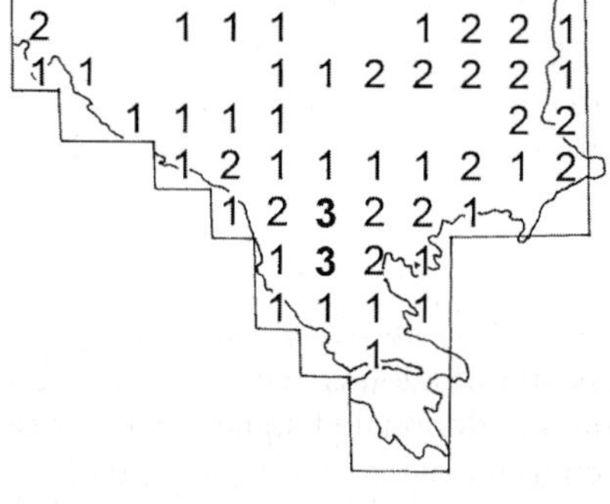

Figure 6. Cumulative number of hot spots in four mammalian groups (insectivores, bats, rodents, and mustelids) in 100 x 100 km squares of the Balkan Peninsula. Hot spots for each group were defined as the upper quartile squares in number of species.

Is the Peloponnisos effectively an island?

A sharp decrease in species density in the Peloponnisos Peninsula occurred in all groups. This pattern is ascribable to the peninsula effect, the generality and the causes of which have been much debated, however (Brown & Lomolino, 1998). As will be demonstrated below (see *Latitudinal pattern*), the Peloponnisos is located at the lowest point of a gradual latitudinal decline in species density, which clearly starts further north (around 40°N) and is not associated with a decrease in area (*cf.* Figure 13). Because of its unique geography it can be hypothesised that the Peloponnisos actually behaves as a biogeographic island. The Peloponnisos is mountainous (the highest peaks >2,000 m a.s.l.) and displays considerable environmental heterogeneity. Due to its narrow connection with the rest of the Balkan Peninsula (the isthmus of Corinth, connecting the Peloponnisos to Attica, is a mere *ca.* 6.5 km wide), however, the entire peninsula is effectively an island. Being isolated from the more species-rich Greek mainland by the Gulf of Corinth, the montane islands of the Peloponnisos are small and thus more prone to local extinctions; added to this, the nature of the local topography prevents re-colonisation events.

The above hypothesis was tested by comparing species richness in 39 East Mediterranean islands, including Cyprus, and in eleven Balkan mainland regions (mainly states) plus Asia Minor. Restricting this comparison to rodents resulted in the most complete species lists; all rodent species were considered, including both rats (*Rattus* spp.) and commensal house mice (*Mus domesticus* and *M. musculus*). These exotic rodents are frequently feral on the islands, and constitute the only rodent species on some small islands.

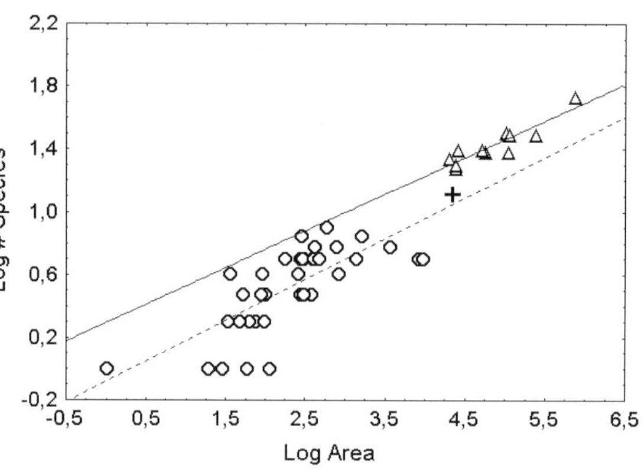

Figure 7. Bivariate plot of species richness against area (in km²) separately for 39 East Mediterranean islands (dots) and 11 mainland regions of the Balkans and Asia Minor. Values were transformed to logarithms (Log_{10}). Best fit regression lines are plotted. Cross – Peloponnisos.

Islands ranged in surface area from 0.05–9,251 km² and contained from 1–8 rodents; comparative values in the mainland were 20,000–755,000 km² for surface area and 19–54 for the species pool. The Peloponnisos (area 21,556 km²) had 13 rodent species. Regression of island and mainland samples showed significant differences in slope (Analysis of covariance; F=6.45, P<0.05), and the mainland slope had a higher intercept value. As is evident from Figure 7, the Peloponnisos sample lies close to the island regression line. Based on the above regressions, 11 species would be predicted for the Peloponnisos as an island and 20 species as part of the mainland. The observed number (*S*= 13) fell well within the 95% confidence limits for the island estimate, and was outside those for the mainland estimate (Table 2). The hypothesis of the Peloponnisos as an effective island has thus not been rejected.

Table 2. Estimated number of rodents for the Peloponnisos (with 95% confidence limits) under the assumption that it is an island, or part of the mainland. Estimates were based on regressions for the 39 East Mediterranean islands and 11 mainland regions, respectively. The observed number in the Peloponnisos (S = 13) is within the 95% confidence limits for the island estimate. See text for further explanation.

	Regression results		Estimated	95% confidence
	intercept	slope	number	limits
Island	0.141	0.186	11	7 – 16
Mainland	0.295	0.234	20	18 – 23

Regional comparison

To put the above study of species richness within the Balkan Peninsula into its broader context, it is necessary to establish whether it is high or low compared to other regions. To this end, species richness was estimated firstly for other European countries (Europe to the west of Russia, Ukraine and Belarus). Only insectivores, bats, rodents and mustelids were considered in order to minimise the effects of human impact on local faunas; island endemics were also excluded. Data were taken from Mitchell-Jones *et al.* (1999). Species richness in the 32 countries concerned varied between 19 and 77 (4-fold increase). The countries, however, differed significantly in area, ranging between 20,000 and 547,000 km^2 (27-fold increase). Because the relation between area and species richness is not linear, species richness was adjusted for the area effect by applying the Arrhenius equation (Rosenzweig 1996; see Appendix). Area-adjusted species richness (given as *c*-values, z=0.132) varied between 4.35 and 17.27 (4-fold increase). A clear geographic pattern emerged when *c*-values were plotted on the map of Europe, with the lowest values in the north and the highest values in the south-east (Figure 8). Area-adjusted richness was significantly higher in the Balkan countries than in the rest of Europe (Kruskal-Wallis H=15.55, P<0.0001). The *c*-values correlated negatively with surface area transformed to logarithms (r=-0.40, P<0.05), indicating that the smaller the area the higher its mammalian richness tended to be. Although this is best ascribed to the small size of the Balkan countries, even by European standards, species richness, as perceived through *c*-values, might also be an artefact of the area.

To overcome this, Balkan species richness was then compared with thirteen regions of the western Palaearctic. The regions were comparatively large and, as such, less sensitive to possible recent modifications of local faunas by direct human impact. Such an approach thus allowed the entirety of mammals to be considered, and, in particular, the larger species. Recently extirpated mammals were also taken into account. Between 62 and 136 mammals occurred in individual regions (2.2-fold increase) and *c*-values (z=0.132) varied between 9.87 and 24.09 (2.4-fold increase). Again, a clear pattern emerged when *c*-values were plotted on the geographic map (Figure 9). The Near and Middle East showed by far the highest *c*-scores, followed by the Mediterranean region (except Algeria). Balkan biodiversity appeared to be of modest magnitude within the western Palaearctic. The results were similar for even broader comparisons (Figure 10). Temperate regions followed a similar pattern in species richness, being strongly surpassed by tropical countries. There were several outliers among temperate regions (the Caucasus, the Levant, southern Africa, and New Mexico), and Sri Lanka had a lower *c*-value than expected for a tropical country of its size (Figure 10); this is evidently the consequence of an island effect. The Balkans is placed close to the regression line for temperate regions, indicating modest biodiversity, as perceived through mammal species richness.

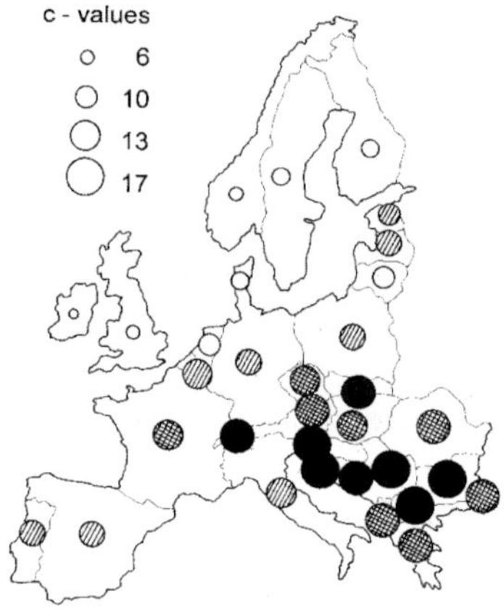

Figure 8. Area-adjusted species densities (c-values, z=0.132) in 32 European countries. Circles indicate limit intervals, whose size corresponds to c-values. The lower quartile circles are white (c≤9.95); the central quartiles are lightly shaded (c=10.20-13.05) or densely shaded (c=13.11-15.23); the upper quartile circles are black (c≥15.73).

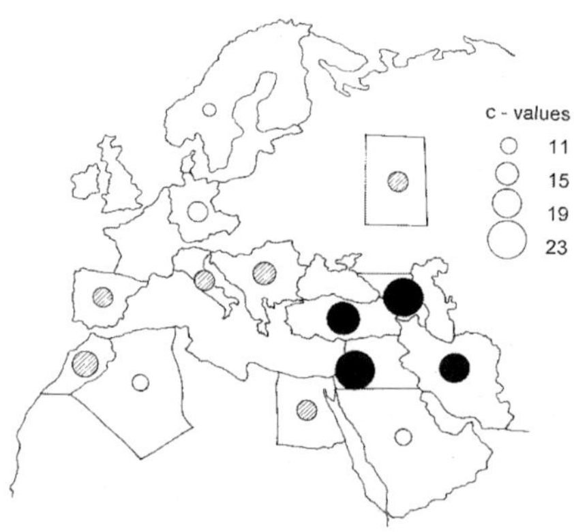

Figure 9. Area-adjusted species densities in the western Palaearctic (c-values, z=0.132). Circle sizes correspond to c-values. The lower quartile circles are white (c=9.87–13.33); the central quartile circles are shaded (c=13.83-17.36); the upper quartile circles are black (c=20.56–24.09). Regions included (for details see Appendix): Pyrenean Peninsula, Italian Peninsula, Balkan Peninsula, Central Europe, northern Europe, eastern Europe, Asia Minor, the Caucasus, the Levant, Iran, Arabia, Egypt, Algeria, and Morocco.

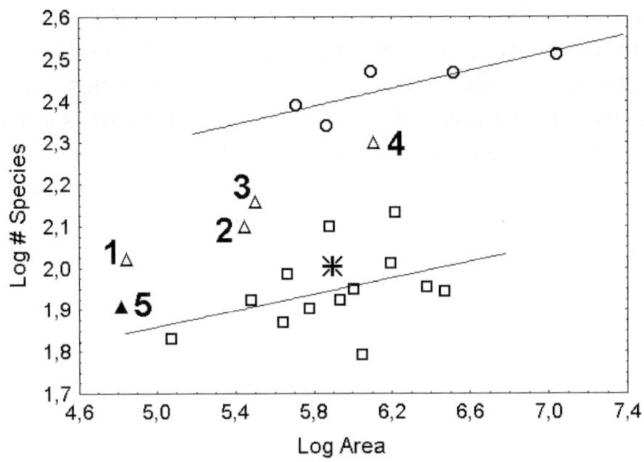

Figure 10. Plot of species richness against area for various regions of the World. Regression lines are plotted separately for temperate (squares) and tropical (circles) countries. Triangles indicate outliers: 1 – the Levant, 2 – the Caucasus, 3 – New Mexico, 4 – southern Africa, 5 – Sri Lanka; star – the Balkans. Values were transformed to logarithms. Regions included (in addition to those in Figure 9): Mongolia, Pennsylvania, North-Central states of the USA, New Mexico, southern Africa, India, Sri Lanka, Thailand, Island of Borneo, Angola, and Bolivia.

Local-scale diversity

Within the study region, the most detailed and comprehensive local-scale study seems to be the biogeographic analysis of insectivores and rodents in the Dinarides (Kryštufek & Griffiths, 1999). The Dinarides are a mountain chain of pronounced biotic heterogeneity, running along the eastern Adriatic coast from the Alps in the north-west as far south-east as the basin of Kosovo and Metohija. Species richness here show a sharp decline in the coastal belt. This reflect a lack of adequate replacement for continental species in coastal habitats, a pattern which appears to be widespread throughout the Mediterranean region (*e.g.* Barbosa & Benzal, 1996, Torre *et al.*, 1996). A zone of rapid faunal change followed the coastal line and coincided with the border between the Mediterranean deciduous and continental vegetation types (as defined by Horvat *et al.*, 1974), rather than with the Mediterranean sclerophyllous evergreen forests (Figure 11). The altitudinal threshold for faunal turnover was at *ca.* 700 ± 100 m in the northern part (the northern Velebit Mts.) and between 900 ± 100 m further south in the Dinaric karst of Bosnia and Dalmatia; this indicates an altitudinal shift of 200 m over a distance of only 250 km. Species density correlated positively and significantly (P<0.0001) with two environmental variables: the number of phytic communities (r=0.343) and the elevation range (r=0.322). Environmental (habitat) diversity is thus a good predictor for species richness in the area.

Widely-distributed species (*Erinaceus concolor, Crocidura suaveolens, C. leucodon, Apodemus sylvaticus* and *Glis glis*) contributed more to local small mammal faunas in the coastal (mean=54.8%; range = 38.5-83.3%) than in the continental zones (mean=25.3%; range=20.8-45.5%). Amongst the six species that were restricted to the Dinaric coastal belt, five were either rock-associated (*Dinaromys bogdanovi, Apodemus mystacinus, Eliomys quercinus*) or subterranean

(*Talpa caeca, Microtus thomasi*). It was suggested (Kryštufek & Griffiths 1999) that in the early Postglacial the small mammal fauna of the coastal regions of the Dinarides may have been essentially similar to that seen inland now. Following habitat deterioration in the last five millennia, the remnant fauna consists mainly of a few widespread ubiquitous species, plus geophilic specialists (which presumably "escaped" adverse drought conditions and possibly also frequent fires) and rock-dwelling species seeking mesic conditions in rocky habitats with abundant fissures and crevices (Mares & Lacher, 1987).

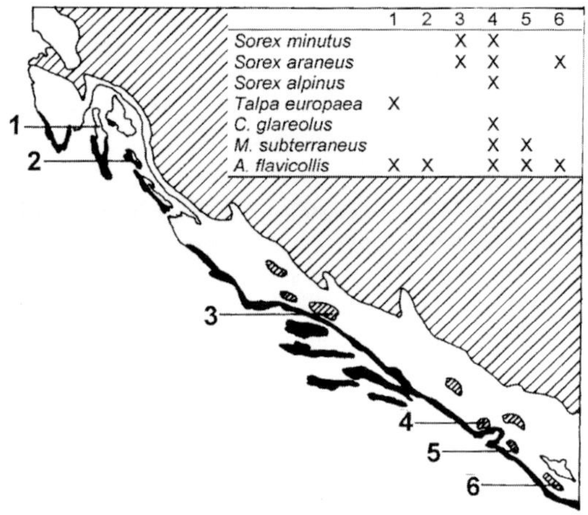

Figure 11. Summary of relict occurrences of insectivores and rodents of mesic continental habitats on the northern Adriatic islands and in the habitat islands of the coastal Dinarides. Vegetation types (modified from Horvat et al., *1974 and Jovanović* et al., *1986): black – dry Mediterranean evergreen; white – dry Mediterranean deciduous; hatched – mesic continental deciduous. 1 – Island of Krk; 2 – Island of Cres; 3 – Mt. Biokovo; 4 – Mt. Orjen; 5 – Mt. Lovćen; 6 – Mt. Sutorman.*

The assumption of a formerly wider distribution of insectivore and rodent species in the coastal belt of the Dinarides seems plausible, taking into account the relict occurrences of several mammals of continental mesic habitats both on the Adriatic islands and in the habitat islands of isolated high mountains along the coast (Figure 11).

SPATIAL PATTERNS

Spatial autocorrelation

Species richness showed a sharp decline in Moran's autocorrelation coefficient I over the initial 500 to *ca.* 1,000 km, depending on the group (Figure 12). This decline was fairly monotonous in rodents, bats, and mustelids, suggesting a simple clinal pattern of spatial variation in species density. The depression, which occurred in the mammals combined and in the insectivores, is interpreted as a circular cline with the approximate radius of 500 – 800 km (*i.e.* the distance of the lowest values of Moran's I). In summary, species richness showed clinal variation in all groups, but the nature of the cline varied between groups.

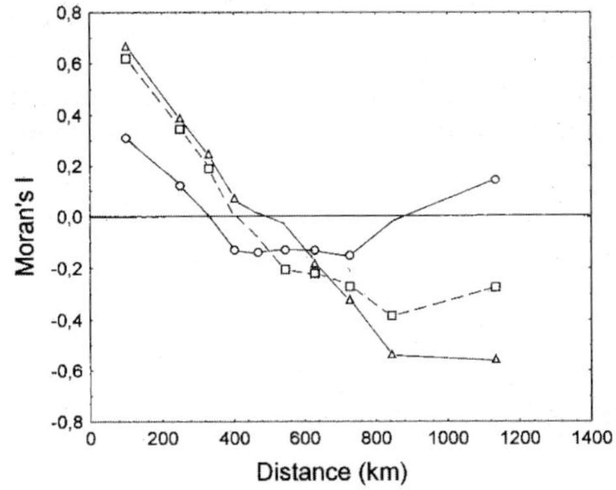

Figure 12. Spatial correlogram of species richness per 100 x 100 km square in three mammalian groups in the Balkan Peninsula. Dots – all mammals (see text for the scope); squares – bats; triangles – mustelids. Insectivores mimicked all the mammals and rodents mimicked the pattern of bats (not shown). Ordinate: Moran's autocorrelation coefficient I. Symbols are placed only for significant autocorrelation coefficients (P<0.05).

Latitudinal pattern

The number of species within a 1° latitudinal belt varied between 31 in the south and 77 in the north (latitudinal belt 43–44° N), *i.e.* a 2.5- fold increase. The curve of species density against latitude was humped rather than linear in all groups, but the latitudinal belt with the maximum number did not correspond exactly between groups. Insectivores peaked (12 species per latitudinal belt) between 41–43° N, bats (29 species) between 42–45° N, and rodents (28 species) and mustelids (10 species) between 43–46° N. The southernmost latitudinal belt contained the lowest number in all the groups but the magnitude (compared to the most species-rich belt) varied and was 1.4-fold in mustelids, 1.6-fold in bats, 3-fold in insectivores, and 5.6-fold in rodents.

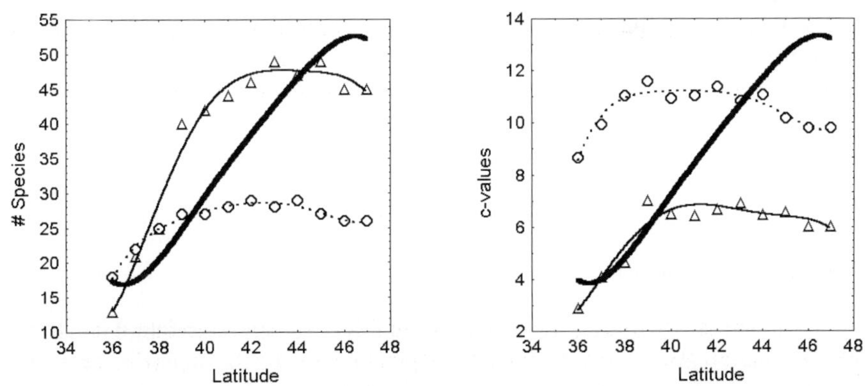

Figure 13. Estimated latitudinal trends (least squares fit) in species richness in bats (circles) and in non-volant mammals (triangles) in the Balkan Peninsula within the intervals of 1° latitudinal belt. Species density was adjusted for surface area (c-values) in the right-hand graph (z=0.082 for bats and z=0.169 for non-volant mammals). Heavy line shows surface area.

Non-volant mammals thus accumulated species more rapidly with increasing latitude than did bats (Figure 13).

Since the surface area of the peninsula shrinks towards the south (Figure 13), the latitudinal pattern might be due to a species – area effect, with larger areas supporting more species (Rosenzweig 1996). In fact, the increase of non-volant mammals' number roughly followed the surface increase from south to north (Figure 13). The plot of adjusted densities (given as *c*-values) showed similar latitudinal responses in both volant and non-volant groups (Figure 13; note that the two groups differ in *z*-values; *cf.* Appendix). Area-adjusted species richness followed the increase in the surface area up to 38-39° N, with a peak at *ca.* 40°N (Figure 13). To the north of 43° N, this broad peak was followed by a shallow decline, which was more pronounced in non-volant mammals.

ENDEMISM

Patterns of endemism

A taxon is deemed endemic to an area if it occurs within it and nowhere else. As a consequence, taxa can be endemic to a geographic location on a variety of spatial scales and at various taxonomic levels (Brown & Lomolino, 1998). Definition of the area is thus of prime importance. Only six species are confined in their distribution to the Balkan Peninsula as defined in this paper and are thus endemic to it ("endemics" in the subsequent text), in addition to two island endemics from the Island of Crete (*Crocidura zimmermanni, Acomys minous*).

Since there is no *a priori* reason for Balkan endemics to be confined strictly in their distribution to the geographical scope of the peninsula used in this study, species with small distributional ranges were also taken into account. An area of *ca.* < 500,000 km^2 was selected arbitrarily (subsequently called "small-range species"). Such an approach resulted in a data-set of nine small-range taxa: three insectivores and six rodents (Table 3). Six small-range taxa were shared with the rest of Europe and three occurred also in Asia Minor. This suggests the greater influence of Europe (rather than Asia Minor) in formation of the recent mammal fauna of the Balkan Peninsula.

Uncertain taxonomy can impose important limitations to any biogeographic study, and the Balkan endemics are no exception. Until recently, for example, the genus *Mesocricetus* was still considered as monospecific (Corbet, 1978), while it is now split into four allopatric species, three of which having very small ranges (Musser & Carleton, 1993). Similarly, the genera *Talpa, Spermophilus, Microtus*, and *Myomimus* have all been revised in the light of fairly recent advances in taxonomic research. Taxonomy still remains obscure in mole rats *Nannospalax*, and *N. leucodon* continues to be a collection of 22 different allopatric and parapatric chromosomal forms. Savić & Soldatović (1984) recognise 13 biological species within *N. leucodon s. lat.* (not including *N. nehringi*), the majority of which are endemic to the Balkans. The blind mole *Talpa caeca* and the common shrew *Sorex araneus* are further represented in the Balkans by endemic chromosomal forms (see Zima, this volume) of unknown taxonomic affinities. For the purposes of this study, two rodents were included which are still in need of taxonomic revision: *Microtus guentheri* and *Apodemus mystacinus epimelas*. The scope of the former is poorly defined; however, in its narrow sense, this vole is restricted to the Balkans, the

Anatolian plateau of Asia Minor and to the coastal Near East; its range as such is <500,000 km^2. The subspecies *A. m. epimelas* which is endemic to the Balkans, is morphologically well differentiated (Storch, 1977) and might represent a species in its own right (Storch, 1999).

Table 3. List of taxa considered in the analysis of Balkan endemism. Small-range species have small distributional areas (< 500,000 km^2) but occur also in the Balkans; endemic taxa are restricted to the Balkans. See text for further explanation.

Taxon	Small range Species	Endemic species	Generic Range
Sorex alpinus	+		Holarctic
Talpa caeca	+		W Palaearctic
T. levantis	+		- " -
T. stankovici		+	- " -
Spermophilus citellus	+		Holarctic
Mesocricetus newtoni		+	SW Palaearctic
Dinaromys bogdanovi		+	Balkans
Microtus felteni		+	Holarctic
M. guentheri s. str.	+		- " -
M. multiplex	+		- " -
M. tatricus	+		- " -
M. thomasi		+	- " -
Spalax graecus		+	SW Palaearctic
Nannospalax leucodon	+		SW Palaearctic
Apodemus mystacinus epimelas		+	Palaearctic
Myomimus roachi	+		SW Palaearctic

Nearly all the Balkan squares contained at least one small-range species and eight species square^{-1} was the maximum (mean$_{90}$=3.2; *S.D.*=1.733). A clear geographic pattern emerged in small-range species densities, with a gradual increase towards the south; again, the Peloponnisos violated this trend (Figure 14). Scores were highest in the southern Dinarides, the northern Šara-Pindhos Mts. and in eastern Thraki. The density of small-range species correlated positively with the total number of mammals square^{-1} although the fit was poor (r=0.40, P<0.0001). The correlation coefficient did not differ significantly from zero when rodents (representing the bulk of narrow-range species) were used instead of the total mammalian species pool (r=0.10, P=0.338).

Endemics were restricted to merely 57 squares (*i.e.* 63% of the total), with densities of up to five species square^{-1} (mean$_{57}$=1.8, *S.D.*=1.020; Figure 15). No endemism was recorded in the north-western corner of the study area, *i.e.* the region most affected by Pleistocene glaciation. The highest density, on the other hand, was in the southern Dinarides and the northern Šara-Pindhos Mts. The peak values (≥4 endemics square^{-1}) coincided with the squares with the highest species richness values (*cf.* Figure 2).

Both the number of endemics and of small-range species showed very similar spatial autocorrelation patterns: a sharp decline of Moran's autocorrelation coefficient *I* from the initial *ca.* 800 km (Figure 16), thus suggesting a cline. A very similar pattern was also seen in rodent and bat species richness (*cf.* Figure 12).

Endemics, as well as small-range taxa, belong mainly to the southwestern Palaearctic group of genera/subgenera, which exhibit pronounced peripheral speciation along their southern distributional borders. Patterns seen in the red-toothed shrews *Sorex*, moles *Talpa* and voles *Microtus* (*Terricola*) are repeated in southern Europe and the Near East, with each glacial refugium having endemic taxa. Chromosomal diversification within *Nannospalax leucodon* is characteristic of the genus across its entire range (Nevo, 1999), while species of hamsters *Mesocricetus* and mouse-tailed dormice *Myomimus* invariably have small ranges (Panteleyev, 1998). The bulk of the Balkan endemics is thus derived from groups which are widespread and species rich in the southwestern Palaearctic. The observed pattern most likely results from broad-scale processes which also operate outside the peninsula.

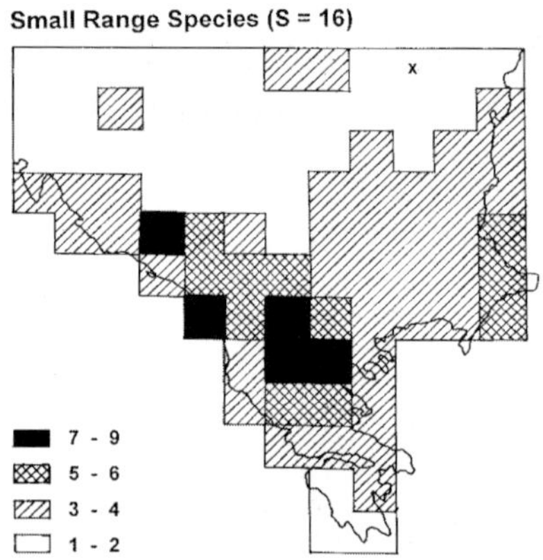

Figure 14. Variation of small-range species richness within 100 x 100 km squares. Scores are grouped into the quartiles. See Table 3 for the taxonomic scope. S – number of small-range species. X – square with no small-range species.

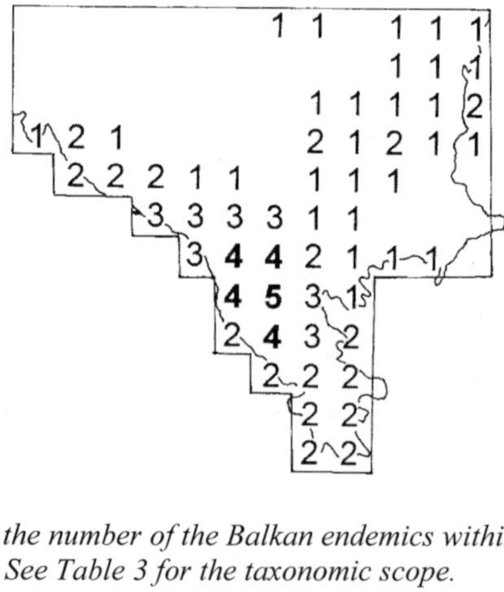

Figure 15. Variation in the number of the Balkan endemics within 100 x 100 km squares. See Table 3 for the taxonomic scope.

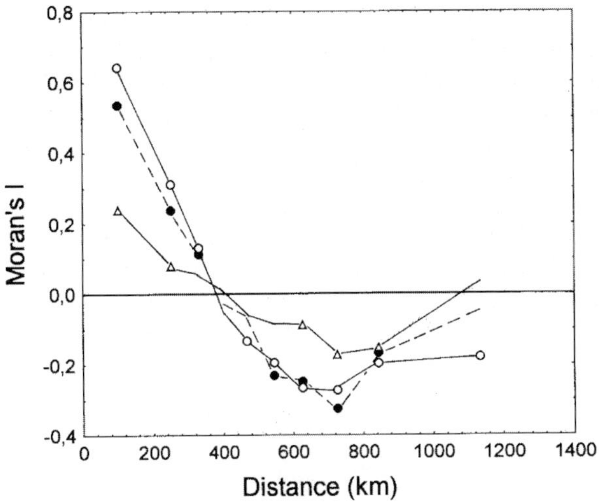

Figure 16. Spatial correlogram of a number of small-range species within 100 x 100 km squares (dots), number of endemics within squares (circles) and Index of Faunistic Change (triangles) in the Balkan Peninsula. Ordinate: Moran's autocorrelation coefficient I. Symbols are placed only for significant autocorrelation coefficients (P<0.05).

Figure 17. Martino's vole Dinaromys bogdanovi *resembles externally the snow vole* Chionomys nivalis. *Although these two rock dwelling voles are only distantly related, they are presumed to be in scramble competition for narrow fissures and caverns on the rocky substrate which provides shelter. (Photograph by Alenka Kryštufek).*

Martino's vole

A unique feature of the Balkan mammal fauna is an ancient vole from the monospecific genus *Dinaromys*. Martino's vole *Dinaromys bogdanovi* (Figure 17) is the only surviving species of the arvicoline subtribe Pliomyi (Gromov & Polyakov, 1992). It is known recently from a small range (<10^5 km^2) in the mountainous regions of the western Balkans.

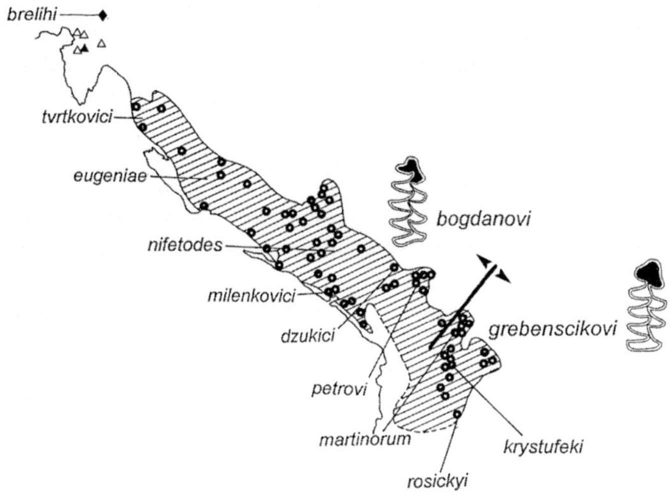

Figure 18. Summary of the zoogeography and variation of Martino's vole Dinaromys bogdanovi *in the north-west Balkans. Records of occurrence are given by circles and recent range is shaded; the broken line in the south-west indicates uncertainty due to lack of data from Albania. Triangles are past records: open triangles are from the Pleistocene and the closed triangle is of the Holocene age. The heavy line indicates a major division of the species into two groups of subspecies (*bogdanovi *and* grebenscikovi*, respectively); the characteristic shape of the grinding surface of the 1st lower molar is shown (anterior loop is black). Subspecies of a specific flea* Ctenophthalmus nifetodes *indicate possible further evolutionary division of the host. Diamond – record of a flea* C. nifetodes brelihi *outside the actual range of Martino's vole which possibly indicate the historic range of the host. Based on Petrov & Todorović (1982), Petrov (1992), Kryštufek & Petkovski (in press), Brelih (1986), Brelih & Trilar (2000), and Todorović (1956).*

As suggested by scarce electrophoretic evidence, the genus *Dinaromys* is a sister taxon to the *Clethrionomys-Microtus* clade (Gill *et al*., 1987), and on the basis of internal structure of the molar enamel, Koenigswald (1980) believes it shows no close relationship with any other living arvicoline. The genus' fossil history dates back to the Late Pliocene (Gliozzi *et al*., 1997). Species which are closely related to the recent Martino's vole are known from northeastern Italy (since the Late Pliocene; Sala, 1996), the western Balkans (since the Early Pleistocene; Malez & Rabaeder, 1984), the Hungarian Carpathians (since the Middle Pleistocene; Gromov & Polyakov, 1992) and from the Ukraine (the Early Pleistocene; Nesin & Skorik, 1989). It is thus evident that the Martino's vole lives only in a small part of *Dinaromys*' historical range. Furthermore, it seems that its range is still shrinking. It was still present in northern Italy, Slovenia and Attica, Greece during the

Pleistocene (Petrov & Todorović, 1982). Its characteristic parasite *Ctenophthalmus nifetodes* survived in western Slovenia, albeit on a different host (*Glis glis*; Brelih, 1986) and Holocene remnants of Martino's vole were excavated recently in southwestern Slovenia *ca.* 130 km outside its current range (Toškan & Kryštufek, in preparation; Figure 18).

Reasons for the shrinkage of the distributional range are not well understood. Kryštufek (1987) suggests that competitive exclusion with the snow vole *Chionomys nivalis,* a species distantly related to Martino's vole but with similar external and cranial morphology and presumably identical habitat requirements, might be responsible. Both voles depend strictly on the availability of a rocky substrate to provide shelter in narrow fissures and caverns. Martino's vole is no doubt an oldtimer in the Balkan fauna, whereas the snow vole's range may be in expansion. Nei's genetic distance (Gill *et al.*, 1987) and dental morphology (Todorović, 1956) suggest profound divergence within Martino's vole and further evolutionary split is suggested by the taxonomic diversification of its specific flea fauna. All this accords with its long fossil record, which, as noted, dates back to the Early Pleistocene. The snow vole was also present in the Balkans during the Pleistocene and the Balkan refuge presumably served as a source for post-Pleistocene colonisation of the Alps (Filippucci *et al.*, 1991). As with Martino's vole, Balkan populations of the snow vole are well differentiated, both morphologically (Kryštufek, 1990) and genetically (Filippucci *et al.*, 1991). However, during the Upper Pleistocene the snow vole was widespread only in the northern parts of the peninsula (*i.e.* to the north of the rivers Sava and Danube; Terzea, 1972, Kryštufek, 1997); records are scarce further south and at least some of them need to be re-evaluated. It is thus plausible to hypothesise that the snow vole penetrated into the range of the Martino's vole in the Dinarides and the northern part of the Šara-Pindhos Mts. at some point during the Holocene, where it established itself as a marginal population. This process is possibly still going on, presumably resulting in the range shrinkage of the more ancient of the two rock-dwelling voles. Ecological and behavioural interactions between these two vole species are entirely unknown, however.

Crete

Crete is an ancient island which has not been connected to the mainland since the Early Pliocene, *ca* 5 Ma. BP (Dermitzakis, 1990). During its long isolation, Crete witnessed a spectacular adaptive radiation, particularly so in deer (Cervidae; Vos, 1984), and supported an unbalanced island fauna until very recently (Reese, 1996). The only surviving endemics are evidently of very different origin. According to the most plausible hypothesis, the ancestor of a recent *Crocidura zimmermanni* reached the island before the Late Villányian *via* rafting either from the Balkans or from Asia Minor (Reumer, 1986, 1996). *Acomys minous*, on the other hand, which is not represented in the fossil record (Jarman, 1996), was evidently introduced by man in antiquity. The species is composed of two maternal lineages, one showing close affinities with *A. nesiotes* from Cyprus and the other with *A. cilicicus* from southern Turkey (Barome *et al.*, 2000). Available molecular evidence alone does not allow a firm conclusion to be reached as to whether the two *A. minous* lineages were derived from distinct stocks or from a single but highly diversified population (Barome *et al.*, 2000). The two surviving Crete endemics evidently evolved independently of historical processes on the Balkan mainland.

PROVINCIALISM

Provincialism is based on the fact that species range borders are not randomly or uniformly distributed, and that endemics tend to concentrate in certain regions. It is thus used to describe the phenomenon of different groups showing similar patterns in endemism (Brown & Lomolino, 1998). This enables the recognition of geographic regions with relatively homogeneous faunas, being separated from other such regions by zones of faunal heterogeneity (Hagmeier & Stults, 1964). It was shown above that hot spots in both species richness and endemism clump in the Balkans rather than exhibiting random distribution. This gives reasonable grounds to explore the differences between regions of faunal homogeneity and those with a high proportion of distributional borders. Two approaches were used for this analysis.

First, a data matrix (rectangular matrix of species presence/absence within a square) was subjected to *K*-means clustering. From two to ten clusters were specified and the results were plotted on the 100 x 100 km grid geographic maps of the region. Two main clusters were recognised in all nine analyses. Eight borders, out of a total of nine, coincided fairly closely (lines A and B, respectively, on Figure 19) and only one single border (eight clusters specified) was shifted further south (line C on Figure 19). The belt between the first two lines (densely shaded on Figure 19) was interpreted as the zone of overlap between two faunas (the southern and the northern one, respectively) and as such corresponded with an area of faunal heterogeneity. This coincided with the Balkan Mts. and the central part of the Dinarides. Hot spots in mammalian diversity were mainly to the south of this belt as were also the distributional ranges of four small-range species (out of a total of nine) and of five endemics (out of a total of seven). Of the 86 mammals analysed, 14 occurred only in the northern cluster and ten were found only in the southern cluster. A further 14 mammals with a "northern" prevalence attained the southern borders of their distribution within the belt of overlapping clusters.

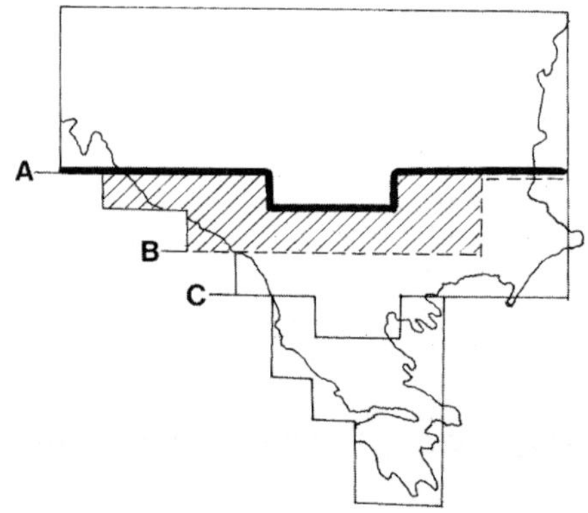

Figure 19. Position of borders separating two main clusters of squares resulting from K-means clustering (from two to ten clusters specified). Lines A and B – coincidence of four borders, respectively; line C – single border. Shaded – belt of overlap between the "northern" and "southern" cluster. See text for further explanation.

Secondly, the Index of Faunistic Change (IFC) varied from 0 to 36.5% ($median_{91}$ = 18.52%) with half the scores being between 14.04 and 23.08%. It is clear from Figure 20 that the region of faunal heterogeneity roughly coincided with the belt of overlap between the two main clusters as they emerged from the above *K*-means clustering (*cf.* Figure 19). The upper quartile squares, however, were not contiguous, although the majority of them occurred within two blocks: in the southern Dinarides and northern Šara-Pindhos Mts., and in Bulgaria, respectively. Autocorrelation analysis of the IFC values showed a sharp decline of Moran's coefficient *I* over the initial *ca.* 800 km, being followed by a depression at *ca.* 800 – 900 km (Figure 16). Since the following increase was not supported by Moran's *I* significantly different from zero, it is difficult to conclude exactly what type of cline was in question.

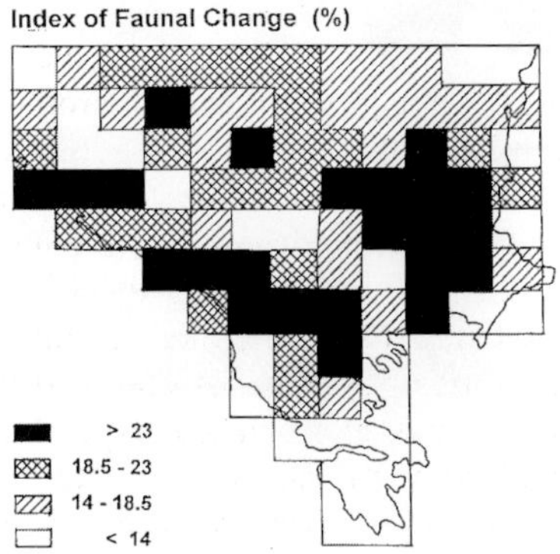

Figure 20. Geographic variation of the Index of Faunistic Change scores (according to quartiles). The higher the score, the larger is the faunal heterogeneity within a square.

BROADER PERSPECTIVES

The Balkan Peninsula links Central Europe and Asia Minor. To assess broader biogeographic affinities, the faunal affinities of the Balkan 100 x 100 km squares were compared with those of Central Europe (*S*=61) and western Asia Minor (*S*=71), respectively. Faunal similarities are given as the Coefficient of Community values (*CC*). Since the distribution of *CC* scores did not deviate from normality (Kolmogorov-Smirnov $d < 0.09$, not significant) a One-way ANOVA was applied to test the differences. The Balkan squares showed higher affinities with the mammal fauna of Central Europe (Table 4) and the difference was highly significant (F=108.69, P<0.000001). The majority of the Balkan squares (88%) achieved high *CC* values (>0.50) with the Central European fauna, while there were only 20% of squares showing such high affinities with western Asia Minor (Figure 21).

Table 4. Descriptive statistics for the Coefficient of Community values of the 91 Balkan 100 x 100 km squares with the mammal faunas of Central Europe and western Asia Minor, respectively. S.D. – standard deviation.

	Mean	S.D.	Range
Central Europe	0.594	0.122	0.315-0.813
Western Asia Minor	0.444	0.062	0.330-0.568

In conclusion, it may be inferred that the majority of the peninsula, except its southernmost parts, has a strong Central European character. The southern border of this area roughly coincided with the southernmost border between the two main clusters as derived from K-means clustering (*cf.* Figure 19). Squares with strong affinities to Asia Minor were located in Thraki, northern Greece, Macedonia (FYROM), Albania, and southern Dalmatia (Figure 21). Eleven squares (*i.e.* 12% of the total) showed *CC*>0.50 with both regions. This is not surprising, since nearly a half of western Asia Minor mammals also occur in Central Europe. The strongest resemblance with Central Europe (*CC*>0.75) was detected in the northwestern corner of the Balkan Peninsula (Figure 21).

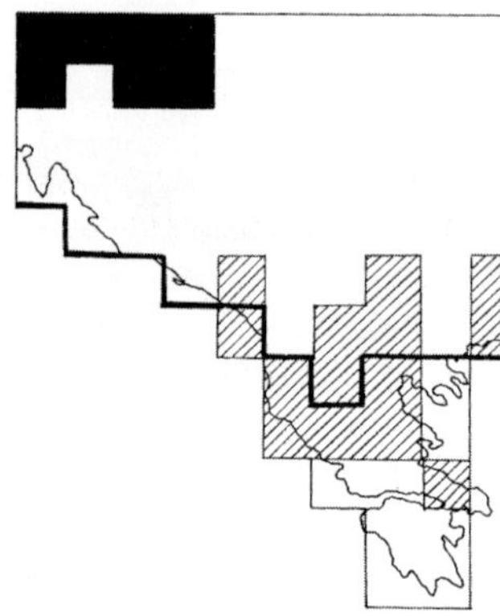

Figure 21. Similarities (Coefficient of Community, CC) between the mammal fauna of the Balkan 100 x 100 km squares and those of Central Europe and western Asia Minor, respectively. The heavy line is the southern border of contiguous squares showing CC≥0.50 with Central Europe; black – squares having CC≥0.75 with Central Europe; shaded – squares showing CC≥0.50 with western Asia Minor.

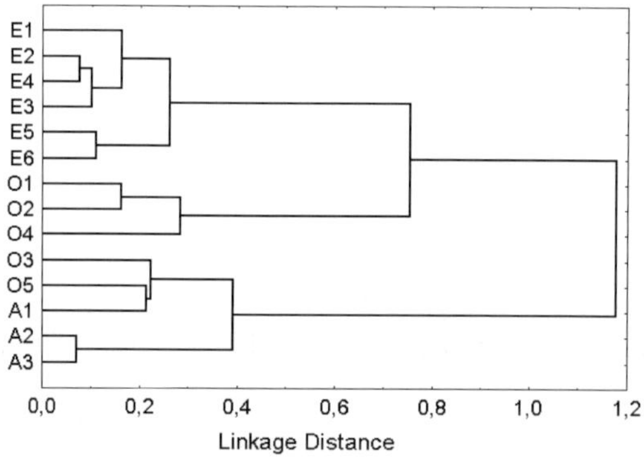

Figure 22. Ward's tree summarising Percent disagreement among 14 mammalian faunas from the western Palaearctic. Cophenetic correlation coefficient r=0.87. E1 – Pyrenean peninsula; E2 - Italian peninsula; E3 – Balkan Peninsula; E4 – Central Europe; E5 – northern Europe; E6 – eastern Europe; O1 – the Caucasus; O2 – Asia Minor; O3 – the Levant; O4 – Iran; O5 – Arabia; A1 – Egypt; A2 – Algeria; A3 – Morocco. See Figure 9 for definition of the regions.

The affinities of the Balkan fauna with the mammalian faunas of the western Palaearctic were assessed by clustering (Percent disagreement among 14 regions based on a presence/absence rectangular matrix; Ward's algorithm). The entire mammalian faunas were taken into account (see *Broader comparison* under *Species richness* for definition). Three clusters emerged which broadly coincided with the continents, namely the European, south-west Asian, and northern African, respectively (Figure 22). This pattern was disturbed by the Levant and Arabia, which both tied to the North African faunas. The remaining mammalian faunas from Asia appeared closer to Europe than to Africa. Central Europe and the three southern peninsulas formed a compact cluster. The Balkan mammal fauna thus has a predominantly European character.

CONCLUSIONS

Between 35 and 60 species of insectivores, bats, rodents, and mustelids were found to occur in individual 100 x 100 km squares superimposed on the Balkan Peninsula. Hot spots in species richness coincided with the southern Dinarides, the Šara-Pindhos Mts., the Rila-Rhodopes, and the Balkan Mts.. Spatial patterns differed between groups, with bats being the most distinct. The more mammal species there were within a square, the lower the proportion of bats tended to be. Species density declined sharply in the Peloponnisos Peninsula, and the number of rodents matched closely the estimate its being an island rather than part of the mainland. Area-adjusted species richness of the Balkan Peninsula was significantly higher than for the rest of Europe, although Balkan mammal diversity was significantly less than that of the Near and Middle East. The latitudinal pattern in species richness showed a humped curve with a peak at *ca.* 40°N. Ranges of six species endemic to the Balkans were confined to 63% of the peninsula's area. Hot spots in endemism, in the southern Dinarides and the northern Šara-Pindos Mts., coincided with those of species richness. K-mean clustering on a species presence/absence matrix suggested two main clusters: a northern and a southern one. The majority of the peninsula (88% of its area) showed close affinities with Central Europe (Coefficient of Community values $CC>0.50$) and only 20% of the area was equally closely tied to Asia Minor. Joining-clustering of the 14 west-Palaearctic mammal faunas placed the Balkans within the European cluster, thus suggesting its prevailing European character.

ACKNOWLEDGEMENTS

I would like to thank Dr. Vladimír Vohralík (Prague) for providing data for the Greek islands and Dr. Franc Janžekovič (Maribor) for help with autocorrelation analyses.

REFERENCES

Anderson, S. 1997. Mammals of Bolivia, taxonomy and distribution. *Bulletin of the American Museum of Natural History* 231, 1-652.

Aulagnier, S. & Thevenot, M. 1986. Catalogue des mammifères sauvages du Maroc. *Travaux de l'Institut scientifique Rabat, Ser. Zoologie* 41, 1-163.

Barbosa, A. & Benzal, J. 1996. Diversity of small mammals in Iberia: peninsula effect or habitat suitability? *Zeitschrift für Säugetierkunde* 61, 236-241.

Barome, P-O., Lymberakis, P., Monnerot, M. & Gautun, J-C. 2000. Cytochrome b sequences reveal *Acomys minous* (Rodentia, Muridae) paraphyly and answer the question about the ancestral karyotype of *Acomy dimidiatus*. *Molecular Phylogenetics and Evolution* 18, 37-46.

Bates, P.J.J. & Harrison, D.L. 1997. *Bats of the Indian subcontinent*. Harrison Zoological Museum, Sevenoaks.

Brelih, S. 1986. Ectoparasitical entomofauna of Yugoslav mammals. II. Siphonaptera from *Dinaromys bogdanovi* and *Chionomys nivalis* (Rodentia: Cricetidae). *Scopolia* 11, 1-47.

Brelih, S. & Trilar, T. 2000. New data on Siphonaptera from *Dinaromys bogdanovi* (Rodentia: Muridae). *Scopolia* 43, 1-22.

Brown, J.H. & Lomolino, M.V. 1998. *Biogeography*. 2nd ed. Sinauer Associates Inc., Sunderland.

Corbet, G.B. 1978. *The mammals of the Palaearctic region: a taxonomic review*. British Museum (Natural History), London.

Corber, G.B. & Hill, J.E. 1992. *The mammals of the Indomalayan region*. Oxford University Press, Oxford.

Dautbegović, A. 1988. [Uz stogodišnjicu Zemljaskog muzeja Bosne i Hercegovine u Sarajevu]. In: *Spomenica stogodišnjice rada Zemljskog muzeja Bosne i Hercegovine 1888-1988* (A. Dautbegović, ed.), pp. 7-34. Zemljski muzej Bosne i Hercegovine, Sarajevo (in Serbo-Croatian).

Dermitzakis, M.D. 1990. Paleogeography, geodynamic processess and event stratigraphy during the Late Cenozoic of the Aegean area. *Atti Convegni Lincei (Acc. Nat. Linc.)* 85, 263-288.

Doflein, F. 1921. *Mazedonien. Erlebnisse und Beobachtungen eines Naturforschers im Gefolge des Deutschen Heeres*. Gustav Fischer, Jena.

Eisenberg, J.F. & McKay G.M. 1970. An annotated checklist of the recent mammals of Ceylon with keys to the species. *Ceylon Journal of Science (Biological Sciences)* 8, 69-99.

Feiler, A. 1990. Distribution of mammals in Angola and notes on biogeography. In *Vertebrates in the tropics* (G. Peters & R. Hutterer, eds.), pp. 221-236. Museum Alexander Koenig, Bonn.

Filippucci, M.G., Fadda, V., Kryštufek, B., Simson, S. & Amori, G. 1991. Allozyme variation and differentiation in *Chionomys nivalis* (Martins, 1842). *Acta Theriologica* 36, 47-62.

Findley, J.S., Harris, A.H., Wilson, D.E. & Jones, C. 1975. *Mammals of New Mexico*. University of New Mexico Press, Albuquerque.

Gaston, K.J. & David, R. 1994. Hotspots across Europe. *Biodiversity Letters* 2, 108-116.

Gerasimov, S. & Minkova, T. 1999. Biodiversity characteristics of small-mammal communities in Bulgaria. *Contributions to the Zoogeography and Ecology of the Eastern Mediterranean Region, Athens* 1, 89-97.

Gill, A., Petrov, B., Živković, S. & Rimsa, D. 1987. Biochemical comparisons in Yugoslavian rodents of the families Arvicolidae and Muridae. *Zeitschrift für Säugetierkunde* 52, 247-256.

Glenny, M. 2000. *The Balkans 1804 – 1999. Nationalism, war, and the great powers*. Granta Books, London.

Gliozzi, E., Abbazzi, L., Argenti, P., Azzaroli, A., Caloi, L., Capasso Barbato, L., Di Stefano, G., Esu, D., Ficcarelli, G., Girotti, O., Kotsakis, T., Masini, F., Mazza, P., Mezzabotta, C., Palombo, M.R., Petronio, C., Rook, L., Sala, B., Sardella, R., Zanalda, E. & Torre, D. 1997. Biochronology of selected mammals, molluscs and ostracods from the Middle Pliocene to the Late Pleistocene in Italy. The state of art. *Riv. Ital. Paleont. Stratigraf.* 103, 369-388.

Gromov, I.M. & Polyakov, I.Ya. 1992. *Voles (Microtinae). Fauna of the USSR. Mammals, Vol. III, No. 8*. Smithsonian Institution Libraries and The National Science Foundation, Washington, D.C.

Hagmeier, E. M. & Stults, C. D. 1964. A numerical analysis of the distributional patterns of North American mammals. *Systematic Zoology* 13, 125-155.

Harrison, D.L. & Bates, P.J.J. 1991. *The mammals of Arabia*. Harrison Zoological Museum Publication, Sevenoaks.

Horvat, I., Glavać, V. & Ellenberg, H. 1974. *Vegetation Südosteuropas*. Gustav Fischer Verlag, Stuttgart.

Hosey, G.R. 1982. The Bosporous Land-bridge and mammal distributions in Asia Minor and the Balkans. *Säugetierkundliche Mitteilungen* 30, 53-62.

Jarman, M.R. 1996. Human influences in the development of Cretan mammalian fauna. In: *Pleistocene and Holocene Fauna of Crete and Its First Settlers* (D.S. Reese, ed.), pp. 211-229. Prehistory Press, Madison, WI.

Jones, J.K. Jr. & Birney, E.C. 1988. *Handbook of mammals of the north-central States*. University of Minnesota Press, Minneapolis.

Jovanović, B., Jovanović, R. & Zupančič, M. 1986. *Natural potential vegetation of Yugoslavia (Commentary to the map 1: 1,000,000)*. Scientific Council of Vegetation Map of Yugoslavia, Ljubljana.

Kent, M. & Coker, P. 1998. *Vegetation description and analysis. A practical approach*. John Wiley & Sons, New York.

Koenigswald, von W. 1980. Schmeltzstruktur und Morphologie in den Molaren der Arvicolidae (Rodentia). *Abteilunden des Senckenberg Naturforsch. Gesellschaft* 539, 1-129.

Kowalski, K.& Rzebik-Kowalska, B. 1991. *Mammals of Algeria*. Polish Academy of Sciences, Warszawa.

Krapp, F. (ed.) 2001. *Handbuch der Säugetiere Europas. Band 4: Fledertiere. Teil I. Rhinolophidae, Vespertilionidae I*. Aula-Verlag, Wiebelsheim.

Kryštufek, B. 1987. Some aspects of sympatric occurrence of two petricolic voles (*Dinaromys bogdanovi* and *Chionomys nivalis* – Rodentia, Mammalia) in Yugoslavia. *4th International Congress on Zoogeography and Ecology of Greece and Adjacent Regions, Kamena Vurla 20-25.IV.1987. Book of Abstracts*. Hellenic Zoological Society, p. 93.

Kryštufek, B. 1990. Geographic variation in *Microtus nivalis* (Martins, 1842) from Austria and Yugoslavia. *Bonner zoologische Beitraege* 41, 121-139.

Kryštufek, B. 1991. *Mammals of Slovenia*. Prirodoslovni muzej Slovenije, Ljubljana (in Slovene with English abstract).

Kryštufek, B. 1997. Small mammals (Insectivora, Chiroptera, Rodentia). In: *Mousterian "bone-flute" and other finds from Divje babe I cave site in Slovenia*. (I. Turk, ed.), pp. 85-98. Založba ZRC, Ljubljana.

Kryštufek, B. & Griffiths, H.I. 1999. Mediterranean *vs.* continental small mammal communities and the environmental degradation of the Dinaric Alps. *Journal of Biogeography* 26, 167-177.

Kryštufek, B. & Griffiths, H.I. 2001. Species richness and rarity in European rodents. *Ecography* 24, 120-128.

Kryštufek, B. & Petkovski, S. 2003. Annotated checklist of the mammals of Macedonia. *Bonner zoologische Beitraege* 51, 229-254.

Kryštufek, B. & Vohralik, V. 2001. *Mammals of Turkey and Cyprus. Introduction, Checklist, Insectivora*. Knjižnica Annales Majora, Koper.

Lay, D.M. 1967. A study of the mammals of Iran resulting from the Street expedition of 1962-63. *Fieldiana: Zoology* 54, 1- 282.

Lekagul, B. & McNeely, J.A. 1988. *Mammals of Thailand*. 2nd ed. Darnsutha Press, Bangkok.

Malez, M. & Rabaeder, G. 1984. Neues Fundmaterial von Kleinsäugern aus der altpleistozänen Spaltenfüllung Podumci 1 in Norddalmatien (Kroatien, Jugoslawien). *Beitr. Paläont. Österr.* 11, 439-510.

Mares, M. & Lacher, T.E. 1987. Ecological, morphological, and behavioral convergence in rock-dwelling mammals. In: *Current mammalogy, Vol. 1*. (H.H. Genoways, ed.), pp. 307-348. Plenum Publishing Corporation, New York.

Markov, G. 1957. *Die Insektenfressenden Säugetiere in Bulgarien*. Izdanie na B'lgarskata Akademija na anukite, Sofija.

Merritt, J.F. 1987. *Guide to the mammals of Pennsylvania*. University of Pittsburgh Press, Pittsburgh.

Miller, G.S. 1912. *Catalogue of the mammals of Western Europe (Europe exclusive of Russia) in the collection of the British Museum*. British Museum (Natural History), London.

Mitchell-Jones, A.J., Amori, G., Bogdanowicz, W., Kryštufek, B., Reijnders, P.J.H., Spitzenberger, F., Stubbe, M., Thissen, J.B.M., Vohralik, V. & Zima, J. 1999. *The atlas of European mammals*. T. & A.D. Poyser, London.

Musser, G.G. & Carleton, M.D. 1993. Family Muridae. In: *Mammal species of the World. A taxonomic and geographic reference*. 2nd ed. (D.E. Wilson & , D.-A.M. Reeder, eds.), pp. 501-755. Smithsonian Institution Press, Washington.

Nesin, V.A. & Skorik, A.F. 1989. First find of the *Dinaromys* vole (Rodentia, Microtinae) in the USSR. *Vestnik Zoologii* 5, 14-17.

Nevo, E. 1999. *Mosaic evolution of subterranean mammals. Regression, progression, and global convergence*. Oxford University Press, Oxford.

Niethammer, J. 1986. Über griechische Nager im Museum A. Koenig in Bonn. *Annalen des Naturhistorisches Museum Wien* 88/89B, 245-256.

Niethammer, J. & Krapp, F. (eds.) 1978. *Handbuch der Säugetiere Europas. Band 1 Rodentia 1 (Sciuridae, Castoridae, Gliridae, Muridae)*. Akademische Verlagsgesellschaft, Wiesbaden.

Niethammer, J. & Krapp, F. (eds.) 1982. *Handbuch der Säugetiere Europas. Bd. 2/1 Rodentia II (Cricetidae, Arvicolidae, Zapodidae, Spalacidae, Hystricidae, Capromyidae)*. Akademische Verlagsgesellschaft, Wiesbaden.

Niethammer, J. & Krapp, F. (eds.) 1983. *Handbuch der Säugetiere Europas. Band 3/1. Insektenfresser – Insectivora, Herrentiere – Primates*. Aula-Verlag, Wiesbaden.

Niethammer, J. & Krapp, F. (eds.) 1986. *Handbuch der Säugetiere Europas. Band 2/II Paarhufer – Artiodactyla (Suidae, Cervidae, Bovidae)*. Aula-Verlag, Wiesbaden.

Ondrias, J.C. 1966. The taxonomy and geographical distribution of the rodents of Greece. *Säugetierkundliche Mitteilungen* 14, 1-136.

Osborn, D.J. & Helmy, I. 1980. The contemporary land mammals of Egypt (including Sinai). *Fieldiana Zoology N. S.* 5, 1-579.

Panteleyev, P.A. 1998. *The rodents of the Palaearctic. Composition and areas*. Russian Academy of Sciences, Moscow.

Payne, J., Francis, C.M. & Phillipps, K. 1985. *A field guide to the mammals of Borneo*. The Sabah Society, Kota Kinabalu.

Petrov, B. 1979. Some questions of the zoogeographical division of the western Palaearctic in the light of the distribution of mammals in Yugoslavia. *Folia Zoologica* 28, 13-24.

Petrov, B. 1985. Numerical elements for zoogeographical division of the Western Palaearctic using the distribution of mammals in Yugoslavia as an example. *Acta Zoologica Fennica* 170, 209-210.

Petrov, B.M. 1992. Mammals of Yugoslavia. Insectivores and rodents. *Natural History Museum in Belgrade, Suppl.* 37, 1-186.

Petrov, B. & Todorović, M. 1982. *Dinaromys bogdanovi* (V. et E. Martino, 1922) – Bergmaus. In: *Handbuch der Säugetiere Europas, Bd. 2/1*. (J. Niethammer & F. Krapp, eds.), pp. 193-208. Akademische Verlagsgesellschaft, Wiesbaden.

Qumsiyeh, M.B. 1985. The bats of Egypt. *Special Publications of the Museum of the Texas Technical University* 23, 1-102.

Qumsiyeh, M.B. 1996. *Mammals of the Holy Land*. Texas Technical University Press, Lubbock.

Reese, D.S. 1996. *Pleistocene and Holocene Fauna of Crete and its first settlers*. Prehistory Press, Madison, WI.

Reumer, J.W.F. 1986. Notes on the Soricidae (Insectivora, Mammalia) from Crete. I. The Pleistocene species *Crocidura zimmermanni*. *Bonner zoologische Beitraege* 37, 161-171.

Reumer, J.W.F. 1996. Shrews (Soricidae) on islands, with special reference to *Crocidura zimmermanni* from Crete. In: *Pleistocene and Holocene Fauna of Crete and Its First Settlers* (D.S. Reese, ed.), pp. 173-179. Prehistory Press, Madison, WI.

Rosenzweig, M.L. 1996. *Species diversity in space and time*. Cambridge University Press, Cambridge.

Sala, B. 1996. *Dinaromys allengranzii* n. sp. (Mammalia, Rodentia) from Rivoli Veronese (northeastern Italy) in a Villanyian association. *Acta zoologica cracoviensia* 39, 469-472.

Savić, I., Soldatović, B. 1984. Karyotype evolution and taxonomy of the genus *Nannospalax* Palmer, 1903, Mammalia, in Europe. *Serbian Academy of Sciences, Dept. Nat. Math. Sci.* 59, 1-104.

Shidlovsky, M.V. 1976. [Opredelitely grizunov Zakavkazya]. Mecniereba, Tbilisi (in Russian).

Skinner, J.D. & Smithers, R.H.N. 1990. *The mammals of the southern African subregion*. University of Pretoria, Pretoria.

Sokal, R.R. 1978a. Spatial autocorrelation in biology. 1. Methodology. *Biological Journal of the Linnean Society* 10, 199-228.

Sokal, R.R. 1978b. Spatial autocorrelation in biology. 2. Some biological implications and four applications of evolutionary and ecological interest. *Biological Journal of the Linnean Society* 10, 229-249.

Sokolov, V.E. & Orlov, V.N. 1980. [*Opredelitely mlekopityoschich Mongolskoi Narodnoi Respubliki*]. Nauka, Moscow (in Russian).

Sokolov, V.E. & Tembotov, A.K. 1989. [*Pozvonochnije Kavkaza. Mlekopitayuschie. Nasekomoyadniye*]. Nauka, Moscow (in Russian).

Storch, G. 1977. Die Ausbreitung der Felsenmaus (*Apodemus mystacinus*): Zur Problematik der Inselbesiedlung und Tiergeographie in der Ägäis. *Natur und Museum* 107, 174-182.

Storch, G. 1999. *Apodemus mystacinus* (Danford & Alston, 1877). In: *The atlas of European mammals* (A.J. Mitchell-Jones *et al.*, eds.), pp. 272-273. T. & A.D. Poyser, London.

Strelkov, P.P. & Iljin, V.Ju. 1992. Bats of easternmost Europe: distribution and faunal status. In: *Prague studies in mammalogy* (I. Horaček & V. Vohralik, eds.), pp. 193-205. Charles University Press, Prague.

Stubbe, M. & Krapp, F. (eds.) 1993a. *Handbuch der Säugetiere Europas. Band 5 Raubsäuger – Carnivora (Fissipedia). Teil I: Canidae, Ursidae, Procyonidae, Mustelidae 1*. Aula-Verlag, Wiesbaden.

Stubbe, M. & Krapp, F. (eds.) 1993b. *Handbuch der Säugetiere Europas. Band 5 Raubsäuger – Carnivora (Fissipedia). Teil II: Mustelidae 2, Viverridae, Hespertidae, Felidae*. Aula-Verlag, Wiesbaden.

Terzea, E. 1972. Remarques sur la morphologie dentaire et la répartition de *Microtus nivalis* Martins dans le Pléistocène de Roumanie. *Travaux de l'Institut de spéologie "Émile Racovitza"* 11, 271-298.

Torre, I., Tella, J.L. & Arrizabalaga, A. 1996. Environmental and geographic factors affecting the distribution of small mammals in an isolated Mediterranean mountain. *Zeitschrift für Säugetierkunde* 61, 365-375.

Todorović, M. 1956. Variability of the endemic genus *Dolomys* Nehring (Microtinae, Mammalia). *Arhiv bioloških nauka, Beograd* 8, 93-109.

Tvrtković, N., Dulić, B. & Mrakovčić, M. 1985. Distribution of Insectivora and Rodentia on the north-east Adriatic coast (Yugoslavia). *Acta Zoologica Fennica* 170, 201-203.

Vohralík, V. 1985. Notes on the distribution and the biology of small mammals in Bulgaria (Insectivora, Rodentia) I. *Acta Universitatis Carolinae – Biologia* 1981, 445-461.

Vohralík, V., Sofianidou, T. 1987. Small mammals (Insectivora, Rodentia) of Macedonia, Greece. *Acta Universitatis Carolinae – Biologia* 1985, 319-354.

Vohralík, V., Sofianidou, T. 1991. Notes on the distribution of small mammals (Insectivora, Rodentia) in Epeirus, Greece. *Bonner zoologische Beitraege* 42, 125-135.

Vohralík, V., Sofianidou, T. 1992. Small mammals (Insectivora, Rodentia) of Thrace, Greece. *Acta Universitatis Carolinae – Biologia* 36, 341-369.

Vos, de J. 1984. *The endemic Pleistocene deer of Crete*. Verhandling der Koninklijke Nederlandse Akademie van Wetenschappen, Afdeling Natuurkunde, Eerste Reeks, Deel 31. North-Holland Publ. Comp., Amsterdam.

Ward, J.H. 1963. Hierarchical grouping to optimize an objective function. *Journal of the American Statistical Association* 58, 236.

Wartenberg, D. 1989. *SAAP –Spatial Autocorrelation Analysis Program*. University of Medicine and Dentary of New Jersey, Placataway, NJ.

Wilson, D.E. & Reeder, D.-A.M. (eds.) 1993. *Mammal species of the World. A taxonomic and geographic reference*. 2nd ed. Smithsonian Institution Press, Washington.

Whittaker, J.O. Jr. 1996. *National Audubon Society field guide to North American mammals*. Chanticleer Press, New York.

APPENDIX

Definition of the study region

Since there is no consensus on the geographic scope of the Balkan Peninsula, the region was defined for the purposes of this study to include the following countries: Slovenia, Croatia, Bosnia and Herzegovina, Federal Republic of Yugoslavia, Macedonia (FYROM), Albania, Greece, Bulgaria, Romania, and European Turkey. This gave an area of 788,689 km^2 which was used for species/area estimates. The bordering regions of Austria, Hungary, Moldova, and Russia were also included in cartographic presentations and corresponding analyses.

Materials

Distribution maps and taxonomy were taken from Mitchell-Jones *et al.* (1999) but, when appropriate, Niethammer & Krapp (1978, 1982, 1983) and other papers on individual species were also consulted, in addition to the author's unpublished data. The aim was to consider only those species whose ranges are not the result of direct human impact. This gave a total of 86 insectivores, bats, rodents, and mustelids (see below). Introduced exotics, synanthropic rodents (both rats *Rattus* spp. and two commensal house mice *Mus* spp.) and three bats (*Nyctalus lasipterus, Vespertilio murinus, Barbastella barbastellus*) were excluded, whose range borders could not have been delimited accurately.

Taxonomic list. – Insectivora: *Erinaceus concolor, Sorex alpinus, S. araneus, S. minutus, Neomys anomalus, N. fodiens, C. leucodon, C. suaveolens, Suncus etruscus, Talpa caeca, T. levantis, T. europaea, T. stankovici.* Chiroptera: *Rhinolophus blasii, R. euryale, R. ferrumequinum, R. hipposideros, R. mehelyi, Myotis bechsteini, M. blythi, M. brandtii, M. capaccinii, M. dasycneme, M. daubentonii, M. emarginatus, M. myotis, M. mystacinus, M. nattereri, Pipistrellus kuhlii, P. nathusii, P. pipistrellus, P. savii, Nyctalus leisleri, N. noctula, Eptesicus nilssonii, E. serotinus,*

Plecotus auritus, P. austriacus, Miniopterus schreibersii, Tadarida teniotis. Rodentia: *Sciurus vulgaris, Spermophilus citellus, Cricetus cricetus, Cricetulus migratorius, Mesocricetus newtoni, Clethrionomys glareolus, Dinaromys bogdanovi, Arvicola terrestris, Microtus agrestis, M. arvalis, M. felteni, M. guentheri, M. multiplex, M. oeconomus, M. rossiaemeridionalis, M. subterraneus, M. thomasi, Chionomys nivalis, Spalax graecus, Nannospalax leucodon, Micromys minutus, Apodemus agrarius, A. flavicollis, A. mystacinus, A. sylvaticus, A. uralensis, Mus macedonicus, M. spicilegus, Glis glis, Muscardinus avellanarius, Eliomys quercinus, Dryomys nitedula, Myomimus roachi, Sicista betulina, S. subtilis*. Mustelidae: *Mustela erminea, M. eversmanii, M. lutreola, M. nivalis, M. putorius, Vormela peregusna, Martes foina, M. martes, Meles meles, Lutra lutra.*

Distribution maps were prepared for each species, on which were superimposed a grid of squares proportionally 100 km. Presence/absence of a species was scored for each square. The square grid was quite crude for a geographic area of 788,689 km^2. However, the aim was to evaluate rough patterns; besides, the 100 x 100 km squares were not too sensitive to gaps in distributional detail, which were particularly evident in bats.

An additional 24 regions were used for broader assessments of the Balkan fauna. Mammal lists for European countries were based on Mitchell-Jones *et al.* (1999); sources for east-European and non-European regions are in parentheses: Pyrenean peninsula (Spain and Portugal); Italian peninsula (Italy); Central Europe (Germany, Czech Republic); northern Europe (Sweden, Norway, Finland); eastern Europe (square between 50-60° N and 40-50° E; Corbet 1978, Strelkov & Iljin, 1992, Panteleyev, 1998); Asia Minor (Asian Turkey; Kryštufek & Vohralík, 2001); the Caucasus (Armenia, Azerbaijan, Georgia, and Russia to the south of 44° N; Corbet, 1978; Shidlovsky, 1976; Panteleyev, 1998; Sokolov & Tembotov, 1989); the Levant (see Qumsiyeh, 1996 for definition and scope); Iran (Lay, 1967); Arabia (Arabian peninsula to the south of 30°N; Harrison & Bates, 1991); Egypt (incl. Sinai; Osborn & Helmy, 1980; Qumsiyeh, 1985); Algeria (Kowalski & Rzebik-Kowalska, 1991); Morocco (Aulagnier & Thevenot, 1986); Mongolia (Sokolov & Orlov, 1980); Pennsylvania (Merritt, 1987; Whittaker, 1996); North-Central states of the USA (Minnesota, Iowa, Wisconsin, Illinois, Michigan, Indiana, Ohio; Jones & Birney, 1988; Whitaker, 1996); New Mexico (Findley *et al.*, 1975); southern Africa (Republic of South Africa, Lesotho, Swaziland; Skinner & Smithers, 1990); India (Corbet & Hill, 1992; Bates & Harrison, 1997); Sri Lanka (Eisenberg & McKay, 1970); Thailand (Lekagul & McNeely, 1988); Island of Borneo (Payne *et al.*, 1985); Angola (Feiler, 1990); Bolivia (Anderson, 1997). Older lists were up-dated using Wilson & Reeder (1993).

Methods

Statistical analyses were performed using STATISTICA (Release 5.5 '99).

Arrhenius equation. Because of the allometric relations between species richness (S) and area (A), direct comparisons in species richness among regions of different sizes are possible only when S is adjusted for the size of the area (A). Adjusted species richness is given as the c-value and was derived from Arrhenius equation:

$$c = S/A^z$$

where z is the slope of the species-area curve (Rosenzweig, 1996). Since the slope of the species-area curve varies between biota, z-values were calculated for the mammals of the Balkans. Species-area curves were obtained by plotting species richness against increased areas; values were transformed to logarithms (Log_{10}). The basic area unit was the 100 x 100 km square; this was doubled in each subsequent step until the entire peninsula was included. Non-volant mammals (insectivores, rodents and mustelids) accumulated new species at a faster rate than did bats (Figure 23), resulting in different z-values: non-volant mammals z=0.169, bats z=0.082, and both groups combined z=0.132.

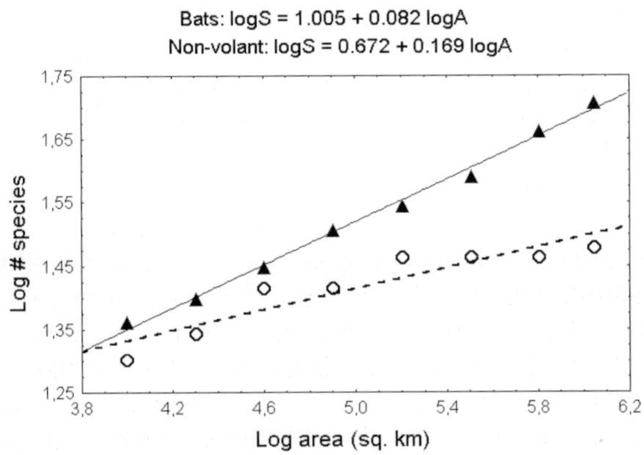

Figure 23. Species-area curves for bats (dots) and non-volant mammals (triangles) of the Balkan Peninsula.

Index of Faunistic Change. The Index of Faunistic Change (*IFC*) was computed from the number of species whose ranges ended within each square:

$$IFC = 100L/n$$

where L is the number of range limits in the square, and n the number of species occurring there, whether their ranges ended or not. The resultant IFC values are a measure of the percentage of species whose ranges end within a 100 km square area (Hagmeier & Stults, 1964).

Coefficient of Community. Coefficient of Community (or Jaccard coefficient; *CC*) was used in comparing different faunas:

$$CC = 100C/\, n_1 + n_2 - C$$

where C represents the number of species common to two areas, n_1 is the number in the smaller fauna, and n_2 the number in the larger (Hagmeier & Stults, 1964).

Associations. Pairwise associations in hot spots between taxonomic groups were tested by the χ^2 test and 2 x 2 contingency tables, using Yate's correction for small samples. To assess the nature of the association an expected frequency was calculated using the formula:

$$(a + b)\ (a + c)\ N^{-1}$$

where a is the number of 100 x 100 km squares being hot spots in i and j, b and c are the numbers of squares being hot spots of i and j, respectively (Kent & Coker, 1992). Lower predicted than observed frequencies denote negative associations and *vice versa*.

Spatial autocorrelation analysis. Spatial autocorrelation analysis was performed to describe and summarise patterns of variation and covariation of species densities and the Index of Faunal Change. Spatial analysis tests whether the observed value of a variable at one locality is significantly dependent on values of the variable at neighbouring localities. The autocorrelation coefficient (Moran's statastic I) varies from –1 (no correlation) to +1 (perfect correlation). See Sokal (1978a, b) and Wartenberg (1989) for methodological considerations.

Joining-clustering. This multivariate statistical technique was used to join together objects (species or regions) into successively larger clusters, using measures of similarity, to produce a clustering tree. The clustering algorithm used was Ward's method, which attempts to minimise the sum of squares of any two (hypothetical) clusters that can be formed at each step (Ward 1963). If the dimensions included into the analysis were categorical in nature, the distance matrix was computed using "percent disagreement":

$$distance\ (x,y) = (number\ of\ x_i \neq y_i)\ /\ i$$

K-means clustering. The method is based on an *a priori* defined number of clusters and thus differs conceptually from tree-clustering. It is actually an analysis of variance "in reverse" with the aim of minimising within-cluster variance and maximising between-cluster variance. In general, the *K*-means method will produce exactly *K* different clusters of greatest possible distinction.

7. KARYOTYPIC VARIATION IN MAMMALS OF THE BALKAN PENINSULA

JAN ZIMA*

INTRODUCTION

This paper investigates chromosomal variation in a well-known animal group - the mammals. Chromosome research in this animal group has had a long-term tradition in the Balkan Peninsula, and has contributed significantly to our understanding of its regional biodiversity. Most studies have examined chromosomes of small mammal species, within the orders of insectivores, chiropterans, rodents and carnivores, whose size does not usually exceed that of the hedgehog or common hamster. This contribution attempts to summarise our current knowledge of the karyotypes of Balkan mammals, and to show the importance of chromosomal data in systematic and phylogeographic research.

Biological diversity at the chromosomal level

The diversity of life can be conceived and studied at three basic levels. The first is that of genes, determined basically by sequences of the base pairs in molecules of DNA. The second is the species level, and the last is that of communities and ecosystems. This contribution deals with the first two levels of biodiversity. Studies of the genetic level of biodiversity are based mainly on molecular techniques of sequencing and restriction analyses of DNA, and those of the species level largely on morphological analyses of phenotypic traits. Between these two fundamental approaches to demonstrating biological diversity lie chromosomal and/or karyotypic studies which attempt to describe the basic features of the gross organisation of the genome, manifested in the number and morphology of chromosomes in the nucleus.

The aim of this contribution is to demonstrate the significance and usefulness of chromosomal data in elucidating the diversity of animals, and to show that data obtained from karyotypic studies can reveal new information that is not redundant in respect both to molecular and to morphological analyses. We will attempt to review the available karyotypic data and to provide a synopsis of the extent of chromosomal variation reported for the mammal fauna of the Balkans. It will be demonstrated that this chromosomal variation represents an important aspect of the pattern of biological diversity. Various implications for systematic, phylogenetic, and biogeographic studies will then be proposed.

Available karyotypic data

Information concerning all the mammal species of the Balkans is summarised in a synopsis of available karyotypic data, in Tables 1-3, below. For the purposes of this paper, the Balkans are defined artificially according to current political borders, to include the countries of Albania, Bosnia and Herzegovina, Bulgaria, Croatia, Greece,

H.I. Griffiths et al. (eds.), Balkan Biodiversity, 109–133.

Macedonia (Former Yugoslav Republic of Macedonia; FYROM), Romania, Slovenia, Serbia, Montenegro and European Turkey. Using this approach, the area covered exceeds the usually accepted geographic limits of the Balkan Peninsula. The faunal list of the mammalian species of the area under consideration and taxonomic nomenclature were derived from the atlas by Mitchell-Jones *et al.* (1999).

This compilation indicates that there are about 123 terrestrial mammal species distributed in the Balkans. Some marginal species occurring in the surrounding mountain ranges of the Alps and the Carpathians are also included in the list (*Erinaceus europaeus, Myotis dasycneme, Marmota marmota, Microtus tatricus, Sicista betulina, Mustela erminea, Capra ibex*). The list comprises several non-native, introduced species (*Oryctolagus cuniculus, Ondatra zibethica, Rattus norvegicus, R. rattus, Myocastor coypus, Nyctereutes procyonoides, Mustela vison, Herpestes auropunctata, Axis axis, Dama dama, Odocoileus virginianus, Ovis musimon*), and other species that became extinct in the area in the past and have since been re-introduced (*Castor fiber, Bison bonasus*). Almost all the species in this list have been studied karyologically, although some of them solely from areas outside the Balkans. For some species, available data are restricted to simple karyotypic characteristics derived from conventional staining, and the chromosomal reports are based on examination of only a few specimens from a restricted geographic area. The only species of the region which has not been studied is the monk seal, *Monachus monachus*.

The present evaluation is based mainly on published data relating to our own research. In the studies performed in our laboratory during the past 25 years, the karyotypes of a total of 290 individuals originating from the Balkans have been examined, 85 of which were also studied in the surrounding geographic regions. These data are combined with the rich information derived from other published papers, with a total of more than 200 publications having been taken into account.

The study encompasses both between- and within-species variation in karyotypic characteristics, with an emphasis on chromosomal differences occurring between different geographic populations and closely related species.

EXTENT OF CHROMOSOMAL VARIATION

The data show that the level of chromosomal variation within individual species is considerable. There are 25 species (20.3%) with a polymorphic karyotype and 24 species (19.5%) are polytypic. Altogether, 39 species of the Balkan mammal fauna (31.7%) reveal intraspecific karyotype variation. In more than 20 species, karyotypic studies contributed significantly to defining their precise taxonomic status. In some cases, these chromosomal studies also allowed morphological sibling species with specific karyotypes to be distinguished. The resultant data on the overall extent of variation provide rich material with which to study biological diversity at the chromosomal level.

Chromosomal polymorphism

In polymorphic species, karyotype variation is recorded within single populations. This variation originates from various kinds of rearrangements (Table 1). The frequent cause of chromosomal polymorphism is the occurrence of supernumerary or B chromosomes. These B chromosomes are accessory and occur in addition to the elements of the standard complement. They are mainly heterochromatic, that is, genetically inactive. Such chromosomes can occur in certain populations of the lesser white-toothed shrew, *Crocidura suaveolens* (Zima *et al.*, 1998) or *Nyctalus leisleri* (Volleth, 1992), and in several species of the genus *Apodemus*. They are also quite common in the yellow-necked wood mouse, *Apodemus flavicollis* (Vujošević & Živković, 1987; Belcheva *et al.*, 1988; Vujošević *et al.*, 1991; Figure 1).

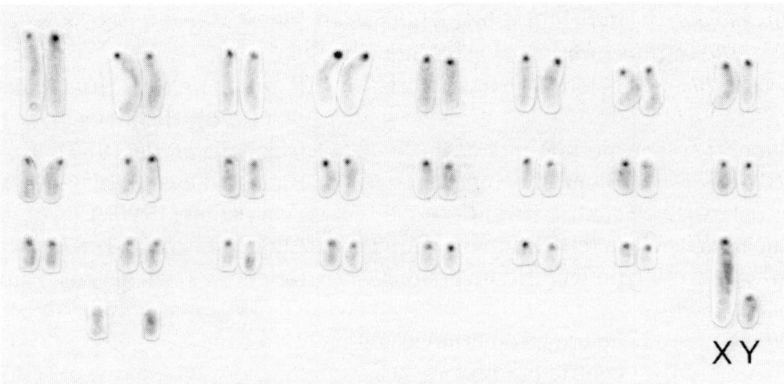

Figure 1. C-banded karyotype of the yellow-necked wood mouse, Apodemus flavicollis, *from Mt. Pelister in Macedonia (FYROM). The complement comprises 50 chromosomes including two supernumerary (B-) chromosomes. From Zima* et al. *(1997b), with permission of the publishers of* Scopolia.

In the yellow-necked wood mouse, studies of the occurrence of supernumerary chromosomes revealed a clinal increase in frequency from south-eastern to Central Europe (Zima & Macholán, 1995). Compared to other regions, the Balkan populations usually show rather low frequency of supernumeraries, but increased frequencies were recorded in mountain populations living at high altitude and under extreme climatic conditions (Vujošević & Blagojević, 2000). The frequency of B chromosomes appears to be relatively stable in individual geographic populations over successive years (Vujošević, 1992) but it varies seasonally and may have a role in population dynamics (Blagojević & Vujošević, 1995). Blagojević & Vujošević (2000) assessed the effect of B chromosomes on certain morphometric characters, indicating the possible adaptive significance of B chromosomes. It does not seem likely that the frequency of B chromosomes in the yellow-necked wood mouse is related to environmental pollution (Zima *et al.*, 1999) as had been suggested previously for Balkan populations (Giagia *et al.*, 1985).

Variation in the amount and the distribution of heterochromatin can also be observed in chromosomes of the standard complement, both in autosomes and sex chromosomes. Such polymorphism has been described in Bulgarian populations of the suslik, *Spermophilus citellus*, the variation pattern of which differed from that found in Central European populations (Belcheva & Peshev, 1979, 1985).

Table 1. Species of mammals with the polymorphic karyotype occurring in the Balkans

Species	Nature of polymorphism	References
Sorex araneus	Robertsonian changes	Searle & Wójcik (1998)
Crocidura suaveolens	B chromosomes	Zima *et al*. (1998)
Nyctalus leisleri	B chromosomes	Volleth (1992), own data
Cricetulus griseus	pericentric inversion, sex chromosomes variation	Zagorodnyuk (1986), own data
Cricetus cricetus	heterochromatin variation	Ružić *et al*. (1975b), Gamperl *et al*. (1976), Belcheva *et al*. (1988)
Clethrionomys glareolus	morphology of the Y, B chromosomes	Belcheva *et al*. (1987), Radosavljević *et al*. (1988)
Microtus agrestis	centric fusion	Zima *et al*. (1990b)
Microtus arvalis	heterochromatin variation	Lungeanu *et al*. (1987)
Microtus rossiaemeridionalis	pericentric inversions	Belcheva *et al*. (1985), Zima *et al*. (1991)
M. guentheri	heterochromatin variation, centric fusion	own data
M. multiplex	the Y morphology	Tvrtković *et al*. (1979), Zima & Král (1984)
M. subterraneus	pericentric inversions	Živković *et al*. (1975a), Král & Mitev (1976), Sablina *et al*. (1989)
Apodemus agrarius	pericentric inversions, B chromosomes	Soldatović *et al*. (1969), Lungeanu *et al*. (1986), Belcheva *et al*. (1988)
A. flavicollis	B chromosomes	Giagia *et al*. (1985),Vujošević & Živković (1987), Zima & Macholán (1995)
A. sylvaticus	B chromosomes, centric fusion	Giagia *et al*. (1985),Vujošević & Živković (1987)
A. mystacinus	pericentric inversion, B chromosomes	Giagia *et al*. (1985), Belcheva *et al*. (1988)
Rattus norvegicus	pericentric inversions	Zima & Král (1984)
R. rattus	pericentric inversions, B chromosomes	Belcheva & Bisserov (1982)
Mus domesticus	Robertsonian changes	Djulić *et al*. (1980), Giagia *et al*. (1987), Tichy & Vucak (1987)
M. musculus	centric fusion	Zima *et al*. (1990a)
Vulpes vulpes	B chromosomes	review in Zima & Král (1984)
Mustela erminea	centric fusion	Mandahl & Fredga (1980)
M. nivalis	heterochromatin variation	Mandahl & Fredga (1980)
Sus scrofa	Robertsonian changes	Živković *et al*. (1971), Mayr *et al*. (1984)
Capreolus capreolus	B chromosomes	Živković & Isaković (1972), Zernahle (1980)

The pine vole, *Microtus (Terricola) subterraneus*, is both a polymorphic and polytypic species, but the mechanism producing variation is different in each case. In polymorphic populations, pericentric inversions occur in certain autosomes; such populations have been recorded both from the Balkans and Central Europe (Král & Mitev, 1976; Král & Zima, 1978). Polymorphism caused by pericentric inversions has also been recorded in some local populations of *Rattus rattus* (Belcheva & Bisserov, 1982) and *Microtus rossiaemeridionalis* (Belcheva *et al.*, 1977; Zima *et al.*, 1981, 1991).

Polytypic karyotypes

In some species, polymorphism can result in fixed differences between populations. This kind of variation is called chromosomal polytypy (Table 2). An example of this process is provided by the Robertsonian variation described for the house mouse, *Mus domesticus*.

This variation results from centric fusions of uniarmed, acrocentric chromosomes that produce biarmed, metacentric chromosomes with varied combinations of arms. As a result of these changes, the diploid number of chromosomes is lowered. Some populations of the house mouse from the Alps with several fixed Robertsonian fusions were previously considered as separate species (for instance the well-known *Mus poschiavinus*). The Robertsonian populations are dispersed throughout the European range of *Mus domesticus* (Winking *et al.*, 1988; Nachmann & Searle, 1995), and have also been described from the eastern Adriatic coastal regions and the Peloponnisos peninsula (Djulić *et al.*, 1980; Giagia *et al.*, 1987; Tichy & Vucak, 1987). It is surprising that such populations appear to be absent in the eastern part of the Balkans, in spite of extensive research on material from Bulgaria, Greece and European Turkey.

Another species with the polytypic karyotype is the Alpine shrew, *Sorex alpinus*. Differences in the diploid number of chromosomes originating after Robertsonian fusions have only been described from southern Switzerland; one of the karyotypic races was also recorded in several isolated populations in Central Europe (Dannelid, 1994, Lukáčová *et al.*, 1996). There are no data from the other parts of the mosaic distribution range of this species, and the absence of data from populations distributed in the Eastern Alps and the Dinarides (Dinaric Alps) should be emphasised.

Table 2. Species of Balkan mammals with a polytypic karyotype

Species	Nature of polytypy	References
Erinaceus concolor	heterochromatin changes	Markov & Dobrijanov (1974), Djulić & Tvrtković (1979), Giagia & Ondrias (1980)
E. europaeus	heterochromatin changes	Mandahl (1978)
Sorex alpinus	centric fusions	Dannelid (1994), Lukáčová *et al.* (1996)
S. araneus	Robertsonian changes, the Y morphology	Djulić (1978), Macholán *et al.* (1994), Zima *et al.*

		(1997c)
Crocidura leucodon	the Y size, centric fusions	Zima *et al.* (1998), Biltueva *et al.* (2001)
Talpa caeca	pericentric inversion	Todorović *et al.* (1972)
T. stankovici	pericentric inversions	Soldatović *et al.* (1986), Todorović *et al.* (1987)
Rhinolophus hipposideros	centric fusions	Belcheva *et al.* (1990), Zima *et al.* (1992)
Myotis mystacinus	NORs distribution	Volleth (1987)
Spermophilus citellus	heterochromatin distribution	Savić *et al.* (1971), Belcheva & Peshev (1979)
Cricetus cricetus	heterochromatin and NORs distribution	Belcheva *et al.* (1988)
Clethrionomys glareolus	the Y morphology	Živković *et al.* (1975b), Zima *et al.* (1997b)
Arvicola terrestris	heterochromatin distribution	Raicu *et al.* (1971), Živković & Petrov (1974), Peshev & Belcheva (1978), Kulijev *et al.* (1978)
Microtus arvalis	pericentric inversions, complex rearrangements	Živković *et al.* (1975c)
M. guentheri	heterochromatin and NORs distribution	Živković & Petrov (1975a), Belcheva *et al.* (1980), own data
M. multiplex	complex rearrangements	Storch & Winking (1977), Tvrtković *et al.* (1979), Graf & Meylan (1980)
M. subterraneus	Robertsonian fusion	Sablina *et al.* (1989), Macholán *et al.* (2001)
M. thomasi	Robertsonian fusions, pericentric inversions	Živković *et al.* (1975a), Giagia-Athanasopoulou *et al.* (1995), Giagia-Athanasopoulou & Stamatopooulos (1997)
Nannospalax leucodon	complex changes	Savić & Soldatović (1979, 1984)
Rattus rattus	complex changes	Yosida (1980)
Eliomys quercinus	complex changes	Murariu *et al.* (1985), Vujošević *et al.* (1993)
Sicista subtilis	complex changes	Sokolov *et al.* (1987)
Mustela nivalis	heterochromatin distribution	Zima & Grafodatskij (1985)
Sus scrofa	Robertsonian changes	Živković et al. (1971), Mayr *et al.* (1984)

The common shrew, *Sorex araneus*, exhibits high karyotypic variation compared to most other mammals. More than 60 different karyotypic races are known, some of which occur in the Balkans (Zima *et al.*, 1996). One of the Balkan races, the Pelister race, was first described from Mt. Pelister, Macedonia (FYROM) (Macholán *et al.*, 1994). This race was distinguished by the prevalence of acrocentric autosomes in the complement, and its unique chromosomal feature is the metacentric position of the centromere on the Y chromosome. The acrocentric character of the complement is considered a plesiomorphic, primitive, feature within the respective species. Similar populations exhibiting the primitive, mostly acrocentric, karyotype have been recorded in the Massif Central and western Alps of France, and on the eastern banks of Lake Baikal, Siberia (Zima *et al.*, 1994). The occurrence of primitive karyotypes in marginal populations distributed at the edges of their geographic range strongly suggests that karyotypic variation originated in central rather than peripheral populations. In this respect, the extant populations of the common shrew in the mountains of the southern Balkans may be considered as relics retaining ancestral features compared to Balkan populations occurring further north (Djulić, 1978; Belcheva & Kolevska, 1986).

The Istranca race, which has been described from European Turkey (Zima *et al.*, 1997c), represents a different case. There are three fixed autosomal fusions in the complement, but the combination of fusions is different from that of other populations in south-eastern, Central and eastern Europe. This peculiar feature indicates that the Istranca race should be considered as endemic rather than ancestral to the populations of areas north of the Balkans.

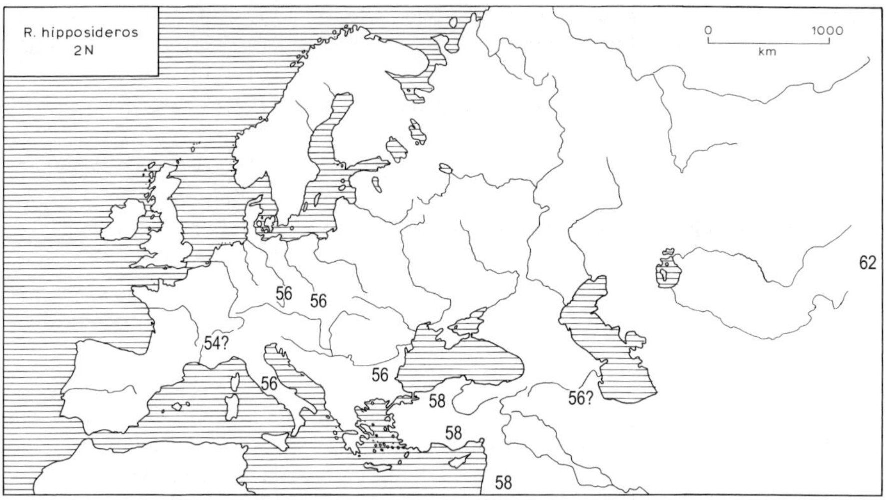

Figure 2. Diploid numbers of chromosomes reported for populations of the lesser horseshoe bat, Rhinolophus hipposideros, *from Europe and the Middle East. Data from Zima* et al. *(1992) and unpublished results.*

The extent of karyotypic variation found within and between species varies between individual mammalian orders. The bats (Chiroptera) are clearly a conservative group which very rarely exhibit intraspecific variation; there is often karyotypic uniformity within individual genera inhabiting the Temperate Zone. One of the few exceptions is the lesser horseshoe bat, *Rhinolophus hipposideros*. Within this species, three karyotypic races have been reported, differing in their respective diploid numbers, that is, 56, 58, and 62 chromosomes (Zima *et al.*, 1992). The karyotype with 62 chromosomes was only found in certain populations from central Asia, and it is likely that these populations belong to a separate species because of considerable morphological differences compared to other populations. The distribution of the other two races with 56 and 58 chromosomes, respectively, is not yet completely clear. The 56-chromosome race has been reported from several parts of Europe including Bulgaria (Belcheva *et al.*, 1990). Our data show that the 58-chromosome race occurs in Asia Minor and the Middle East, and it is possible that the Bosphorus Strait is the boundary between the ranges of these two races (Figure 2).

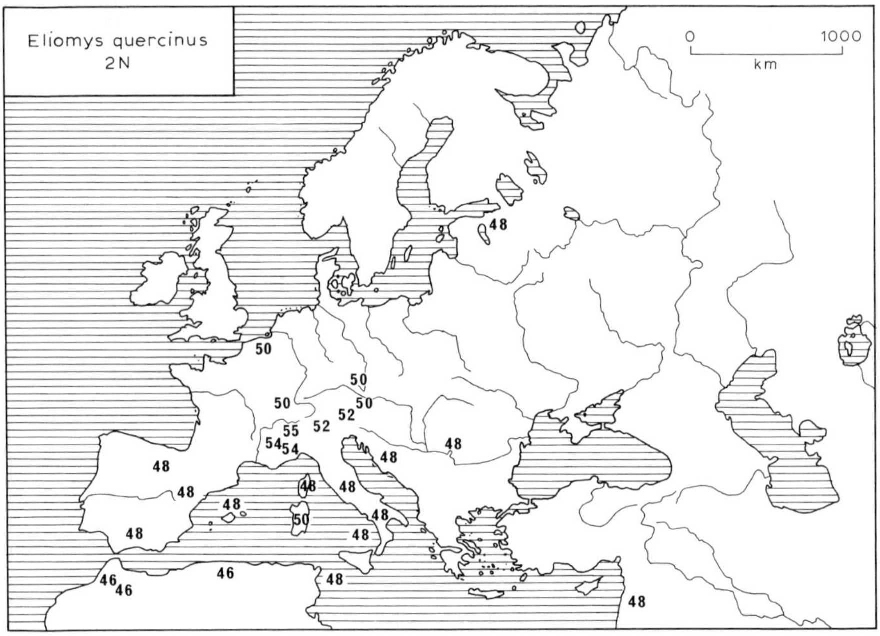

Figure 3. Diploid numbers of chromosomes reported for European populations of the garden dormouse, Eliomys quercinus. *Data from Zima* et al. *(1997a) and unpublished results.*

Another species with an extremely variable karyotype is the garden dormouse, *Eliomys quercinus*. Mediterranean populations of this species (including some parts of the Balkans) mostly possess a karyotype with 48 chromosomes (Murariu *et al.*, 1985; Vujošević *et al.*, 1993). The centre of chromosomal variation appears to be the Alps, and populations with divergent diploid numbers have also been found in western and Central Europe. A surprising result was reported from the Saint Petersburg region, Russia, whereby the karyotype of the local population was the same as that of the Mediterranean (Graphodatsky & Fokin, 1993). This indicates that the karyotype with 48 chromosomes can be considered primitive, and karyotypes with higher diploid numbers advanced. The advanced karyotypes found in populations from western and Central Europe have apparently been derived from those of the Alps rather than the Mediterranean (Zima *et al.*, 1997a; Figure 3).

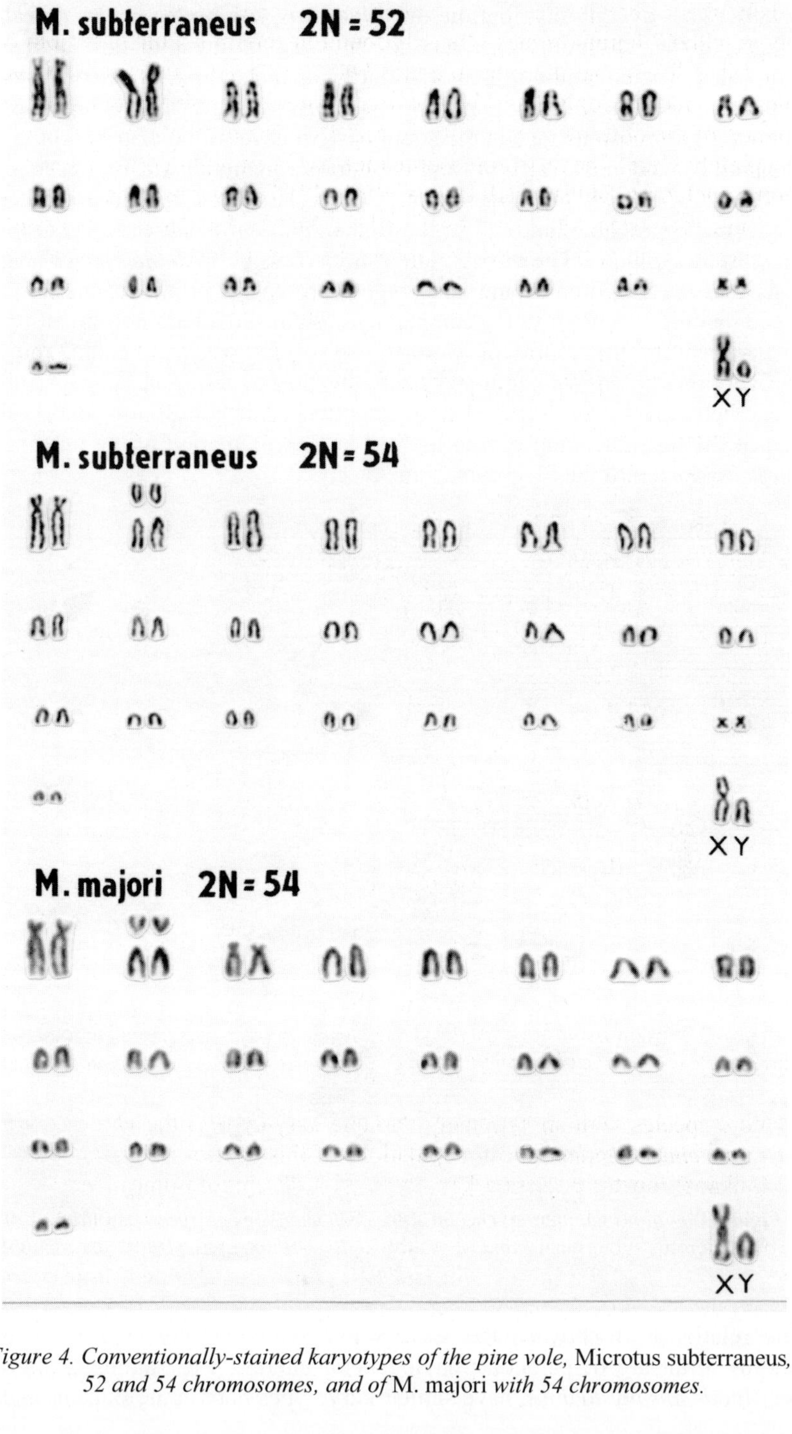

Figure 4. Conventionally-stained karyotypes of the pine vole, Microtus subterraneus, *with 52 and 54 chromosomes, and of* M. majori *with 54 chromosomes.*

Differences between geographical populations of the pine vole, *Microtus subterraneus*, are caused by a single centric fusion of two autosomal pairs (Figure 4). As a result, populations with 52 or 54 chromosomes have been reported (Sablina *et al.*, 1989). The overall distribution of these two karyotypic races reveals the occurrence of the 54-chromosome race in the marginal areas of the range: in the southern Alps, western and north-eastern Europe and also in Asia Minor, where the occurrence of a related species, *Microtus majori*, was previously supposed. We found the 52-chromosome race at the south-eastern edge of the distribution range in Mt. Pelister, Macedonia (FYROM) and the Istranca mountains of European Turkey (Kryštufek *et al.*, 1994; Macholán *et al.*, 2001). Therefore, the occurrence of *M. majori* seems improbable not only in the Balkan parts of south-eastern Europe, but also in western Anatolia. The distribution of both races of *M. subterraneus* (Figure 5) suggests that the 54-chromosome karyotype represents a primitive, plesiomorphic state, and that the 52-chromosome complement originated after a new mutation from somewhere within the centre or eastern part of its range, spreading outwards successively thereafter. We can assume that the Bosphorus Strait is again a natural boundary between the two races. The interruption of the Bosphorus land bridge at the end of the last glaciation can be used as an approximation of the timescale for the range expansion of the 52-chromosome race.

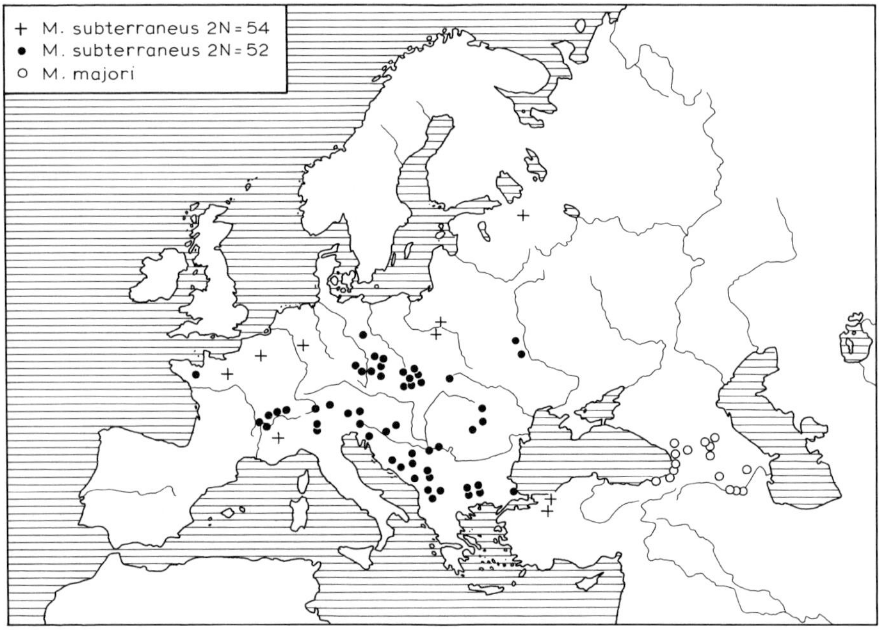

Figure 5. Diploid numbers of chromosomes reported for European populations of the pine vole, Microtus subterraneus *and* M. majori. *After Sablina* et al. *(1989) and Macholán* et al. *(2001).*

The relationships between the species distributed on the either side of the Bosphorus Strait are of particular phylogenetic interest. Populations of the rocky mouse, *Apodemus mystacinus*, have similar karyotypes both in the Balkans and Asia

Minor; extraordinarily deep genetic divergence was found between them, however (Filippucci *et al.*, 2002). Another species distributed both in the Balkans and Asia Minor is the Guenther's vole, *Microtus guentheri.* The basic karyotype with a diploid number of 54 chromosomes seems to be same throughout the range (Živković & Petrov, 1975a; Belcheva *et al.*, 1980; Figure 6). We have found differences between populations from Macedonia (FYROM) and Bulgaria and from Asia Minor, respectively, in the amount and distribution of the C-heterochromatin segments, and in the distribution of the nucleolar organiser regions visualised by silver staining. These slight but distinct chromosomal differences between geographical populations can hardly be interpreted as of definite taxonomic significance, but they do demonstrate some extent of divergence and show how sensitive cytogenetic markers can be.

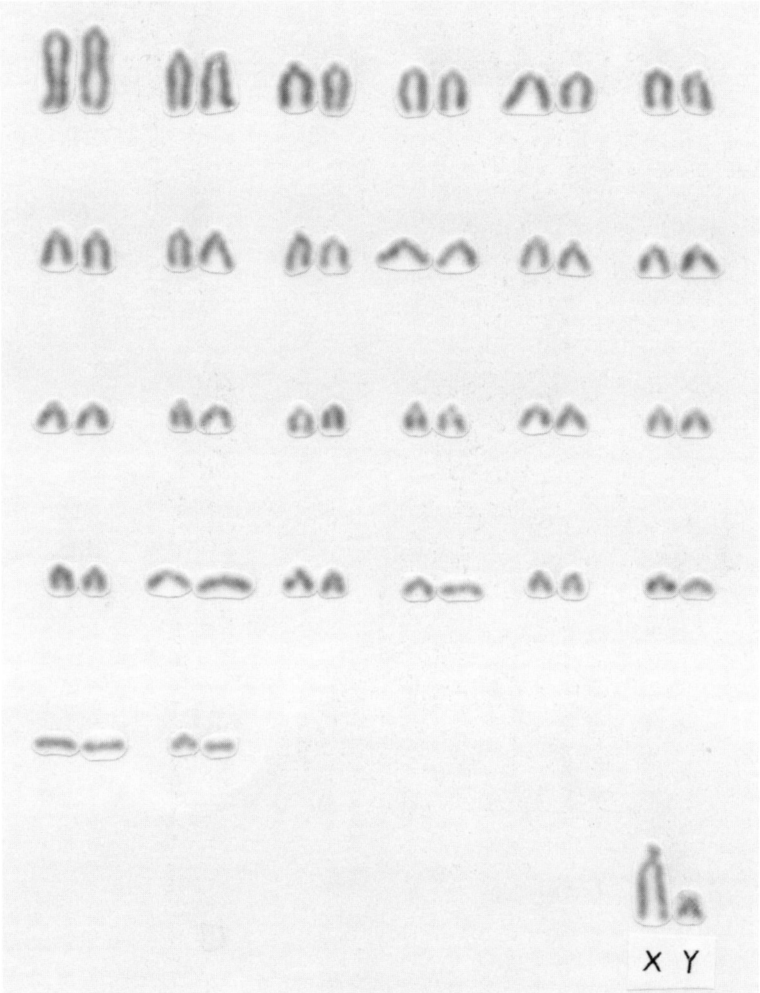

Figure 6. Karyotype of the Guenther's vole, Microtus guentheri.

Extensive polytypic variation has been recorded in the Thomas' pine vole, *Microtus thomasi*, from Greece (Giagia & Ondrias, 1973; Giagia, 1985; Giagia-Athanasopoulou *et al.*, 1995). Giagia-Athanasopoulou & Stamatopoulos (1997) differentiated four distinct karyotypic groups with varying diploid chromosome numbers (2N) and numbers of chromosome arms (FN): 2N=44, FN=44 (*thomasi* cytotype), 2N=44, FN=46 (atticus cytotype), 2N=42, FN=42 (subalpine cytotype), and 2N=40, FN=42 (Rb-subalpine cytotype). The mechanism responsible for this interesting variation is the combination of pericentric inversions and centric fusions, with additional tandem fusions and heterochromatin changes.

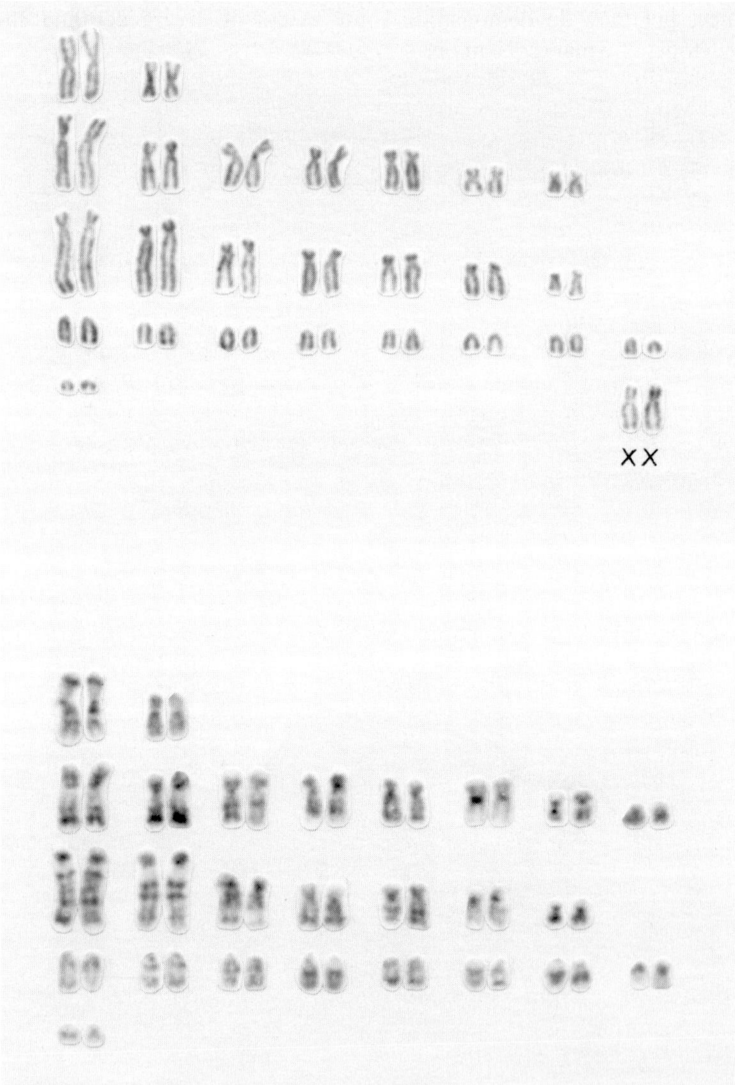

Figure 7. Karyotype of the lesser mole rat, Nannospalax leucodon *from Bistra Planina, Macedonia (FYROM). Conventional staining and G-banding (below).*

A species showing remarkable chromosomal variation within its range, which is confined mainly to the Balkan Peninsula, is the lesser mole-rat, *Nannospalax leucodon* (Raicu & Duma, 1969; Raicu *et al.*, 1973; Peshev, 1981, 1983; Giagia *et al.*, 1982; Savić & Soldatović, 1974, 1978a,b, 1984; Figure 7). The extraordinary extent of within- and between-population variation in the karyotype stimulated the treatment of this mammal as a superspecies consisting of (perhaps many) cryptic species with different karyotypes (Savić & Soldatović, 1979, 1984). The question of their separate taxonomic status is still open, since additional studies using genetic markers other than the karyotype are necessary to demonstrate the possible limitations of gene flow between populations with different karyotypes.

Polytypic karyotypes can even occur within carnivores. Observed chromosomal variation in the weasel, *Mustela nivalis,* provides new insights into the status of Balkan populations. Previously, we found that the number of large heterochromatic arms in autosomes differed between weasel populations from Central and northern Europe, respectively (Zima & Grafodatskij, 1985). These data showed that the karyotype reported from Scandinavia also occurred in populations from Siberia and Alaska, that is, in the northern boreal sector of the Holarctic range, whereas a different karyotype, rearranged after two translocations, was reported from Japan (see Zima & Král, 1984 for a review). This raised the question as to whether the karyotype found in Central European weasels (possessing only six large heterochromatic arms in contrast to northern weasels with seven arms) was confined to this geographic area or occurred elsewhere. A subsequent chromosome banding study of weasels from European Turkey confirmed that the karyotype with the six arms also occurs in the Balkans, and may be widespread in south-eastern Europe. We can conclude that the Balkan and Central European weasels probably belong to the same phylogenetic lineage.

Karyotypic sibling species

The karyotype has often been used as a marker to differentiate between otherwise morphologically identical species, or *siblings*. An example is the whiskered bat, *Myotis mystacinus*. This species is karyotypically quite similar to other species of the large genus *Myotis*. Detailed studies of the distribution of nucleolar organiser regions (Volleth, 1987), however, indicated the existence of separate phylogenetic groups in south-eastern Europe. One of these lineages has recently been described as a new species, *Myotis alcathoe* (Helversen *et al.*, 2001).

Table 3. Species of Balkan mammals which can be distinguished according to their karyotypes

Species	Potentially or actually occurring sibling species	References
Erinaceus concolor	*E. europaeus*	Mandahl (1978), Giagia & Ondrias (1980)
Sorex araneus	*S. satunini*	Şimşek *et al.* (1986), Zima *et al.* (1997c)
S. minutus	*S. volnuchini*	Zima *et al.* (1997c)
Talpa caeca	*T. europaea* and other moles	Todorović *et al.* (1972)
T. levantis	*T. europaea* and other moles	review in Zima &

		Král (1984)
T. stankovici	*T. europaea* and other moles	Todorović *et al.* (1987)
Myotis alcathoe	*M. mystacinus*	Volleth (1987)
Sciurus vulgaris	*S. anomalus*	review in Zima & Král (1984)
Spermophilus citellus	*S. xanthoprymnus*	Živković *et al.* (1968)
Mesocricetus newtoni	*M. auratus*	Voiculescu (1974)
Microtus arvalis	*M. rossiaemeridionalis*	Král (1975), Ružić *et al.* (1975a)
M. felteni	*M. subterraneus*	Petrov *et al.* (1976), Živković & Petrov (1975b)
M. multiplex	*M. subterraneus*	Tvrtković *et al.* (1979)
M. subterraneus	*M. majori*	Kryštufek *et al.* (1994), Macholán *et al.* (2001)
M. tatricus	*M. subterraneus*	Zagorodnyuk & Zima (1992)
Nanospalax leucodon	*Spalax graecus*	Raicu *et al.* (1968)
Apodemus sylvaticus	*A. vohlynensis*	Nadjafova *et al.* (1993), Orlov *et al.* (1996)
A. microps (uralensis)	*A. mosquensis, A. ciscaucasicus*	Orlov *et al.* (1996), Reutter *et al.* (2001)
Mus macedonicus, M. spicilegus	*M. domesticus, M. musculus*	Ivanitskaya *et al.* (1997)
Sicista subtilis	*S. severtzovi*	Sokolov *et al.* (1987)
Mustela putorius	*M. eversmanni*	Graphodatsky *et al.* (1976)
Capreolus capreolus	*C. pygargus*	Zernahle (1980)

The results of cytogenetic research on Bulgarian populations of the wood mouse, *Apodemus sylvaticus*, by Nadjafova *et al.* (1993) indicated the separation of a new species, *A. vohlynensis* (Orlov *et al.*, 1996); the range of the new species is not well defined, and it is not known whether it includes the Balkans. It is debatable, however, whether chromosomal variation resulting from changes in the amount and the distribution of the heterochromatin is a reliable indicator of taxonomic separation. The status of chromosomal sibling species in *A. sylvaticus* and other *Apodemus* species should be re-examined (Reutter *et al.*, 2001; Nová *et al.*, 2002).

Genetic methods (other than chromosomal) have enabled several species of the genus *Mus* to be distinguished in Europe (Boursot *et al.*, 1993). The ranges of four of these species (two indoor and two outdoor in an ecological sense) meet in the Balkan Peninsula, so the species richness of this genus is exceptionally high in the Balkans compared the other parts of the Palaearctic region. The outdoor and indoor mouse species can also be distinguished according to size of the Y chromosome (Ivanitskaya *et al.*, 1996).

Separate specific status has not always been indicated by cytogenetic studies. In the early 1970s, for example, populations of the harvest mouse from the Danube Delta were considered as a separate species, *Micromys danubialis* (Simionescu, 1974). However, the karyotypic data revealed no differences between populations from the Danube Delta and other parts of the species range (Zima, 1983). The validity of the description of *Micromys danubialis* was also brought into doubt by morphometric studies.

Figure 8. Marginal geographic records of Microtus rossiaemeridionalis *in the Balkans and surrounding regions. After Král* et al. *(1980) with additions from unpublished data.*

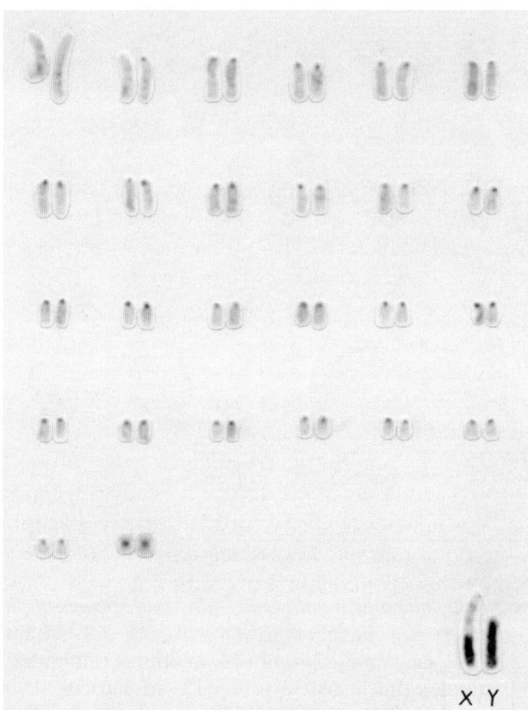

Figure 9. Karyotype of the sibling vole, Microtus rossiaemeridionalis (*C-banding*).

Probably the most well-known "chromosomal" sibling species amongst mammals of the Balkans and eastern Europe is the sibling vole, *Microtus rossiaemeridionalis* (formerly known as *Microtus subarvalis* or *M. epiroticus*). Populations of this species were not previously distinguished from those of the common vole, *Microtus arvalis*, in spite of the fact that these latter are common and abundant and occur over an extensive geographical range. The sibling vole was originally discovered in cytogenetic studies made in Russia and, subsequently, its occurrence was reported from various parts of the Balkans (Král, 1975; Ružić *et al.*, 1975a; Živković *et al.*, 1975c; Belcheva *et al.*, 1977; Zima *et al.*, 1981). The distribution range extends further south than that of *Microtus arvalis*, which is restricted to the mountains of the southern Balkans (Figure 8). The karyotype of *Microtus rossiaemeridionalis* comprises 54 mostly acrocentric chromosomes (Figure 9). This is in sharp contrast with the complement of *Microtus arvalis* that includes only 46 chromosomes, most of them being biarmed. Recently, European populations of the common vole were split again into two separate species differing in the number of acrocentrics in their karyotypes: *Microtus arvalis* with four acrocentric pairs and *M. obscurus* with ten acrocentric pairs (Zagorodnyuk, 1991). *Microtus arvalis*, distributed in the mountains and northern Balkans, has thus become a European endemic species.

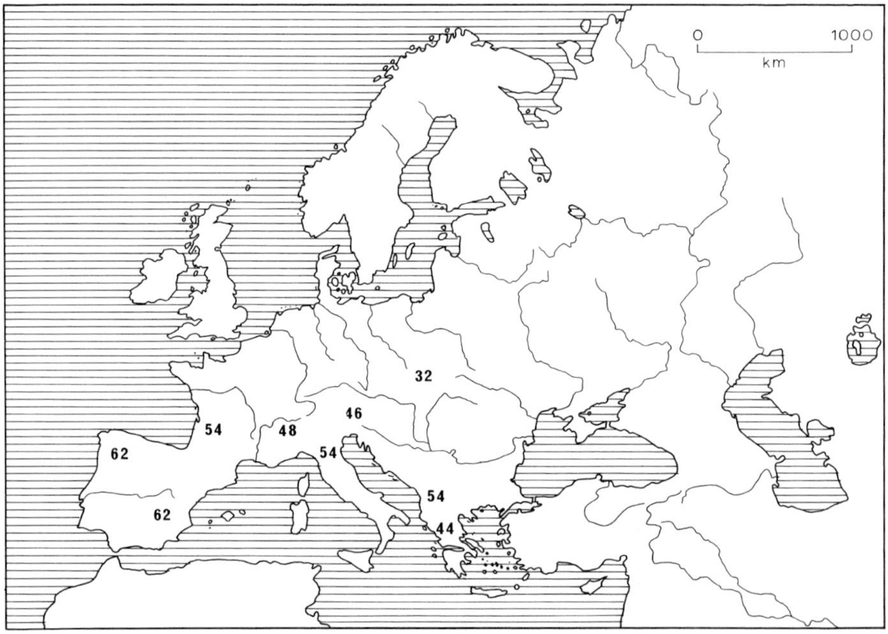

Figure 10. Diploid chromosome numbers in individual species of pine voles (subgenus Terricola*) in southern Europe. From west to east: 2N=62 -* M. lusitanicus, M. duodecimcostatus*; 2N=54 -* M. gerbei*; 2n=48 -* M. multiplex (multiplex s.str.)*; 2N=54 -* M. savii*; 2n=46 -* M. multiplex (liechtensteini)*; 2N=32 -* M. tatricus*; 2N = 54 -* M. felteni*; 2N=44 -* M. thomasi*. See Zima & Král (1984) for a review of the original data.*

Chromosomal research has made a substantial contribution to the understanding of the phylogeny of another species group, the voles of the subgenus, *Terricola* (formerly *Pitymys*). Different lineages of this genus have specific diploid numbers, and the karyotypic characteristics are also useful for distinguishing between individual species within each lineage (Kratochvíl & Král, 1974; Figure 10).

PHYLOGENETIC AND BIOGEOGRAPHIC IMPLICATIONS

Endemism: unique karyotypes

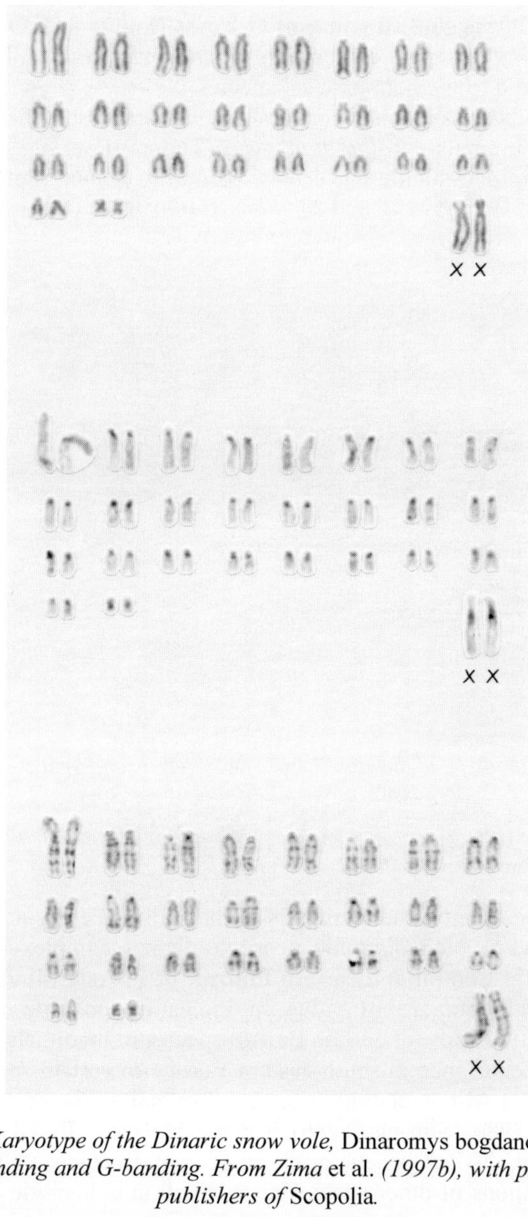

Figure 11. Karyotype of the Dinaric snow vole, Dinaromys bogdanovi. *Conventional staining, C-banding and G-banding. From Zima* et al. *(1997b), with permission of the publishers of* Scopolia.

The determination of karyotypic characteristics has a special significance in the case of Balkan endemic species. The Dinaric snow vole, *Dinaromys bogdanovi*, for example, is a Balkan endemic genus and species. The karyotype characteristics (54 mostly acrocentric chromosomes) indicate ancient roots for this species, because the complement is similar to several other arvicolid species belonging to phylogenetically well separated lineages (Savić *et al.*, 1966, 1967; Djulić *et al.*, 1971; *cf.* Figures 6 and 9). This suggests the plesiomorphic nature of the chromosomal complement. At the same time, our study of banded chromosomes revealed some unique features of the karyotype compared to these other species. These features were demonstrated particularly by the amount and distribution of heterochromatin and the fine structure of G-bands (Figure 11). Our study involved individuals belonging to separate subspecies *D. b. grebenscikovi*, but no differences were found compared to the nominate subspecies.

Karyotypic studies have also confirmed a separate specific status for other Balkan endemic mammals - *Talpa stankovici* (Todorović *et al.*, 1972; Figure 12), *Mesocricetus newtoni* (Voiculescu, 1974), *Microtus felteni* (Petrov *et al.*, 1976; Petrov & Živković, 1977, 1979), and *M. thomasi* (Giagia & Ondrias, 1973).

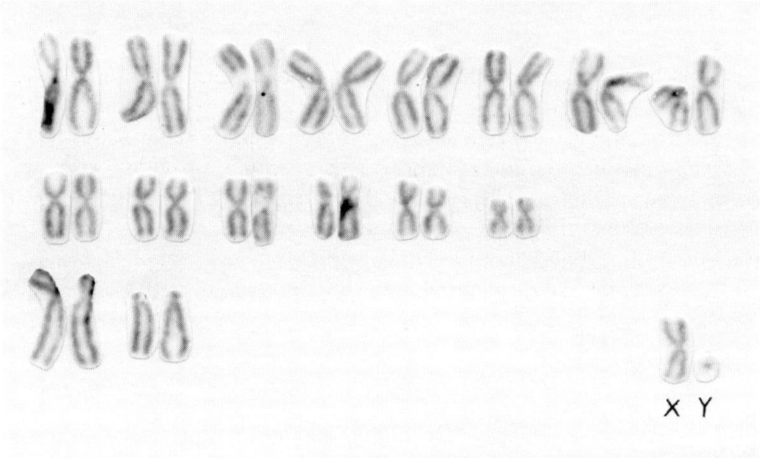

Figure 12. Karyotype of the mole Talpa stankovici. *From Zima* et al. *(1997b), with permission of the publishers of* Scopolia.

Patterns of chromosomal variation

Other than endemic species, the mammals of the Balkan Peninsula can be related in a phylogenetic and/or biogeographic sense to faunal complexes with a modern distribution either in Central and eastern Europe, or in Asia Minor and the Middle East. Chromosomal data are of value in indicating possible phylogenetic and biogeographic relationships for certain Balkan species of mammals.

The marginal occurrence in south-eastern Europe of certain species whose main range is within Asia Minor or the Caucasus Mountains (*Sorex satunini, Microtus majori* - Kıvanç, 1986; Şimşek, 1986) has not been confirmed by chromosomal studies. Conversely, distinct genetic or chromosomal divergence was found between conspecific populations of other species distributed on either side of the Bosphorus

Strait (*Crocidura leucodon, Rhinolophus hipposideros, Apodemus mystacinus, Microtus guentheri, Nannospalax leucodon*). Karyotypic and other genetic data thus do not support a major recent influence on the Balkan mammal fauna from the eastern areas of Asia Minor and the Middle East. Similarly, there are only a few examples of chromosomal identity between populations from the Balkans and Central Europe belonging to single but polytypic species (*e.g. Microtus subterraneus, Mustela nivalis*). In the pine vole, *M. subterraneus*, the karyotypic data indicate that the currently occurring 52-chromosome race invaded the Balkans from an area situated somewhere in Central or eastern Europe during the Holocene.

In other polytypic species such as the common shrew, *Sorex araneus*, or the garden dormouse, *Eliomys quercinus,* the current pattern of chromosomal diversification supports the occurrence of endemic populations in the Balkans, characterised by ancestral features different from those reported from advanced populations distributed in the more northerly parts of the range. At the species level, a similar conclusion can be derived for mammalian faunas occurring recently in other Mediterranean regions of southern Europe, particularly in the Iberian and Apennine peninsulas.

Postglacial colonisation routes

Bilton *et al.* (1998) have recently raised the question of whether the Balkan Peninsula is an area of endemism or a source for northward postglacial colonisation. It is possible that this can be elucidated by chromosomal data. Karyotypic data on Balkan populations of the common shrew, for example, provided some of the first evidence that the Mediterranean region was a centre of endemism rather than simply a source for postglacial colonisation of Europe. The patterns of karyotypic variation in the garden dormouse and the pine vole are similarly indicative. However, it would be premature to make broad generalisations at this stage, and it is not difficult to imagine that the postglacial history of individual plant and animal species has been highly variable.

CONCLUSIONS

The chromosomal data-set for small mammals from the Balkans is almost complete, at least with regard to basic karyotype characteristics. At the same time, these data provide evidence for extensive chromosomal variation within individual species, manifested either as polymorphism or polytypy. Interspecific karyotypic variation is also frequent, and several chromosomal sibling species are known. Where there is distinct variation within and between species, we have been able to derive some taxonomic, phylogenetic or biogeographic conclusions from karyotypic data. Small mammals are thus a unique group of organisms which can serve as a model for future research.

The value of chromosome studies in taxonomic research can be exemplified by their application to resolve issues surrounding the distribution of various morphological sibling species, where the karyotype appeared to be the most important character for differentiation between two species (*e.g. Sorex araneus vs. S. satunini, Sorex minutus vs. S. volnuchini, Microtus arvalis vs. M. rossiaemeridionalis, Microtus subterraneus vs. M. majori*). Local populations of several polytypic species were shown to possess unique karyotypes (*e.g. Sorex araneus, Nannospalax leucodon*). These findings could indicate the existence of new sibling species, and/or the endemic character of their respective populations, which may also be related to habitat fragmentation consequent upon long-term human impact. Chromosomal variation in certain species can be further employed in efforts to reveal phylogenetic relationships between the fauna of the Balkans and that of either Asia Minor (*e.g. Crocidura leucodon, Rhinolophus hipposideros, Microtus guentheri*) or Central Europe (*e.g. Apodemus flavicollis, Mustela nivalis*).

The results of cytogenetic research can thus contribute substantially to our understanding of the endemic status of certain populations or species, and the past routes of their dispersal. In this respect, chromosome studies of karyotypic variation have been shown to be relevant to the recent debate over whether the Balkans is an area of endemism of small mammals, or a source of northwards postglacial colonisation. At present, it is still not clear whether inferences based on data on mammalian cytogenetics are more generally applicable to other groups of plants and animals.

ACKNOWLEDGEMENTS

My thanks are due to H. I. Griffiths and B. Kryštufek for their encouragement to set up this study and for fruitful discussions. I am also obliged to Jane M. Reed for English revision of the earlier draft of the manuscript. This study was supported in part by the grant no. GA CR 206/01/0562.

REFERENCES

Belcheva, R.G. & Bisserov, B.Y. 1982. B-chromosomes in the black rat (*Rattus rattus*) from Bulgaria. *Comptes rendus de l'Académie bulgare des Sciences* 35, 1577-1580.

Belcheva, R.G. & Kolevska, N.G. 1986. Cytogenetic studies of the common shrew *Sorex araneus* L. (Soricidae, Insectivora) from the Vitosha Mountain. *Comptes rendus de l'Académie bulgare des Sciences* 39, 115-118.

Belcheva, R.G. & Peshev, D. 1979. Intersubspecific sex chromosome differences in *Citellus citellus* L. (Rodentia, Sciuridae). *Experientia* 35, 595-596.

Belcheva, R.G. & Peshev, D. 1985. Constitutive heterochromatin in the ground squirrel *Citellus citellus* L. (Sciuridae, Rodentia) from Bulgaria. *Zoologischer Anzeiger* 215, 385-390.

Belcheva, R.G., Peshev, C.Ch. & Radjabli, S.I. 1977. [Analysis of the chromosome set in the Bulgarian population of *Microtus arvalis*]. *Zoologicheskii zhurnal* 56, 315-317 (in Russian with English abstract).

Belcheva, R.G., Peshev, T.H. & Peshev, D.T. 1980. Chromosome C- and G-banding patterns in a Bulgarian population of *Microtus guentheri* Danford et Alston (Microtidae, Rodentia). *Genetica* 52/53, 45-48.

Belcheva, R.G., Topaschka-Ancheva, M.N. & Gerassimov, S. 1985. Karyological characteristics of *Microtus epiroticus*, Ondr. in Bulgaria. *Comptes rendus de l'Académie bulgare des Sciences* 38, 1555-1557.

Belcheva, R.G., Topaschka-Ancheva, M.N., Peshev, D.C. & Gerasimov, S. 1987. [Karyological studies of some species of rodents in Bulgaria]. *Savremenni postizheniya na balgarskata zoologiya, Sofia*, 376-379 (in Bulgarian).

Belcheva, R.G., Topaschka-Ancheva, M.N. & Atanassov, N. 1988. Karyological studies of five species of mammals from Bulgarian fauna. *Comptes rendus de l'Académie bulgare des Sciences* 42, 125-128.

Belcheva, R.G., Topaschka-Ancheva, M.N. & Bisserov, V. 1990. Karyological studies of *Rhinolophus ferrumequinum* and *Rhinolophus hipposideros* (Rhinolophidae, Chiroptera) from Bulgaria. *Comptes rendus de l'Académie bulgare des Sciences* 43, 81-83.

Bilton, D.T., Mirol, P.M., Mascheretti, S., Fredga, K., Zima, J. & Searle, J.B. 1998. Mediterranean Europe as an area of endemism for small mammals rather than a source for northwards postglacial colonization. *Proceedings of the Royal Society London B* 265, 1219-1226.

Biltueva, L.S., Rogatcheva, M.B., Perelman, P.L., Borodin, P.M., Oda, S.-I., Koyasu, K., Harada, M., Zima, J. & Graphodatsky, A.S. 2001: Chromosomal phylogeny of certain shrews of the genera *Crocidura* and *Suncus* (Insectivora). *Journal of Zoological Systematics and Evolutionary Research* 39, 69-76.

Blagojević, J. & Vujošević M. 1995. The role of B-chromosomes in population dynamics of yellow-necked wood mice *Apodemus flavicollis* (Rodentia, Mammalia). *Genome* 38, 472-478.

Blagojević J. & Vujošević M. 2000. Do B chromosomes affect morphometric characters in yellow-necked mice *Apodemus flavicollis* (Rodentia, Mammalia)? *Acta Theriologica* 45, 129-135.

Boursot, P., Auffray, J.-C., Britton-Davidian, J. & Bonhomme, F. 1993. The evolution of house mice. *Annual Review of Ecology and Systematics* 24, 119-152.

Dannelid, E. 1994. Chromosome polymorphism in *Sorex alpinus* (Mammalia, Soricidae) in the western Alps and the Swiss Jura. *Zeitschrift für Säugetierkunde* 59, 161-168.

Djulić, B. 1978. Chromosomenmorphologie bei Waldspitzmausen, *Sorex araneus* Linné, 1758, aus einigen Gegenden Jugoslawiens. *Säugetierkundliche Mitteilungen* 26, 184-190.

Djulić, B. & Tvrtković, N. 1979. On some mammals from the Centraladriatic and Southadriatic islands. *Acta Biologica (Zagreb)* 43, 15-35.

Djulić, B., Savić, I. & Soldatović, B. 1971. The chromosomes of two rodents species, *Dolomys bogdanovi* (V. and E. Martino 1922) and *Glis glis* (Linnaeus 1766) (Mammalia, Rodentia). *Caryologia* 24, 299-305.

Djulić, B., Soldatović, B. & Dunđerski, Z. 1980. Distribution of karyotypes in *Mus musculus* Linnaeus, 1758 (Rodentia, Muridae) in some regions of Yugoslavia. *Biosistematika* 6, 203-210.

Filippucci, M.G., Macholán, M. & Michaux, J.R. 2002. Genetic variation and evolution in the genus *Apodemus* (Muridae: Rodentia). *Biological Journal of the Linnean Society* 75, 395-419.

Gamperl, R., Vistorin, G. & Rosenkranz, W. 1976. A comparative analysis of the karyotype of *Cricetus cricetus* and *Cricetulus griseus*. *Chromosoma* 55, 259-265.

Giagia, E.B. 1985. Karyotypes of "44-chromosomes" *Pitymys* species (Rodentia, Mammalia) and their distribution in southern Greece. *Säugetierkundliche Mitteilungen* 32, 169-173.

Giagia, E. & Ondrias, J.G. 1973. Karyological analysis of the vole *Pitymys atticus* (Rodentia, Mammalia) from Greece. *Biologia Gallo-Helenica* 4, 205-212.

Giagia, E.B. & Ondrias, J.C. 1980. Karyological analysis of eastern European hedgehog *Erinaceus concolor* (Mammalia, Insectivora) in Greece. Mammalia 41, 59-72.

Giagia, E., Savić, I. & Soldatović, B. 1982. Chromosomal forms of the mole rat *Microspalax* from Greece and Turkey. *Zeitschrift für Säugetierkunde* 47, 213-236.

Giagia, E., Savić, I., Soldatović, B. & Zimonjić, D. 1985. Karyotype study of the genus *Apodemus* (Kaup, 1829) populations from the Balkan peninsula. *Acta Veterinaria (Beograd)* 35, 289-298.

Giagia, E.B., Fraguedakis-Tsolis, S.E. & Chondropoulos, B.P. 1987. Contribution to the study of the taxonomy and zoogeography of wild house mouse, genus *Mus* L. (Mammalia, Rodentia, Muridae) in Greece. 2. Karyological study of 2 populations from southern Greece. *Mammalia* 51, 111-116.

Giagia-Athanasopoulou, E.B., Chondropoulos, B.P. & Fraguedakis-Tsolis, S.E. 1995. Robertsonian chromosomal variation in subalpine voles *Microtus* (*Terricola*), (Rodentia, Arvicolidae) from Greece. *Acta Theriologica* 40, 139-143.

Giagia-Athanasopoulou, E.B. & Stamatopoulos, C. 1997. Geographical distribution and interpopulation variation in the karyotypes of *Microtus (Terricola) thomasi* (Rodentia, Arvicolidae) in Greece. *Caryologia* 50, 303-315.

Graf, J.-D. & Meylan, A. 1980. Polymorphisme chromosomique et biochimique chez *Pitymys multiplex* (Mammalia, Rodentia). *Zeitschrift für Säugetierkunde* 45, 133-148.

Graphodatsky, A.S. & Fokin, I.M. 1993. [Comparative cytogenetics of Gliridae (Rodentia)]. *Zoologicheskii zhurnal* 72, 149-152 (in Russian with English abstract).

Graphodatsky, A.I., Ternovsky, D.V., Isajenko, A.A. & Radjabli, S.I. 1976. [G-banding pattern of the chromosomes in seven species of Mustelidae (Carnivora)]. *Zoologicheskii zhurnal* 55, 1704-1711 (in Russian with English abstract).

Helversen von, O., Heller, K.G., Mayer, F., Nemeth, A., Volleth, M. & Gombkoto, P. 2001. Cryptic mammalian species: a new species of whiskered bat (*Myotis alcathoe* n. sp.) in Europe. *Naturwissenschaften* 88, 217-223.

Ivanitskaya, E., Gorlov, I., Gorlova, O. & Nevo, E. 1996. Chromosome markers for *Mus macedonicus* (Rodentia, Muridae) from Israel. *Hereditas* 124, 145-150.

Kıvanç, E. 1986. *Microtus (Pitymys) majori* Thomas, 1906 in der europäischen Türkei. *Bonner zoologische Beiträge* 37, 39-42.

Král, B., 1975: A species of the genus *Microtus* (Microtidae, Mammalia), new for the fauna of Bulgaria. *Zoologické listy* 24, 353-360.

Král, B. & Mitev, D.B., 1976. Karyotypes of *Pitymys subterraneus* (Microtidae, Rodentia) from the mountains of southern Bulgaria. *Zoologické listy* 25, 27-31.

Král, B. & Zima, J. 1978. Chromosomal polymorphism in *Pitymys subterraneus* (Microtidae, Rodentia). *Folia Zoologica* 27, 13-24.

Král, B., Bel'anin, A.N., Zima, J., Malygin, V.M., Gajčenko, V.A. & Orlov, V.N., 1980: Distribution of *Microtus arvalis* and *M. epiroticus*. *Acta Sc. nat. Brno* 14(9), 1-31.

Kratochvíl, J. & Král, B. 1974. Die Karyotypenforschung als Weg zur Erkenntnis der Evolution europäischer Arten der Gattung *Pitymys*. In: *Symposium Theriologicum II* (J. Kratochvíl & R. Obrtel, eds.), pp. 299-311. Academia, Praha.

Kryštufek, B., Filippucci, M.G., Macholán, M., Zima, J. Vujošević, M. & Simson, S. 1994. Does *Microtus majori* occur in Europe? *Zeitschrift für Säugetierkunde* 59, 349-357.

Kulijev, G.N., Kulijev G.K. & Radjabli, S.I. 1978. [Karyotypical differences between different populations of *Arvicola terrestris* (Rodentia, Cricetidae)]. *Zoologicheskii zhurnal* 57, 1409-1411 (in Russian with English abstract).

Lukáčová, L., Dannelid, E., Hausser, J., Macholán, M. & Zima, J. 1996. G-banded karyotype of the Alpine shrew, *Sorex alpinus* (Mammalia, Soricidae) from the Šumava Mts. *Folia Zoologica* 45, 223-226.

Lungeanu, A., Gavrila, L., Ciresica, S. & Murariu, D. 1987. [Characterization of the G and C banding pattern in the genome of species *Microtus arvalis* (Pallas, 1779) (Mammalia, Rodentia, Arvicolidae)]. *Travaux du Muséum d'Histoire naturelle "Grigore Antipa"* 29, 319-322 (in Romanian with English abstract).

Lungeanu, A., Gavrila, L., Murariu, D. & Ciresica, S. 1986. The distribution of the constituent heterochromatin and the G-banding pattern in the genome of *Apodemus agrarius* (Pallas, 1771) (Mammalia, Muridae). *Travaux du Muséum d'Histoire naturelle "Grigore Antipa"* 28, 267-270.

Macholán, M., Filippucci, M.G., Zima, J., Kryštufek, B. & Simson, S. 1994. Karyological and allozyme survey of the common shrew, *Sorex araneus*, from Macedonia. *Zeitschrift für zoologische Systematik und Evolutions-forschung* 32, 129-136.

Macholán, M., Filippucci, M.G. & Zima, J. 2001. Genetic variation and zoogeography of pine voles of the *Microtus subterraneus/majori* group in Europe and Asia Minor. *Journal of Zoology, London* 255, 31-42.

Mandahl, N. 1978. Variation in C-stained chromosome regions in European hedgehogs (Insectivora, Mammalia). *Hereditas* 89, 107-128.

Mandahl, N. & Fredga, K. 1980. A comparative chromosome study by means of G-, C-, and NOR-banding of the weasel, pygmy weasel and the stoat (*Mustela*, Carnivora, Mammalia). *Hereditas* 93, 75-83.

Markov, G. & Dobrijanov, D. 1974. Karyologische Analyse der Weissbrust oder Ostigel (*Erinaceus rumanicus* Barr.-Ham.) in Bulgarien. *Zoologischer Anzeiger* 193, 181-188.

Mayr, B., Schweizer, D. & Geber, G. 1984. NOR activity, heterochromatin differentiation, and the Robertsonian polymorphism in *Sus scrofa* L. *Journal of Heredity* 75, 79-80.

Mitchell-Jones, A.J., Amori, G., Bogdanowicz, W., Kryštufek, B., Reijnders, P.J.H., Spitzenberger, F., Stubbe, M., Thissen, J.B.M., Vohralík, V. & Zima J. 1999. *Atlas of European mammals.* Poyser Natural History - Academic Press, London.

Murariu, D., Lungeanu, A., Gavrila, K. Gavrila, K. & Stepan, C. 1985. Preliminary data concerning the study of the karyotype of *Eliomys quercinus* (Linnaeus, 1766) (Mammalia, Gliridae). *Travaux du Muséum d'Histoire naturelle "Grigore Antipa"* 27, 325-327.

Nachman, M.W. & Searle, J.B. 1995. Why is the house mouse karyotype so variable? *Trends in Evolution and Ecology* 10, 397-402.

Nadjafova, R.S., Bulatova, N.S., Chasovlikarova, Z. & Gerassimov, S. 1993. Karyological differences between two *Apodemus* species in Bulgaria. *Zeitschrift für Säugetierkunde* 58, 232-239.

Nová, P., Rábová, M., Reutter, B. & Zima, J. 2002. Sex chromosome heterochromatin variation in the wood mouse, *Apodemus sylvaticus*. *Cytogenetic and Genome Research* 96, 186-190.

Orlov, V.N., Bulatova, N.S., Nadjafova, R.S. & Kozlovsky, A.I. 1996. Evolutionary classification of European wood mice of the subgenus *Sylvaemus* based on allozyme and chromosome data. *Bonner zoologische Beiträge* 46, 191-202.

Peshev, D., 1981. On the karyotypes in some populations of the mole rat (*Spalax leucodon* Nordmann) in Bulgaria. *Zoologischer Anzeiger* 206, 129-133.

Peshev, D., 1983. New karyotype forms of the mole rat, *Nannospalax leucodon* Nordmann (Spalacidae, Rodentia) in Bulgaria. *Zoologischer Anzeiger* 211, 65-72.

Peshev, T. & Belcheva, R. 1978. Banding patterns of the water vole (*Arvicola terrestris* L., Rodentia, Mammalia) chromosomes. *Zoologischer Anzeiger* 200, 132-140.

Petrov, B. & Živković, S. 1977. Some questions of the actual systematics in the light of use of karyological methods. *Biosistematika* 3, 13-22.

Petrov, B. & Živković, S. 1979. Present knowledge on the systematics and distribution of *Pitymys* (Rodentia, Mammalia) in Yugoslavia. *Biosistematika* 5, 113-125.

Petrov, B., Živković, S. & Rimsa, D. 1976. Über die Artengeständingkeit der Kleinwühlmause *Pitymys felteni* (Mammalia: Rodentia). *Senckenbergiana biologica* 57, 1-10.

Radosavljević, J., Vujošević, M. & Živković, S. 1988. Intrapopulation variability of the Y chromosome of the bank vole (*Clethrionomys glareolus*, Schreber 1880). *Arhiv bioloških nauka (Beograd)* 40, 3P-4P.

Raicu, P. & Duma, D. 1969. Cytogenetical study in *Spalax leucodon* Nordm. in Moldavia (Rumania). *Genetical Research Cambridge* 13, 99-104.

Raicu, P., Bratosin, S. & Hamar, M. 1968. Study on the karyotype of *Spalax leucodon* Nordm. and *S. microphthalmus* Guld. *Caryologia* 21, 127-135.

Raicu, P., Duma, D., Kirillova, M. & Tuta, A. 1971. Chromosomal polymorphism in the water vole (*Arvicola terrestris* L.). *Revue Roumaine de biologie - zoologie* 16, 293-296.

Raicu, P., Duma, D. & Torcea, S., 1973. Chromosomal polymorphism in the lesser mole rat, *Spalax leucodon*. *Chromosomes Today* 4, 383-386.

Reutter, B., Nová, P., Vogel, P. & Zima, J. 2001. Karyotypic variation between wood mouse species: banded chromosomes of *Apodemus alpicola* and *A. microps*. *Acta Theriologica*. 46, 353-362.

Ružić, A., Živković, S. & Rimsa, D. 1975a. On the species independence of the 54-chromosome vole *Microtus epiroticus* Ondrias 1966 (Mammalia, Rodentia), its distribution, ecology and importance as a pest in the west part of Balkan peninsula. *Journal of Sciences of Agricultural. Research* 28, 153-160.

Ružić, A., Živković, S. & Rimsa, D. 1975b. Karyotype of the common hamster (*Cricetus cricetus*) from Yugoslavia. *Arhiv bioloških nauka (Beograd)* 27, 3P-4P.

Sablina, O.V., Zima, J., Radjabli, S.I., Kryštufek, B. & Goleniščev, F.N. 1989. New data on karyotypic variation in the pine vole, *Pitymys subterraneus* (Rodentia, Arvicolidae). *Věstník československé Společnosti zoologické* 53, 295-299.

Savić, I. & Soldatović, B., 1974. Die Verbreitung der Karyotypen bei Blindmaus *Spalax* (*Mesospalax*) in Jugoslawien. *Arhiv bioloških nauka (Beograd)* 26, 115-122.

Savić, I. & Soldatović, B., 1978a. Karyotypes of some populations in the genus *Spalax* (*Mesospalax*) in Bulgaria and Turkey. *Säugetierkundliche Mitteilungen* 26, 252-256.

Savić, I. & Soldatović, B., 1978b. Studies on the karyotype and distribution range of the mole rat (*Spalax leucodon* Nordmann) in Greece. *Caryologia* 31, 63-73.

Savić, I. & Soldatović, B., 1979. Distribution range and evolution of chromosomal forms in Spalacidae of the Balkan peninsula and bordering regions. *Journal of Biogeography* 6, 363-374.

Savić, I. & Soldatović, B., 1984. *Karyotype evolution and taxonomy of the genus* Spalax *in Europe*. Separate Editions of Serbian Academy of Sciences and Arts, Department of Natural and Mathematical Sciences, Beograd.

Savić, I., Soldatović, B. & Rimsa, D. 1966. Zur Kenntnis der Chromosomen einer Population der Gattung *Dolomys* Nehring, 1898 (Microtinae, Rodentia). *Arhiv bioloških nauka (Beograd)* 18, 43P-44P.

Savić, I., Soldatović, B. & Rimsa, D. 1967. Zur Kenntnis der Chromosomen einer Population der Gattung *Dolomys* Nehring, 1898 (Microtinae, Rodentia) aus Mazedonien. *Arhiv bioloških nauka (Beograd)* 19, 15P-16P.

Savić, I., Milošević, M. & Živković, S. 1971. [Die Chromosomen des Ziesels (*Citellus citellus* Linnaeus, 1766) aus Jugoslawien]. *Arhiv bioloških nauka (Beograd)* 23, 35-37 (in Serbian with German abstract).

Searle, J.B. & Wójcik, J.M. 1998. Chromosomal evolution: the case of *Sorex araneus*. In: *Evolution of shrews* (J.M. Wójcik & M. Wolsan, eds.), pp. 219-268. Mammal Research Institute, Polish Academy of Sciences, Bialowieza.

Simionescu, V. 1974. Revision de la systématique de genre *Micromys* Dehne, 1841 (Rodentia) du Paléarctique, fondée sur les critères morphologiques. In: Symposium Theriologicum II (J. Kratochíl & R. Obrtel, eds.), pp. 139-154. Academia, Praha.

Şimşek, N. 1986. A new subspecies of *Sorex caucasicus* (Mammalia: Insectivora) from Turkey. *Turkish Journal of Biology* 10, 206-208.

Sokolov, V.E., Koval'skaya, J.M. & Baskevič, M.I. 1987. Review of karyological research and the problems of systematics in the genus *Sicista* (Zapodidae, Rodentia, Mammalia). *Folia Zoologica* 36, 35-44.

Soldatović, B., Djulić, B., Savić, I. & Rimsa, D.1969. Chromosomes of two species of the genus *Apodemus* (*A. agrarius* and *A. mystacinus* - Mammalia, Rodentia) from Yugoslavia. *Arhiv bioloških nauka (Beograd)* 21, 27-32.

Soldatović, B., Dunderski, Z. & Todorović, M. 1986. A new mole karyotype from the Balkans. *Arhiv bioloških nauka (Beograd)* 38, 9P-10P.

Storch, G. & Winking, H. 1977. Zur Systematik der *Pitymys multiplex - Pitymys liechtensteini* - Gruppe (Mammalia: Rodentia). *Zeitschrift für Säugetierkunde* 42, 78-88.

Tichy, H. & Vucak, I. 1987. Chromosomal polymorphism in the house mouse (*Mus domesticus*) of Greece and Yugoslavia. *Chromosoma* 95, 31-36.

Todorović, M., Soldatović, B. & Dundjerski, Z. 1972. [Characteristics of the karyotype of the populations of the genus *Talpa* from Macedonia and Montenegro]. *Arhiv bioloških nauka (Beograd)* 24, 131-139 (in Serbian with English abstract).

Todorović, M., Dundjerski, Z. & Soldatović, B. 1987. Distribution of the mole species (*Talpa*) in the south-west region of the Balkans. *Biologica Gallo-Hellenica* 13, 119-126.

Tvrtković, N., Djulić, B. & Kryštufek, B. 1979. Methods of distinction of *Pitymys liechtensteini* Wettstein, 1927 and *Pitymys subterraneus* de Sélys-Longchamps, 1836 and their distribution in northwestern Yugoslavia. *Biosistematika* 5, 201-204.

Voiculescu, I., 1974. A comparative study of the chromosome banding pattern of *Mesocricetus newtoni* and *Mesocricetus auratus. Zeitschrift für Säugetierkunde* 39, 211-219.

Volleth, M. 1987. Differences in the location of nucleolus organizer regions in European vespertilionid bats. *Cytogenetics Cell Genetics* 44, 186-197.

Volleth, M. 1992. Comparative analysis of the banded karyotypes of the European *Nyctalus* species (Vespertilionidae; Chiroptera).). In: *Prague studies in mammalogy* (I. Horáček & V. Vohralík, eds.), pp. 221-226. Charles University Press, Praha.

Vujošević, M. 1992. B-chromosome polymorphism in *Apodemus flavicollis* (Rodentia, Mammalia) during five years. *Caryologia* 49, 347-352.

Vujošević, M. & Blagojević, J., 2000. Does environment affect polymorphism of B chromosomes in the yellow-necked mouse *Apodemus flavicollis*? *Zeitschrift für Säugetierkunde* 65, 313-317.

Vujošević, M. & Živković, S. 1987. Numerical chromosome polymorphism in *Apodemus flavicollis* and *A. sylvaticus* (Mammalia: Rodentia) caused by supernumerary chromosomes. *Acta Veterinaria (Beograd)* 37, 81-92.

Vujošević, M., Blagojević, J., Radosavljević, J. & Bejaković, D. 1991. B chromosome polymorphism in populations of *Apodemus flavicollis* in Yugoslavia. *Genetica* 83, 167-170.

Vujošević, M., Filippucci, M.G., Blagojević, J. & Kryštufek, B. 1993. Evolutionary genetics and systematics of the garden dormouse, *Eliomys* Wagner, 1840. 4. Karyotype and allozyme analyses of *E. quercinus* from Yugoslavia. *Bolletino di Zoologica* 60, 47-51.

Winking, H., Djulić, B. & Bulfield, G. 1988. Robertsonian karyotype variation in the European house mouse, *Mus musculus*. Survey of present knowledge and new observations. *Zeitschrift für Säugetierkunde* 53, 148-161.

Yosida, T.H. 1980. *Cytogenetics of the black rat. Karyotype evolution and species differentiation.* University of Tokyo Press, Tokyo.

Zagorodnyuk, I.V. 1986. [A case of chromosome polymorphism in the grey hamster, *Cricetulus migratorius,* from Ukraine]. *Vestnik Zoologii* 1986 (5), 45 (in Russian with English abstract).

Zagorodnyuk, I.V. 1991. [Karyotypic variation of 46-chromosome forms of the vole group of *Microtus arvalis* (Rodentia) – a taxonomic evaluation]. *Vestnik Zoologii* 1999 (1), 36-45 (in Russian with English abstract).

Zagorodnyuk, I.V. & Zima, J. 1992. *Microtus tatricus* (Kratochvíl, 1952) in the eastern Carpathians: cytogenetic evidence. *Folia Zoologica* 41, 123-126.

Zernahle, K. 1980. Zytogenetische Untersuchungen am europäischen Rehwild (*Capreolus c. capreolus* L., 1758), sibirischen Rehwild (*Capreolus c. pygargus* Pallas, 1771) und deren Bastarden. *Beiträgen zum Jagd- und Wildforschung* 9, 304-309.

Zima, J. 1983. Chromosomes of the harvest mouse, *Micromys minutus*, from the Danube delta (Muridae, Rodentia). *Folia Zoologica* 32, 19-22.

Zima, J. & Grafodatskij, A.S. 1985. C-heterochromatin arm variation in the weasel, *Mustela nivalis* (Mustelidae, Carnivora). *Folia Zoologica* 34, 125-132.

Zima, J. & Král, B. 1984. Karyotypes of European mammals. I, II, III. *Acta Scientiarum Naturalium Academiae Scientiarum Bohemoslovacae Brno* 18 (7), 1-51; 18 (8), 1-62; 18 (8), 1-51.

Zima, J. & Macholán, M. 1995. B chromosomes in the wood mice (genus *Apodemus*). *Acta Theriologica* (Supplement 3), 75-86.

Zima, J., Červený, J., Hrabě, V., Král, B. & Šebela, M. 1981. On the occurrence of *Microtus epiroticus* in Rumania (Arvicolidae, Rodentia). *Folia Zoologica* 30, 139-146.

Zima, J., Gaichenko, V.A., Macholán, M., Radjabli, S.I., Sablina, O.V. & Wójcik, J.M., 1990a. Are Robertsonian variations a frequent phenomenon in mouse populations in Eurasia? *Biological Journal of the Linnean Society* 41, 229-233.

Zima, J., Lukš, D. & Macholán, M., 1990b: Unusual karyotypes in *Apodemus cf. flavicollis* and *Microtus agrestis* (Mammalia, Rodentia). *Acta Societatis Zoologicae Bohemoslovacae* 54, 146-149.

Zima, J., Zagorodnyuk, I.V., Gaichenko, V.A. & Zhezherina, T.O. 1991. [Polymorphism and chromosomal variability in *Microtus rossiaemeridionalis*]. *Vestnik Zoologii* 1991, 48-53 (in Russian with English abstract).

Zima, J., Volleth, M., Horáček, I., Červený, J., Červená, A., Průcha, K. & Macholán, M. 1992: Comparative karyology of rhinolophid bats (Chiroptera: Rhinolophidae). In: *Prague studies in mammalogy* (I. Horáček & V. Vohralík, eds.), pp. 229-236. Charles University Press, Praha.

Zima, J., Macholán, M., Filippucci, M.G., Reiter, A., Andreas, M., Lípa, M. & Kryštufek, B. 1994. Karyotypic and biochemical status of certain marginal populations of *Sorex araneus*. *Folia Zoologica* 43 (Supplement 1), 43-51.

Zima, J., Fedyk, S., Fredga, K., Hausser, J., Mishta, A., Searle, J.B., Volobouev, V.T. & Wójcik, J.M. 1996. The list of chromosome races of the common shrew (*Sorex araneus*). *Hereditas* 125, 97-107.

Zima, J., Macholán, M., Anděra, M. & Červený, J. 1997a. Karyotypic relationships of the garden dormouse (*Eliomys quercinus*) from central Europe. *Folia Zoologica* 46, 105-108.

Zima, J., Macholán, M., Kryštufek, B. & Petkovski, S. 1997b. Karyotypes of certain small mammals (Insectivora, Rodentia) from Macedonia. *Scopolia* 38, 1-15.

Zima, J., Slivková, L., Andreas, M., Benda, P. & Reiter, A. 1997c. Karyotypic status of shrews (*Sorex*) from Thrace, European Turkey. *Zeitschrift für Säugetierkunde* 62, 315-317.

Zima, J., Lukáčová, L. & Macholán, M. 1998. Chromosomal evolution in shrews. In: *Evolution of shrews* (J.M. Wójcik & M. Wolsan, eds.), pp. 175-218. Mammal Research Institute, Polish Academy of Sciences, Bialowieza.

Zima, J., Ieradi, L.A., Allegra, F., Sartoretti, A., Wlosoková, E. & Cristaldi, M. 1999. Frequencies of B chromosomes in *Apodemus flavicollis* are not directly related to mutagenetic environmental effects. *Folia Zoologica* 48 (Supplement 1), 115-119.

Živković, S. & Isaković, I. 1972. [Karyotype analyses of game as a method of genetical testing in the hunting exploitation]. *Simposium o lovstvu, Beograd* 104-112 (in Serbian with English abstract).

Živković, S. & Petrov, B. 1974. Das Chromosomenkomplement der Wasserwühlmaus (*Arvicola terrestris stankovici* Petrov, 1949) und die taxonomische Stellung dieser Form. In: *Symposium Theriologicum II* (J. Kratochvíl & R. Obrtel, eds.), pp. 343-346. Academia, Praha.

Živković, S. & Petrov, B. 1975a. The karyotype of *Microtus guentheri* Danford et Alston, 1880 from Yugoslavia and taxonomic status of that vole (Mammalia, Rodentia). *Arhiv bioloških nauka (Beograd)* 27, 15-16.

Živković, S. & Petrov, B. 1975b. Record of a vole of the *Microtus arvalis* group with 54 chromosomes (*Microtus subarvalis* Meyer, Orlov et Scholl, 1972) in Yugoslavia and comparison of its karyotype with that of *Pitymys felteni* Malec et Storch, 1963 (Rodentia, Mammalia). *Genetika (Beograd)* 27, 15-16.

Živković, S., Soldatović, B., Milošević, M. & Savić, I. 1968. Analyse der Chromosomen der drei Populationen von Ziesel (*Citellus citellus* L.) aus Serbien. *Zoologischer Anzeiger* 181, 181-185.

Živković, S., Jovanović, V., Isaković, I. & Milošević, M., 1971. Chromosome complement of the European wild pig (*Sus scrofa* L.). *Experientia* 27, 224-226.

Živković, S., Petrov, B. & Rimsa, D. 1975a. [New data on the taxonomy of Balkan *Pitymys* representatives in the light of karyological analysis]. *Biosistematika* 1, 31-42 (in Serbian with English summary).

Živković, S., Petrov, B., Soldatović, B. & Savić, I. 1975b. Two morphologically different Y chromosomes in the bank vole (*Clethrionomys glareolus* Schreb.) from Serbia (Yugoslavia). *Acta veterinaria (Beograd)* 25, 241-246.

Živković, S., Rimsa, D., Ružić, A. & Petrov, B. 1975c. Cytogenetical characteristics, taxonomic status and distribution of the voles with 46 and 54 chromosomes of the *Microtus arvalis* group in Yugoslavia (Rodentia, Mammalia). *Arhiv bioloških nauka (Beograd)* 26, 123-134.

8. LATE PLEISTOCENE RODENT DISPERSAL IN THE BALKANS

GERHARD STORCH*

INTRODUCTION

The paper is focused on the zoogeographical relations of Late Pleistocene mammalian faunas from the Balkans, and in particular on the issue of whether direct mammalian dispersal between Anatolia and the Balkan Peninsula across the Aegean island arcs and the Bosphorus-Dardanelles region made a significant contribution to Balkan biodiversity. The study is based on micromammals, and more precisely on rodent species from selected local faunas which date from around the penultimate glacial cycle. During this period the Recent fauna of the Balkans became essentially established and rodent taxa can be mostly referred to the extant species. The paper does not deal with fossil sites from the northernmost Balkan region (Slovenia to northern Romania).

The study of fossil micromammal distributions, migrations, and diversity dynamics across the area under consideration is somewhat hampered by taxonomic problems at the species level and by problems with biochronological correlation of fossil sites. Faunas from the penultimate glacial, the Last Interglacial, and the Last Glacial periods are rather similar to each other in species composition and do not mirror distinct cold and warm phases as in regions further to the north. Evolutionary processes within rodent lineages are quite often not easy to substantiate over this rather short Late Pleistocene time interval. Moreover, faunal renewals during this period by immigrants are slight, and turnovers caused by extinction and subsequent immigration, a common feature among Central European Late Pleistocene faunal successions, appear to be slight or absent in most parts of the Balkans and western Anatolia. Actually, Late Pleistocene local rodent faunas from many regions do not differ essentially from extant ones in species composition.

LATE PLEISTOCENE DISPERSAL AND FAUNAL PROVINCIALISM

Only a rather limited number of Late Pleistocene local faunas from the Balkans and western Anatolia are known for being rich in micromammalian fossils. Rodent faunas are selected which most likely date from the penultimate glacial cycle (around 200 ka.). These fossil sites are located in the eastern half of the Balkan Peninsula, extending from Dobrugea (south-east Romania) in the north to Athens in the south. A single local fauna of corresponding age is available from western Anatolia (Figure 1). The localities are: (1) Cave "La Adam", layers α-11 (Terzea, 1972) and the upper layers of Gura Dobrogei-4 (Radulescu & Samson, 1996), both located in Dobrugea, south-east Romania. - (2) Morovitsa cave in northern Bulgaria (Popov, 1989). - (3) Arnissa, northern Greece (Mayhew, 1978). - (4) Varkiza 1-2 near Athens, Greece (van de Weerd, 1973). - (5) Latomi-1, island of Chios; Chios was part of the Anatolian mainland at the time the faunal remains were deposited (Storch, 1975).

H.I. Griffiths et al. (eds.), Balkan Biodiversity, 135–145.

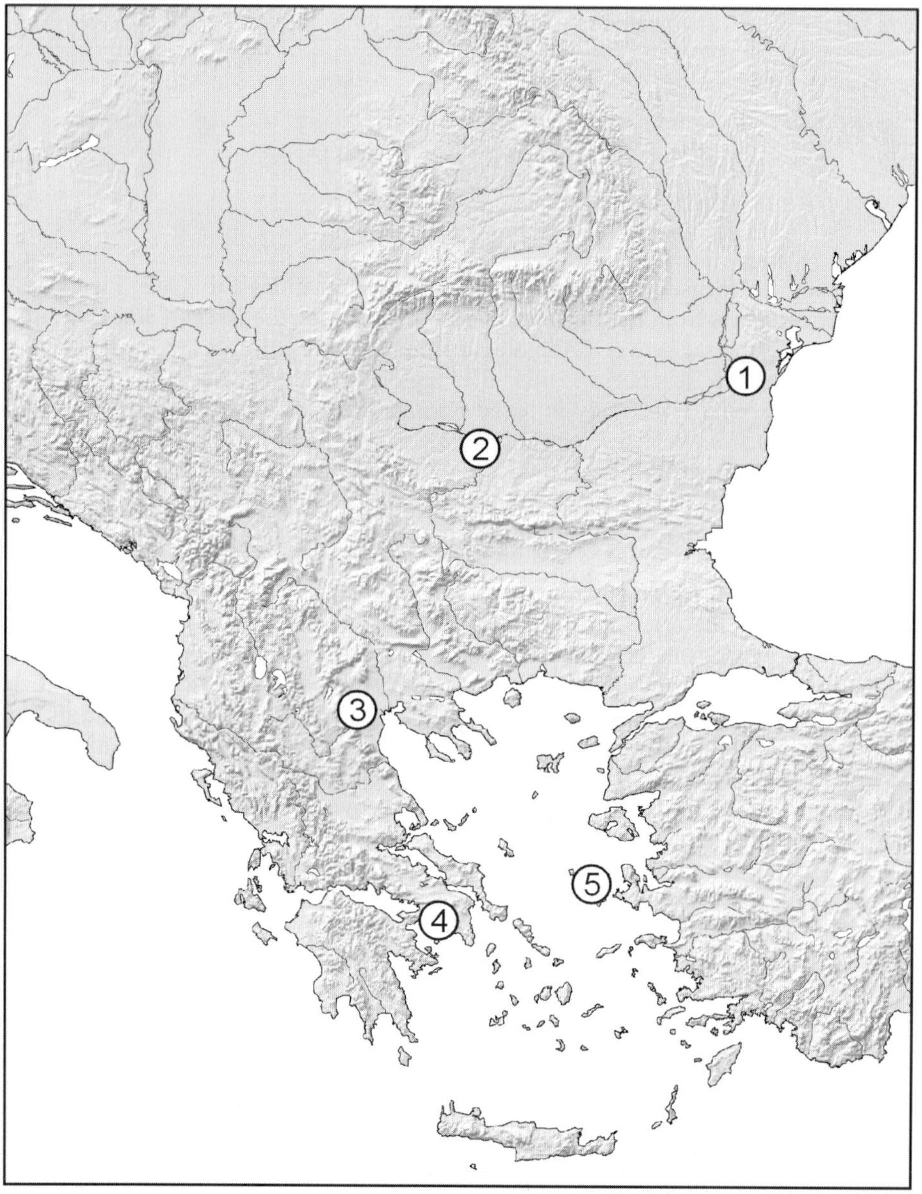

Figure 1. Location of fossil sites from the penultimate glacial cycle. – 1, Cave La Adam and Gura Dobrogei; 2, Morovitsa Cave; 3, Arnissa; 4, Varkiza near Athens; 5, Latomi, island of Chios.

Table 1. Composition of rodent faunas from the localities shown in Figure 1. A few of the original specific assignments are changed. Sites: 1 – La Adam, Gura, Dobrogei (Terzea, 1972; Radulescu & Samson, 1996); 2 – Morovitsa Cave, N Bulgaria (Popov, 1989); 3 – Arnissa, N. Greece (Mayhew, 1978); 4 – Athens, Greece (van de Weerd, 1973); Chios, W Anatolia (Storch, 1975).

Site	1	2	3	4	5
Microtus oeconomus	+				
Microtus agrestis	+				
Microtus gregalis	+				
Pygeretmus pumilio	+				
Allactaga major	+				
Allocricetulus eversmanni	+				
Ellobius talpinus	+				
Cricetus cricetus	+	+			
Eolagurus luteus	+	+			
Lagurus lagurus	+	+	+		+
Sicista subtilis	+	+	+		+
Microtus guentheri	+		+	+[1]	+
Myomimus roachi				+	+
Spermophilus citellus	+	+	+		
Spermophilus xanthoprymnus					+
Mesocricetus newtoni	+	+	+		
Mesocricetus brandti					+
Nannospalax leucodon	+	+	+		
Nannospalax nehringi					+
Apodemus epimelas			+	+	
Apodemus mystacinus					+
Allactaga euphratica					+
Kritimys sp.					+
Apodemus hermonensis					+
Dinaromys bogdanovi				+	
Cricetulus migratorius	+	+	+	+	+
Arvicola mosbachensis / terrestris	+	+	+		+
Microtus arvalis / rossiaemeridionalis	+	+	+		+
Microtus nivalis	+	+	+		+
Apodemus flavicollis/ sylvaticus	+	+	+	+	
Clethrionomys glareolus	+	+	+		
Microtus subterraneus	+	+	+[1]	+[2]	
Castor fiber	+				
Rattus dobrogicus	+				
Rattus rattus					+
Mus sp.				+	+
Glis glis		+			
Dryomys nitedula			+		+

[1] *Microtus* spp.; [2] *Microtus* sp.

The occurrences of rodent species indicate moderate provincialism among the fossil faunas (Table 1). There appears to be a decrease in species number from north to south within the Balkans. In particular, the location of Dobrugea, at the crossroads of eastern, northern, and southern influences, is obvious. Based on their present distribution areas, we can subdivide the rodent species of the fossil sites from the Balkans and western Anatolia roughly into six categories.

(1) Boreal species. Root vole (*Microtus oeconomus*) and field vole (*Microtus agrestis*) have wide Palaearctic ranges. The field vole still occurs in Dobrugea to the south, and the southernmost records of the root vole are isolated relict populations from Hungary. Both species were restricted to the northern Balkans during the Late Pleistocene.

(2) Eurasian steppe species. Their numbers among the Late Pleistocene localities decrease from northeast to south. The species are presently distributed in the steppe zone ranging from Ukraine through northern Kazakhstan to western Mongolia and northwestern China. The hamster (*Cricetus cricetus*) and southern birch mouse (*Sicista subtilis*) have extended ranges into western and Central Europe and they are the only steppe species still present in the Balkans. The yellow steppe lemming (*Eolagurus luteus*) became extinct recently in Kazakhstan and is now confined to north-west China and western Mongolia. The ranges of the great jerboa (*Allactaga major*), mole lemming (*Ellobius talpinus*), and Eversmann´s hamster (*Allocricetulus eversmanni*) do not reach Mongolia and China and extend east only to the regions of Lake Balkhash and Lake Zajsan. The narrow-skulled vole (*Microtus gregalis*) is included in this group of eastern European/middle Asian steppe dwellers. Late Pleistocene invasions into the Balkan area most likely originated from the narrow-skulled vole´s central Asian segment, while the expansion into Central Europe most likely originated from its high Arctic segment.

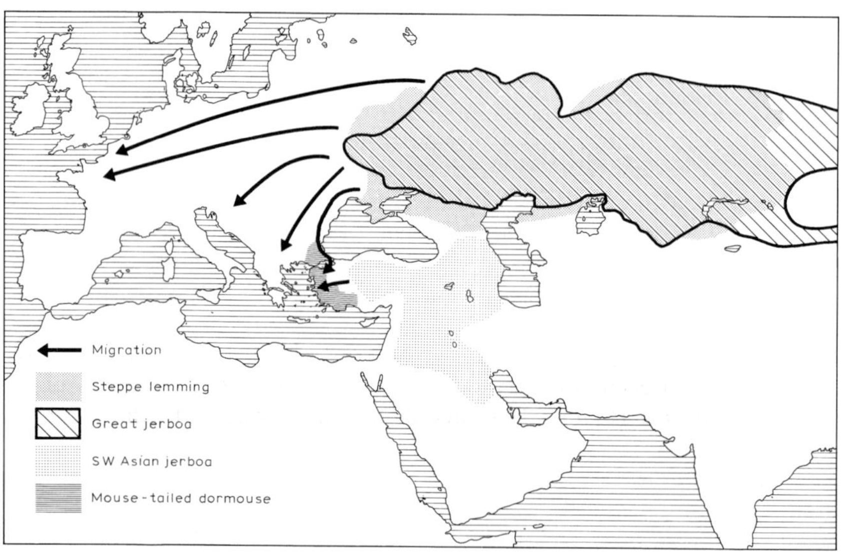

*Figure 2. Recent ranges of the Steppe lemming (*Lagurus lagurus*), Great jerboa (*Allactaga major*), South-west Asian jerboa (*Allactaga euphratica*), and mouse-tailed dormouse (*Myomimus roachi*). Arrows indicate Late Pleistocene expansions.*

The Steppe lemming (*Lagurus lagurus*) is a very able coloniser and exhibits the widest Late Pleistocene range extensions, reaching west to England and western France and south to northern Greece and western Asia Minor. The fossil record shows that the differentiation of the species *Lagurus lagurus* took place in eastern Europe and western Siberia, well before its first appearance in the Balkans during the penultimate glacial. A second immigration wave occurred around the early Last Glacial. These migrations are more or less synchronous with the expansion of the range of the great jerboa into Central Europe (Figure 2). It appears that the Steppe lemming reached Chios, then part of the western Anatolian mainland, from the northeastern Balkan region by crossing the Bosphorus/Dardanelles area rather than by migration from the plains of Ukraine and Kazakhstan through Caucasus and Anatolia. The species is lacking in the micromammal association of Emirkaya-2 in central Anatolia (Montuire *et al.*, 1994), being more or less contemporary with the Chios assemblage. It is also absent from the southern Anatolian rodent succession of Karain near Antalya, spanning the Last Interglacial to Holocene periods (Storch, 1988 and unpublished data).

(3) Rodents which are distributed both in southeastern Europe and Anatolia. They occur in both regions either as a single species or as two different yet apparently closely-related species. Based on the current state of knowledge, the timing and location of initial dispersal and possible faunal exchange between the Balkans and Anatolia is hard to assess. Judging from their present ranges, the mouse-tailed dormouse (*Myomimus roachi*) (Figure 2) and Guenther´s vole (*Microtus guentheri*) may appear to have invaded the Balkan region via the Bosphorus/Dardanelles area. However, we have to consider that their Pleistocene ranges were much more extensive and that both appear rather early in the fossil record, the genus *Myomimus* being present as early as the Miocene in Europe and western Asia, and Guenther´s vole during the Elster glacial in Romania (Terzea, 1995). Thus, alternative palaeobiogeographical patterns for the origin of the Balkan populations cannot be excluded. Fossil evidence suggests that the genus *Mesocricetus* originated in the Mio/Pliocene of western Asia. It is known from Late Pleistocene localities of eastern Europe and ancestral populations of the extant Romanian hamster (*Mesocricetus newtoni*) may have reached the Balkans from southern Ukraine along the Black Sea coastal plains. The present ranges of the lesser mole rat (*Nannospalax leucodon*) and the European souslik (*Spermophilus citellus*), for example, still include southern Ukraine. Thus, there appears to be no strong evidence either for direct rodent migrations between Asia Minor and the Balkans during the Pleistocene, or for speciation processes after the breakdown of a direct land corridor.

(4) Species occurring in Anatolia and the Near/Middle East. In the Balkans, this group lacks any Recent or Pleistocene records. The South-west Asian jerboa *Allactaga euphratica* (Figure 2) is distributed in steppes and semi-deserts from central Anatolia east to the Caspian Sea and northern Saudi Arabia. It occurred on Chios during the Late Pleistocene but is absent, however, from the present fauna of western Anatolia. The wood mouse *Apodemus hermonensis* is known from western Anatolia to Israel. It is possibly synonymous with *Apodemus fulvipectus* (Filippucci *et al.*, 1996) and would thus range east into the Middle East. The murid *Kritimys* sp. was originally referred to as cf. *Rattus* sp. (Storch, 1975). It shows no close relationships, however, to known Pleistocene *Rattus* taxa from the northern Balkans, *R. casimcensis* Radulesco & Samson 1973 and *R. dobrogicus* Terzea 1973, which

appear to be closely related to the black rat (*R. rattus*). The taxon *Kritimys* sp. from Chios most likely has Asian affinities. Its Aegean palaeobiogeography is discussed below.

(5) Endemic species. The Balkan snow vole (*Dinaromys bogdanovi*) is an archaic endemic to the Balkans. Its Late Pleistocene range was obviously more extended to the northwest and southeast than it is today.

(6) Species which are presently widely distributed and occur in the major natural zones of at least parts of the area under consideration. Examples are the water vole (*Arvicola terrestris*) and gray hamster (*Cricetulus migratorius*) which have wide Palaearctic ranges and show considerable habitat plasticity.

The Late Pleistocene micromammalian fauna of the Balkans comprises different zoogeographical sources, including a distinct Eurasian steppe component. Faunal innovations in the Balkans essentially started from adjacent northeastern and northern areas; invasions from Anatolia have only played an insignificant role. The main part of the Balkan Peninsula was obviously not a major glacial refuge and hence did not act as a centre of dispersal during warmer intervals.

CASE STUDIES

Palaeobiogeographical data on rodent species from the above groups (3) and (4) are discussed below in some detail to assess the probability of direct Anatolian-Balkan micromammalian migration routes.

Rock mice, Apodemus mystacinus *and* A. epimelas

Extant Aegean island faunas are impoverished and rather uniform in species composition. Most often, they are dominated by commensal species such as the black rat and house mouse. Unilateral zoogeographic relations such as the occurrence of the jird *Meriones tristrami* on the island of Kos are the exception. This species occurs in Asia Minor but not in the Balkans, and Kos became separated from the Anatolian mainland only rather recently. However, the present and known fossil ranges of rock mice can provide conclusions about the zoogeographical origin of Aegean island populations and the nature of island colonisation. Its present occurrence on various Aegean islands even suggests the possibility of a trans-Aegean dispersal between the Balkans and Asia Minor via Aegean island arcs.

Rock mice have an extended fossil record. The earliest known records are from the latest Tertiary (= latest Villanyian, around 2 Ma) of the isle of Kos (van der Meulen & Kolfschoten, 1986), Tuscany, Italy (Kotsakis *et al*., 2001), and Greece (Koufos, 2001). They are followed by Early Pleistocene (= Biharian, within the Biharian mainly Bavelian) records from the island of Kalymnos, southeastern Aegean Sea, close to Anatolia (Kuss & Storch, 1978), Greece (Koufos, 2001), Croatia (Malez & Rabeder, 1984), Italy (Kotsakis *et al*., 2001), southern France (Pasquier, 1974), southeastern Spain (Agusti *et al*., 1987), and northwestern Romania (Terzea, 1992, 1995). Aside from the Romanian locality Betfia, all fossil sites are located more or less near the Mediterranean Sea. There are no records from the Plio/Pleistocene of Ukraine north of the Black Sea (Nesin & Storch, in preparation).

The oldest known European fossil remains already exhibit the morphology of the first upper molar which is diagnostic of the living species *A. epimelas* from the

Balkans. The fossil Kalymnos population, on the other hand, is already distinct in having the typical occlusal pattern of the extant Rock mouse species *A. mystacinus* from Asia Minor and the Near East (Figure 3). The fossil record suggests that the speciation process between rock mice from Europe and Asia Minor occurred in pre-Pleistocene times. In *A. mystacinus*, the small distal cusp t12 on the first upper molar is usually connected directly both with t8 and t9. The t12-t9 connection is stronger than that between t8-t9 and the latter may be totally lacking. As distinguished from *mystacinus*, in *epimelas* t12 is only connected to t8, the t12-t9 connection is absent, and there is a strong ridge connecting t8-t9 (Storch, 1977).

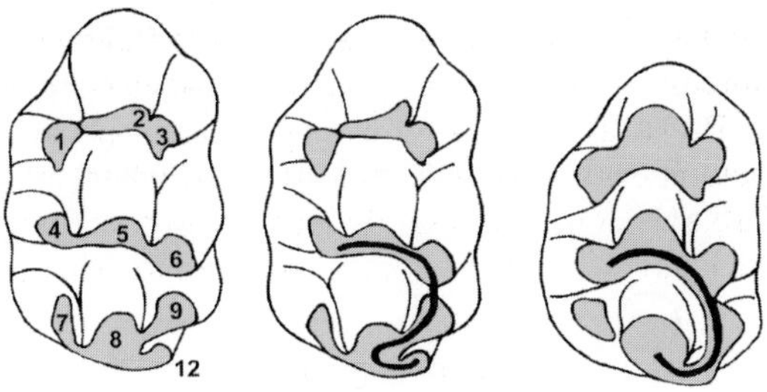

Figure 3. Occlusal pattern of the first upper molar of rock mice. – Left, numbering of cusps; middle, cusp connections typical of Apodemus epimelas*; right, cusp connections typical of* A. mytacinus. *After Storch (1977).*

In regard to zoogeographical analyses, rock mice show several highly favorable preconditions:

(1) Early fossil remains from the Balkans and Asia Minor are distinguishable from each other. The tooth morphology thus indicates the origin of island populations regardless whether they originated from the breakup of a populated island from the mainland or else from immigration from the mainland into an unpopulated island.

(2) Rock mice have wide extant and fossil ranges on the Balkans and in Asia Minor. They could thus take advantage of any intermittent dispersal opportunity. Colonisation of island arcs from one mainland only cannot be explained by the absence of rock mice on the other land mass.

(3) Rock mice are typical rock dwellers, which prefer rocks and rock debris that are rich in crevices and show some cover of grasses and shrubs. They are thus qualified for crossing barren rocky filter corridors into islands and even to compete with black rats. On the other hand, they are not good candidates for an unintended introduction into an island by humans.

(4) Rock mice are not able to cross marine barriers, either by active swimming or passive drifting. Swimming over short distances would lead to death by exposure and exhaustion. Drifting on a tree trunk would require dry shelter and the provision of food, an unlikely situation for a specialised rock dweller (yet possibly feasible for rodents such as dormice which can fall into hypothermy).

The analyses of extant populations (Figure 4) show that islands close to the coasts of the Balkans and Asia Minor are inhabited by their respective mainland species. The eastern Aegean island chain from Lesbos in the north to Rhodos in the south became separated from the western Anatolian mainland only quite recently (*e.g.* Besenecker *et al.*, 1972) and they are inhabited accordingly by *A. mystacinus*. The records of *A. mystacinus* from Karpathos and Crete in the South Aegean island arc, however, are rather unexpected. They are inconsistent with most previous zoogeographic conclusions which favor a demarcation line between Karpathos and Rhodos and unilateral faunal relations of islands west to this line with the Peloponnisos (see Strid, 1971). The records of the eastern rock mouse species along the South Aegean island arc strongly suggest a Pleistocene dispersal route from Asia Minor into Crete that was strongly filtered ecologically and only allowed the passage of very few micromammal species. The dispersal of *A. mystacinus* was probably more or less contemporary with the immigration of an ancestral population of the Pleistocene endemic murid genus *Kritimys* to Crete. Specifically unidentified remains of *Apodemus* sp. are included in Late Pleistocene faunas dominated by *Kritimys* (Mayhew, 1977).

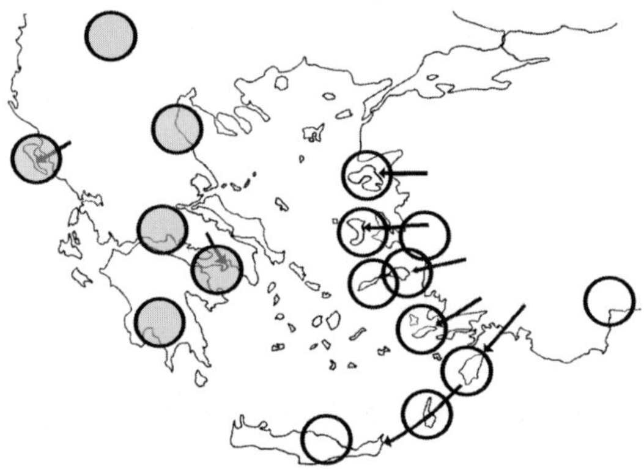

Figure 4. Distribution and geographic origin of island populations of Apodemus epimelas *(hatched circles) and* A. mystacinus *(blank circles). After Storch (1977).*

Rock mice, which provide the best preconditions for a trans-Aegean dispersal, did not manage to cross the Aegean Sea during the Pleistocene by means of island arcs, in any direction whatsoever, and the same is obviously the case for all non-commensal rodents occurring in the region under discussion.

Kritimys*, endemic Pleistocene murid from Crete*

In general, the Pleistocene mammalian faunas of Aegean islands are poorly documented (Kotsakis *et al.*, 1980) and this is particularly true for micromammals. A succession of extinct endemic rodent species of the genera *Kritimys* and *Mus* from various sites on the island of Crete is the best documented. However, their phylogenetic and zoogeographic origins are still poorly understood. Thus far, two hypotheses relating to *Kritimys* are discussed: 1) it is a Tertiary relic that originated on the Greek mainland (Kuss, 1973), or, 2) it is closely related to or even a descendant of the African genus *Praomys* (Kuss & Misonne, 1968; Mayhew, 1977).

Praomys includes various extant African species and the Pleistocene *Praomys pomeli* from the Maghreb.

A series of islands and submarine ridges form the South Aegean island arc that separates the Aegean from the Libyan Sea. The postulated migration routes of *Kritimys* ancestors are mainly with the Balkans (Sondaar & Boekschoten, 1967); in addition, dispersal via a direct connection between Crete and Africa has been suggested. A further possible immigration to Crete from Rhodos via Kasos and Karpathos has been suggested, based on the resemblance of a second lower murid molar from the Pliocene (Late Ruscinian) of Rhodos with corresponding teeth of the Pleistocene *Kritimys* aff. *kiridus* from Crete (Mayhew, 1977). This tooth was later assigned to cf. *Kritimys* sp. by Kotsakis *et al.* (1980) but was never figured or described.

There is no plausible fossil candidate for *Kritimys* ancestry known from the Balkans or elsewhere in Europe to corroborate a western migration route to Crete. An eastern route, however, is strongly suggested by a fossil specimen from the island of Chios. A second upper molar dating from the penultimate glacial of Chios was originally referred to as cf. *Rattus* sp. (Storch, 1975). Subsequent close comparisons with abundant *Kritimys* fossils revealed the close resemblance between the two taxa in several diagnostic characters such as an elevated crown height, an inflated cusp t1 and a faint crest-like cusp t3, the gable-shaped structure of t5-t6, and the rather isolated t4 (Figure 5). Both taxa must be considered closely related, the Chios specimen being most likely referrable to *Kritimys*. The enigmatic *Kritimys* thus appears to have colonised Crete from Asia Minor via an eastern migration route. It obviously did not succeed in crossing the South Aegean island arc to reach the Balkan mainland.

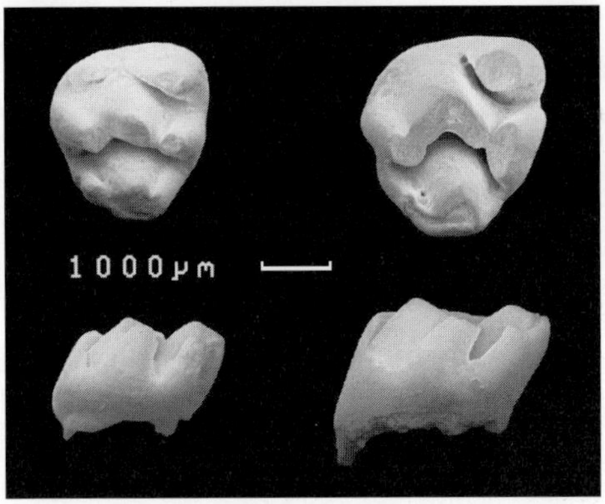

Figure 5. Occlusal (above) and labial (below) views of second upper molars. - Left, Kritimys *sp. from the Late Pleistocene of Chios; right,* K. catreus, *Pleistocene of Crete.*

CONCLUSIONS

There is no evidence from Late Pleistocene fossils that direct faunal exchanges between Asia Minor and the Balkans played a major role in present Balkan rodent biodiversity. If present at all, such an exchange must have been limited to a few species and is not indicative of a widely accessible land corridor.

The Balkans are usually considered an important refuge for European mammalian faunas during Pleistocene glacial episodes. The northern Balkan regions may well have contributed to the re-establishment of the present central European temperate rodent faunas after arctic and continental Asian associations had vanished during the dramatic faunal turnover of the Late Glacial and the Pleistocene-Holocene boundary about 10,000 yr BP. Adjacent to the south, however, most of the Balkan region mirrors rather continuous autochthonous evolutionary processes during the Late Quaternary, resulting occasionally in taxa indigenous to these areas only.

REFERENCES

Agusti, J., Moyà-Solà, S. & Pons-Moyà, J. 1987. La sucesión de mamiferos en el Pleistoceno inferior de Europa: proposición de una nueva escala biostratigráfica. *Paleontologia i Evolució. Memòria Especial* 1, 287-295.

Besenecker, H., Spitzenberger, F. & Storch, G. 1972. Eine holozäne Kleinsäuger-Fauna von der Insel Chios, Ägäis (Mammalia: Insectivora, Rodentia). *Senckenbergiana biologica* 53, 145-177.

Filippucci, M.G., Storch, G. & Macholan, M. 1996. Taxonomy of the genus *Sylvaemus* in western Anatolia – morphological and electrophoretic evidence (Mammalia: Rodentia: Muridae). *Senckenbergiana biologica* 75, 1-14.

Kotsakis, T., Abbazzi, L., Angelone, C., Argenti, P., Barisone, G., Fanfani, F., Marcolini, F. & Masini, F. 2001. Plio-Pleistocene biogeography of Italian mainland micromammals. *Distribution and migration of Tertiary mammals in Eurasia. A three-day international conference in honour of Hans de Bruijn*, Utrecht, Abstract volume, 26-27.

Kotsakis, T., Petronio, C. & Sirna, G. 1980. The Quaternary vertebrates of the Aegean islands: Palaeogeographical implications. *Annales Géologiques des Pays Helléniques* 30, 31-64.

Koufos, G.D. 2001. The Villafranchian mammalian faunas and biochronology of Greece. *Bollettino della Società Paleontologica Italiana* 40, 217-223.

Kuss, S. 1973. Die pleistozänen Säugetierfaunen der ostmediterranen Inseln. Ihr Alter und ihre Herkunft. *Berichte der Naturforschenden Gesellschaft Freiburg i.Br.* 63, 49-71.

Kuss, S. & Misonne, X. 1968. Pleistozäne Muriden der Insel Kreta. *Neues Jahrbuch für Geologie und Paläontologie, Abhandlungen* 132, 55-69.

Kuss, S. & Storch, G. 1978. Eine Säugetierfauna (Mammalia: Artiodactyla, Rodentia) des älteren Pleistozäns von der Insel Kalymnos (Dodekanés, Griechenland). *Neues Jahrbuch für Geologie und Paläontologie, Monatshefte* 1978, 206-227.

Malez, M. & Rabeder, G. 1984. Neues Fundmaterial von Kleinsäugern aus der altpleistozänen Spaltenfüllung Podumci 1 in Norddalmatien (Kroatien, Jugoslawien). *Beiträge zur Paläontologie von Österreich* 11, 439-510.

Mayhew, D.F. 1977. The endemic Pleistocene murids of Crete. I-II. *Proceedings of the Koninklijke Nederlandse Akademie van Wetenschappen* (B) 80, 182-214.

Mayhew, D.F. 1978. Late Pleistocene small mammals from Arnissa (Macedonia, Greece). *Proceedings of the Koninklijke Nederlandse Akademie van Wetenschappen* (B) 81, 302-321.

Montuire, S., Sen, S. & Michaux, J. 1994. The Middle Pleistocene mammalian fauna from Emirkaya-2, Central Anatolia (Turkey): Systematics and paleoenvironment. *Neues Jahrbuch für Geologie und Paläontologie, Abhandlungen* 193, 107-144.

Pasquier, L. 1974. *Dynamique évolutive d'un sous-genre de Muridae, Apodemus (Sylvaemus). Étude biométrique des charactères dentaires de populations fossiles et actuelles d'Europe occidentale.* Unpublished PhD thesis. Université des Sciences et Techniques du Languedoc, Montpellier.

Popov, V.V. 1989. Middle Pleistocene small mammals (Insectivora, Lagomorpha, Rodentia) from Morovitsa Cave (North Bulgaria). *Acta zoologica cracoviensia* 32, 561-588.

Radulescu, C. & Samson, P.-M. 1996. On some Middle and Late Pleistocene rare small mammal elements from the karst deposits of central Dobrogea (Romania). *Theoretical and Applied Karstology* 8, 163-173.

Sondaar, P.Y. & Boekschoten, G.J. 1967. Quaternary mammals in the South Aegean island arc; with notes on other fossil mammals from the coastal regions of the Mediterranean. I-II. *Proceedings of the Koninklijke Nederlandse Akademie van Wetenschappen* (B) 70, 556-576.

Storch, G. 1975. Eine mittelpleistozäne Nager-Fauna von der Insel Chios, Ägäis (Mammalia: Rodentia). *Senckenbergiana biologica* 56, 165-189.

Storch, G. 1977. Die Ausbreitung der Felsenmaus (Apodemus mystacinus): Zur Problematik der Inselbesiedlung und Tiergeographie in der Ägäis. *Natur und Museum* 107, 174-182.

Storch, G. 1988. Eine jungpleistozäne/altholozäne Nager-Abfolge von Antalya, SW-Anatolien (Mammalia, Rodentia). *Zeitschrift für Säugetierkunde* 53, 76-82.

Strid, A. (ed.) 1971. Evolution in the Aegean. Proceedings of a symposium held at the Department of Plant Taxonomy, Lund, Sweden on 22-24 January, 1971. *Opera Botanica* 30, 1-83.

Terzea, E. 1972. Remarques sur la morphologie dentaire et la répartition de *Microtus nivalis* Martins dans le Pleistocène de Roumanie. *Traveaux de l'Institut de Spéologie "Émile Racovitza"* 11, 271-298.

Terzea, E. 1992. *Apodemus mystacinus* (Danford & Alston) (Rodentia, Mammalia) dans le Pleistocène inférieur final de Betfia-VII (Bihor, Roumanie). *Traveaux de l'Institut Spéologie "Émile Racovitza"* 31, 83-94.

Terzea, E. 1995. Mammalian events in the Quaternary of Romania and correlations with the climatic chronology of Western Europe. *Acta zoologica cracoviensia* 38, 109-120.

van der Meulen, A.J. & van Kolfschoten, T. 1986. Review of the Late Turolian to Early Biharian mammal faunas from Greece and Turkey. *Memorie della Società Geologica Italiana* 31, 201-211.

van de Weerd, A. 1973. Rodentia from two Pleistocene fissure fillings near Athens. *Proceedings of the Koninklijke Nederlands Akademie van Wetenschappen* (B) 76, 148-166.

9. EARLY HOMINIDS IN THE BALKANS

NENA GALANIDOU*

INTRODUCTION

The aim of this paper is to introduce the human factor, in the shape of our Palaeolithic past, into the study of Balkan biodiversity and to identify those parts of the story of Pleistocene hominids in the Balkans that are most relevant to this study. The paper's temporal perspective will thus be a long one, since this period encompassed many thousands of years, witnessing the alternation of cold and warm stages and the geographic restriction and expansion of various plant and animal populations. Against this periodically changing Ice Age background there took place the arrival of ancient hominids in the Balkans, the extinction of certain species and the arrival and subsequent expansion of our species, *Homo sapiens*, through even the most remote of upland regions, formerly inaccessible. The Balkan Peninsula, the south-eastern entrance to Europe, is a region of critical concern in the process of reconstructing ancient European history. As one of the many pieces in the European hominid mosaic, the Balkan record will be discussed here in conjunction with the major issues of early European prehistory.

The story of ancient hominids in this part of the world is woven from many different threads: emergence and extinction, arrival and colonisation, social life, subsistence patterns, settlement patterns, demography and more. All of these threads have to do with the past, yet they are woven together by means of techniques and motivations shaped by the present: the methodological tools developed by the archaeological and biological disciplines and the questions that are currently regarded as of particular interest. I have chosen to follow the thread of the history of hominid presence in this area, a theme that will be dealt with within three subheadings: speciation, variation and adaptation.

SPECIATION

Many of you must be familiar with scenarios of human evolution such as the one represented in Figure 1. This is a typical example of what was until very recently considered to be a scientifically correct visual reconstruction of the sequence of major human evolutionary events. Such images are representative of the main paradigms that have informed research into early humans. This approach does not ignore biodiversity altogether, but reduces it to a neat linear sequence consisting of a limited number of species. The image overlooks the possibility that more than one hominid species may have existed at the same time and assigns the business of evolution exclusively to our white male ancestors, female or non-white hominids generally being conspicuous by their absence. The scheme of Figure 1, the product of a western, male-dominated tradition of palaeoanthropological research, conveniently obscures the considerable difficulties faced by taxonomists of human fossils, who must distinguish morphological variation within a single species from variation between different species and assign fragmentary elements of ancient

H.I. Griffiths et al. (eds.), Balkan Biodiversity, 147–165.

skeletons to discrete species according to hard-tissue characteristics alone (Tattersall, 1986). All the skeletal and cranial remains found on the Balkan Peninsula have been identified as belonging to one of three hominid taxa: *Homo heidelbergensis*, *H. neanderthalensis* and *H. sapiens*.

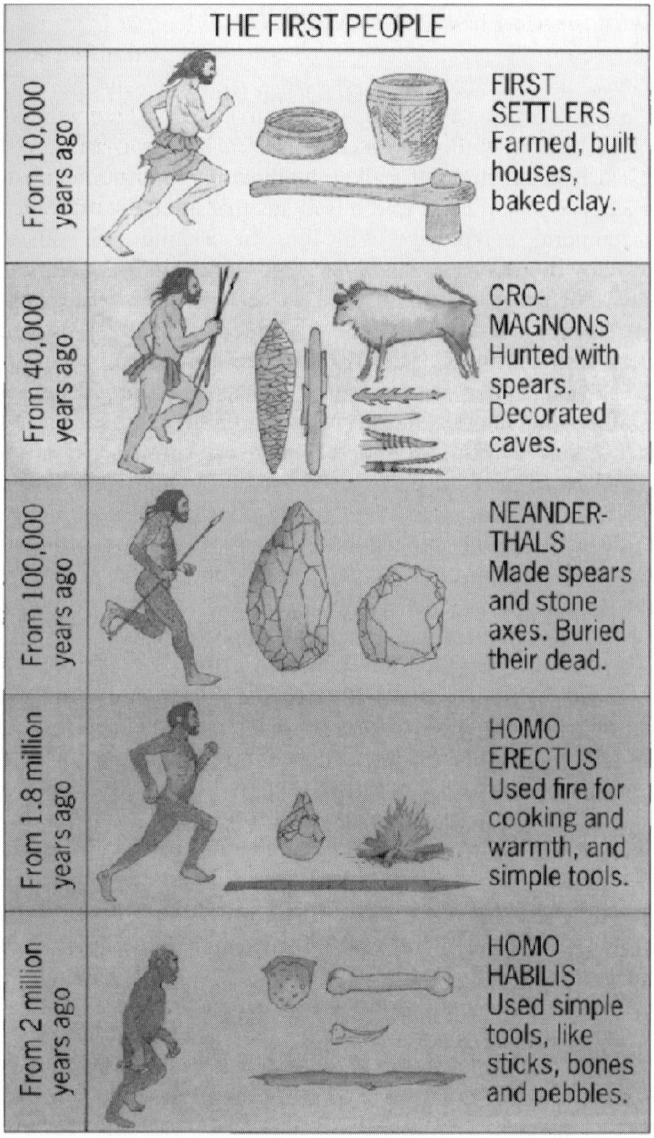

Figure 1. An outdated model of hominid evolution (reproduced from Maynard, 1996, by permission of Kingfisher Plc. All rights reserved).

In phylogenetic terms the earliest human species present in the Balkans was *Homo heidelbergensis,* represented by the Petralona skull (Figure 2). This specimen was discovered accidentally by a villager in the homonymous cave in northern

Greece in 1960. The skull was well protected by the stalagmitic material that had accumulated on its surface, and is as a result one of the best preserved European Middle Pleistocene crania we have. Despite claims to the contrary, no postcranial remains that can safely be attributed to the same hominid have been recovered from the cave. The Petralona site is an impressive karstic formation, 300 m a.s.l., at the foot of Mount Katsika in Chalkidiki. Excavations carried out in various parts of the cave long after the discovery of the skull revealed a long stratified sequence rich in macro-faunal and micro-faunal remains, with some lithic and bone artefacts. The thoroughness with which the cave's fauna have been documented (Kretzoi & Poulianos, 1981; Kurtén & Poulianos, 1977, 1981) is in marked contrast to the treatment of the artefacts, concerning which we have as yet no detailed publication, although photographs and illustrations have appeared sporadically (Poulianos, 1971, 1978, 1982).

For many years the species, the context and the period to which the skull should be assigned were some of the most bitterly debated questions in Balkan prehistory. This controversy, which divided the scientific community, students and the general public, damaged careers and aroused bitter resentment, was not confined to the boundaries of scientific meetings and journals, but frequently overflowed into the magistrates' courts. Today it has been established beyond all doubt that the Petralona specimen should be classified as *Homo heidelbergensis*, a species whose fossils have been found in both Africa and Europe (previously referred to in some literature as 'European *Homo heidelbergensis*', 'African archaic *Homo sapiens*' or 'archaic *Homo sapiens*'). The European sample consists of cranial and skeletal remains from Arago, Atapuerca, Boxgrove, Bilzingsleben, Mauer, Petralona, Steinheim and Swanscombe. The cranial remains suggest an average cranial capacity of 1100cc, but the fragmentary nature of the postcranial finds does not permit detailed reconstruction of *H. heidelbergensis*' anatomy.

Deriving its name from the specimen (a lower jaw) found in a sand quarry at Mauer, near Heidelberg, in 1907 (Schoetensack, 1908), *Homo heidelbergensis* is widely considered to have evolved from *H. erectus* about 500 ka. *Homo erectus* was as far as we know the first species to leave Africa, at some time not longer than 1.7 Ma, and spread as far afield as China and Java. It is not clear, however, whether *erectus* made it as far as Europe and there formed an early population that later evolved into *heidelbergensis* (a question that applies equally to the Balkans) or whether *heidelbergensis* itself evolved in Africa, then became the first hominid to populate Europe (Gowlett, 1999). A. Templeton's recent analysis of genetic data identifies two other major hominid expansions outwards from Africa after the original spread of *H. erectus* (Templeton, 2002). This new model of human evolution, which places the second expansion between 840 and 420 ka, is compatible with the significant changes that appear in both the archaeological and the fossil record during this period (Gowlett, 1999): the first emergence of Acheulean technology in Eurasia and an increase in average cranial capacity.

We know that both Africa and Europe had *heidelbergensis* populations about 400 ka, but do not know how long ago the Petralona hominid lived. Various Lower and Middle Pleistocene dates, estimated on the basis of either the faunal evidence or the speleothems (Ikeya, 1980; Poulianos, 1982; Shen & Yokoyama, 1986), have been proposed for strata excavated in various areas of the cave some time after the removal of the skull. Useful though these dates may be, they cannot be taken as reliable measures of the age of the hominid remains, since the skull was recovered in a manner that left the original context of its deposition and discovery extremely uncertain and thus impeded any subsequent attempt to date it. The most reliable dates that have been obtained for the Petralona hominid are based on calculations of the age of the calcitic crust on the skull's surface. This approach places the Petralona hominid in the later Middle Pleistocene (Grün, 1996; Hennig *et al.*, 1982; Latham & Schwarcz, 1992).

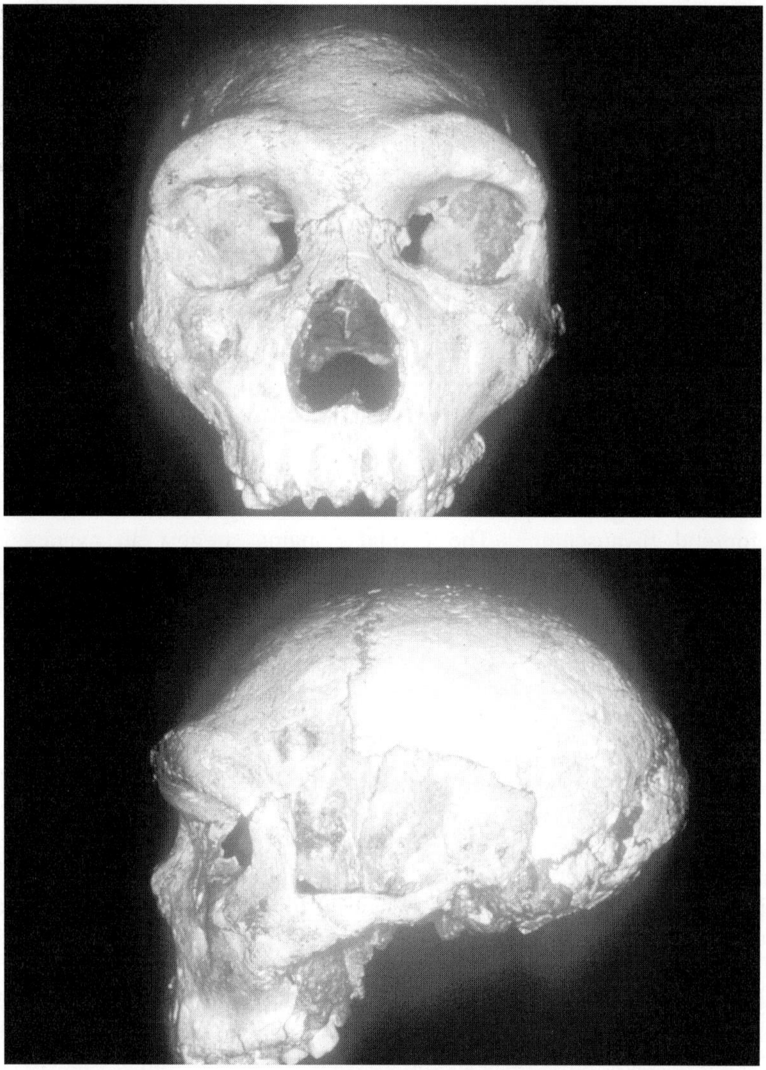

Figure 2. The Petralona specimen (front view and profile).

The proper dating of the Petralona hominid is linked to the question of early hominid arrival in Europe. This issue has given rise to two hypotheses, popularly termed the 'long' and the 'short chronology' (Roebroeks & van Kolfschoten, 1994). The former claims that there is sufficient evidence, both archaeological and palaeontological, to prove that hominids were present in Europe prior to the Brunhes/Matuyama boundary (777 ka) and in some places from before 1.5 Ma. The latter raises three separate objections to the theory of a long chronology, querying the real ages of certain allegedly early sites, the authenticity of the artefacts found at some sites and the actual species of the various remains claimed to be those of hominids. This theory asserts that the first unequivocal evidence of hominid presence only appears in the record from *ca.* 500 ka onwards. Although its criticisms are valid for most areas of Europe, the short chronology model fails to account for the Iberian Peninsula, where uncontested evidence has been found (at Gran Dolina, Atapuerca and possibly Fuente Nueva 3) for hominid presence dating from around 800 k yr. BP. The data we have suggest that hominids did indeed reach the SW gateway to Europe early on. The question of whether or not something similar may have happened at the other end, the SE gateway, must for the moment remain open. It is nonetheless worth mentioning the argument, unsupported by any hard evidence, that the earliest hominids to reach Iberia could have done so by crossing the straits of Gibraltar. This would of course mean that early dates in Iberia did not necessarily imply early dates elsewhere in Europe.

Amongst the Balkan fossils sharing a number of anthropometric and morphological characteristics with the Petralona specimen is one of the two skulls found in Apidima Cave A, one of the many karstic formations on the steep coast near Areopolis in southern Greece. The skulls, ΛΑΟ1/Σ1 and ΛΑΟ1/Σ2, were embedded in a single block of hard breccia that filled a recess in the inner cave. This block was extracted from the site in one piece so that the skulls might be worked free of it under laboratory conditions. The heavy overlying sediments have damaged and distorted the skulls. Of the two, ΛΑΟ1/Σ2 is the better preserved, lending itself to comparison with other pre-*sapiens* cranial fossils. It exhibits a number of affinities with the Petralona skull, but also several differences in the cranial vault and facial skeleton (Pitsios & Liebhaber, 1995). In view of these differences, pending the complete extraction of the second skull from the surrounding breccia, the excavator has proposed that ΛΑΟ1/Σ2 should be classified as belonging to an archaic group with traits transitional between those of *Homo erectus* and *H. sapiens* (Pitsios, 2000) that he has provisionally named *H. (sapiens) taenarius*. On the basis of the geological context and the morphology of the better-preserved fossil, he has suggested a date of between 300 and 100 ka BP (Pitsios & Liebhaber, 1995). Radiometric techniques will be required to evaluate the age of these hominids more precisely. Clarification of their species will have to wait until the second specimen has been worked clear of the breccia, and must of course take into account the latest archaeological and genetic evidence, which does not support the multiregional hypothesis of modern human evolution (see below for a more detailed discussion).

In phylogenetic terms the second earliest hominid species found in the Balkans is *Homo neanderthalensis*, represented by the fossil collections from the caves of Krapina and Vindija, both in the Hrvatsko Zagorje, NW Croatia. This hominid, whose remains are restricted to Europe and western Asia, and possibly North Africa (*e.g.* Haua Fteah), is known for its distinctive anatomy (fairly short stature and robust build). Neanderthals lived in Europe between about 130 ka and 30 ka. They are generally considered to have evolved through regional differentiation. According to this view the Sahara desert acted as a natural barrier promoting hominid isolation and speciation, *heidelbergensis* populations north of the Sahara giving rise to the Neanderthals, and those south of the Sahara evolving into modern humans

(Beaumont *et al.*, 1978; Stringer & MacKie, 1996). According to Gamble the traits of the fossil human crania found at Sima de los Huesos, Atapuerca, in Spain support the view that the Neanderthals in Europe evolved locally from *heidelbergensis* groups (Gamble, 1999). The hominids found at La Sima have an anatomy that combines African *erectus* features with others that show some degree of adaptation to the colder European climate. This combination of elements is suggestive of a species that is developing into the Neanderthal.

Recent genetic research has lent further support to this theory by extracting and analysing mtDNA sequences from three Neanderthal samples: the original specimen found in the Feldhofer cave in Germany (Krings *et al.*, 1997), a child found in the Mezmaiskaya cave in the Caucasus (Ovchinnikov *et al.*, 2000) and a specimen from the Vindija cave (Krings *et al.*, 2000). These three have given us some idea of the Neanderthals' genetic makeup. Like modern humans, they seem to have exhibited very little genetic variation, suggesting that the original population from which they developed was small (Krings *et al.*, 1997, 2000). The same studies have shown the Neanderthal genome to have differed considerably from that of modern humans. Although modern humans and Neanderthals share a common ancestry, the Neanderthals are thought to have become separated from the ancestors of modern humans about 600 to 700 ka, although this date is based on the DNA clock and should therefore be treated with caution.

The largest collection of Neanderthal remains ever found at a single European site was discovered over a century ago in the Krapina rockshelter, at the foot of Hušnjak, a sandstone cliff 120m above sea level in northern Croatia. Excavations began in the last year of the 19th century and were completed in 1905 by D. Gorjanović-
Kramberger (1902, 1906). His field technique, which aimed to expose the horizontal distribution of the finds following the rockshelter's natural stratigraphy, was accompanied by detailed recording of the position and depth of each fossil find (Karavanić, 2000). 5000 archaeological finds were recovered. Between the early Gorjanović-Kramberger publications and the 1990s these finds received little archaeological attention (but see Miracle in press; Simek, 1991; Simek & Smith, 1997). Krapina, nonetheless, immediately became well known for its palaeoanthropological remains. A recent study has shown that these are part of the cranium of an adult female and about 800 fragments belonging to between 12 and 28 individuals aged from infancy to adulthood (Radovčić *et al.*, 1988; Smith, 1982). Krapina is unique amongst sites containing human fossils in having yielded the remains of what was almost certainly a biological population of Neanderthals. Mousterian artefacts and faunal remains were also found in the hominid zone. ESR and U-series dates obtained using tooth enamel from the Krapina hominids suggest that this population lived over a period of up to fifty thousand years, with a mean date of *ca.* 130 ka (Rink *et al.*, 1995).

The state of preservation of the Krapina sample has raised many questions about the conditions and context of its deposition. The fossils are highly fragmented and disarticulated; some have been exposed to fire and others bear on their surfaces what appear to be scratch or cut marks. It is, moreover, impossible to associate any of the postcranial bones with cranial or dental remains (Kricun *et al.*, 1999) or indeed to assign any two anatomical parts to the same individual. Although finds from all over the Neanderthal distribution area have made it clear that some Neanderthal groups buried their dead in caves, there is no indication that the Krapina hominids were buried intentionally. The questions of how so many bones came to be incorporated into the Krapina cave sediments and why they are so fragmented thus demand an answer. Gorjanović-Kramberger responded to these questions by hypothesising that the Krapina Neanderthals practised cannibalism, a suggestion that has received both support and severe criticism. Alternative hypotheses attribute the condition of the bones either to taphonomic factors (Trinkaus, 1985) or to postmortem treatment of the corpses in preparation for a secondary burial ritual (Russell, 1986a, b). This issue remains a murky one, but the Krapina finds have shed light on several other aspects of Neanderthal economic and social life. Miracle's study of subsistence practices at Krapina finds that the Merck's rhino bones found in the cave, some with signs of surface modification or burning, demonstrate an age distribution that suggests systematic selection of young-juvenile animals, which would have had to be separated from their mothers by means of some well planned strategy. His study thus argues for advanced ways of procuring meat that would have required considerable co-ordination and collective effort by the Neanderthal groups (Miracle, in press).

The second major Balkan sample of Neanderthal fossils is from Vindija, a limestone cave in the Hrvatsko Zagorje, 50 km from Krapina. This site contained sediments that were rich in archaeological material and human fossils. Of the Vindija hominid sample, 40 specimens from layer G3 and 4 specimens from layer G1, all of which are postcranial or cranial fragments or teeth, have been identified as belonging to *Homo neanderthalensis* (Malez *et al.*, 1980). Despite their fragmentary condition, these fossils have told us much about the life and dietary habits of the Vindija Neanderthals and about the time at which they lived. This population was more lightly built (gracile) than the Neanderthals in other parts of Europe (Smith, 1982) and most of the protein in its diet was of animal origin; in other words, these groups subsisted largely on meat (Richards *et al.*, 2000).

The lowermost Vindija sequence (level K) overlaps in time with some of the Krapina Mousterian, but the G complex in which the Neanderthal remains were found has produced dates that seem rather recent. The most reliable dates obtained are those for the upper level, G1: 33,000±400 yr. BP, obtained from a bear bone, and 29,080±400 yr. BP and 28,020±360 yr. BP, obtained directly from Neanderthal bones (Karavanić, 1995; Smith *et al.*, 1999). The Vindija population would seem from this to have been amongst the last surviving Neanderthal groups, along with the late Neanderthals of the Zafarraya cave in southern Spain (Tattersall, 1999) and those of the Mezmaiskaya cave in the northern Caucasus (Golovanova *et al.*, 1999). It is interesting to note that Wolpoff's reaction to the results of ancient DNA analysis was to question on morphological grounds whether the Mezmaiskaya infant was really a Neanderthal, or a modern human (Hawks & Wolpoff, 2001). Observing the conjunction of these apparently late dates with this population's overall morphology and lack of robustness compared to other western European Neanderthals, some researchers suggested that the Vindija fossils belonged to a group in a state intermediate between that of the majority of central European Neanderthals and that of early Upper Palaeolithic humans (Smith, 1982). In other words, the hypothesis was advanced that the type represented by the Vindjia population was a transitional stage between Neanderthal and modern humans (Malez *et al.*, 1980). This hypothesis was tenable within the context of the multiregional theory of modern human evolution, founded partly upon empirical data concerning the Vindija Neanderthals. According to this theory, modern human populations were descended from local populations of *Homo erectus* or *H. ergaster* that spread out of Africa into Eurasia at least 1 Ma and evolved locally and independently into the various archaic *H. sapiens* of the Old World (Thorne & Wolpoff, 1992). Modern European populations thus evolved from Neanderthal ones. More recent research using genetic (*e.g.* Cann *et al.*, 1987) and archaeological (*e.g.* Yellen *et al.*, 1995) evidence indicated, however, that it was more probable that modern humanity was of African origin. This suggested that Stringer and Gamble (1993) might be right in seeing the Vindija population's lack of robustness as the result either of intra-species variation (in other words, this particular group of Neanderthals was simply physically smaller on average than most other groups) or of interbreeding with contemporaneous modern humans. The latter hypothesis was made more tenable by the fact that level G1, in which some hominid specimens were found, also contained an industry that was distinctively Early Upper Palaeolithic (Aurignacian) and a few bone artefacts. This created a case for a possible association of Neanderthal groups with material culture created by modern humans (Karavanić, 1995; Karavanić & Smith, 1998). Too many unresolved issues surrounded Vindija's stratigraphy to allow this matter to be decided on an archaeological or anthropological basis.

The question of the Vindija hominids' species was clarified by analysing the mtDNA sequence extracted from a bone found in level G3. This sequence differed in only nine respects from that extracted from the Feldhofer cave specimen, but showed 35 differences from sequences extracted from modern humans, leaving us in no doubt that the Vindija hominids were definitely Neanderthals (Krings *et al.*, 1997, 2000). The morphological variation observed in the Vindija fossils is therefore probably best explained not as the outcome of a speciation process, but as an instance of just such intra-species variation as is clearly visible in modern human populations (compare, for example, the anatomy of an Inuit Eskimo with that of a Masai).

Table 1. Upper Pleistocene Homo sapiens *remains from the Balkans.*

Country	Site	Hominid remains	Cultural material date	Source
Bulgaria	Bacho Kiro layer 11	Fragment of juvenile left mandible with 1 molar	39,000 - >43,000 yr. BP	Kozlowski, 1982
Croatia	Vindija level D	34 cranial and postcranial remains		Smith, 1976, 1982
Greece	Apidima cave Γ	1 female skeleton	A few artefacts of Aurignacian character	Mompherratou & Pitsios, 1995; Darlas, 1995
Greece	Klithi	2 teeth	16.5-13 k yr. BP	Bailey, 1997; Gamble, 1997
Greece	Theopetra	Calvaria and fragments of long bones	13,723 $\pm$ 60 yr. BP	Stravopodi & Manolis, 2000

Genetic investigation has likewise contributed to a better understanding of the ancestry of modern humanity. Molecular research on nuclear and mtDNA shows that of modern African populations to have the greatest variability, suggesting that the modern human genome first evolved in this area of the world (Cann *et al.*, 1987; Goldstein *et al.*, 1995; Harpending *et al.*, 1993; Rogers & Jorde, 1995). Although the process by which our species evolved has yet to be worked out in greater detail by means of fossil, archaeological and molecular analysis (see Brooks, 1996 for discussion), current thinking is therefore basically that *Homo sapiens* evolved from a small founder population in sub-Saharan Africa and from this region spread outwards to the rest of the world. Templeton's model assumes that interbreeding, rather than replacement, should be regarded as the key to the human evolutionary process, gene flow and territorial expansion leading to genetic interchange between human populations (Templeton, 2002). His analysis places the third hominid expansion out of Africa between 150 and 80 ka.

Figure 3. A map of the Balkans showing the sites with hominid remains referred to in the text.

It is not clear when modern humans first arrived in the Balkans, partly because so few fossil hominid remains have been found there (Table 1, Figure 3) and partly because the correct dating of this region's early modern human record is problematic. The ages calculated for Bacho Kiro layer 11 are too imprecise to be of much use in this connection; we have no radiometric dates for other sites such as Apidima cave Γ, while in the case of other sites such as Temnata layer 4 the dates obtained using different techniques (TL and ^{14}C) are inconsistent (Kozlowski, 1996). Amongst the earliest human remains found to exhibit distinctively modern traits are a fragment of a mandible with molar, found in Bacho Kiro layer 11 (Bulgaria), that dates from between 37 and >43 k yr. BP (Table 1). The ages of the stratum in which modern human remains have been found and of those containing the remains of the Vindija Neanderthals do, however, make it clear that Neanderthal and modern human populations co-existed in the Balkans for some part of the period between approximately 40 and 28 k yr. BP.

The Balkan *Homo sapiens* sample dating from the Pleistocene is smaller than the Neanderthal sample and tells us little about past pathology, demography or diet (Table 1); incidentally, the number of modern human remains increases exponentially in sites that date from the early Holocene, when formal disposal of the dead in caves, in settlements or on the peripheries of settlements became a more regular and widespread practice. Despite the limitations of these fossil remains, however, our picture of the modern human groups that lived in the Balkans during the Pleistocene is complemented by the richer archaeological record of the period between 40 and 10 k yr. BP. This brings us to the next issue under discussion: variation in cultural remains.

VARIATION

The picture of human evolution that we see in Figure 1 demonstrates another implicit assumption that has informed Palaeolithic research almost from its inception. This assumption was based upon the premise that the morphological variation seen in the lithic industries recovered was directly related to hominid variation. A lithic industry is the set of stone tools, together with cores and other waste products of the knapping process, recovered from a site. Certain human species were associated with particular lithic industries and the archaeological record of each of the three major subdivisions of the Palaeolithic was in effect assigned to a different hominid, the Aurignacian and Epigravettian industries of the Upper Palaeolithic being bestowed upon *Homo sapiens*, the Mousterian industries of the Middle Palaeolithic upon *H. neanderthalensis* and the Oldowan and Acheleuan industries of the Lower Palaeolithic upon *H. habilis* and *H. erectus*. In recent years, however, developments in genetic research, refinements in archaeological recovery and dating techniques and the discovery of a large number of human fossils in Africa and the Near East have changed the picture radically. It has become clear that many more hominid species evolved during the Pleistocene than had previously been thought, that some of these co-existed in time and space and that some of them shared the same technology and material culture.

Biological events of speciation and extinction do not by any means fall neatly within the traditional Palaeolithic subdivisions (*i.e.* Lower, Middle, Upper), instead frequently crossing their boundaries. For instance, Vindija shows us that Neanderthal groups survived in what is today Croatia until after 30 k yr. BP, well into the Upper Palaeolithic. By this time modern humans had been established elsewhere in the Balkans (in the caves of Bacho Kiro and Temnata, in what is today Bulgaria, and in Kleisoura cave 1, in Greece, for example) for many millennia (Kozlowski 1996; Koumouzelis *et al.*, 2001). Likewise, if the late dates obtained for the Petralona hominid are valid (Grün, 1996), these together with the early dates calculated for the Krapina fossils (Rink *et al.*, 1995) permit the hypothesis that during the last part of the Middle Pleistocene *H. heidelbergensis* and *H. neanderthalensis* may have lived side by side on different parts of the peninsula.

The deconstruction of the idea that biological variation equates to variation in material culture has rendered the Lower–Middle–Upper Palaeolithic scheme inadequate to the task of signifying industrial diversity. This scheme, developed within the intellectual milieu of the late 19th century, today fails to account for the temporal and regional peculiarities of ancient hominid manifestations. Dispensing with the old scheme is, of course, easier than replacing it: a daunting task, since the generic and global character of the old scheme made it both flexible and widely applicable. At present a more effective and refined, though not purely archaeological, framework for discussing the archaeology of the Palaeolithic in the northern hemisphere is the Marine Oxygen Isotope Record, which acts as the global standard for reconstructing Quaternary chronology and climate. This record replaces the old geological scheme of Glacials and Interglacials rather than the scheme linking industries with hominid types. The Isotope Record is only as effective or as ineffective as our ability to correlate with it isolated fragments of the terrestrial record, or cultural and hominid remains.

The demise of the old scheme has also challenged the ways in which we conceptualise and analyse lithic industries. These, the most abundant and enduring remains of ancient hominids, are a dynamic expression of cognitive abilities, of technical traditions and of the social context within which they were produced. In recent years Palaeolithic archaeology has opened up its agenda, moving on from its previous purely descriptive typological approach to examine operational sequences as a means to the better understanding of the conceptual, technical and economic issues involved in the production and use of stone tools (Perlès, 1992). Along with these developments has come the recognition that either the lithic record or our approach to it is insufficiently refined for any answer to the question of which hominid was responsible for which part of the record to be possible. This is true both of the European and Near Eastern industries previously described as 'Lower' or 'Middle' Palaeolithic and of the Balkan industries.

The African record offers a basis for discussion of the earlier part of the Palaeolithic. Broadly speaking, two main technocomplexes can be identified here. The earlier Oldowan technocomplex consists of core and flake tools of types that began to be manufactured about 2.5 Ma and remained in production for at least a million years. The more recent Acheulean technocomplex, which appeared about 1.5 Ma, consists of core tools of another kind: bifaces, in the form of hand axes or cleavers. Both of these technological traditions are found in the Balkans. Choppers and flakes have been found on the banks of the river Peneios in Thessaly, Greece (Runnels & van Andel, 1999), in the Yarimburgaz cave near Istanbul, which dates from the middle of the Middle Pleistocene (Arsebük & Özbaşaran, 1999), in the Gajtan cave in northern Albania and on the fluvial terrace of Baran in the same region (Fistani, 1993). The occasional chopper or chopping tool has been reported at the Korrission lagoon, on the island of Corfu in the Ionian sea (Kourtessi-Philippakis, 1999), at the Sandalja cave, near Pula in Croatia (Malez, 1974), and in the fluviolacustrine deposits of the river Oltet (a tributary of the Olt) in south Romania (Valoch, 1995). Elsewhere in Europe, Oldowan-like industries have been found at (early) Orce in Spain and (later) Isernia in Italy, at Bilzingsleben in Germany and at Vértesszöllös in Hungary.

So far two Balkan sites have produced hand axes that fall within the Acheulean tradition: Palaiokastro, in western Macedonia, and Kokkinopilos, an old polje that is now an impressive terra-rosa formation on the left bank of the Louros river in NW

Greece. Two hand axes have been reported at this site (Runnels & van Andel, 1993; Ammerman et al., 1999). A few more, smaller bifaces, possibly deriving from Middle Palaeolithic contexts, have also been found: one at Krovili, in the Rhodope province of Aegean Thraki (Ammerman *et al.*, 1999), some small ovate tools at Tripotamia and Kastro, in the Peloponnisos (Reisch, 1982), and others at Punikve in Croatia (Malez, 1979).

Judging by archaeological finds in other parts of Europe such as Germany and Britain, *H. heidelbergensis* used tools made of wood, antler and stone. The stone industries associated with this hominid generally fall within the Acheulean tradition, whereas choppers and chopping tools are traditionally considered the hallmarks of hominids that preceded *Homo heidelbergensis* in phylogenetic terms. Few choppers and chopping tools have been found in the Balkans and more often than not those we do have lack properly dated stratigraphic provenience. Certainly they cannot be considered a reliable data set permitting the hypothesis that *H. heidelbergensis* was not the earliest hominid to inhabit the peninsula. The questions that naturally arise are whether such tools were used by other hominid species that are not represented in the fossil record, whether the tools represent the survival of old technological traditions into a later period or whether tools of this sort were later developed anew and used by other species in response to the nature of the lithic raw materials available to them. None of these possibilities can be excluded, since the Balkan data, like those from the rest of Europe, are sparse and inconclusive. The matter is further complicated by the existence of lithic assemblages that contain elements of more than one technological tradition. Many Mousterian assemblages, for example, contain large numbers of bifaces and chopping tools (Panagopoulou, 1999). In the light of recent discussion of biological and cultural variation, it is now widely accepted that neither the biface nor the chopper can be used as a type fossil (in other words, a tool characteristic of a single species, culture or date). The Acheulean industries in this part of the world cannot, moreover, be presumed to have been preceded by those consisting of simple flakes, choppers and cores, since early dates have been assigned to some bifaces found in this area, while certain choppers are of later date (Gowlett, 1999). Similar conclusions have been reached in other parts of Europe. In Britain, for example, Clactonian industries have recently been shown both to pre-date and to post-date Acheulean biface industries (Roberts *et al.*, 1995). It is thus not improbable that Acheulean and chopper-flake industries should have existed side by side in time and space in the Balkans, and the question of whether or not they were manufactured and used by *H. heidelbergensis* alone remains open. What we are perhaps seeing in this early record is what Gowlett has called 'dancing of industries' (Gowlett, 1999: 48). Many more modern excavations of clearly stratified sites are needed to clarify the picture.

From at least 130 k yr. BP onwards the Balkans were inhabited by human groups that used Mousterian artefacts, as is shown by the early Krapina dates that we have already discussed and by the sequence from Asprochaliko, in NW Greece, whose deepest strata have been TL dated to 102±14 k yr. BP (Huxtable *et al.*, 1992). It is not clear which type of hominid produced these industries at some sites. Recent evidence from the Near East suggests that for some thousands of years Neanderthals co-existed with anatomically modern humans. We do not know whether or not the two interacted, but they certainly shared the same industrial technocomplex. Sites such as Krapina and Vindija, where Mousterian artefacts have been found in contexts containing Neanderthal bones, give us good reason to hypothesise that

Neanderthals created the material culture recovered there. The Mousterian lithic assemblages recovered from some other sites cannot be assigned to either species. Mousterian signatures have been found in numerous cave sites in Greece (Asprochaliko, Kalamakia and Theopetra), in Croatia (Krapina, Mujina pećina and Vindija), in northern Bosnia, in Slovenia, in Bulgaria, in Turkey and in Romania (Bordul Mare and Curata) that have not necessarily yielded hominid remains.

To this sample we may add a large number of unstratified or poorly stratified open-air Mousterian sites where dating and preservation of organic remains are problematic. Elements of earlier technological traditions (such as bifaces) or of later ones (such as backed bladelets) are often unearthed at these sites (at Kokkinopilos, in NW Greece, for example [Dakaris *et al.*, 1964]). Our picture of these open sites suffers from interference. The fundamental question that remains unanswered is whether they are merely palimpsests of temporally separate events of use by various hominids or whether they are evidence of different technical traditions coexisting in a single lithic industry.

ADAPTATION

It is widely agreed that the Balkans and Gibraltar must have been the gateways to Europe used by early human migrants. The Balkan Peninsula is an area of extremely diverse landscapes, with climatic conditions varying from the maritime climate of the Mediterranean coast to the continental conditions of the north and east. Within this general trend may be found numerous microclimates whose temperatures and rainfall differ markedly despite their physical proximity to one another (Bailey & Gamble, 1990). The mountain ranges in the Balkans, which reach heights of about 3000 m a.s.l., may have delayed the expansion of early humans to the west and north, but they certainly did not prevent it; the mosaic habitats encountered in this area offered conditions that would probably have been favourable to the survival and continuity of the various hominid species.

An important marker of how well a species has adapted to a habitat is population density. Out of all the periods of human prehistory, however, the Palaeolithic is the one over whose population densities there hangs the largest question mark. Although at one time or another various formulae and numbers have been proposed (*e.g.* Sturdy & Webley, 1988), these have been based on generic models of environmental sustainability borrowed from the biological sciences. This approach neglects humanity's most vital weapon: social organisation. Social organisation cannot be quantified; forming any estimate whatsoever of the population densities represented by the early record thus remains an inherent difficulty.

We know very little about how well *Homo heidelbergensis* adapted to the Balkans. This is because many uncertainties still surround the material culture associated with this species. Forty years after the cave of Petralona was discovered, deficiencies in the publication of its archaeological finds have left it merely a palaeontological site, rather than an archaeological one. The number of Mousterian sites found and the diversity of their locations lead us to suspect that the Neanderthals and early modern humans may have been better adapted to the Balkan landscape than was *H. heidelbergensis*. The anatomically modern human populations of the Pleistocene probably were too. It should be remembered, however, that neither here nor elsewhere did adaptation develop linearly, with numbers of settlements and population density increasing steadily over the course of time. The climatic changes of the last 100,000 years and the loss of former habitats such as the productive Adriatic plain that today lies submerged under the Adriatic Sea, to which human groups had become accustomed and adapted over the course of many millennia, must have had a significant impact upon human population densities and survival strategies.

As we have already said, the Balkan landscape is of considerable variety, essential to biodiversity. Its mosaic of heterogeneous biotopes, some naturally isolated and thus protected, may explain how different hominid species could have managed to coexist (initially *Homo heidelbergensis* with *H. neanderthalensis*, if the late Middle Pleistocene dates derived for the Krapina and Petralona hominids are valid, and later *H. neanderthalensis* with *H. sapiens*, as we observed in the previous section).

It must also have favoured the adaptation of individual hominid species. This is best seen in the record left by anatomically modern humans, who survived challenges such as periods of climatic change by making use of the many alternative habitats offered by this landscape. During the process of de-glaciation, for example, when the resources exploited by modern human groups before the Last Glacial Maximum changed significantly, modern humans were able to expand into previously unexplored mountainous areas. During this period new camps associated with specialised ibex and chamois hunting sprang up in previously inaccessible upland locations. Klithi and Megalakkos, two sites in the Voidomatis river valley in NW Greece, are a good example of this sort of adaptation (Bailey, 1997).

CONCLUSIONS

The discussion of ancient hominid biodiversity has two closely related parts: the reconstruction of ancient hominid morphology and the reconstruction of social, economic and symbolic behaviour. Bringing together these two different lines of study demands a continuous interaction between archaeologists, palaeontologists, geneticists, and palaeoecologists. Here Palaeolithic archaeology meets biology. Unlike biologists, who work directly with living or extinct species, archaeologists research by proxy. We come to grips with past human societies by excavating and studying the material culture they have left behind: their artefacts, their settlement structures and their food residues, along with the occasional skeletal remain.

For most of the 20th century the two components of ancient hominid biodiversity were researched more or less independently of one another, biologists exploring pure phylogeny and pure morphology while archaeologists studied the various hominid species' behavioural hallmarks. Since the 1990s, however, discoveries and developments in palaeoanthropological research have raised many questions about the fundamental nature of the differences and similarities between species (Trinkaus & Shipman, 1994). An interdisciplinary and multidisciplinary approach in which archaeologists, palaeontologists, geneticists and palaeoecologists all cooperate is vital if the two parts of the human biodiversity puzzle are to be put together. Our aim must be to build a global view out of detailed local records. In line with this aim, the present contribution has pieced together the main lines of evidence for the presence of ancestral hominid populations in the Balkans and highlighted the questions that have arisen from these.

For many years biologists and archaeologists, and through them the world at large, assumed that the various hominid species developed linearly through processes of progressive change, extinction and replacement. It has now been recognised that in most instances the linear view is not correct. The continual discovery of new finds and of new methodological refinements makes palaeoanthropology one of the fastest changing areas in science. The palaeoanthropology of the Balkans is no exception. Our uncertainties about this area's record are part of a far wider uncertainty as to how events of this sort took place throughout the world.

I began this section by pointing out the main difference between Palaeolithic archaeologists and other researchers in biodiversity. I shall end it with the features that they share.

Firstly, present-day national borders are irrelevant to our research. More often than not, indeed, they are an obstacle to a thorough understanding of the regional patterns of the human past. The populations we study, the human groups of the Pleistocene, were highly mobile throughout the vast expanses of southeast Europe. It was only during the last part of the Stone Age, the Neolithic, that human groups settled down and ethnic traits began to show up in the archaeological record. Although collaboration and the transmission of information across borders are absolutely essential to our research, we are hampered by our different archaeological traditions and schools of thought, by socio-political factors to do with modern Balkan history and of course by language barriers.

The second similarity has to do with our objects and angles of research. Like other animal populations, the human groups of the Palaeolithic spent many thousands of years foraging. They remained highly dependent on the ecosystem in which they lived and must have developed various strategies to manage the resources they lived off. The process of actively modifying natural resources by clearing vegetation for large-scale agriculture and by domesticating and systematically exploiting what were originally wild plants and animals appeared later on in Balkan prehistory, as the Holocene advanced. Biologists and archaeologists alike are thus centrally concerned with adaptation in populations highly dependent on environmental conditions and change. We share, moreover, an interest in the relations between human and animal, between human and plant and between predator and prey. Just what impact, for instance, did human strategies have upon the natural environment of the Pleistocene? Although the traditional view in both archaeology and theoretical biology would hold that Pleistocene humans imposed no undue pressure on natural communities, recent computer modelling of human ecodynamics during the Palaeolithic does not support this view (Winder, 1997). It appears that there is open ground for an interchange of ideas, hypotheses and methods between the two disciplines.

Finally, just as most biologists in this area are actively concerned with conserving the local fauna and flora and with the future of the Balkans' natural heritage, archaeologists here are concerned with conserving a cultural heritage. The fact that certain threatened natural resources have been saved because they are close to significant archaeological sites and *vice versa* is a potent symbol. The Palaeolithic record affords us a long perspective on human history. Learning from the past, archaeologists are able to appreciate the fragility of the ecosystem in which we live. Let us not forget that the Pleistocene witnessed the rise and extinction of many different hominid, animal and plant species. Communication between prehistorians and those whose interest is biodiversity can only have positive results.

REFERENCES

Ammerman, A.J., Efstratiou, N. & Adam, E. 1999. First evidence for the Palaeolithic in Aegean Thrace. In: *The Palaeolithic of Greece and adjacent areas. Proceedings of the ICOPAG conference* (G.N. Bailey, E. Adam, E. Panagopoulou, C. Perlès & K. Zachos, eds.), pp. 211-214. British School at Athens Studies 3, London.

Arsebük, G. & Özbaşaran, M. 1999. Pleistocene archaeology at the Cave of Yarimburgaz in Eastern Thrace/Turkey: preliminary results. In: *The Palaeolithic of Greece and adjacent areas. Proceedings of the ICOPAG conference* (G.N. Bailey, E. Adam, E. Panagopoulou, C. Perlès & K. Zachos, eds.), pp. 59-72. British School at Athens Studies 3, London.

Bailey, G.N. 1997. Klithi: a synthesis. In: *Klithi: Palaeolithic settlement and Quaternary landscapes in northwest Greece. Klithi in its local and regional setting* (G. Bailey ed.), pp. 655-677. MacDonald Institute for Archaeological Research, Cambridge.

Bailey, G.N & Gamble, C. 1990. The Balkans at 18,000 BP: the view from Epirus. In: *The world at 18,000*, (C.S. Gamble & O. Soffer, eds), pp. 148-167. Unwin-Hyman, London.

Beaumont, P., de Villiers, H. & Vogel, J. 1978. Modern man in Sub-Saharan Africa prior to 49,000 years BP: a review and evaluation with particular reference to Border Cave. *South African Journal of Science* 74, 409-19.

Brooks, A.S. 1996. Behavioral perspectives on the origin of modern humans: another look at the Africal evidence. In: *The Lower and Middle Palaeolithic. Colloquia of the XIII International Congress of Prehistoric and Protohistoric Sciences Forli 8-14 September 1996* (O. Bar-Yosef, L.L. Cavalli-Sforza, R.J. March & M. Piperno, eds.), pp. 157-166. A.B.A.C.O, Forli.

Cann, R., Stoneking M. & Wilson, A.C. 1987. Mitochondrial DNA and human evolution. *Nature* 325, 31-36.

Dakaris, S.I., E.S. Higgs & Hey, R.W., 1964. The climate, environment and industries of Stone Age Greece: Part I. *Proceedings of the Prehistoric Society* 30, 199-244.

Darlas, A. 1995. The lithic industry found with the skeleton LAO 1/S3 (Apidima, Mani) (in greek). *Acta Anthropologica* 1, 59-62.

Fistani, A.B. 1993. Human evolution in Albania for the Quaternary period. In: *Before the Wall Fell: the Science of Man in Socialist Europe* (B.A. Sigmon, ed.), pp.141-178. Canadian Scholars' Press, Toronto.

Gamble, C. 1997. The animal bones from Klithi. In: *Klithi: Palaeolithic settlement and Quaternary landscapes in northwest Greece.Excavation and intra-site analysis at Klithi* (G. Bailey ed.), pp. 207-244. MacDonald Institute for Archaeological Research, Cambridge.

Gamble, C. 1999. *The Palaeolithic Societies of Europe*. Cambridge University Press, Cambridge.

Goldstein, D.A., Linares, A.R., Cavalli-Sforza, L.L. & Feldman, M.W. 1995. Genetic Absolute dating based on microsatellites and the origin of modern humans. *Proceedings of the National Academy of Sciences* 92, 6723-6727.

Golovanova L.V., Hoffecker, J.F., Kharitonov, V.M. & Romanova, G.P. 1999. Mezmaiskaya Cave: A Neanderthal Occupation in the Northern Caucasus. *Current Anthropology* 40, 77-86.

Gorjanović-Kramberger, D. 1902. Der paläolitische Mensch und seine Zeitgenossen aus dem Diluvium van Krapina in Kroatien II, *Mitteilungen der Anthropologischen Gesellschaft in Wien* 31, 189-216.

Gorjanović-Kramberger, D. 1906. *Der diluviale Mensch von Krapina in Kroatien: ein Beitrag zur Paläoanthropologie.* Studien über die Entwicklungsmechanik des Primatenskelletes, Wiesbaden.

Gowlett, J.A.J. 1999. The Lower and Middle Palaeolithic, transition problems and hominid species: Greece in broader perspective. In: *The Palaeolithic of Greece and adjacent areas. Proceedings of the ICOPAG conference* (G.N. Bailey, E. Adam, E. Panagopoulou, C. Perlès & K. Zachos, eds.), pp. 43-58. British School at Athens Studies 3, London.

Grün, R. 1996. A re-analysis of electron spin resonance dating results associated with the Petralona hominid. *Journal of Human Evolution* 30, 227-241.

Harpending, H.C., Sherry, S.T., Rogers A. R. &. Stoneking, M. 1993. Genetic structure of ancient human populations. *Current Anthropology* 34, 483-496.

Hawks, J. & Wolpoff, M.H. 2001. Paleoanthropology and the population genetics of ancient genes. *American Journal of Physical Anthropology* 114, 269-272.

Hennig, G.J., Herr, W., Weber, E. & Xirotiris, N.I. 1982. Petralona cave dating controversy. *Nature* 299, 281-282.

Huxtable, J., Gowlett, J.A.J., Bailey, G.N., Carter, P.L. & Papaconstantinou V. 1992. Thermoluminescence dates and a new analysis of the early Mousterian from Asprochaliko. *Current Anthropology* 33, 109-114.

Ikeya, M. 1980. ESR dating of carbonates at Petralona Cave, *Anthropos* 7, 143-151.

Karavanić, I. 1995. Upper Palaeolithic occupation levels and late-occuring Neandertal at Vindija cave (Croatia) in the context of central Europe and the Balkans, *Journal of Anthropological Research* 51, 9-35.

Karavanić, I. 2000. Les recherches de Dragutin Gorjanović-Kramberger à Krapina et les débuts de la préhistoire en Croatie, *Praehistoria* 1, 25-29.

Karavanić, I. & Smith, F.H. 1998. The Middle/Upper Paleolithic interface and the relationship of Neanderthals and early modern humans in the Hrvatsko Zagorje, Croatia. *Journal of Human Evolution* 34, 223-248.

Koumouzelis. M., Ginter, B., Kozlowski, J.K., Pawlikwski, M., Bar-Yosef, O., Albert, R.M., Litynska-Zajac, M., Stworzwicz, E., Wojtal, P., Lipecki, G., Tomek, T., Bochenski, M. & Pazdur A. 2001. The Early Upper Palaeolithic in Greece: The excavations in Klisoura Cave. *Journal of Archaeological Science* 28 (5), 515-539.

Kourtessi-Philippakis, G. 1999. The Lower and Middle Palaeolithic in the Ionian islands. In: *The Palaeolithic of Greece and adjacent areas. Proceedings of the ICOPAG conference* (G.N. Bailey, E. Adam, E. Panagopoulou, C. Perlès & K. Zachos, eds.), pp. 282-287. British School at Athens Studies 3, London.

Kozlowski, J.K. 1982. *Excavation in the Bacho Kiro cave (Bulgaria) Final Report.* Panstwowe Wydawnictwo Naukowe, Warszawa.

Kozlowski, J.K. 1996. Cultural context of the last Neanderthals and Early Modern Humans in Central-Eastern Europe. In: *The Lower and Middle Palaeolithic. Colloquia of the XIII International Congress of Prehistoric and Protohistoric Sciences Forli 8-14 September 1996* (O. Bar-Yosef, L.L. Cavalli-Sforza, R.J. March & M. Piperno, eds.), pp. 205-218.A.B.A.C.O., Forli.

Kretzoi, M. & Poulianos, N.A. 1981. Remarks on the Middle and Lower Pleistocene vetebrate fauna in the Petralona cave. *Anthropos* 8, 57-72.

Kricun, M., Monge, J., Mann, A., Finkel, G., Lampl, M. & Radovčić, J. 1999. *The Krapina Hominids, A Radiographic Atlas of the Skeletal Collection.* Croatian Natural History Museum, Zagreb.

Krings, M., Stone, A., Schmitz, R.W., Krainitzki, H., Stoneking, M., & Paabo, S. 1997. Neandertal DNA sequences and the origin of modern humans, *Cell* 90, 19-30.

Krings, M., Capelli, C., Tschentscher, F., Geisert, H., Meyer, S., von Haesler, A. Grossschmidt, K., Possnerrt, G., Paunovic, M. & Paabo, S. 2000. A view of Neandertal diversity, *Nature Genetics* 26, 144-146.

Kurtén, B. & Poulianos, N.A. 1977. New stratigraphic and faunal material from Petralona cave with special reference to the Carnivora. *Anthropos* 4, 47-130.

Kurtén B. & Poulianos N.A. 1981. Fossil carnivora of Petralona cave (status 1980). *Anthropos* 8, 9-56.

Latham, A.G. & Schwarcz, H. 1992. The Petralona hominid site: uranium-series re-analysis of "Layer 10" calcite and associated palaeomagnetic analyses. *Archaeometry* 34, 135-140.

Malez, M. 1974. Noviji rezultati istraživanja paleolitika u Velikj pećini, Veternici I Šandalji. *Arheološki radovi i rasprave Jugosl. Akad. Znan. Umjet.* 7, 7-44

Malez, M. 1975. Die Höhle Vindija: Eine neue Fundstelle fossiler Hominiden in Kroatien. *Bull. Sci.Conceil Acad. Sci. Arts RSF Yougoslavie*, Section A 20, 139-141.

Malez, M. 1979. Nalažista paleolitskog I mezolitskog doba u Hrvatskj. In: *Praistorija jugoslavenskih zemalja 1*, (A. Benac, ed.), pp. 227-276. Svjetlst, Sarajevo.

Malez, M., Smith, F.H., Radovčić, J. & Rukavina, D. 1980. Upper Pleistocene Hominids from Vindija, Croatia, Yugoslavia. *Current Anthropology* 21, 365-367.

Maynard, C. 1996. *Stone Age Times*. Kingfisher, London.

Miracle, P.T. in press. *The Krapina Neandertal Site: Zooarchaeology, Taphonomy, and Catalog of the Faunal Remains.*

Mompherratou, E. & Pitsios, T.1995. Apidima Cave Gamma (Γ): The burial of female skeleton ΛΑΟ1/Σ3. *Acta anthropologica* 1, 27-51 (in Greek with an English abstract).

Ovchinnikov, I.V., Gotherstrom, A., Romanova, G.P., Kharitov,V.M., Liden, K. & Goodwin, W. 2000. Molecular analysis of Neanderthal DNA from the northern Caucasus. *Nature* 404, 490.

Panagopoulou, E. 1999. The Theopetra Middle Palaeolithic assemblages: their relevance to the Middle Palaeolithic of Greece and adjacent areas. In: *The Palaeolithic of Greece and adjacent areas. Proceedings of the ICOPAG conference* (G.N. Bailey, E. Adam, E. Panagopoulou, C. Perlès & K. Zachos, eds.), pp. 252-265. British School at Athens Studies 3, London.

Perlès C. 1992. In search of lithic strategies: a cognitive approach to Prehistoric chipped stone assemblages. In: *Representations in Archaeology* (J-C. Gardin & C.S. Peebles eds.), pp. 223-47. Indiana University Press, Bloomington Indianapolis.

Pitsios, T.K. 2000. Ο Ταινάριος Άνθρωπος: *Ανθρωπολογικές Έρευνας στη Μέσα Μάνη.* Ελληνικά Γράμματα, Αθήνα [in Greek].

Pitsios, T.K. & Liebhaber, B. 1995. Research conducted in Apidima and the surrounding region – Taenarios Man-, *Acta Anthropologica* 1, 171-179 [in Greek with an English abstract].

Poulianos,A.N. 1971. Petralona A Middle Pleistocene Cave in Greece, *Archaeology* 24, 6-11.

Poulianos, A. N. 1978. Τα αρχαιότερα μέχρι σήμερα λίθινα εργαλεία στα Πετράλωνα. *Anthropos* 5, 74-80 [in Greek].

Poulianos, A.N. 1982. *Το σπήλαιο του αρχανθρώπου των Πετραλώνων*. Βιβλιοθήκη της Ανθρωπολογικής Εταιρείας της Ελλάδας 3, Αθήνα [in Greek].

Radovčić, J., F.H. Smith, E. Trinkaus, M.H. Wolpoff 1988. *The Krapina Hominids: An Illustrated Catalog of Skeletal Collections*. Mladost, Zagreb.

Reisch, L. 1982. The transition from Lower to Middle Palaeolithic in Greece and the Southern Balkans. In: *The Transition from Lower to Middle Palaeolithic and the Origin of Modern Man* (A. Ronen, ed), pp. 223-231. British Archaeological Reports International Series 151, Oxford.

Richards, M.P., Pettitt, P.B., Trinkaus, E., Smith, F.H., Paunović, M. & Karavanić I. 2000. Neanderthal diet at Vindija and Neanderthal predation: the evidence from stable isotopes. *Proceedings of the National Academy of Sciences USA* 97, 7663-7666.

Rink, W. J., Swarcz, H. P., Smith, F.H., Radovčić, J. 1995. ESR ages for Krapina hominids. *Nature* 378, 24.

Roberts, M.B., Gamble, C.S. & Bridgland D.R. 1995. The earliest occupation of Europe: the British Isles. In: *The Earliest occupation of Europe: Proceedings of the European Science Foundation Workshop at Tautavel (France) 1993. Analecta Praehistorica Leidensia 27.* (W. Roebroeks & T. van Kolfschoten, eds.). University of Leiden, Leiden.

Roebroeks, W. & van Kolfschoten T. 1994. The earliest occupation of Europe: a short chronology. *Antiquity* 68, 489-503.

Rogers, A. R. & Jorde, L.B. 1995. Genetic evidence on modern human origins. *Human Biology* 7, 1-36.

Runnels, C. & van Andel, T.H. 1993. A handaxe from Kokkinopilos, Epirus and its implications for the Palaeolithic of Greece. *Journal of Field Archaeology* 20, 191-203.

Runnels, C. & van Andel, T.H. 1999. The Palaeolithic in Larisa, Thessaly. In: *The Palaeolithic of Greece and adjacent areas. Proceedings of the ICOPAG conference* (G.N. Bailey, E. Adam, E. Panagopoulou, C. Perlès & K. Zachos, eds.), pp. 215-220. British School at Athens Studies 3, London.

Russell, M.D. 1986a. Bone breakage in the hominid collection. *American Journal of Physical Anthropology* 72, 373-379.

Russell, M.D. 1986b. Mortuary practices at the Krapina Neanderthal site. *American Journal of Physical Anthropology* 72, 381-397.

Schoetensack, O. 1908. *Der Unterkiefer des* Homo heidelbergensis *aus den Sanden von Mauer bei Heidelberg*. Wilhelm Engelmann, Leipzig.

Shen, G. & Yokoyama, Y. 1986. T-230/U-234 Dating of the Petralona Speleothems, *Anthropos* 11, 23-32.

Simek, J.F. 1991. Stone tool assemblages from Krapina (Croatia, Yugoslavia). In: *Raw Material Economies among Prehistoric Hunter-Gatherers*, (A. Montet-White & S. Holen, eds.), pp. 58-71. Publications in Anthropology 19, University of Kansas, Lawrence.

Simek, J.F. & Smith, F.H. 1997. Chronological changes in stone tool assemblages from Krapina (Croatia). *Journal of Human Evolution* 32: 561-575.

Smith, F.H. 1976. A fossil hominid frontal from Velika Pecina (Croatia) and a consideration of Upper Pleistocene hominids from Yugoslavia. *American Journal of Physical Anthropology* 44, 127-134.

Smith, F.H. 1982. Upper Pleistocene Hominid Evolution in South-Central Europe: a review of the evidence and analysis of trends. *Current Anthropology* 23, 667-703.

Smith, F.H., Trinkaus, E., Pettitt, P., Karavanić, I. & Paunović M. 1999. Direct radiocarbon dates from Vindija G1 and Velika Pecina Late Pleistocene hominid remains, *Proceedings of the National Academy of Sciences, U.S.A.* 96, 12281-12286.

Stringer, C. & Gamble, C. 1993. *In Search of the Neanderthals*. Thames and Hudson, London.

Stringer, C. & MacKie, R. 1996. *African Exodus. The Origins of modern humanity*. Jonathan Cape, London.

Stravopodi, E. & Manolis, S.K. 2000. The bioarchaeological profile of the Theopetra Cave: a pilot project in Greece. In: *Theopetra Cave: Twelve years of excavation and research 1987-1998. Proceedings f the Internatinal Conference*, Trikala 6-7 November 1998 (N. Kyparissi-Apostolika, ed.), pp.93-108. Athens.

Sturdy, D.A. & Webley, D.P. 1988. Palaeolithic geography: or where are the deer? *World Archaeology* 19, 262-280.

Tattersall, I. 1986. Species recognition in human paleontology. *Journal of Human Evolution* 15, 165-175.

Tattersall, I. 1999. *The Last Neanderthal*. Westview Press, Boulder - Colorado.

Templeton, A. 2002. Out of Africa again and again. *Nature* 416, 45-51.

Thorne, A.G. & Wolpoff, M.H. 1992. The multiregional evolution of humans. *Scientific American* 266, 76-83.

Trinkaus, E., 1985. Cannibalism and Burial at Krapina. *Journal of Human Evolution* 14, 203-216.

Trinkaus, E. & Shipman, P. 1994. *The Neandertals: Changing the Image of Mankind.* Pimlico, London.

Valoch, K. 1995. The earliest occupation of Europe: Eastern Central and Southeastern Europe. In: *The Earliest occupation of Europe: Proceedings of the European Science Foundation Workshop at Tautavel (France) 1993. Analecta Praehistorica Leidensia 27.* (W. Roebroeks & T. van Kolfschoten, eds.), pp.165-191. University of Leiden, Leiden.

Winder, N. 1997. Dynamic modelling of an extinct ecosystem: refugia, resilience and the overkill hypothesis in Palaeolithic Epirus. In: *Klithi: Palaeolithic settlement and Quaternary landscapes in northwest Greece. Klithi in its local and regional setting* (G. Bailey ed.), pp. 615-636. MacDonald Institute for Archaeological Research, Cambridge.

Yellen, J.E., Brooks, A. S., Cornelissen, E., Mehlman, M.J. & Stewart, K. 1995. A Middle Stone Age Worked Bone Industry from Katanda, Upper Semliki Valley Zaire. *Science* 268, 553-556.

10. THE BIODIVERSITY OF AMPHIBIANS AND REPTILES IN THE BALKAN PENINSULA

GEORG DŽUKIĆ AND MILOŠ L. KALEZIĆ*

"The northern Adriatic region, the south-eastern region from the Alps to the Carpathian Mts., Dalmatia and Albania, as well as their neighbouring seas, have a particular biogeographic significance, which I will explain in more detail elsewhere along with the Pendulation theory. Here, we have an ancient corner of Europe and one of the most interesting parts of the whole World." (Simroth, 1907; our translation)

INTRODUCTION

Biologists have long recognised that the circum-Mediterranean region, and the Balkans in particular, contains an extraordinary variety of life forms in both plant and animal groups (Gaston & David, 1994), which is expressed in terms of biogeographical diversity, including that of amphibians and reptiles. The Mediterranean region of Europe comprises three large southern peninsulas and their archipelagoes. The amphibians and reptiles in this part of Europe exhibit complex biogeographic patterns, with large numbers of endemic taxa, dynamism, and the presence of certain general biological phenomena (see below). Distinguished by a high level of heterogeneity and of mosaicism, and natural fragmentation due to mountain ranges, the Balkan Peninsula is particularly intriguing in this respect.

The biogeographical uniqueness of the peninsula is indisputable. The Balkans are delimited by a strong boundary (which occurs between two regions with significantly different biotas) separating them from the rest of Europe in regard to the distributional pattern of lizards (Ramirez *et al.*, 1992), for example, but are less clearly delimited in terms of the quantitative biogeography of amphibians (Real *et al.*, 1992). It is therefore remarkable that this is the least-studied batrachofauna and herpetofauna of Europe. Publications which deal with particular groups of amphibians and reptiles of the entire Balkan Peninsula are scarce (*e.g.* Radovanović & Martino, 1950), as are general remarks on the Balkan batrachofauna and herpetofauna (Džukić, 1993; Džukić *et al.*, 2001; Borkin, 1998).

The intention of this paper is to assess current evidence for batrachological and herpetological diversity in the Balkans. Because of their limited dispersal ability and temperature dependence, terrestrial and freshwater amphibians and reptiles can be highly sensitive indicators for the study of biogeographic pattern and process. The specific aim of this paper is to provide a holistic treatise on amphibians and reptiles of the Balkan Peninsula, rather than focusing on zoogeographical issues *per se*. Various aspects of diversity are discussed in comparison with statistics for other European faunas of these vertebrates, from the Iberian and Apennine peninsulas in particular. The origins of the present day biodiversity of the Balkan batrachofauna and herpetofauna are also considered. Finally, it is demonstrated that in some general biological phenomena such as hybridisation and hybridogenesis, interspecific and intraspecific heterochronic phenomena, and switching of

H.I. Griffiths et al. (eds.), Balkan Biodiversity, 167–192.

reproductive modes within a single species, Balkan amphibians and reptiles are distinct in comparison with other European areas.

PHYSICAL GEOGRAPHY OF THE STUDY REGION

In contrast to the Iberian and Apennine peninsulas, the Balkan Peninsula is not separated physically from the European mainland by mountain barriers. As a result, the northern and, especially, western boundaries of the Balkan Peninsula have never been demarcated precisely. Some suggest that the tributaries of the Black Sea hydrological system which are closest to the Adriatic shore (running in a west-east direction) should be followed. The Kupa-Sava-Danube border is constructed in this way, wherein Rijeka, on the Adriatic coast, is taken to be the westernmost mark. The resultant designated area covers 490,000 km^2. Others define a dividing line between the Alps and Dinarides (Dinaric Alps) as a northwestern border. This border follows the Sava River up to the mouth of the Krka and then runs up the Krka to its source. Thereafter, the Vipava and Soča Rivers are followed to their mouths in the Bay of Trieste. The area of the Balkan Peninsula defined in this way comprises 520,000 km^2.

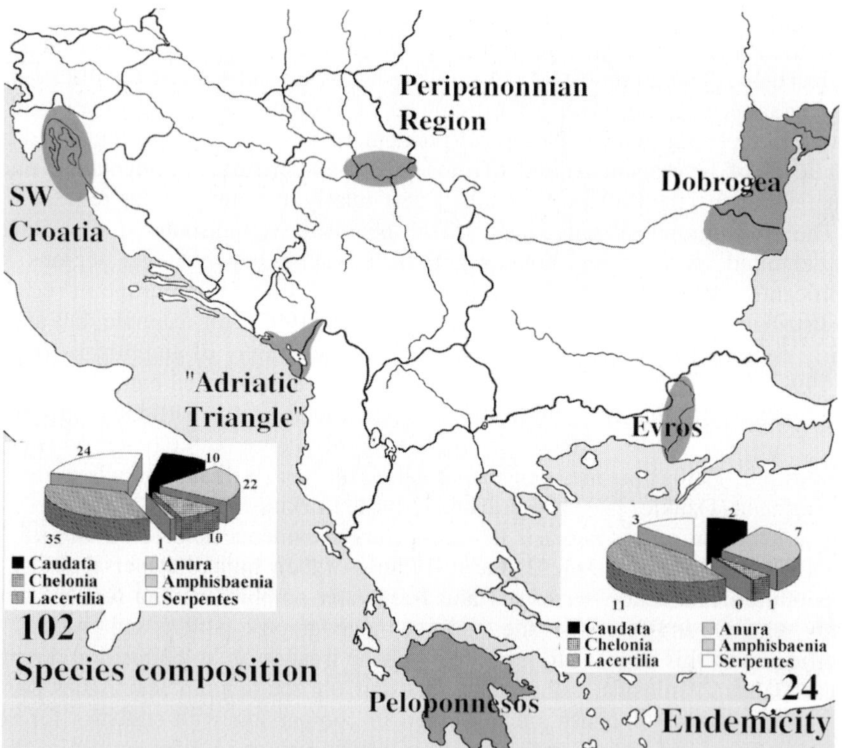

Figure 1. Map showing the location of the six Balkan study regions, and pie diagrams of the presence of Balkan amphibian and reptile species, and their endemicity.

In order to assess differences in species richness and levels of endemicity, the following six regions were selected on a rather arbitrary basis: SW Croatia, the Peripannonian region, Dobrugea, the "Adriatic Triangle", the Peloponnisos, and Evros. It was assumed that these regions were large enough to include all relevant attributes of diversity across the Balkans as a whole (Figure 1). The designation of regions was based mainly on their geography, and orography in particular; the altitudinal range of an area was considered to be highly significant. Additional criteria included the variability of habitats and biomes, biogeographical connections with other regions, and previous proposals that some of them be designated hot spots of diversity in this part of Europe and/or set aside as biogenetic reserves (Simroth, 1907; Corbett, 1989). It should be noted that these regions are not zoogeographic entities in any sense. Some of the regions are physiognomically well separated, such as Dobrugea and the Peloponnisos, whilst others (SW Croatia and the Peripannonian, "Adriatic Triangle", and Evros regions) are closely adjacent.

The largest of these six areas is the Peloponnisos region (Peloponnisos and its nearest islands), with an area of 30,200 km^2. Dobrugea includes parts of Romania and Bulgaria and has an area of 25,800 km^2. With an area of 15,200 km^2, SW Croatia consists of the Kvarner area, the eastern part of Istria, and the Velebit and Kapela Mountains. As the northernmost part of the Balkans, the Peripannonian region includes the Central-Balkan edge of the Pannonian Plain, as well as the region's hilly areas and river valleys, all of which cover 10,000 km^2. The "Adriatic Triangle" consists of the Prokletije Mountain Massif and the lowland Lake Skutari (= Skadar) area, with an area of 7,500 km^2. The Evros is the smallest region (6,700 km^2), comprising the lower valley of the Marica River and adjacent areas, including the Evros reservation. With the exception of the Peripannonian region, each of these regions includes a maritime coastal zone.

TAXONOMIC DIVERSITY AND PATTERNS OF SPATIAL DISTRIBUTION

Amphibians and reptiles occur throughout the Balkan Peninsula, regardless of altitude and latitude. Their wide distribution is a function both of the moderate climate (whose characteristics are well within the tolerance limits of the fauna) and the high adaptive ability of amphibians and reptiles which allows them to colonise a variety of biomes. They live either cryptically in secluded places on land or visibly in the open, and some species alternate between these two modes of existence during their life cycle. Several are known to be troglophilous and trogloxenous, including those which are fossorial.

Species and subspecies diversity

The Balkans harbour many phylogenetic lineages of reptiles and amphibians, some of which have undergone extensive radiation, which has probably occurred *in situ*. Our current annotated taxonomic list (Appendix) includes 104 species: 33 amphibians and 71 reptiles. Lizards are the most species-rich group, followed by the snakes, tailless amphibians, turtles, and tailed amphibians; amphisbaenians are very rare.

To the best of our knowledge, two species have been introduced to the Balkan Peninsula (a frog *Rana catesbeiana* and a turtle *Trachemys scripta*), probably around 30 years ago or more. It has also been speculated that some taxa in the

Mediterranean region, especially on islands and probably including the Balkan Peninsula, are of anthropochorous origin (Corti *et al.*, 1997; 1999). The African chameleon (*Chamaeleo africanus*), for example, was probably introduced into Greece (the Peloponnisos) during the Mycenaean era between 3200 - 3500 B.C. (Böhme *et al.*, 1998), being kept as pets by citizens and kings. The Romans might also have had an influence on native herps, by introducing the Aesculapian snake (*Elaphe longissima romana*) and possibly others for religious rituals in temples. This snake was worshipped by followers of the cult of the god Aesculapius during Roman times all over the Balkans, especially in eastern Serbia, where specimens with the characteristics of *E. l. romana* have been found unexpectedly (see Appendix).

As would be expected, species diversity is not distributed evenly across the peninsula. In regard to overall species richness (amphibians + reptiles), the "Adriatic Triangle" appears to be the most species-rich, with 15 amphibians and 39 reptiles (=54 species; Figure 2; Appendix). The southernmost part of the Balkans, the Peloponnisos, has 11 amphibians and 38 reptiles (49 species). Southwestern Croatia has similar species richness, but with more amphibians (16 species) and fewer reptiles (33 species). The Evros area has 12 amphibians and 33 reptiles (45 species). The least species-richness region is the Peripannonian, where amphibians outnumber reptiles (19 and 16 species, respectively), and the most northeastern part of the Balkans (Dobrugea), with 12 amphibians and 23 reptiles (35 species; Figure 2; Appendix).

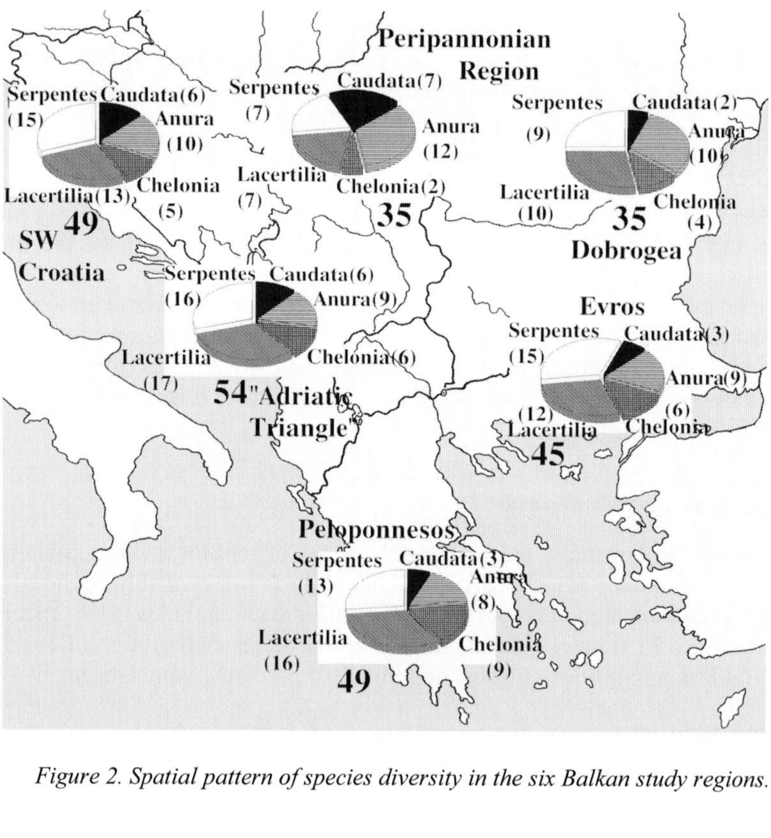

Figure 2. Spatial pattern of species diversity in the six Balkan study regions.

There is no correlation between number of species and surface area. The Spearman's nonparametric correlation coefficients between these two variables are negative, but statistically insignificant (rs = -0.36 for amphibians, rs = -0.01 for reptiles, $P > 0.05$). The greatest disproportion between number of species and region size appears in the case of Peloponnisos for the amphibians, where the largest region has the lowest number of species. This relationship is probably due to ecological factors, since the region is climatically unfavorable for amphibians. For reptiles, the greatest disproportion occurs in the case of the "Adriatic Triangle", the fifth region in size but with, as noted, the largest number of reptile species. This is again attributable to environmental factors, with a complex range of habitats varying from low-altitude Mediterranean, to medium-altitude temperate and high-altitude boreal habitats.

The taxonomic richness of the Balkans is especially apparent in a number of sister species (*Triturus cristatus* superspecies, *Rana ridibunda* superspecies, *Podarcis* spp., *Archeolacerta* spp., *Lacerta* spp.). Equally, there is substantial subspecific diversity in the Balkans. As many as 239 autochthonous subspecies of amphibians and reptiles are known, with an average of 1.03 subspecies per amphibian species, and 2.94 subspecies per reptile species. The most extensive subspecific differentiation occurs among lizards (156 subspecies, Table 1), especially in *Cyrtodactylus kotschyi*, *Podarcis erhardii*, *Podarcis melisellensis* and *Podarcis sicula*. It should be noted that these figures must be taken cautiously, however, since the taxonomic validity of many subspecies has not been established with rigour, and important aspects of their distributions have not yet been resolved. A good example is the alpine newt (*Triturus alpestris*), for which the number of subspecies has been reduced substantially following recent taxonomic revision (Sotiropoulos *et al.*, 2001a). Our list includes all described subspecies that have not been disputed to date.

Table 1. Number of autochthonous Balkan amphibian and reptile taxa, listed according to taxonomic category of the main groups, and the number of species (spp.) which have the marginal zone of their distributional range in the Balkans. (sspp. = subspecies)

Order	Families	Spp.	Sspp.	Endemic spp.	Endemic sspp.	Range limits
AMPHIBIA						
Caudata	2	10	21	2	9	8
Anura	5	22	12	7	3	11
REPTILIA						
Testudines	4	10	8	1	2	4
Amphisbaenia	1	1	1	0	0	1
Lacertilia	6	35	156	11	65	19
Serpentes	4	24	41	3	14	17
Total	**22**	**102**	**239**	**24**	**93**	**60**

The available evidence suggests that new species emerged during the Tertiary (mostly during the Miocene and Pliocene), while the origin of subspecies was confined to Pleistocene glacial stages (*e.g.* *Triturus* and *Salamandra* taxa: Bolkay, 1928; Wallis & Arntzen, 1989; Oosterbroek & Arntzen, 1992; Crnobrnja-Isailović *et al.*,

1997; Grossenbacher, 1994; Weisrock *et al.*, 2001; Denoël *et al.*, 2001; lacertids: Böhme & Corti, 1993; amphibians and reptiles in general: Paunović, 1990).

Although biogeographical importance does not necessarily correlate directly with the size of a peninsula (as is clear in Scandinavia, where species richness is low), it is generally argued that the potential number of species is proportional to area if geographic locations are similar. This does not hold true in southern Europe, however, wherein the Balkan Peninsula (with 104 amphibian and reptile species) has the highest species richness, and yet is only the second largest peninsula. For purposes of comparison, the Italian Peninsula has 76 amphibians and reptiles within an area of 150,000 km^2 (OECD, 1994, cited in European Centre for Nature Conservation, 2001), while the largest European peninsula (the Iberian) has only 78 within an area of 581,000 km^2 (World Conservation Monitoring Centre, 1992, cited in European Centre for Nature Conservation, 2001) (for the number of species see also Corbett, 1989; Astudillo & Arano, 1995; World Conservation Monitoring Centre, 1997). As an aside, it is interesting to note that both these peninsulas contain more ancient phylogenetic lineages of some animal groups than in the Balkans (Oosterbroek & Arntzen, 1992).

Taxonomic peculiarities

The list of Balkan amphibian and reptile taxa is sure to increase in the near future. Ongoing detailed morphological research has revealed new evidence for intraspecific differentiation, within species such as the alpine crested newt *Triturus carnifex* (Kalezić *et al.*, 1997), or the eastern spadefoot toad *Pelobates syriacus* (Ugurtas *et al.*, 2002). There is also evidence that a number of morphologically-cryptic taxa may exist, since, in regard to the taxonomy of many amphibians and reptiles (such as the *Archeolacerta* and *Zootoca* lizard groups, or the brown frogs - '*Ranae fuscae*'), the Balkan Peninsula is a highly perplexing region. Future genetically-based taxonomic studies may reveal further species-level diversity amongst existing taxa. Genetic data have already thrown light on old controversies over the origin and evolutionary diversification of some taxa, such as the crested newts (*Triturus cristatus* superspecies; Litvinchuk *et al.*, 1999), viviparous lizards (*Zootoca vivipara*; Mayer *et al.*, 2000), or the fire-bellied toads (*Bombina variegata* group; Radojičić, unpublished data).

Another taxonomic peculiarity of Balkan amphibians and reptiles is the high degree of phenotypic and genotypic variation in their populations, such that there is often a problem in applying diagnostic traits used to differentiate taxa in western and northern Europe, especially near hybrid zones. This is particularly the case where population affinities appear to be highly sensitive to the number of specimens examined from each population, such that many more samples need to be analysed statistically in order to establish firm taxonomical relationships (*e.g.* Arntzen & Wallis, 1999).

In addition, genetic differentiation among congeneric species would appear to be lower on the Balkan Peninsula than in more northern parts of Europe. For example, maximum hybrid index scores between *Bombina bombina* and *B. variegata* in Croatia reach a value of eight, while in Poland this measure of relative genetic distinctness can be as high as 12 (Szymura, 1993). Among *Triturus* species, the smallest genetic difference was found among Balkan congeneric populations (Kalezić & Hedgecock, 1980; Crnobrnja *et al.*, 1989) in comparison with those from western and Central Europe (Rafinski & Arntzen, 1987), and the former USSR (Litvinchuk *et al.*, 1994). Greater genetic differentiation was found in Greek populations of *Triturus* species than in the northern Balkans, Central and Northern Europe (Sotiropoulos *et al.*, 2001b), however.

ENDEMICITY

The number of endemic taxa present is a very important component of assessing the importance of any fauna. The circum-Mediterranean region, including the Balkans, is already well known as a region of high endemicity (Oosterbroek & Arntzen, 1992; Jong, 1998). Levels of endemicity are especially high for amphibians and reptiles (Sara, 1996).

The Balkan Peninsula has 24 endemic amphibian and reptile species: nine amphibians (28.1% of the total number of autochthonous species) and 15 reptiles (21.4%) (Table 1, Figure 1). Within the most well-represented groups, the highest degree of endemicity occurs in tailless amphibians (31.8%) and the next highest in lizards (31.4%), while snakes have the lowest number of Balkan endemics (12.5%). In general, endemicity at the subspecific level is very high, ranging from 42.9% in Caudata to 25.0% in Testudines, with an average value of 38.9%.

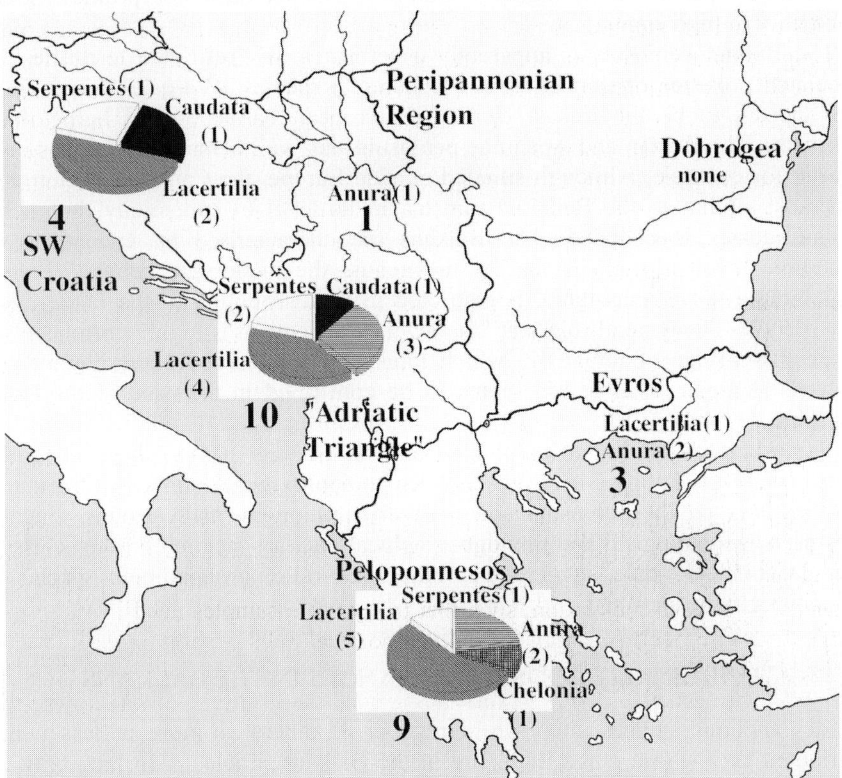

Figure 3. Incidence of Balkan endemics in the six Balkan study regions.

In respect to spatial patterns of Balkan endemicity (for regions where diversity has been compared), the "Adriatic Triangle" appears to have the highest concentration of Balkan endemics (ten out of 54 species; 18.5%), followed by the Peloponnisos region (nine out of 49 species; 18.4%) (Figure 3). The "Adriatic Triangle" is part of the Dinaric karst, a unique landscape formation that evolved on the limestone substrate of the Dinarides. However, despite a high number of species, the northern part of the Dinaric karst (SW Croatia) has a very low proportion of the Balkan endemics (four out of 49 species; 8.2%). The same is true of the Evros region (three endemics out of 45 species; 6.7%). Interestingly enough, the northernmost parts of the Balkans (species-poor areas) lack Balkan endemic species. The Peripannonian region has only one Balkan endemic out of a total of 35 species (2.9%), while the Dobrugea region is completely without Balkan endemics (Figure 3).

It is well known that centres of species richness tend to coincide with those of endemism (Anderson, 1994; Varga, 1995). This has been demonstrated clearly for Balkan amphibians and reptiles, in regard to the six Balkan regions defined herein. The southern part of the Balkans is also characterised by a few instances of microinsular endemism (some lizards and one snake, *Macrovipera schweizeri*, on the Greek islands). However, in the Balkans most endemic species are not in general restricted to islands, so a higher number of islands does not promote greater probability of high endemism.

The Balkan Peninsula is apparently a core area, of "restricted territories with cumulated presence of stenochorous ("endemic") species" (Varga, 1995, p. 71). It also appears to be the richest centre of European batracho- and herpetofaunal endemism. The Iberian and Apennine peninsulas lag well behind the Balkans in this respect, for example. Although situated outside Europe, the Caucasus region is the area most similar to the Balkan Peninsula in its level of endemicity, but only if Trans-Caucasia is included. The Balkans are characterised by endemism at a taxonomic level as high as that of the genus: the monotypic genus *Proteus*, a troglobiontic tailed amphibian, is restricted in its distribution to the Dinaric karst (Sket, 1997). The general rule that "the level of endemism is highly correlated with the geological age of refuges in which relict-like species could survive and have evolved" (Varga, 1995, p. 85) seems to be confirmed in the case of the Balkan Peninsula.

The relictness of Balkan amphibian and reptile taxa has not been adequately investigated. It is impossible to present even an approximate estimate of the number of relict taxa on the peninsula and their distribution by main groups, since the relevant palaeontological and phylogeographical data are lacking. Nevertheless, the data generated to date suggest that numerous relict groups, both glacial and preglacial, can be expected.

MARGINALITY OF SPECIES RANGES IN THE BALKANS

Besides endemic species, there are a number of otherwise more or less widely-distributed taxa whose range limits are in the Balkans (Table 1). In fact, only four amphibian species and nine reptile species have ranges that extend well into the Balkan region without ending on the Peninsula. Species with the marginal zone of their distributional range in the Balkans comprise more than half of the total number of autochthonous species (19 amphibians and 41 reptiles; see Appendix and Table 1).

Among Balkan amphibian species, range borders orientated towards southern Europe appear to be dominant (52.5%), followed by eastern borders (26.5%), in contrast to southeastern and northwestern borders, which are least numerous (10.5% each). Central Europe and, to a lesser extent, western Europe, appear to be the main sources for the Balkan batrachofauna. In contrast to amphibians, the most numerous range borders for reptiles are in the west, north, and northwest (24.4%, 24.4%, and 22.0%, respectively), followed by southern and eastern borders (12.2% each), while southwestern and southeastern range borders are virtually negligible (2.5% each). These facts confirm without doubt that reptiles of the Balkans carry the "seal" of southern and southeastern herpetofaunas.

Marginal zones of distribution ranges are much less frequent in Central and Northern Europe than in the Balkans. This seems to support the Rapoport rescue hypothesis as to the origin of latitudinal gradients in richness, which holds that low-latitude localities have relatively more species near the edge of their range than do high-latitude zones (see Stevens, 1992). In addition, the high level of diversity of amphibians and reptiles in the Balkans is at least partly due to the fact that communities composed of a high percentage of species at the edge of their geographical range are presumed to be more species-rich than communities composed primarily of species at the centre of their distribution. The evolutionary importance of range margins lies in the fact that although most peripheral populations are not going to evolve into distinct entities, there is still an astonishing potential for further diversification of the Balkan batrachofauna and herpetofauna.

BIOGEOGRAPHICAL DIVERSITY

The Balkan batrachofauna and herpetofauna have varied biogeographical origins and connections with other regions. More specifically, the Balkans appear to have been a centre of speciation within some taxa, such as the crested newts (*Triturus cristatus* superspecies; Crnobrnja-Isailović *et al.*, 1997), the alpine newt (*Triturus alpestris*; Arano & Arntzen, 1987), the "true salamanders" (*Salamandra* spp. and *Mertensiella* spp; Weisrock *et al.*, 2001), brown frogs ("*Ranae fuscae*"; Oosterbroek & Arntzen, 1992), the western Palaearctic water frog complex (*Rana* spp.; Plötner & Ohst, 2001); *Anguis* spp. (Grillitsch & Cabela, 1990). In addition, intense migrations of amphibians and reptiles from other zoogeographical areas have occurred over several migration routes. In the first place, there is a large front of contact with steppe in the east and northeast (Pontic area), while the Morava River corridor provides contacts (faunal exchange) in a south-north direction. Finally, the Aegean islands also constituted a connection with the faunas of Africa and Asia. Extensive faunal exchange between Asia Minor and the Balkans might have occurred due to the existence of a Bosphorus land bridge between 20,000 and 10,000 yr. B.C. (Hosey, 1982). Thus, multiple connections followed by vicariant separations have existed between Anatolia and the Balkans (Weisrock *et al.*, 2001). A number of taxon-area cladograms for diverse groups including some amphibians (*Rana* spp. and *Triturus* spp.) revealed that species and subspecies from the Balkans and Western Asia Minor belong to Asia—Transmediterranean phylogenetic lineages (Oosterbroek & Arntzen, 1992).

In contrast to the Iberian and Apennine peninsulas, the Balkan Peninsula is not isolated by any mountain barriers from the European mainland. It is thus broadly connected with Central Europe and the rest of the continent. This has been crucial for the biogeography of the European batrachofauna and herpetofauna. Namely, it has been demonstrated that Central and North Europe were populated by an expansion from Balkan refuges during postglacial warming periods (Hewitt, 1999). Also, certain groups have withdrawn from Northern and Middle European latitudes, some of them being mostly or partially restricted within southern refuge areas (Venczel, 2000).

At the level of demes, changes in ranges after successive Pleistocene glacial episodes involved much colonisation, extinction, dispersal, and adaptation. For many groups, it has been assumed that Balkan populations were the main source for the recolonisation of Europe; *e.g. Bombina* spp. (Szymura, 1993), *Triturus* spp. (Oosterbroek & Arntzen, 1992; Crnobrnja-Isailović *et al.*, 1997), *Pelobates* spp. (Eggert, 2000), *Rana temporaria* (Pidancier *et al.*, 2001), *Zootoca* spp. (Guillaume *et al.*, 1997), or *Natrix tessellata* (Guicking *et al.*, 2001). In addition, a considerable number of lineages, subspecies, and species of Balkan amphibians and reptiles are crowded parapatrically into the southern parts of their range, close to where their ice age distribution would presumably have been (*e.g. Triturus* spp., Wallis & Arntzen, 1989; Litvinchuk *et al.*, 1997; Litvinchuk & Borkin, 2000; the western Palaeartic water frog complex, Plötner & Ohst, 2001). If, as it has been argued, northerly interglacial expansions were extinguished by subsequent glacial period(s), then it seems likely that the richness of southern taxa has been generated by para-allopatric divergence over several repeated glacial stages.

Serial recolonisation from south to north, west, and east involved a series of bottlenecks for the colonising genome, which led to a loss of alleles and a tendency toward homozygosity (although not in every case; see Carson, 1990). This has been demonstrated for a number of groups (Hewitt, 1996). Also, clear geographical trends in heterozygosity within European amphibians and reptiles have been discovered: northern populations, whose present ranges largely include previously-glaciated terrain and permafrost zones, are less variable than are southern populations. Thus, northern populations of *Bombina bombina* and *B. variegata* show little or no variation at the studied gene loci, while southern populations exhibit numerous polymorphisms (Szymura, 1988). Among European newts, there is a general trend towards greater variation in Balkan populations, in terms of gene locus polymorphism, number of alleles, and levels of heterozygosity, than in conspecific populations from western and Central Europe (Kalezić & Hedgecock, 1980; Rafinski & Arntzen, 1987; Sotiropoulos *et al.*, 2001b; Frelow *et al.*, unpublished data). Also, mtDNA data indicate that southern populations of crested newt (*Triturus cristatus* superspecies), especially *T. carnifex* and *T. karelinii*, exhibit greater variation than more northerly populations (Wallis & Arntzen, 1989). As far as the European pond terrapin (*Emys orbicularis*) is concerned, a number of different mtDNA haplotypes can be recognised in Balkan populations (Lenk *et al.*, 1998; Fritz, personal communication). The same relation appears to exist in the case of the viviparous lizard (*Zootoca vivipara*; Mayer *et al.*, 2000).

Many Balkan amphibians and reptiles show discontinuous patterns of geographical distribution, as this part of Europe is characterised by the presence of many mountain blocks and marine trenches. These geographical barriers contribute to range fragmentation and genetic isolation resulting from long-term extrinsic barriers to gene flow. Low mobility and poor genetic exchange appear to be strong enough to evoke genetic distinctness and prevent gene pools from amalgamating.

This has been documented for crested newts in the Balkans (Crnobrnja & Kalezić, 1990).

In terms of zoogeography, the Balkan Peninsula is characterised by numerous discontinuities. The zoogeographical map is very complex, with elaborate subdivisions and a wealth of preglacial faunal elements (Hadži, 1935; Matvejev, 1961; Haxhiu, 1994). The biogeographical diversity of the so-called "Adriatic Triangle" aroused great interest among biologists in the first half of the 20th century (Simroth, 1907), and comprises nearly the entire richness of the Balkan Peninsula. It is interesting to note that a popular and long-standing hypothesis of Simroth (1907) linked the biogeographical "wealth" of the Balkans with the general theory of pendulation, which has been proposed to explain climatic change and the origin of Quaternary glaciation. This view has been accepted by many authors, including the famous Russian ichthyologist Berg, the Serbian ecologist Stanković, the Czech hydrobiologist Komarek and the Austrian botanist Hanausek.

As shown above, the "Adriatic Triangle" is an area within the Balkans with highly specious amphibian and reptile groups and a high degree of endemism. As such, it is acknowledged to be one of the main European biogenetic reserves containing rare and endemic taxa. Many of these rare and endemic taxa are under threat (Džukić, 1995; Crnobrnja-Isailović & Džukić, 1995; Haxhiu, 1998).

In general, the genesis of diversity among Balkan amphibians and reptiles has (as elsewhere) been dictated by many proximate factors, such as habitat heterogeneity, topographic diversity (mostly influenced by Pleistocene glaciation), great climatic variation, differences of immigration rates, etc. However, it can also be assumed that historical events had a fundamental role to play in the generation of this versatility (Humphries & Parenti, 1986). This could be especially true in the case of the Balkan Peninsula, as its geological history and palaeogeography have undergone complex perturbations, especially during the Tertiary. The Mesozoic and Caenozoic geological development of the Mediterranean in general can be regarded tectonically as the evolution of a continental collision zone that originally separated Eurasia (the ancient Laurasian continental landmass) from Africa (Gondwanaland). During its unique and dynamic geological history, small terrestrial regions made up the margin of the growing area termed the Balkans, which became a part of Eurasia much later than western parts of the circum-Mediterranean region (Hadži *et al.*, 1974; Hsu & Bernoulli, 1978; Oosterbroek & Arntzen, 1992; Steininger & Rogl, 1996; Rage, 1997). Thus, the southern margins of the Balkan Peninsula represent a collision zone between ancient Gondwanan tectonic fragments and are geologically active. This may have been crucial to the development Balkan faunistic diversity, as tectonic plates introduced new faunal elements (Macey *et al.*, 2000), while mountain building in the suture zone is important in causing high levels of genetic diversity among faunal elements.

Two models of speciation apparently occurred on the Balkan Peninsula: speciation due to climatic-ecological fluctuations during the Tertiary and Quaternary, and speciation due to tectonic-palaeogeographical change. The latter is demonstrated indirectly by the correspondence of zoogeographical boundaries with that of deep lithospheric fractures, or crustal megablocks (Sikošek, 1986) which has been noted on several occasions (*e.g.* Hadži, 1935; Szyndlar, 1984). Geological events might have profoundly affected the evolutionary history of at least some Balkan amphibians. This has been suggested for the origin of the crested newts (*Triturus cristatus* superspecies, Crnobrnja-Isailović *et al.*, 1997).

HYBRIDISATION AND HYBRIDOGENESIS

Hybrid zones are predominantly narrow regions in which genetically-distinct populations meet, mate, and produce hybrids. Because the hybridising populations exchange genes and yet remain distinct, and because of striking patterns of variation along the contact zone, hybrid zones have constantly attracted the attention of taxonomists and evolutionary biologists (*e.g.* Barton & Hewitt, 1989; Harrison, 1993). This is especially true in the Balkans, where (as in other southern regions) hybrid zones appear to be more numerous than in the rest of Europe, a situation that can be attributed to both population genetics and geographic factors (Hewitt, 1999), since populations that are at present close to each other in space may have quite distant geographical origins and divergent genomes.

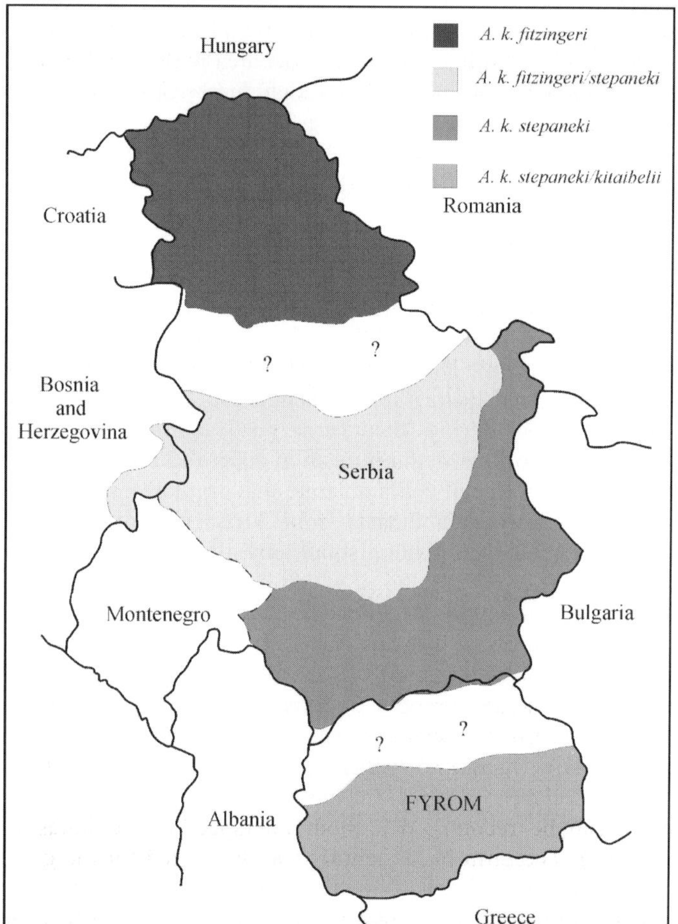

*Figure 4. Spatial pattern of the intergradation zones between subspecies of the snake-eyed skink (*Ablepharus kitaibelii*) in the central Balkans (modified from Ljubisavljević* et al.*, 2002, with permission of the Managing Editor of Herpetozoa).*

Several species show western and eastern forms in Europe, with hybrid zones running roughly north-to-south from Scandinavia down to the Alps, marking the southwestern and southeastern glacial refugia from which they have expanded. On the Balkan Peninsula, hybridisation and intergradation zones are differently orientated in space, including N-S, NW-SE, and W-E distributions. Thus, rather broad intergradations of subspecies zones occur in bands running mainly in a W-E direction, between populations of the snake-eyed skink (*Ablepharus kitaibelii*; Ljubisavljević *et al*., 2002; Figure 4). Similar W-E clines of morphological characters and allozymes are found among *Triturus cristatus* superspecies in Greece (Sotiropoulos *et al*., 1999). Multivariate analysis of morphological characters indicates that the transitional zone between two subspecies of sand viper (*Vipera ammodytes*) in Kosovo and Metohija is orientated in a NW-SE direction (Tomović & Džukić, *in press*).

Individuals of some taxa that usually form distinct taxonomic clusters reproduce without recombination. Hybridogenetic hybrids in the western Palaearctic water frog complex are of general biological interest because this complex appears to be a suitable model for studying genetic diversity as an integral part of biodiversity (Plötner & Schmeller, 2001), as well as because of the unique population genetics of the *Rana* synklepton *esculenta*. Studies of hybridogenesis, particularly those which deal with different hybridogenetic population systems on the Balkan Peninsula (which is characterised by the strong evolutionary divergence of water frog taxa; Plötner & Ohst, 2001), can be of special importance due to several peculiarities. In addition to the fact that those individuals of *R. ridibunda* (the common parent of all lineages of hybridogenetic taxa) which are of Balkan origin are known to be resistant to hybridogenesis (Guerrini *et al*., 1997), it is also noteworthy that the genomes of two Balkan water frogs (*R. epeirotica* and *R. shqiperica*), and most likely *Rana balcanica*, are at least partially resistant to germ line exclusion in hybrids. The overall pattern of reproductive relationships of water frogs in the Balkans is therefore very complex. Unfortunately, studies on such relationships in this part of Europe have only been anecdotal to date (Gavrilović, 1997; Spasić-Bošković *et al*., 1999).

HETEROCHRONY

Heterochrony, *i.e.* a shift in the timing and/or rates of processes underlying development in a descending ontogeny, has long been recognised as constituting a major class of developmental dissociation during evolution (*e.g.* Raff & Kaufman, 1983). Some heterochronic changes lead to paedomorphosis. Paedomorphic descendants pass through fewer stages of ontogenetic development than do their ancestors, so descendant adults resemble larvae or juveniles of their ancestors. Paedomorphosis can be recognised at both interspecific and intraspecific levels (Reilly *et al*., 1997). The term paedogenesis was suggested for the general process that produces an intraspecifically truncated pattern. Both paedomorphosis and paedogenesis occur in tailed amphibians of the Balkans: paedomorphosis in the olm (*Proteus anquinus*) and paedogenesis in the newts (*Triturus* spp.) (Kalezić, 2002).

Paedogenesis appears to be rather common in the genus *Triturus*, especially in the smooth newt and alpine newt, and is particularly limited to the southern margin of the distribution area, *i.e.* to meridional subspecies on the Italian and Balkan peninsulas (Džukić *et al*., 1990; Denoël *et al*., 2001). The current distribution of paedogenetic populations is linked to Holocene colonisation. It is in fact restricted to the Italian and Balkan peninsulas, which are thought to have acted as refuges during Quaternary glacial events. Whether paedogenesis first appeared during the harsh conditions of the Pleistocene or earlier remains unknown. In the alpine newt, intraspecific differentiation occurred between the Upper Miocene and the Pleistocene (Arano, 1988). Thus, paedogenesis might have appeared as early as the Miocene in one or several of the subspecies. The Balkan and Italian peninsulas were connected during the Würm (Furon, 1950), during which time exchange of genes would have been possible, and might explain the occurrence of paedogenesis on the two peninsulas.

The simultaneous occurrence of paedogenesis in syntopic populations of two *Triturus* species (*T. carnifex* and *T. vulgaris*) has only been recorded in one pond of the Submediterranean part of Montenegro (Džukić *et al.*, 1990), although the ranges of these newts overlap significantly.

INTRASPECIFIC VARIATION IN REPRODUCTIVE MODES (OVIPARITY *VS.* VIVIPARITY)

Reproductive bimodality within a single species is a very rare phenomenon among amphibians and reptiles and is of general interest for the study of the evolution of viviparity. Among reptiles, it has been reliably reported for a few lizards and snakes (Shine, 1985). These include the viviparous lizard (*Zootoca vivipara*), which has allopatric oviparous and viviparous populations. Viviparity here appears to be the much more common reproductive mode, occurring over the greatest part of this lizard's range, whereas oviparity is restricted to northern Spain, the Pyrenees, southern France, and a small area in the northwestern Balkans.

Oviparous and viviparous strains differ in many respects, including the presence of genetic markers (*i.e.* diagnostic gene loci and different mtDNA haplotypes). It has been suggested that viviparity as a derived reproductive mode (oviparity is ancestral for the vertebrates in general) evolved recently in *Z. vivipara*, probably during ecological changes of the last Pleistocene glaciations (60,000-20,000 years ago) (Heulin *et al.*, 1993; Arrayago *et al.*, 1996). According to this scenario, viviparity first appeared and was subsequently propagated as an adaptation to cold climatic conditions in eastern populations (somewhere between the Balkan Peninsula and the Altai Mountains), while oviparity remained unchanged in isolated southern refuges (Heulin *et al.*, 1993; Guillaume *et al.*, 1997). Oviparity has thus far been confirmed in samples of five Slovenian populations and one population from Croatia (Böhme *et al.*, 1999; Heulin *et al.*, 2000; Mayer *et al.*, 2000). There is not yet any firm evidence indicating that oviparous populations might be more common on the Balkan Peninsula, which marks the southernmost extent of this lizard's range and where it is considered to be a glacial relict. The only viviparous populations to have been found thus far are in Bulgaria and Serbia (Guillaume *et al.*, 1997; Heulin, personal communication, respectively). It should be noted that the oviparous populations from Slovenia and Austria differ substantially from other conspecific populations. It follows that a new subspecies of viviparous lizard has been described, whose range extends to the northwestern part of the Balkan Peninsula (Mayer *et al.*, 2000).

Among European amphibians, reproductive bimodality has been reported for the fire salamandar (*Salamandra salamandra*), viviparous specimens of which have been found on the Iberian Peninsula (in the Pyrenees and Cantabrian Mountains) and whose evolution can be linked with glaciation events in the region (Alcobendas *et al.*, 1996). The presence of viviparity has also been reported in Balkan populations (in Greece; Sotiropoulos, personal communication).

OUTLOOK FOR FUTURE RESEARCH

It is not yet possible to reconstruct the evolutionary history of the Balkan batrachofauna and herpetofauna or even to decide which model best fits the present data. All sets of data, both modern and historical, show many gaps and are frequently subject to conflicting interpretations. However, some points may help to guide future discussions.

There is a need for a great deal of research to be done in order to understand of history of the Balkan Peninsula and the origins of its high biodiversity. In the first place, we need to know more about taxonomy (even at the level of basic taxonomical questions) and require more detailed faunal surveys. There is still a paucity of data on long-standing taxonomical problems, the population structure of many species, and the distributional characteristics of taxa - including size, shape, boundaries, overlaps, and the definition of geographic ranges. The latter applies in particular to those taxa which can be considered as "zoogeographic paradoxes", *i.e.* where they occur as isolated populations far from the the known limits of their geographical ranges; examples include exotic species such as *Blanus strauchi*, *Laudakia stellio*, *Chamaeleo* spp., and *Macrovipera* spp. Secondly, the biogeographical patterns of many taxa remain virtually unexplored to date, and matters such as phylogenetic relationships and taxon-area cladograms have escaped analysis. Finally, studies of the age of present taxa based mainly on DNA techniques are needed in order to estimate a time span that can be correlated with possible causes of cladogenetic events. Our aim in the future is essentially to reconstruct the biogeographic history of differentiation of populations (*i.e.* centres of origin, patterns of relictness, refuge areas during glacial phases, postglacial recolonisation routes, *etc.*). In summary, we are confident that surprises, wonders, and even a few dangers await discovery in the study of the batrachofauna and herpetofauna of the Balkans.

CONCLUSIONS

Balkan amphibians and reptiles are diverse in many respects. Species richness is very high, with a total of 104 species (33 amphibians and 71 reptiles), two of which are introduced. It is likely that a number of morphologically cryptic taxa also exist. The highest level of species diversity is recorded in the region of the "Adriatic Triangle" (the Prokletije Mountain Massif and adjacent areas), while the Peripannonian area and Dobrugea are the most species-poor regions.

The Balkans are the centre of endemicity in Europe, with *ca.* 28% of amphibians and 21% of reptiles being endemic. In addition, more than half of the rest of the autochthonous species have the marginal zone of their distributional range in the Balkans, with marked differences between amphibians and reptiles in regard to the orientation of dominant range borders.

The Balkan batrachofauna and herpetofauna have varied biogeographical origins and connections with other regions. The Balkans appear to have been a centre of intense speciation within some taxonomic groups. Also, there has been a significant number of migrations of amphibians and reptiles from other zoogeographical areas. Central and Northern Europe appear to have been populated repeatedly by an expansion from Balkan refugia during Pleistocene interglacials. Certain groups have withdrawn from northern and mid-European latitudes, some of them being restricted to southern refuge areas. The zoogeographical map of the Balkan Peninsula is very complex, with elaborate subdivisions and numerous discontinuities, as well as a wealth of preglacial faunal elements.

In terms of general biological phenomena such as hybridisation and hybridogenesis, interspecific and intraspecific heterochronic phenomena (paedomorphosis *vs.* paedogenesis), and switching of reproductive modes (oviparity *vs.* viviparity) within a single species, amphibians and reptiles on the Balkan Peninsula show numerous distinctive features in comparison with other European areas.

ACKNOWLEDGEMENTS

We thank Ljiljana Tomović, Katarina Ljubisavljević, and Imre Križmanić for their help in many phases of paper preparation. The constructive comments of Leo Borkin, Basil Chondropoulos, Dan Cogalniceanu, Ana Djorović, Spartak Litvinchuk, and Konstantinos Sotiropoulos sharpened the focus of the study and are greatly appreciated. Ismail Ugurtas and Kurtulus Olgun were consulted for information on the presence of amphibians and reptiles in Turkish Thrace. The clarity of this article was greatly improved by comments from editors. The authors claim sole responsibility for any errors of fact, interpretation, logic, citation, or omission. Our research is partly supported by the Serbian Ministry of Science, Technology and Development (project "Integrative study of amphibians and reptiles of the central Balkans", No. 1623).

REFERENCES

Alcobendas, M., Dopazo, H. & Alberch, P. 1996. Geographic variation in allozymes of populations of *Salamandra salamandra* (Amphibia: Urodela) exhibiting distinct reproductive modes. *Journal of Evolutionary Biology* 9, 83-102.

Anderson, S. 1994. Area and endemism. *The Quarterly Review of Biology* 69, 451-471.

Arano, B. 1988. *Aspectos filogenéticos del género Triturus con especial consideración a la evolución del complejo Triturus apestris*. Unpublished PhD thesis, Madrid University, Madrid.

Arano, B. & Arntzen, J.W. 1987. Genetic differentiation in the alpine newt, *Triturus alpestris*. In: *Proceedings of the 4th Ordinary General Meeting of the Societas Europea Herpetologica* (J. J. van Gelder, H. Strijbosch & P.J.M. Bergers, eds.), pp. 21-24. Faculty of Sciences, Nijmegen.

Arntzen, J.W. & Wallis, G.P. 1999. Geographic variation and taxonomy of crested newts (*Triturus cristatus* superpecies): morphological and mitochondrial DNA data. *Contribution to Zoology* 68, 181-203.

Arrayago, M.-J., Bea, A. & Heulin, B. 1996. Hybridization experiment between oviparous and viviparous strains of *Lacerta vivipara*: a new insight into evolution of viviparity in reptiles. *Herpetologica* 52, 333-342.

Astudillo, G. & Arano, B. 1995. Europa y su Herpetofauna: responsabilidades de cada país a lo referente a su conservación. *Boletin del Asociación Herpetología Española* 6, 14-45.

Barton, N.H. & Hewitt, G.M. 1989. Adaptation, speciation and hybrid zones. *Nature* 341, 97-503.

Bolkay, S.J. 1928. Die Schadel der Salamandrinen mit besonderer Rucksicht auf ihre Systematische Bedeutung. *Zeitschrift für Anatomie und Entwicklungsgeschichte (Munchen, Berlin)* 86, 259-319.

Böhme, W. & Corti, C. 1993. Zoogeography of the lacertid lizards of the western Mediterranean basin. In: *Lacertids of the Mediterranean Region* (E.D. Valakos, W. Böhme, V. Perez-Mellado, P. Maragou, eds.), pp. 17-33. Hellenic Zoological Society, Athens, Bonn, Alicante.

Böhme, W., Bonetti, A. & Chiras, G. 1998. The chameleons of the Greek mainland: taxonomic allocation and conservation needs of a second European species (Squamata: Sauria: Chamaeleonidae). *Herpetozoa* 11, 87-91.

Böhme, W., Heulin, B. & Bischoff, W. 1999. First data on an oviparous population of the viviparous lizard, *Zootoca vivipara* Jacquin 1787 (Squamata: Lacertidae) from Slovenia. In: *Abstracts of the 10th Ordinary General Meeting of the Societas Europaea Herpetologica*, pp. 34-35. Natural History Museum of Crete, Erakleio.

Borkin, L.J. 1998. Distribution of amphibians in North Africa, Europe, Western Asia, and the Former Soviet Union. In: *Patterns of Distribution of Amphibians: A Global Perspective* (W.E. Duellman, ed.), pp. 329-420. The John Hopkins University Press. Baltimore and London.

Bour, R. 1997. *Testudo hermanni* Gmelin, 1789. In: *Atlas of Amphibians and Reptiles in Europe* (J.-P. Gasc *et al.*, eds.), pp. 178-179. Societas 1Europae Herpetologica, Museum National d'Histoire Naturelle, Paris.

Bringsoe, H. 1988. Comments on: Chondropoulos, B.P. 1986. A checklist of the Greek reptiles. I. The lizards. *Amphibia-Reptilia* 7, 217-235. *Amphibia-Reptilia* 9, 201-204.

Carson, H.L. 1990. Increased genetic variance after a population bottleneck. *Trends in Ecology and Evolution* 5, 228-230.

Corbett, K. 1989. *The Conservation of European Reptiles and Amphibians*. Christopher Helm, London.

Corti, C., Masseti, M. & Delfino, M. 1997. The herpetofauna of anthropochorous origin of the Mediterranean islands. In: *Abstracts of the Third World Congress of Herpetology* (Z. Rocek & S. Hart, eds.), p. 44. Prague.

Corti, C., Masseti, M., Delfino, M. & Perez-Mellado, V. 1999. Man and herpetofauna of the mediterranean islands. *Revista Española de Herpetología* 13, 83-100.

Crnobrnja, J., Kalezić, M.L. & Džukić. G. 1989. Genetic divergence in the crested newt (*Triturus cristatus* Complex) from Yugoslavia. *Biosistematika (Beograd)* 15, 81-92.

Crnobrnja, J. & Kalezić, M.L. 1990. The organization of the genetic variation in the crested newt (*Triturus cristatus* complex) from Yugoslavia. *Genetika (Belgrade)* 22, 133-146.

Crnobrnja-Isailović, J. & Džukić, G. 1995. First report about conservation status of herpetofauna in the Skadar lake region (Montenegro): current situation and perspectives. *Scientia Herpetologica*, pp. 373-380.

Crnobrnja-Isailović, J., Džukić, G., Krstić, N. & Kalezić, M.L. 1997. Evolutionary and paleogeographical effects on the distribution of the *Triturus cristatus* superspecies in the central Balkans. *Amphibia-Reptilia* 18, 321-332.

Delfino, M. 1995. *Blanus* (Reptilia, Amphisbenia) from the Pleistocene of Southern Italy: another small tessera from a big mosaic. In: *Abstracts of the 10th Ordinary General Meeting of the Societas Europaea Herpetologica*, p. 49. Bonn.

Delfino, M. 1997. Italian Paleoherpetofauna Database: Neogene-Quaternary. In: *Abstracts of the 3rd World Congress of Herpetology*, pp. 51-52. Prague 1997.

Denoël, M., Duguet, R., Džukić, G., Kalezić, M. L. & Mazzotti, S. 2001. Biogeography and ecology of paedomorphosis in *Triturus alpestris* (Amphibia, Caudata). *Journal of Biogeography* 28, 1271-1280.

Džukić, G. 1975. Contribution to knowledge of *Elaphe longissima* Laurenti, 1768 (Colubridae, Reptilia). *Biosistematika (Beograd)* 1, 137-141 (in Serbian with English summary).

Džukić, G. 1993. *Tailed Amphibians (Caudata) of Serbia. A faunistical and zoogeographical study, regarding protection and conservation*. Unpublished PhD thesis, University of Belgrade, Belgrade.

Džukić, G. 1995. "Diverzitet vodozemaca (Amphibia) i gmizavaca (Reptilia) Jugoslavije sa pregledom vrsta od medjunarodnog značaja". In: "*Biodiverzitet Jugoslavije sa pregledom vrsta od medjunarodnog znacaja*" (V. Stevanović & V. Vasić, eds.), pp. 447-469. Biološki fakultet i Ecolibri, Beograd (in Serbian).

Džukić, G., Kalezić, M.L., Tvrtković, N. & Djorović, A. 1990. An overview of the occurrence of paedomorphosis in Yugoslav newt (*Triturus*, Salamandridae) populations. *British Herpetological Society Bulletin* 34, 16-22.

Džukić, G., Kalezić, M.L, Petkovski, S. & Sidorovska, V. 2001. General remarks on batracho - and herptetofauna of the Balkan Peninsula. In: *75 years Maced. Mus. of Nat. Hist.* (T. Boskova, ed.), pp 195 - 204. Prirodonaučen Muzej na Makedonija, Skopje.

Eggert, C. 2000. *Le declin du pelobate brun (*Pelobates fuscus*, amphibien anoure): Apport de la phylogéographie et de la dynamique de population a sa comprehension. Implications pour sa conservation.* Unpublished PhD thesis, Université de Savoie, Savoie.

European Centre for Nature Conservation [ECNC]. 2001. Richness in Europe. (*http://www.ecnc.nl)*.

Furon, R. 1950. Les grandes lignes de la paléogéographie de la méditerranée (tertiaire et quaternaire). *Vie et Milieu* 1: 131-162.

Gaston, K.J. & David, R. 1994. Hotspots across Europe. *Biodiversity Letters* 2, 108-116.

Gavrilović, V. 1997. *Interspecific hybridization in Balkan water frogs.* Unpublished masters thesis, Faculty of Biology, Belgrade.

Gavrilović, V., Cvetković, D.D., Džukić, G. & Petkovski, S. 1999. Comparative morphological study of *Rana balcanica* and *Rana ridibunda. Contributions to the Zoogeography and Ecology of the Eastern Mediterranean Region* 1, 205-210.

Grillitsch, H. & Cabela, A. 1990. Zum systematischen Status der Blindschleichen (Squamata: Anguidae) der Peloponnes und der sudlichen Ionischen Inseln (Griechenland). *Herpetozoa* 2, 131-153.

Grossenbacher, K. 1994. Zur Systematik und Verbreitung der Alpensalamander (*Salamandra atra atra, Salamandra atra aurorae, Salamandra lanzai*). *Abhandlungen und Berichte für Naturkunde* 17, 75-81.

Guerinii, F., Bucci, S., Ragghianti, M., Mancino, G., Hotz, H., Uzzell, T. & Berger, L. 1997. Genomes of two water frog species resist germ line exclusion in intraspecific hybrids. *Journal of Experimantal Zoology* 279, 163-176.

Guicking, D., Joger, U. & Wink, M. 2001. Phylogeography of the viperine snake *Natrix maura* and the dice snake *Natrix tessellata*: first results. *Biota*, 2, (Supplement: 12), 27.

Guillaume, C.-P., Heulin, B. & Beshkov, V. 1997. Biogeography of *Lacerta* (*Zootoca*) *vivipara*: reproductive mode and enzyme phenotypes in Bulgaria. *Ecogeography* 20, 240-246.

Hadži, J. 1935. Kurze zoogeographisce Übersicht Jugoslaviens. *Verhandlungen der Internationalen Vereinigung für theoretische und angewandte Limnologie* 7, 36-45.

Hadži, E., Pantić, N., Aleksić, V. & Kalenić, M. 1974. "Alpi jugoistožne Europe u svetlu tektonskih ploča". In: *Metalogenija i koncepcije geotektonskog razvoja Jugoslavije* (S. Janković, ed.), pp. 231-260. Rudarsko-geoloski fakultet, Beograd (in Serbian).

Harrison, G. R. 1993. *Hybrid zones and the evolutionary process*. Oxford University Press, New York.

Haxhiu, I. 1994. The herpetofauna of Albania. Amphibia: species composition, distribution, habitats. *Zoologische Jahrbuecher Systematik (Jena)* 12, 321-334.

Haxhiu, I. 1998. The reptiles of Albania: composition, distribution, habitats. *Bonner Zoolgische Beitrage* 48, 35-57.

Heulin, B., Guillaume, C., Arrayago, M.-J. & Bea, A. 1993. Interpretation biogéographique de la bimodalité de reproduction du lézard *Lacerta vivipara*: un modele pour l'étude de l'évolution de la viviparité. *Biogeographica* 69, 1-11.

Heulin, B., Guillaume, C., Vogrin, N., Surget-Groba, Y & Tadić, Z. 2000. Further evidence of the existence of oviparous populations of *Lacerta* (*Zootoca*) *vivipara* in the NW of the Balkan Peninsula. *C.R. Academie des Sciences (Paris), Sciences de la vie/Life Sciences* 323, 461-468.

Hewitt, G.M. 1996. Some genetic consequences of ice ages, and their role in divergence and speciation. *Biological Journal of the Linnean Society* 58, 247-276.

Hewitt, G.M. 1999. Post-glacial re-colonization of European biota. *Biological Journal of the Linnean Society* 68, 87-112.

Hirtz, M. 1937. "Nova vrsta otrovnih zmija" *Priroda (Zagreb)* 27, 221-222 (in Croatian).

Hosey, G.R. 1982. The Bosphorous land-bridge and mammal distributions in Asia Minor and the Balkans. *Säugetierkundliche Mitteilungen* 30, 53-62.

Hsu, K.J. & Bernoulli, D. 1978. Genesis of the Tethys and the Mediterranean. In: *Initial Reports of the Deep Sea Drilling Project*, Vol. 42(1), pp. 943-949. U.S. Government Printing Office, Washington.

Humphries, C.J. & Parenti, L.R. 1986. *Cladistic biogeography*. Clarendon Press, Oxford.

Jong, H. de 1998. In search of historical biogeographic patterns in the western Mediterranean terrestrial fauna. *Biological Journal of the Linnean Society* 65, 99-164.

Kalezić, M.L. 2002. Heterochrony in the Balkan newts (*Triturus*, Caudata): An overview. In: *Genetics, Ecology and Evolution* (B.P.M. Ćurćić & M. Andjelković, eds.), pp. 65-76. Institute of Zoology, Belgrade.

Kalezić, M.L. & Hedgecock, D. 1980. Genetic variation and differentiation of three common European newts (*Triturus*) in Yugoslavia. *British Journal of Herpetology* 6, 49-57.

Kalezić, M.L., Džukić, G., Mesaroš, G. & Crnobrnja-Isailović, J. 1997. The crested newt (*Triturus cristatus* Superspecies) in ex-Yugoslavia: morphological structuring and distribution patterns. *University Thoughts (Pristina)* 4, 39-46.

Karaman, S. 1921. Beitrage zur Herpetologie von Jugoslawien. *Glasnik Hrvatskog Prirodoslovnog društva* 33, 194-209.

Kincel, F. 1962. Eine *Vipera lebetina* L. aus der Bucht von Cattaro. *Mitteilungen der Abteilung für Zoologie und Botanik am Landesmuseum "Joanneum" (Graz)* 14, 10-11.

Kincel, F. 1969. Noch einmal die *Vipera lebetina* L. aus der Bucht von Cattaro. *Mitteilungen der Abteilung für Zoologie und Botanik am Landesmuseum "Joanneum" (Graz)* 31, 18-19.

Križmanić, I., Mesaroš, G., Džukić, G. & Kalezić, M.L. 1997. Morphology of the smooth newt (*Triturus vulgaris*) in former Yugoslavia: taxonomical implications and distribution patterns. *Acta Zoologica Academiae Scientiarum Hungaricae* 43, 345-357.

Lapini, L., A. dall'Asta, Bressi, N., Dolce, S. & Pellarini, P. 1999. *Atlante Corologico: Degli anfibi e dei rettili del Friuli-Venezia Guiula.* Comune di Udine, Edizioni del Museo Friulano di Storia Naturale, Pubblicazione no. 43.

Lenk, P., Joger, U., Fritz, U., Heidrich, P. & Wink, M. 1998. Phylogeographic patterns of the mitochondrial cytochrone b gene of the European pond turtle (*Emys orbicularis*): first report. *Mertensiella* 10, 159-175.

Litvinchuk, S., Sokolova, T.M. & Borkin, L.J. 1994. Biochemical differentiation of the crested newt (*Triturus cristatus* group) in the territory of the former USSR. *Abdhandlungen und Berichte für Naturkunde* 17, 67-74.

Litvinchuk, S., Rosanov, J.M. & Borkin, L.J. 1997. A contact zone between the newts *Triturus cristatus* and *Triturus dobrogicus* in the Ukrainian Transcarpathians: distribution and genome size variation. In: *Herpetologia Bonnensis* (W. Bohme *et al*., eds.), pp. 229-235.

Litvinchuk, S.N., Borkin, L.J., Džukić, G., Kalezić, M.L., Khalturin, M.D. & Rosanov, J.M. 1999. Taxonomic status of *Triturus karelinii* on the Balkans. *Russion Journal of Herpetology* 6, 153-163.

Litvinchuk, S. & Borkin, L.J. 2000. Intraspecific taxonomy and nomenclature of the Danube crested newt, *Tritururs dobrogicus*. *Amphibia-Reptilia* 21, 419-430.

Ljubisavljević, K., Džukić, G. & Kalezić, M.L. 2002. Morphological differentiation of the snake-eyed skink *Ablepharus kitaibelii* (Bibron & Bory, 1833), in the north-western part of the range: systematic implications. *Herpetozoa* 14, 107-121.

Macey, J.R., Schulte, J.A., II. Larson, A., Ananjeva, N.B., Wang, Y., Pethiyagoda, R., Rastegar-Pouyani, N. & Papenfuss, T.J. 2000. Evaluating trans-Tethys migration: an example using acrodont lizard phylogenetics. *Systematic Biology* 49, 233-256.

Matvejev, S. 1961. *Biogeography of Yugoslavia.* Biološki institut N.R. Srbije, Posebna izdanja 9, Beograd (in Serbian with English summary).

Mayer, W., Bohme, W., Tiedemann, F. & Bischoff, W. 2000. On oviparous populations of *Zootoca vivipara* (Jacquin, 1787) in south-eastern Central Europe and their phylogenetic relationship to neighbouring viviparous and South-west European oviparous populations (Squamata: Sauria: Lacerilia). *Herpetozoa* 13, 59-69.

Mertens, R. & Wermuth, H. 1960. *Die Amphibien und Reptilien Europas.* Verlag Waldemar Kramer, Frankfurt am Main.

Mršić, N. 1978. A contribution to the knowledge of te taxonomy, zoogeography and ecology of the reptiles of the Velebit mountain-chain. *Academia Scientiarum et Artium Slovinica, Classis IV: Historia Naturalis Diss.* 21, 1-43.

Oosterbroek, P. & Arntzen, J.W. 1992. Area-cladograms of Circum-Mediterranean taxa in relation to Mediterranean palaeogeography. *Journal of Biogeography* 19, 3-20.

Paunović, M. 1990. Grundzüge der Herkunft und das Alter der Herpetofauna Jugoslawiens. *Rad JAZU 449, Razred za prirodne znanosti* 24, 309-317 (in Croatian with German summary).

Pidancier, N., Miaud, C. & Taberlet, P. 2001. Phylogeography of the common frog *Rana temporaria* inferred from mitochondrial DNA sequences. *Biota*, 2, (Supplement: 12), 101.

Plötner, J. & Ohst, T. 2001. New hypotheses on the systematics of the western Palearctic water frog complex (Anura, Ranidae). *Mitteilungen aus dem Museum für Naturkunde in Berlin* 77, 5-21.

Plötner, J. & Schmeller, D. 2001. Some remarks on current research on western Palearctic water frogs. *Mitteilungen aus dem Museum für Naturkunde in Berlin* 77, 3-4.

Radovanović, M. & Martino, K. 1950. *"Zmije Balkanskog poluostrva".* Srpska Akademija Nauka, Institut za ekologiju i biogeografiju, Beograd (in Serbian).

Raff, R.A. & Kaufman, T.C. 1983. *Embryos, genes, and evolution.* MacMillan, New York.

Rafinski, J. & Arntzen, J.W. 1987. Biochemical systematics of the Old World newts, genus *Triturus*. *Herpetologica* 43, 446-457.

Rage, J.-C. 1997. Palaeobiological and palaeogeographical backrounds of the European herpetofauna. In: *Atlas of Amphibians and Reptiles in Europe* (J.-P. Gasc *et al*., eds.), pp. 23-28. Societas Europae Herpetologica, Museum National d'Histoire Naturelle, Paris.

Ramirez, J.M., Vargas, J.M. & Guerrero, J.C. 1992. Distribution patterns and diversity in European reptiles. In: *Proceedings of the 6th Ordinary General Meeting of the Societas Europaea Herpetologica* (Z. Korsos & I. Kiss, eds.), pp. 371-376. Budapest.

Raxworthy, C.J. 1990. A review of the smooth newt (*Triturus vulgaris*) subspecies, including an identification key. *The Herpetological Journal* 1, 481-492.

Real, R., Antunez, A. & Vargas, J.M. 1992. A biogeographic synthesis of European amphibians. In: *Proceedings of the 6th Ordinary general Meeting of the Societas Europaea Herpetologica* (Z. Korsos & I. Kiss, eds.), pp. 377 -380. Budapest.

Reilly, S.M., Wiley, E.O. & Meinhardt, D.J. 1997. An integrative approach to heterochrony: the distinction between interspecific and intraspecific phenomena. *Biological Journal of the Linnean Society* 60, 119-143.

Sara, M. 1996. *Uomo e Fauna Nelle Isole Mediterranee*. Medical Books, Palermo.

Schulz, K.-D. & Entzeroth, A. 1996. *A Monograph of the Colubrid Snakes of the genus* Elaphe *Fitzinger*. Koeltz Scientic Books, Havličkuv Brod.

Shine, R. 1985. The evolution of viviparity in reptiles: an ecological analysis. In: *Biology of Reptilia* (C. Gans & F. Billet, eds.), Vol. 15, pp. 605-694. John Wiley and Sons, New York.

Sikošek, B. 1986. Seismotectonical composition and earthquake source regions in Yugoslavia. In: *Collection of Scientific Works dedicated to the Academician Zarija M. Besić* (Lj. Pavičević, ed.), pp. 165-185. Montenegrin Academy of Science and Arts, Vol. 2. Titograd.

Simroth, H. 1907. III. "O puževima golačima Crne Gore". *Glasnik Zemaljskog muzeja u BiH* 19, 509-519 (in Serbo-Croatian).

Sket, B. 1997. Distribution of *Proteus* (Amphibia: Urodela: Proteidae) and its possible explanation. *Journal of Biogeography* 24, 263-280.

Sofianidou, T. S., Schneider, H. & Sinsch, U. 1994. Comparative electrophoretic investigation on *Rana balcanica* and *Rana ridibunda* from northern Greece. *Alytes* 12, 93-108.

Sotiropoulos, K., Polymeni, R.-M. & Legakis, A. 1999. Allozyme clines and population differentation in *Triturus cristatus* complex (Amphibia: Urodela) from Greece: preliminary results. In: *Abstracts of the 10th Ordinary General Meeting of the Societas Europaea Herpetologica*, p. 240. Natural History Museum of Crete, Erakleio.

Sotiropoulos, K., Tomović, Lj., Džukić, G. & Kalezić, M.L. 2001a. Morphological differentiation of the alpine newt (*Triturus alpestris*) in the Balkans: taxonomical implications. *The Herpetological Journal* 11, 1-8.

Sotiropoulos, K., Polymeni, R.M. & Legakis, A. 2001b. *Triturus alpestris* in Greece: extensive genetic variability and population differentiation. *Biota*, 2, (Supplement: 12), 112.

Spasić-Bošković, O., Križmanić, I. & Vujošević, M. 1999. Population composition and genetic variation of water frogs (Anura: Ranidae) from Yugoslavia. *Caryologia* 52, 9-20.

Steininger, F.F. & Rogl, F. 1996. Palaeogeography and palinspatic reconstructions of the Neogene of the Mediterranean and Paratethys. In: *The Geological Evolution of the Eastern Mediterranean* (J.E. Dixon & A.H.F. Robertson, eds.), pp. 659-668. The Geological Society, London.

Stevens, G.C. 1992. The elevational gradient in altitudinal range: an extension of Rapoport's latitudinal rule to altitude. *The American Naturalist* 140, 893-911.

Szymura, J.M. 1988. Regional differentiation and hybrid zones between the fire-bellied toads *Bombina bombina (L.) and Bombina variegata (L.) in Europe*. Rozprawy Habilitacyjne No. 147. Cracow: Uniwersytet Jagiellonski (in Polish with English abstract).

Szymura, J.M. 1993. Analysis of hybrid zones within *Bombina*. In: *Hybrid Zones and the Evolutionary Process* (R.G. Harrison, ed.), pp. 261-287. Oxford University Press, New York.

Szymura, J.M., Uzzell, T. & Spolsky, C. 2000. Mitochondrial DNA variation in the hybridizing fire-bellied toads, *Bombina bombina* and *B. variegata*. *Molecular Ecology* 9, 891-899.

Szyndlar, Z. 1984. Fossil snakes from Poland. *Acta Zoologica Cracoviensia* 28, 1-156.

Tomović, Lj. & Džukić, G. in press. Geographic variability and taxonomy of the nose-horned viper, *Vipera ammodytes* (L. 1758), in the central and eastern parts of the Balkans: a multivariate study. *Amphibia-Reptilia*.

Ugurtas, I., Ljubisavljević, K., Sidorovska, V., Kalezić, M.L. & Džukić, G. 2002. Morphological differentiation of the eastern spadefoot toad (*Pelobates syriacus*) populations. *Israel Journal of Zoology* 48, 13-32.

Varga, Z. 1995. Geographical patterns of biological diversity in the Palaeartic region and the Carpathian basin. *Acta Zoologica Academiae Scientiarum Hungaricae* 41, 71-92.

Venczel, M. 2000. *Quaternary Snakes from Bihor (Romania)*. Tarii Crisurilor Museum, Oradea.

Wallis, G.P. & Arntzen, J.W. 1989. Mitochondrial DNA variation in the crested newt superspecies: limited cytoplasmic gene flow among species. *Evolution* 43, 88-104.

Weisrock, D.W., Macey, J.R., Ugurtas, I.H., Larson, A. & Papenfuss, T.J. 2001. Molecular phylogenetics and historical biogeography among salamandrida of the "true" salamander clade: rapid branching of numerous highly divergent lineages in *Martensiella luschani* associated with the rise of Anatolia. *Molecular Phylogenetics and Evolution* 18, 434-448.

World Conservation Monitoring Centre. 1997. *Red data book of European vertebrates*. Council of Europe, Strasbourg.

APPENDIX

Annotated checklist of Balkan amphibians and reptiles with taxonomic inferences. Amphibian and reptile species affiliations to the six regions defined for the Balkans are given in parentheses (see text for explanation). Regions are abbreviated as follows: AT - "Adriatic Triangle", P - Peloponnisos, SWC - South-west Croatia, E - Evros, PP - Peripannonian region, D - Dobrugea. Other abbreviations and symbols: * - endemic taxon, Rb - range border (taxon with marginal populations in the Balkans), E - eastern range border (marginal part of the species range on the Balkans is orientated towards Eastern Europe), N - northern range border, NW - northwestern range border, S - southern range border, SE - southeastern range border. A question mark (?) denotes that the presence of a taxon in the Balkans is probable, but not confirmed.

CAUDATA

1. *Proteus anguinus** (AT, SWC)
   Subspecies: *anguinus, parkelj*
2. *Salamandra atra* Rb.S. (AT, SWC)
   Subspecies: *atra, prenjensis**
3. *Salamandra salamandra* Rb.E. (AT, P, SWC, E, PP)
   Subspecies: *beshkovi*, salamandra, werneri**
4. *Mertensiella luschani**
   Subspecies: *helverseni*
5. *Triturus alpestris* Rb.SE. (AT, P, SWC, PP)
   Subspecies: *alpestris, serdarus*, veluchiensis**
6. *Triturus cristatus* Rb.S. (PP)
7. *Triturus dobrogicus* Rb.S. (PP, D)
   Subspecies: *dobrogicus, macrosomus*
8. *Triturus carnifex* Rb.E. (AT, SWC, PP)
   Subspecies: *carnifex, macedonicus**
9. *Triturus karelinii* Rb.NW. (E, PP)
   Subspecies: *arntzeni*, ?karelinii*
10. *Triturus vulgaris* Rb.S. (AT, P, SWC, E, PP, D)
    Subspecies: *graecus*, meridionalis, tomasinii*, vulgaris*
    Comment: To judge from male qualitative characteristics of many populations, the separate subspecific status of *tomasinii* appears to be confirmed (Križmanić *et al.*, 1997), but for different opinion see Raxworthy (1990).

ANURA

11. *Bombina bombina* Rb.S. (E, PP, D)
    Subspecies: *bombina*
12. *Bombina variegata* Rb.SE. (SWC, PP)
    Subspecies: *variegata*
13. *Bombina scabra** (AT, E)
    Comment: Taxonomic status uncertain. To judge from allozyme analysis, the distinctiveness of *scabra* at the specific level appears to be conclusively proved

(Radojičić *et al.*, unpublished data). Szymura *et al.* (2000) would disagree, however.

14. *Pelobates fuscus* Rb.S. (SWC, PP, D)
    Subspecies: *fuscus*
15. *Pelobates syriacus* Rb.NW. (P, E, PP, D)
    Subspecies: *balcanicus**, unnamed subspecies*
    Comment: Our morphological analysis shows that populations from Serbia are distinct from the Macedonian, Bulgarian, and Turkish Thrace (Edirne) ones, which belong to the subspecies *P. s. balcanicus* (Ugurtas *et al.*, 2002).
16. *Bufo bufo* (AT, P, SWC, E, PP, D)
    Subspecies: *bufo,* cfr. *spinosus*
17. *Bufo viridis* (AT, P, SWC, E, PP, D)
    Subspecies: *viridis*
18. *Hyla arborea* (AT, P, SWC, E, PP, D)
    Subspecies: *arborea*, *kretensis**
19. *Hyla intermedia* Rb.E.
20. *Rana* kl. *esculenta* Rb.S. (SWC, PP, D)
21. *Rana lessonae* Rb.S. (PP, D)
22. *Rana ridibunda* (AT, P, SWC, E, PP, D)
23. *Rana shqiperica** (AT)
24. *Rana balcanica* (=kurtmuelleri*)
    Comment: The specific status of this taxon has been refuted recently (Plötner & Schmeller, 2001). However, bioacoustic data and electrophoretic investigations (Sofianidou *et al.*, 1994), as well as morphological study (Gavrilović *et al.*, 1999) clearly distinguished it from nearby populations of *R. ridibunda.*
25. *Rana epeirotica** (P)
26. *Rana cerigensis**
27. *Rana cretensis**
28. *Rana latastei* Rb.E.
29. *Rana dalmatina* Rb.E. (AT, P, SWC, E, PP, D)
30. *Rana temporaria* Rb.S. (AT, SWC)
    Subspecies: *temporaria*
31. *Rana graeca** (AT, P, E, PP)
32. *Rana arvalis* Rb.S. (SWC)
    Subspecies: *wolterstorffi*

? 33. *Rana catesbeiana* Introduced

CHELONIA

1. *Caretta caretta* (AT, P, SWC, E, D)
   Subspecies: *caretta*
2. *Chelonia mydas* (AT, P, SWC, E, D)
   Subspecies: *mydas*
3. *Eretmochelys imbricata*
4. *Dermochelys coriacea* (AT, P, SWC)
5. *Emys orbicularis* (AT, P, SWC, E, PP, D)
   Subspecies: *orbicularis, hellenica**
6. *Mauremys caspica* Rb.NW.(AT, P, E)
   Subspecies: *rivulata*

7. *Trachemys scripta* Introduced
   Subspecies: *elegans*
8. *Testudo graeca* Rb. NW. (P, E, D)
   Subspecies: *ibera*
9. *Testudo hermanni* Rb.N. (AT, P, SWC, E, PP)
   Subspecies: *hermanni, boettgeri**
   Comment: Subspecies distinctness and its distribution seem to be much more complex than was previously thought (Bour, 1997). The presence of ssp. *hermanni* in the northwestern part of the Balkans is suggested (*e.g.* Lapini *et al.*, 1999).
10. *Testudo marginata* Rb.N. (P)
11. *Testudo weissingeri** (P)

AMPHISBAENIA

12. *Blanus strauchi* Rb.W.
    Subspecies: *strauchi*
    Comment: Distribution is restricted to one Adriatic island (see Karaman, 1921). It was previously considered an introduced species, but palaeontological and palaeogeographic data (Delfino, 1995, 1997) suggest it is autochthonous.

LACERTILIA

13. *Laudakia stellio* Rb.W.
    Subspecies: *daani, stellio*
14. *Anguis cephallonicus** (P)
15. *Anguis fragilis* Rb.S. (AT, SWC, E, PP, D)
    Subspecies: *colchicus, fragilis*
16. *Pseudopus apodus* Rb.NW. (AT, P, SWC, E, D)
    Subspecies: *thracius*
17. *Chamaeleo africanus* Rb.N. (P)
18. *Chamaeleo chamaeleo* Rb.N
19. *Cyrtodactylus kotschyi* (AT, P, E)
    Subspecies: *adelphiensis*, bartoni*, beutleri*, bibroni, bileki*, buchholzi*, danilewskii*, fuchsi*, kalypsae*, kotschyi*, oertzeni*, rarus*, rumelicus*, saronicus*, schultzewestrumi*, skopjensis*, solerii*, stepaneki*, stubbei, tinensis*, unicolor*, weltsteini**
20. *Hemidactylus turcicus* Rb.N. (AT, P, SWC, E)
    Subspecies: *H. t. turcicus*
21. *Tarentola mauritanica* Rb.N. (AT, P, SWC)
    Subspecies: *mauritanica, fascicularis**
22. *Algyroides moreoticus** (P)
23. *Algyroides nigropunctatus** (AT, SWC)
    Subspecies: *nigropunctatus, kephallithacius*
24. *Eremias arguta* Rb. SW. (D)
    Subspecies: *deserti*
25. *Lacerta agilis* Rb.S. (AT, SWC, PP, D)
    Subspecies: *agilis, argus, bosnica*, chersonensis*
26. *Lacerta bilineata* Rb.E. (SWC)

27. *Lacerta trilineata* Rb.NW. (AT, P, SWC, E, D)
Subspecies: *cariensis*, citrovittata*, diplochondrodes, dobrogica*, hansschweizeri*, major*, polylepidota*, subcellata*, trilineata*
28. *Lacerta viridis* (AT, E, PP, D)
Subspecies: *citrovittata*, meridionalis*, viridis*
29 *Iberolacerta horvathi* (SWC)
30. *Darevskia praticola* (E, PP, D)
Subspecies: *?hungarica, pontica, praticola*
31. *Archaeolacerta graeca** (P)
32. *Archaeolacerta mosorensis** (AT)
33. *Archaeolacerta oxycephala** (AT)
Subspecies: *oxycephala, tomasinii*
34. *Zootoca vivipara* Rb.SE. (AT, SWC)
Subspecies: *pannonica, vivipara, carniolica*
35. *Ophisops elegans* Rb.W. (E)
Subspecies: *ehrenbergii*, macrodactylus*
36. *Podarcis erhardii** (P, E)
Subspecies: *amorgensis, biinsulacola, buchholzi, cretensis, elaphonisi, erhardii, kinarensis, levithensis, leukaorii, livadiaca, makariaisinii, megalophthenae, mykonensis, naxensis, ophidusae, pachiae, phytiusae, punctigularis, rechingeri, riveti, ruthveni, schiebeli, subobscura, syrinae, thermiensis, thessalica, werneriana, zafranae*
37. *Podarcis melisellensis** (AT, SWC)
Subspecies: *aeoli, bokicae, caprina, curzolensis, digenea, fiumana, galvagnii, gigantea, gigas, jidulae, kammereri, kornatica, lissana, lupa, melisellensis, mikavicae, plutonis, pomoensis, thetidis, traguriana*
38. *Podarcis milensis**
Subspecies: *adolfjordansi, gaigeae, gerakuniae, milensis, schweizeri, weigandi*
39. *Podarcis muralis* Rb.E. (AT, P, SWC, E, PP, D)
Subspecies: *albanica*, maculiventris, muralis*
40. *Podarcis peloponnesiaca** (P)
Subspecies: *lais, peloponnesiaca, thais*
41. *Podarcis sicula* Rb.E. (AT, SWC)
Subspecies: *adriatica*, astorgae*, bagnolensis*, bolei*, campestris*, dupenici*, fiumanoidae*, flavigula*, hadzii*, hieroglyphica*, insularum*, kattaroi*, kolombatovici*, kurtklari*, laganjensis*, nikolici*, pelagosae*, pirosoensis*, polenci*, pohlibensis*, premudensis*, premudana*, pretneri*, radovanovici*, ragusae*, samogradi*, veseljuhi*, zeii**
42. *Podarcis taurica* Rb.NW. (AT, P, E, PP, D)
Subspecies: *ionica*, taurica, thasopulae**
43. *Ablepharus kitaibelii* Rb.W. (AT, P, E, PP, D)
Subspecies: *fabichi*, fitzingeri, kitaibelii, stepaneki*
44. *Chalcides moseri**
Comment: dubious taxon (Mertens & Wermuth, 1960; Bringsoe, 1988)
45. *Chalcides ocellatus* Rb.N. (P)
Subspecies: *ocellatus**
46. *Mabuya aurata* Rb.W.
Subspecies: *fellowsii**
47. *Ophiomorus punctatissimus* (P)

SERPENTES

48. *Typhlops vermicularis* Rb.NW. (AT, P, SWC, E)
49. *Eryx jaculus* Rb.N. (P, E, D)
   Subspecies: *turcicus*
50. *Coluber caspius* Rb.W. (AT, E, PP, D)
51. *Coluber gemonensis** (AT, P, SWC)
52. *Coluber najadum* Rb.NW. (AT, P, SWC, E)
   Subspecies: *dahlii*
53. *Coluber nummifer (=revergieri)* Rb.W.
   Subspecies: *nummifer*
54. *Coluber rubriceps* Rb.W.
   Subspecies: *thracius**
55. *Coluber viridiflavus* Rb.E. (SWC)
   Subspecies: *carbonarius, cfr. viridiflavus*
   Comment: see Mršić (1978) for the presence of two morphotypes in the Velebit Mountains.
56. *Coronella austriaca* Rb.S. (AT, P, SWC, E, PP, D)
   Subspecies: *austriaca*
57. *Eirenis modestus* Rb.W.
   Subspecies: *modestus*
58. *Elaphe longissima* (AT, P, SWC, E, PP, D)
   Subspecies: *longissima, rechingeri*, romana*
   Comment: Džukić (1975) pointed out the morphological characteristics of *E. longissima* from Serbia, which according to Schulz & Entzeroth (1996), should be assigned to *E. l. romana*. See text for possible introduction of *E. l. romana* in the Balkans.
59. *Elaphe quatuorlineata* Rb.N. (AT, P, SWC, E, D)
   Subspecies: *quatuorlineata, muenteri*, sauromates*
60. *Elaphe situla* Rb.NW. (AT, P, SWC, E)
61. *Malpolon monspessulanus* Rb.N. (AT, P, SWC, E)
   Subspecies: *fuscus, insignitus*
62. *Natrix natrix* (AT, P, SWC, E, PP, D)
   Subspecies: *helvetica, ?dalmatina, natrix, ?persa, schweizeri**
63. *Natrix tessellata* (AT, P, SWC, E, PP, D)
   Subspecies: *tessellata, heinroth**
64. *Telescopus fallax* Rb.NW. (AT, P, SWC, E)
   Subspecies: *fallax, intermedius*, multisquamatus*, pallidus*, rhodicus**
65. *Vipera ammodytes* (AT, P, SWC, E, PP, D)
   Subspecies: *ammodytes*, gregorwallneri, illyrica*, meridionalis, montandoni*
66. *Vipera aspis* Rb.E.
   Subspecies: ?*atra, francisciredii*
67. *Vipera berus* Rb.S. (AT, SWC, E, PP)
   Subspecies: *berus, bosniensis**

? 68. *Macrovipera* sp.* (AT)
   Comment: There has long been controversy about the existence of a unusual viper species in Boka Kotorska (Hirtz, 1937). Kincel (1962, 1969) assigned this viper to *lebetina* species. In the absence of recently-captured specimens (this viper might be

extinct now), we still do not know to which species of the genus *Macrovipera* this snake should be assigned.

69. *Macrovipera schweizeri**
70. *Vipera ursinii* Rb.S. (AT, SWC, D)
    Subspecies: *macrops**, *rakosiensis, graeca**
71. *Vipera xanthina* Rb.W. (E)
    Subspecies: *xanthina*

11. A ZOOGEOGRAPHICAL REVIEW OF THE SPIDERS (ARANEAE) OF THE BALKAN PENINSULA

CHRISTO DELTSHEV*

INTRODUCTION

The Balkan Peninsula has an extremely rich and diversified invertebrate fauna. The group of spiders is characterised by considerable species richness and also by the presence of many endemics, both at the level of the genus and the species. The spider fauna has been comparatively well studied due to the efforts of many araneologists from different countries. The earliest data were published by workers such as: C.L. Koch, Lucas, Keyserling, L. Koch, Schiodte, Cambridge, Simon, Kulczynski, Pavesi, Nosek, Absolon, Roewer, Fage, Giltay, Wiehle, Stoichevich, Reimoser, Kolosvary, Kratochvil, Miller and Caporiacco. The first major work concerning the spiders of the region as a whole, however, was the work of Drensky (1936), who reported 1,066 species from 35 families in a review of all the currently available literature. Some years later Hadjissarantos (1940) compiled the currently available faunistic data on the spiders of continental Greece. More recently, Nikolić & Polenec (1981) combined the data concerning the spiders of Former Yugoslavia and estimated a total of 1,022 species for the region. More recent publications list the fauna of Bulgaria, Greece, Serbia, Macedonia, Montenegro and parts of Turkey (Brignoli, 1968, 1971, 1972, 1974a, 1974b, 1976, 1977, 1979, 1984, 1986; Deeleman-Reinhold, 1976, 1978, 1988, 1993; Deltshev, 1979a, 1979b, 1983a, 1983b, 1985, 1988, 1990, 1993, 1996, 1997a, 1997b; Deltshev & Blagoev, 2000; Deltshev & Ćurčić, 1997; Deltshev & Paraschi, 1990; Thaler, 1996; Thaler & Knoflach, 1991, 1993, 1995; Wunderlich, 1980, 1985, 1994a, 1994b, 1994c). These contributions are the result of intensive faunistic research, and the accumulation of new data is now sufficient to allow the critical zoogeographical analysis of the distribution of spiders within the territory of the Balkan Peninsula.

STUDY AREA AND MATERIAL

The Balkan Peninsula is situated in southeastern Europe. As defined here, its northern border follows the rivers Danube (including its delta), Sava and Soča, to the Gulf of Trieste *via* the regions of Gorizia and Monfalcone. Its western border follows the line of Adriatic and Ionian coast including the islands. The eastern border passes to the east of the Aegean Islands Sirina, Astipalea, Amorgos, Miconos, Tinos, Andros, Skiros, Limnos and Imros, continues along the Dardanelles, goes across the Marmara Sea and reaches the Black Sea coast *via* the Bosphorus. The southernmost point of the peninsula is Crete and the islands of Gavdos, Aiduronisi and Kufonisi (Figure 1).

H.I. Griffiths et al. (eds.), Balkan Biodiversity, 193–200.

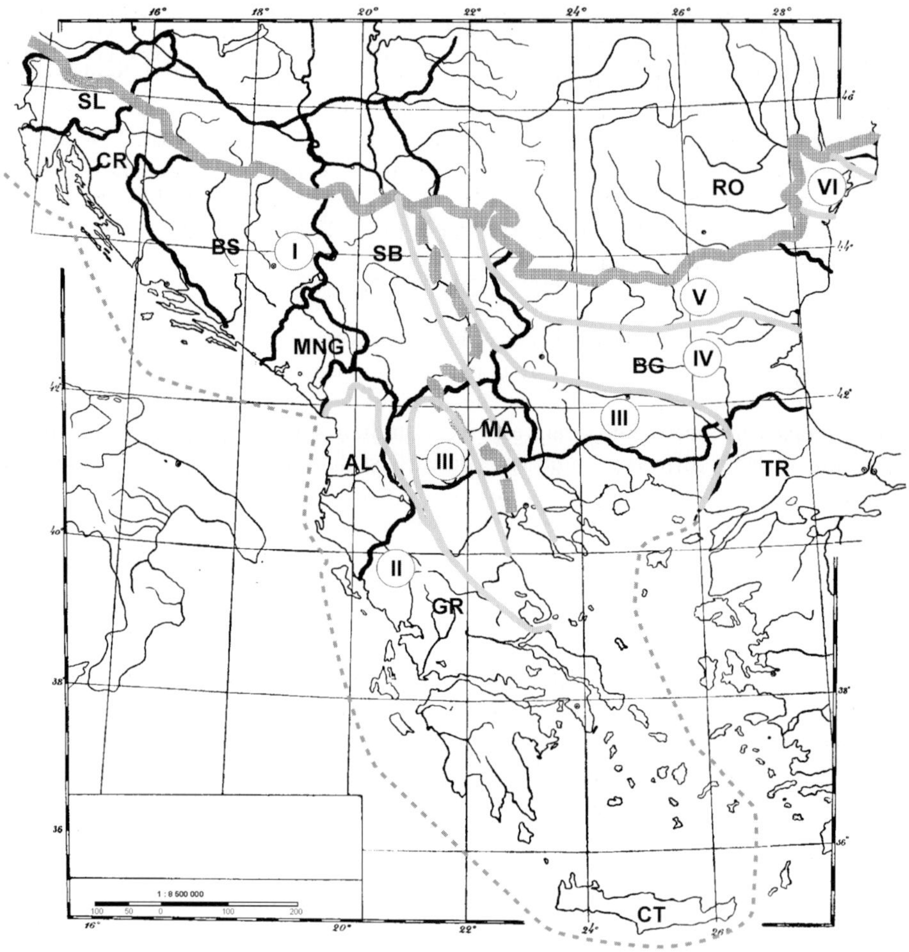

Border between Eastern and Western part of the Balkan peninsula
Borders of main geographic regions
Borders of Balkan countries
Borders of Balkan peninsula aquatory

Figure 1. Geographic division of the Balkan Peninsula. I: Dinaric region; II: Pindus region; III: Thrace-Macedonian region; IV: Balkanid region; V: Danubian Plains region; VI: North Dobrugea region. SL = Slovenia; CR = Croatia; BS = Bosnia; MNG = Montenegro; AL = Albania: GR = Greece; CT = Crete; SB = Serbia; MA = Macedonia; BG = Bulgaria; RO = Romania; TR = Turkey.

The geographical areas and associated abbreviations used in the text are shown in Figure 1. Data on general zoogeographical distribution are taken mainly from Platnick (1989, 1993, 1997). The zoogeographical categories used and their abreviations are as follows: WD - Widely distributed; COS - Cosmopolitan; PPT - Palaearctic-Palaeotropic; H- Holarctic; OW - Old World; P - Palaearctic; WP - West Palaearctic; ECA - European-Central Asian; E - European; HE - Holoeuropean; MEE - Middle East- European; MSE - Middle South European; MSEE - Middle Southeast European; EE - East European; SE - South European; SEE - Southeast European; PO - Pontic; BK - Balkans; M - Mediterranean; HM - Holomediterranean; EM - East Mediterranean; NM - North Mediterranean; NEM - Northeast Mediterranean; SEM - Southeast Mediterranean; BKMA - Balkan-Asia Minor; POM - Pontic-Mediterranean.

RESULTS AND DISCUSSION

Zoogeographical analysis

The spider fauna of the Balkan Peninsula is represented by 1409 species, included in 337 genera and 48 families. The number of species is high compared with the number of spiders recorded from other parts of Europe, which are as follows: France = 1400 (Jones *et al.*, 1990); Russian Plain = 1001 (Michailov, 1997); Alps = 1000 (Thaler, 1980); Germany = 925 (Coponen, 1993); Switzerland = 875 (Maurer & Hanggi, 1990); England & Wales = 624 (Roberts, 1987). According to their current distribution the established 1409 species can be classified into 24 zoogeographical categories, grouped into four main groups (Figures 2 & 3).

The complex of widely distributed species (WD = COS + PPT + H + OW + P + WP + ECA) is most well represented, with 533 species (38.1%). Within the WD complex, Palaearctic species are dominant (75.4%), followed by Holarctic (19.9%), Cosmopolitan (3.8%) and Palaearctic-Palaeotropic (0.2%). The complex includes especially widespread species asociated with lowlands, buildings, woodlands and high altitude zones of mountains.

The complex of Balkan Endemics (BK), forms the second largest group and comprises 379 species (26.9%). The established number is high and reflects the local character of the fauna. The question about the status and distribution of endemic spiders found in the Balkan Peninsula is a complex one. Some of them are found only in restricted areas (even in a single cave) while the others show wider distributions, sometimes even over the whole peninsula. Certainly, some of the widespread Balkan endemics can be found in neighbouring territories as well and can be placed within the Balkans, Asia Minor or the southeast European spider fauna, or beyond.

The most characteristic families are: Linyphiidae s. l. (31.3% of all spider species found in the Balkans), Dysderidae (21.0%) and Agelenidae (8.3%). The genus *Troglohyphantes* is the most numerous of endemics and can be regarded as a faunistic phenomenon since, of the total of 53 species established within the territory of the Balkans, 52 are endemics, distributed mainly in caves. Deeleman-Reinhold (1978) concluded that the present distribution and morphological diversity of *Troglohyphantes* in the Balkan Peninsula represents a repeated processes of expansion and contraction of its range. The representation of the genera *Dysdera* (28 endemics from 38 species), *Lepthyphantes* (18 endemics from 49 species) and *Tegenaria* (17 endemics from 31 species) is also due to expansion in caves, woodlands and highlands. Present-day examples of cave penetration are the species *Lepthyphantes centromeroides* and *L. spelaeorum*, which are widespread in the

Balkan Peninsula. They occur in caves but also in humus and ground detritus and indicate active subterranean colonisation (Deeleman-Reinhold, 1978). It should be emphasised that from the established 14 endemic genera (*Antrohyphantes*, *Barusia*, *Cryphoecina*, *Fageiella*, *Folkia*, *Icariella*, *Lasconia*, *Macedoniella*, *Minotauria*, *Protoleptoneta*, *Parastalita*, *Rhodera*, *Stalagtia*, *Sulcia*) for the Balkan Peninsula, only three (*Anthrohyphantes*, *Macedoniella* and *Protoleptoneta*) occur in the east of the region. The distribution of the genera *Antrohyphantes* and *Fageiella* is especially interesting. *Antrohyphantes* is restricted to high altitude zones and caves of the eastern part of the region (Bulgaria), while *Fageiella* is endemic to the caves of the western part of the peninsula (in Bosnia and Montenegro). The two genera are closely related, and their allopatric distribution indicates that they had been already separated before the establishment of the Vardar tectonic zone (Oligocene-Miocene period) (Deltshev, 1996). This suggests that these two genera may be classed as palaeoendemics.

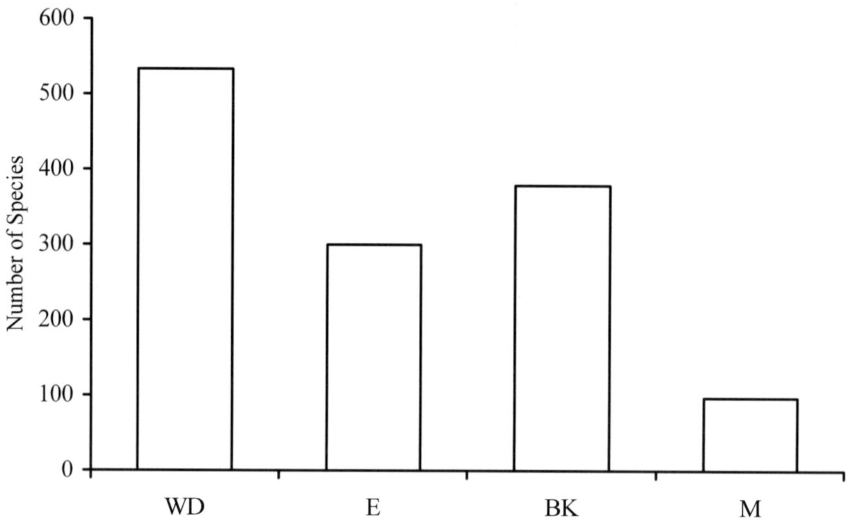

Figure 2. Distribution of Balkan spider species according to four main zoogeographic groups. WD = widely distributed; E = European; BK = Balkan; M = Mediterranean.

The highest number of endemic species is recorded for the territories of Greece (156), Croatia (68), Bulgaria (55) and Crete (46) (Figure 4). Variability in the distribution of endemics in the main geographic systems of the Balkan Peninsula shows that they are most well represented in the Pindus region (150 species), the Dinaric region (145), Thraki-Macedonian region (52), the Balkan Mts. (14), Danube region (4) and North Dobrugea (4) (Figures 4 & 5). The largest proportion of endemics was encountered mainly in mountains and islands, where they inhabit caves (159 species), woodlands (139), coastal sites (48) and high altitude zones (20 species; Figure 5). Within the group of cave spiders, 51 are troglobites (blind or semiblind) with the most species-rich genera being as follows: *Troglohyphantes* (15 species), *Folkia* (6), *Stalagtia* (5), and *Leptonetella*, *Nesticus* and *Stalita* (3 species each). The recent cave spider fauna was formed after gradual changes in the fauna of the ancient humid Mediterranean forests during theTertiary. However, due to a current sparsity of data it is difficult to determine with certainty which of the Balkan endemic cave spiders are Tertiary and which are Quaternary elements. The extreme species richness of endemic cave spiders in the Dinaric region (96 species) leads to the assumption that this was a major centre of speciation and diversification. The same can be concluded for the woodlands of the Pindus region (with 74 species) and for the highest mountains (Rila, Pirin) of the Thraki-Macedonian region, where the greatest number of high altitude elements (15 species) is found.

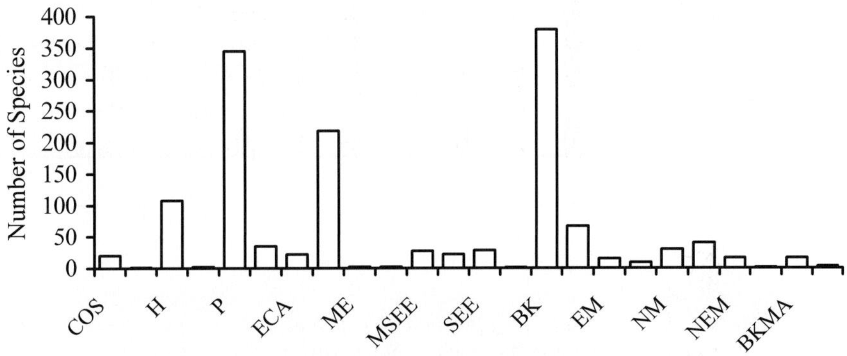

Figure 3. Zoogeographical types in the spider fauna in Balkan Peninsula. See text for abbreviations.

According to their ranges, the endemic spiders of the Balkan Peninsula belong to two different faunal complexes: Mediterranean and European. The Mediterranean elements are distributed in caves, forests, coastal sites and single species at high altitudes, while the European elements are distributed mainly in forests and high altitude sites. This phenomenon can be regarded as a result of the relative isolation of the mountains compared to the lowlands, in the context of palaeoenvironmental changes since the Pliocene (Deltshev, 1996).

The European zoogeographical complex (E = HE + MEE + MSE + MSEE + EE + SEE + PO) includes 300 species (21.3% of the total number of spiders). Within this complex, Holoeuropean species are dominant (72.7% of the European complex), being widespread mainly in the mountains. Middle Southeast European (9.0%), Southeast European species (9.0%) and East European species (7.4%) are comparatively well represented. The complex comprises spiders which occur in Europe and the Balkan Peninsula, and which inhabit both lowlands and mountains. The group of European mountain species is worthy of note, being most well represented in the forest and subalpine belts.

The last complex comprises 195 species occurring in the Mediterranean area (M = HM + EM + WM + NM + SE + NEM + SEM + BKMA + POM). This complex forms 13.8% of the total spider fauna of the Balkan Peninsula, but the real percentage is probably much higher because a large proportion of Balkan endemics are of Mediterranean origin and are widely distributed in the Mediterranean region. The mountain-Mediterranean species (*e.g. Aculepeira talishia, Pardosa incerta*) are of particular interest, since they may be regarded as ancient elements in the high mountains.

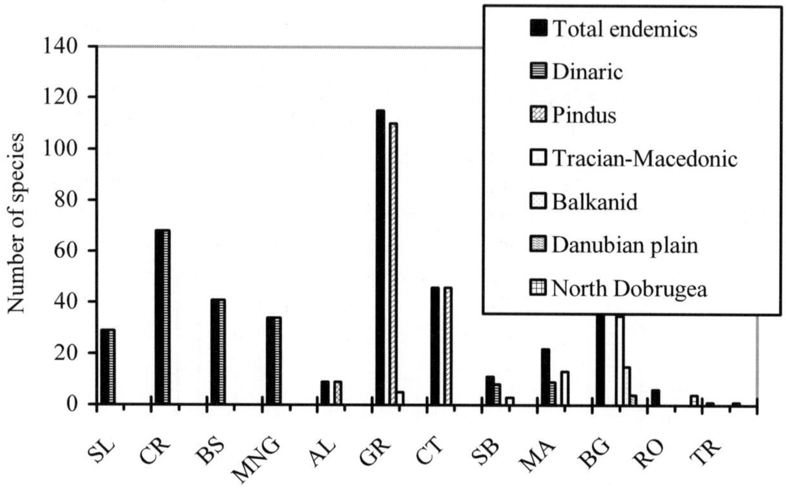

Figure 4. Zoogeographic composition of endemic spiders according to countries and main geographic regions. See Figure 1 for abbreviations.

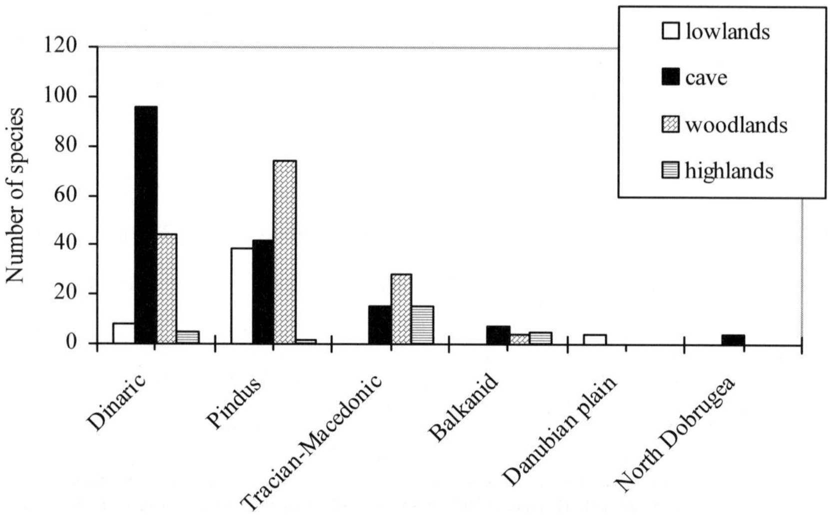

Figure 5. Ecological preferences of endemic spiders according to main geographic regions. See Figure 1 for definition of regions.

CONCLUSIONS

The faunistic diversity of the 1409 spider species shows that the Balkan Peninsula is a territory of considerable species richness. This conclusion is supported also by the existence of 379 endemic species. The uneven species richness in different parts of the Balkan Peninsula is partly a function of the degree of exploration by researchers. In terms of zoogeography, the widely distributed spiders (WD) are dominant, but the most characteristic faunal elements are the Balkan endemics (BK). Their number is high, and their faunistic composition reflects the local character of the fauna. This phenomenon can be explained by the relative isolation of the mountains compared with the lowlands, in the context of palaeoenviromental changes since the Pliocene. The high percentage of Balkan endemics (26.9%) suggests an important process of speciation *in situ*. Thus, the existing data suggest that the Balkan Peninsula represents one of the main centres of speciation in Europe.

ACKNOWLEDGEMENTS

I am especially indebted to my colleagues G. Blagoev, D. Dobrev and S. Lazarov for the discussions and help with the graphics.

REFERENCES

Brignoli, P.M. 1968. Ueber grechische Leptonetidae. *Senkenbergiana biologica* 49, 259-264.

Brignoli, P.M. 1971. Su alcuni *Leptyphantes* di Creta. *Fragmenta Entomologica* 7, 231-241.

Brignoli, P.M. 1972. Su alcuni ragni cavernicoli di Corfu. *Revue suisse de Zoologie* 79, 861-869.

Brignoli, P.M. 1974a. Ragni di Grecia VI. Specie nuove o interessanti delle isole Ionie e della Morea. *Revue suisse de Zoologie* 81, 155-175.

Brignoli, P.M. 1974b. Ragni di Grecia VII. Raccolte in grotte dell'Attica del Dr. P. Strinati. *Revue suisse de Zoologie* 81, 493-499.

Brignoli, P.M. 1976. Ragni di Grecia IX. Specie nuove o interessanti delle famiglie Leptonetidae, Dysderidae, Pholcidae ed Agelenidae (Araneae). *Revue suisse de Zoologie* 83, 539-578.

Brignoli, P.M. 1977. Ragni di Grecia X. Nuovi dati sulla Grecia continentale ed insulare. *Revue suisse de Zoologie* 84, 937-954.

Brignoli, P.M. 1979. Ragni di Grecia XI. Spaecie nuove o interessanti, cavernicole et epigee. *Revue suisse de Zoologie* 86, 181-202.

Brignoli, P.M. 1984. Ragni di Grecia XII. Nuovi dati su varie famiglie (Araneae). *Revue suisse de Zoologie* 91, 281-321.

Brignoli, P.M. 1986. Rapports Biogéographiques entre les Araignées des Balkans et du Moyen Orient (Araneae). *biologia Gallo-hellenica* 12, 93-101.

Coponen, S. 1993. On the biogeography and faunistics of European spiders: latitude, altitude and insularity. *Bull. Soc. neuchatel. Sci. nat.* 116, 141-152.

Deeleman-Reinhold, C.L. 1976. Distribution patterns in European cave spiders. *First International Symposium. Cave biology. Cave paleontology. Cape Town,* pp. 25-35.

Deeleman-Reinhold, C.L. 1978. *Revision of the cave-dwelling and related spiders of the genus Troglohyphantes* Joseph (Linyphiidae), *with special reference to the Yugoslav species.* Academia Scientiarum et Artium Slovenica, Ljubljana.

Deeleman-Reinhold, C.L. 1993. The genus *Rhode* and harpacteine genera *Stalagtia*, *Folkia*, *Minotauria* and *Kaemis* (Araneae, Dysderidae) of Yugoslavia and Crete, with remarks on the genus *Harpactea*. *Revue Arachnologique* 10, 105-135.

Deltshev, C. 1979a. The origin, formation and zoogeography of troglobitic spiders of Balkan peninsula. *Sympozia of the Zoological Society of London* 42, 345-351.

Deltshev, C. 1979b. Contribution to the study of cave spiders (Araneae) in Greece. Four new species of Crete and Thera. *Acta zoolgica bulgarica* 13, 78-92.

Deltshev, C. 1983a. A contribution to the taxonomical study of *sylvaticus* group of genus *Centromerus* F. Dahl (Araneae, Linyphiidae) in Bulgaria. *Acta zoologica bulgarica* 21, 53-58.

Deltshev, C. 1983b. A contribution to the taxonomical and faunistical study of genus *Lepthyphantes* Menge (Araneae, Linyphiidae) from Pirin Mountain. *Acta zoologica bulgarica* 23, 25-32.

Deltshev, C. 1985. New data concerning cave spiders (Araneae) in Grece with description of a new *Leptonetela* (Araneae, Leptonetidae). *Acta zoologica bulgarica* 27, 41-45.

Deltshev, C. 1988. The genus *Fageiella* and *Antrohuphantes* in cave of Balkan peninsula. *TUB-Documentation* 38, 293-299.

Deltshev, C. 1990. A critical review of genus *Coelotes* Blackwall in Bulgaria with description of a new species (*Coelotes drenskii* sp. n.) (Araneae, Agelenidae). *Acta zoolgica bulgarica* 40, 29-43.

Deltshev, C. 1993. The genus *Tegenaria* Latreille in Bulgaria: A critical review with description of two sibling species (Arachnida, Araneae: Agelenidae). *Berichte naturwissenschaftlich-medizinischen Vereins in Innsbruck* 80, 167-174.

Deltshev, C. 1996. The origin, formation and zoogeogaphy of endemic spiders of Bulgaria (Araneae). *Revue suisse de Zoologie,* hors serie, 141-151.

Deltshev, C. 1997a. A New *Cybaeus* from the Mountains of Balkan Peninsula. *Reichenbachia*, 32, 1-4.

Deltshev, C. 1997b. *Crypoecina deelemanae* g. n., sp. n., a fascinating spider of Balkan Peninsula. *Revue suisse de Zoologie* 104, 485-489.

Deltshev, C. & Blagoev, G. 2000. A critical check list of Bulgarian spiders (Araneae). *Bulletin of the British Arachnological Society* 12, 110-138.

Deltshev, C. & Ćurćić, B.P.M. 1997. Contribution to the knowledge of the group *europaeus Centromerus* Dahl (Linyphiidae). *Revue suisse de Zoologie* 104, 49-55.

Deltshev, C. & Paraschi, L. 1990. A contribution to the study of spiders (Araneae) in Greece, with a description of a new species (*Malthonica spinipalpis* Deltshev, sp. n., Agelenidae). *Biologia Gallo-hellenica* 17, 3-12.

Drensky, P. 1936. Katalog der echten Spinnen (Araneae) der Balkanhalbinsel. *Sbornik na Bulgarskata Akademia na naukata* 32, 1-223.

Hadjissarantos, H. 1940. *Les araignees de l'Attique*. Athens (no publisher listed).

Jones, D.J., Ledoux, C. & Emerit, M. 1990. *Guide des Araignées et Opiliones d'Europe.* Delachaux et Niestlé, France, pp. 1-384.

Maurer, R. & Haenggi, A. 1990. Katalog der Schweicerischen spinnen. *Documenta faunistica Helvetiae* 12, 1-33.

Michailov, K.G. 1997. *Katalogue of the spiders of the territories of the former Soviet Union (Arachnida, Aranei)*. Zoological Museum of Moskow State University, Moscow, pp. 1-416.

Nikolić, F. & Polenec, A. 1981. *Catalogus faunae Jugoslaviae. III/4 Aranea.* Academia Scientiarum et Artium Slovenica, Ljubljana.

Platnick, N. 1989. *Advances in spider taxonomy 1981-1987. A Supplement to Brignoli's A Catalogue of the Araneae described between 1940 and 1981.* Manchester University Press, New York.

Platnick, N. 1993. *Advances in spider taxonomy 1988-1991. With synonimies and transfers 1940-1980.* New York Entomological Society, American Museum of Natural History, New York..

Platnick, N. 1997. *Advances in spider taxonomy 1992-1995. With redescription 1940.* New York Entomological Society, American Museum of Natural History, New York.

Roberts, M.J. 1987. *The Spiders of Great Britain and Ireland 2.* Harley Books, Colchester. pp. 1- 204.

Thaler, K. 1980. *Die Spinnenfauna des Alpen: Ein Zoogeograpfischer Versuch.* Verhandlungen 8 International Arachnological Kongress. Wien, pp. 389-404.

Thaler, K. 1996. Three *Walckenaeria* species from Peloponnese, Greece (Araneae: Linyphiidae). *Bulletin of the British Arachnological Society* 10, 156-160.

Thaler, K. & Knoflach, B. 1991. Eine neue *Amaurobius*-Art aus Griechenland (Arachnida: Araneae, Amaurobiidae). *Mitteilungen Schweizerische Entomologische Geselschaft* 64, 265-268.

Thaler, K. & Knoflach, B. 1993. Two new *Amaurobius* species (Araneae: Amaurobiidae) from Greece. *Bulletin of the British Arachnological Society* 9, 132-136.

Thaler, K. & Knoflach, B. 1995. Uber Vorkommen und Verbreitung von *Amaurobius*-Arten in Peloponnes und Aegaeis (Araneida: Amaurobiidae). *Revue suisse de Zoologie* 102, 41-60.

Wunderlich, J. 1980. Linyphiidae aus Sued-Afrika (Arach. : Araneae). *Verhandlungen Naturwissenschaftliche Verein Hamburg* 23, 319-337.

Wunderlich, J. 1985. *Pachygnatha clerckoides* n. sp. aus Jugoslavien (Arachnida: Araneae: Tetragnathidae). *Senkenbergiana biologica* 65, 325-328.

Wunderlich, J. 1994a. Zwei bisher unbekannte Mediterrane Arten der Gattung *Pholcus* Walckenaer 1805 (Arachnida: Araneae: Pholcidae). *Beitrage Araneologie* 4, 625-628.

Wunderlich, J. 1994b. Beschreibung bisher unbekanter Arten dre Baldachinspinnen aus der Oestlichen Mediterraneen (Arachnida: Araneae: Linyphiidae). *Beitrage Araneologie* 4, 655-686.

Wunderlich, J. 1994c. Beschreibung einer bisher unbekanten Art der Gattung *Amaurobius* C. L. Koch 1837 von Kreta (Arachnida: Araneae: Amaurobiidae). *Beitrage Araneologie* 4, 729-730.

PART 2: AQUATIC

12. DISTRIBUTION PATTERN OF THE AQUATIC FAUNA OF THE BALKAN PENINSULA

PETRU M. BĂNĂRESCU*

INTRODUCTION

The Danube River and its tributary, the Drava, define the northern limits of the Balkan Peninsula. This peninsula is drained by many rivers flowing in all directions, and has many lakes, the largest being Ohrid, Prespa, Skutari (or Skadar), and Ioannina. The aquatic fauna of the peninsula is not simple, as each river basin, lake or mountain range has its own faunal peculiarities. On the basis of fish distribution, seven geographic regions can be defined, which correspond to river catchments, together with the lakes of Ohrid and Prespa within one of these regions (Figure 1). This contribution deals primarily with the fish, hydrobioid snail and malacostracean elements of marine Tethyan origin.

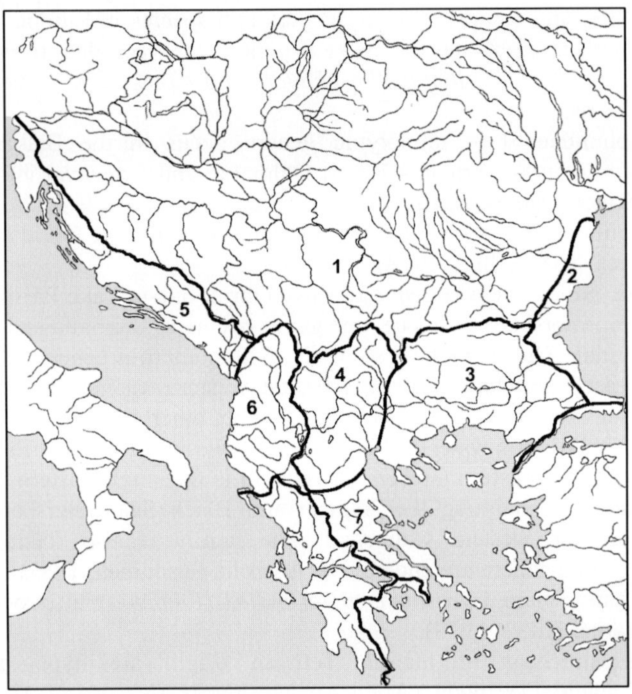

Figure 1. Map showing the Balkan regions used in this study, based on fish fauna and drainage networks. 1 = The Danube River Basin; 2 = The Eastern Balkan watershed (the Black Sea); 3 = The Thraki (Thrace) region; 4 = The Macedonia-Thessaly region; 5 = The Dalmatian region; 6 = The South Adriatic-Ionian region; 7 = The Attiko-Beotia region.

H.I. Griffiths et al. (eds.), Balkan Biodiversity, 203–217.

Summary of the aquatic fauna

All genuine freshwater fish in the peninsula belong to Holarctic or Eurasian families or subfamilies, and most come from the Cyprinidae group; recent marine intruders are not considered here. Five genera in the western watershed are endemic: three from the Cyprinidae and two from the Salmonidae.

The hydrobioids are small prosobranchiates, inhabiting mainly fresh water, but also found in brackish or marine conditions. Most authors include numerous families and subfamilies in Hydrobioidea, which are found throughout the world (see Bănărescu, 1990: 337-342, 347-349). Radoman (1983, 1985) restricts the superfamily to ten families, eight of which are present in the Balkan Peninsula. These are as follows:

(1) Hydrobiidae. Mainly euryhaline, with a wide Holarctic range; about eight species are western Balkan endemics;

(2) Lithoglyphidae. Predominantly North American, but a single genus, *Lithoglyphus*, occurs in Europe, and three species are endemic to the southwestern tributaries of the Danube, whilst one occurs throughout the Danube Basin and in other rivers of the Black Sea watershed;

(3) Emericiidea. One genus, with four species, three of which are endemic to the Dalmatian region, and one of which extends from Italy to Dalmatia.

(4) Radoman's Bythinellidae. This corresponds approximately to the Amnicolidae of other authors, and comprises many genera and species in Europe, Anatolia and North America; eight species in two genera are recorded from the Balkan Peninsula, most notably *Bythinella schmidti* whose range covers the Adriatic and Danube Basins.

(5) Lithoglyphulidae. Two monotypic genera occur in the Balkan Peninsula; *Lithoglyphulus* in the Zrmanja River, Dalmatia, and *Dabriana* in a tributary of the Sava River, Danube Basin.

(6) Micropyrgulidae. One genus, with two species: one in Lake Ohrid and another in the Pyrenees, Spain (Illies, 1978).

(7) Pyrgulidae. Nine genera endemic to Lake Ohrid, one to Lake Prespa, and one to the western watershed of Greece; one genus occurs in Dalmatia and Italy as well.

(8) Orientallinidae. This is the largest family, with numerous genera endemic to four Balkan areas: Lake Ohrid, Lake Prespa, the western watershed of the peninsula, and the southwestern tributaries of the Danube; other genera have ranges which extend to the Adriatic watershed and the southwestern Danube tributaries, or the Vardar Basin and eastern Greece. *Grossuana* is the single eastern Balkan genus of hydrobioids, inhabiting the Aegean and Black Sea watersheds and being absent from the western watershed. The family is also found outside the peninsula: some genera are endemic to Anatolia (Radoman, 1983), while at least three Balkan genera (*Graziana, Belgrandia, Hauffenia*) also occur in other European areas (Illies, 1978).

All malacostraceans of marine Tethyan origin are hypogean. Families represented include Decapoda (Atyidae), Mysidacea (Mysidae), Therosbaenacea (Monodellidae), Isopoda (Microparasellidae, Stenasellidae, Cirolanidae, Sphaeromatidae, Microcerberidae) and Amphipoda (Typhlogammaridae, Pseudoniphargidae, Hadziidae, Ingolfiellidae). All these families have wide distributions beyond the Balkan Peninsula. Five genera (*Spelaeocaris*, *Troglomysis* and three genera of the Typhlogammaridae) are endemic to Dalmatia, *Protelsonia* is endemic to the tributaries on the left bank of Danube from Hungary to Bulgaria, and *Limnosbaena* to the Bosna River (southwestern Danube Basin). *Bulgarocerberus* is endemic to Thraki (Thrace), and *Hadzia* to the western Balkan watershed and the Vardar Basin. The range of *Balkanostenasellus* includes the Danube, Vardar and Maritza (Thraki) basins. Five other genera occur both in Balkan regions and in other European or circum-Mediterranean locations.

Aquatic zoogeography by region

The regions defined and used in this paper (Figure 1) have the following characteristics.

1. The Danube River Basin

Only the southern and southwestern tributaries of the Danube (the largest being the Drava and Sava) belong to the Balkan Peninsula. This basin is considered the main, or possibly the only, refugium in which the preglacial aquatic fauna (especially fish) survived Quaternary glaciation, following which it repopulated central, western, and parts of northern Europe (Thienemann, 1950; Bănărescu, 1960, 1992). Most species of the Danube Basin have wide ranges in Europe, Euro-Siberia and beyond. Taken as a whole, the fish fauna of the Danube has marked western and Central European ties; almost all genuine freshwater fish species from western and Central Europe are present in the Danube Basin. The Danube Basin also supports species with an exclusive Aralo-Caspo-Pontic distribution (some of them are genuine fresh water, others of marine-brackish water origin), and species shared with only a few neighboring rivers (*e.g.* the Vardar), along with ten endemic species.

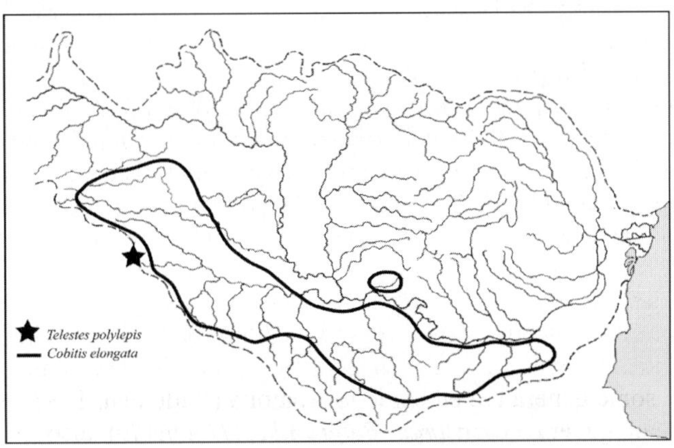

Figure 2. Map showing the distributional ranges of two fish species endemic to the southern (Balkan) sector of the Danube Basin.

One of these endemic species, *Gymnocephalus schraetser,* is widely distributed throughout the Danube catchment area, including the southwestern tributaries. Others have restricted ranges, and two of them occur in the Balkans. *Telestes polylepis* is confined to a few tributaries of the Sava River near Zagreb, and *Cobitis elongata* occurs in most southern tributaries of the Danube, from Slovenia to Bulgaria, and one northern tributary, the river Nera in the Banat, Romania (Figure 2). The subspecies *Zingel streber nerensis* is not fully mapped, but appears to occur mainly in the southern tributaries of the Danube, at least from the Drava River to the Morava River in Serbia, and may also be present in the tributaries flowing from Bulgaria. It is recorded in one northern tributary, the river Nera.

In spite of the existence of some local endemics, the fish fauna of the Danube Basin is essentially uniform, at least in zoogeographic terms. Most species are widely distributed throughout the basin, including the southwestern tributaries. Like the fish, many unionacean mussels, large snails, and the crayfish *Astacus astacus* have wide ranges, occurring throughout the entire Danube drainage area and in other parts of Europe (Illies, 1978). The fish fauna of the Danube Basin differs strongly from that of the Dalmatian region (see section 5 below), at both specific and generic levels.

The zoogeography of the hydrobioid snails is quite different. Hydrobioids are numerous in the western watershed of the Balkan Peninsula, especially in the Dalmatian region, and in the drainage areas of the southwestern tributaries of the Danube, despite being almost absent from other parts of the Danube Basin, even the southeastern tributaries from Bulgaria.

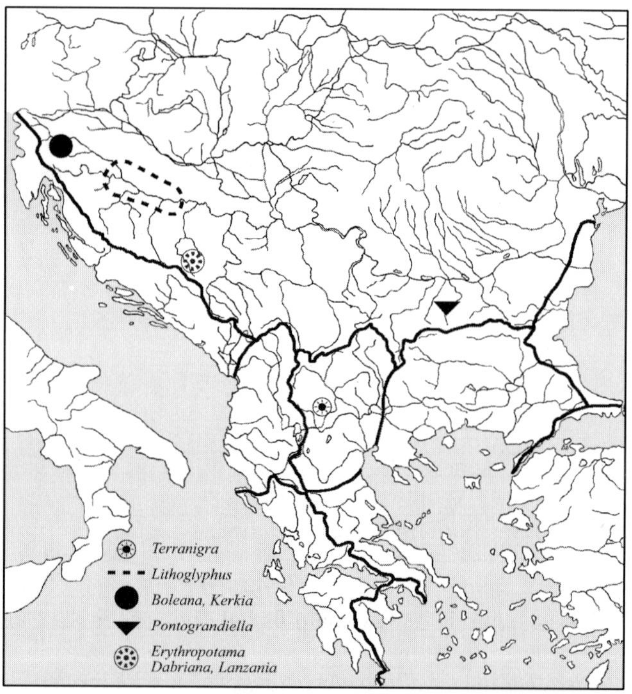

Figure 3. Map showing the distributional ranges of some genera of hydrobioid snails endemic to the Danube Basin or common to the Danube and Vardar basins (Terranigra).

The genus *Dabriana* of Lithoglyphulidae is endemic to the southwestern tributaries of the Danube (and a second genus occurs in Dalmatia). Other endemics include three species of *Lithoglyphus* (a fourth occurs in the northern watershed of the Black Sea) and four monospecific genera of Orientalinidae (Figure 3). Eight genera occur in both the southwestern tributaries of the Danube and in rivers of the western Balkan watershed, and three of these are mainly found in the Danube Basin (*Sadleriana, Graziana* and *Sarajana*) (Figure 4) (Radoman, 1985). Some species are present only in Danube tribuatries and western Balkan rivers (*e.g. Sadleriana sadleriana, S. fluminenesis, Graziana lacheineri,* and *Sarajana apfelbecki*). A single hydrobioid species, *Bythinella dacica*, is present in a northern tributary of the Danube beyond the Balkans, the Cerna River in southern Banat, along with the widely distributed *Lithoglyphus naticoides*, and two monotypic genera are endemic to the southeastern (Bulgarian) tributaries of the Danube (*Pontobelgrandiella* and *Cavernisa*).

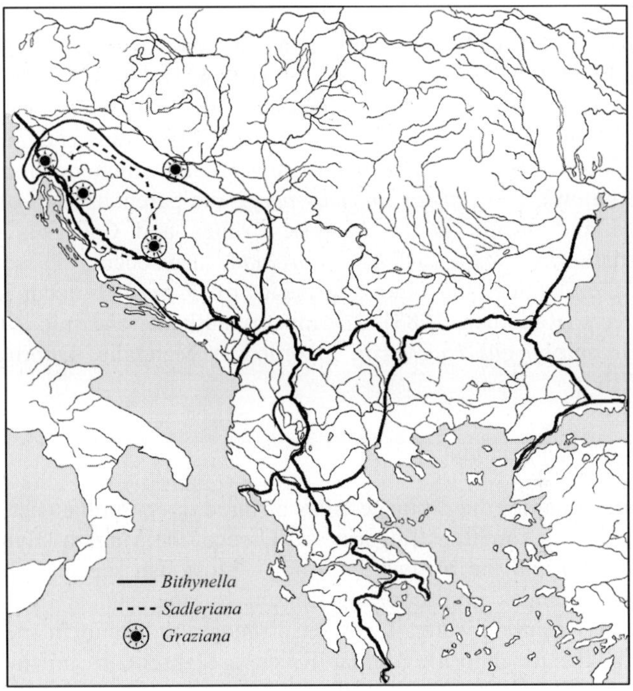

Figure 4. Map showing the distributional ranges of three genera of hydrobioid snails common to the southwestern tributaries of the Danube and the western Balkan watershed.

Radoman (1985) distinguishes two main faunal divisions in the Balkan Peninsula on the basis of hydrobioid distribution; a western one including the western watershed of the peninsula and the southwestern tributaries of the Danube, and an eastern division comprising the remaining part of the Danube Basin in the Balkans, the Thraki region and most of Macedonia-Thessaly. The boundary between divisions is not the hydrographic divide between the Danube and Adriatic watersheds, as in the case of fishes, but lies along the rivers Morava (tributary of the Danube), and Vardar (corresponding to an old marine arm). If the entire aquatic fauna of the Danube Basin is considered, three main subdivisions can be recognized:

(1) a northern, non-Balkanic one; (2) the southwestern tributaries; (3) the southeastern, Bulgarian tributaries. The latter subdivision is, in faunal terms, intermediate between the first two, and is inhabited by *Cobitis elongata* and two endemic hydrobioid genera.

The range of a non-hydrobioid snail genus, *Holandriana*, is worthy of note; it occurs in the western Balkan watershed, the southwestern and probably also southeastern Danube tributaries, and two northern Danube tributaries, the Caras and the Nera in southern Banat, Romania. Three rivers of the Banat are inhabited by taxa which occur mostly in the southwestern Danube tributaries, the Nera (*Cobitis elongata, Zingel streber nerensis, Holandriana*), the Caras (*Holandriana*) and the Cerna (*Bythinella dacica*).

2. The Eastern Balkan watershed (the Black Sea)

This small region encompasses many brackish and freshwater lagoons in the Romanian province of Dobrugea and in eastern Bulgaria, and some eastern Bulgarian rivers. The aquatic fauna consists mainly of brackish water species of Ponto-Caspic origin. There are also European species and some endemic subspecies. Examples include *Cobitis* (*Bicanestrinia*) *strumicae*, which occurs here and in Thraki (see 3 below), and *Barbus tauricus bergi*, the parent species of which has a disjunct range and occurs in the Vistula, Crimea and Caucasus regions. The hydrobioid snail *Grossuana serbica codreanui* also occurs in some southern tributaries of the Danube, and other subspecies of *G. serbica* occur in the Struma and Vardar rivers (Radoman, 1983). The most remarkable endemic is the hypogean amphipod *Pontoniphargus racovitzai*, recorded at Mangalia, Dobrugea (Dancău, 1976).

3. The Thraki (Thrace) region

This region encompasses the drainage areas of three rivers on the eastern watershed of the Aegean Sea in southern Bulgaria and Greece, the Maritza (Evros in Greek), the Mesta (Nestos) and the Struma (Strymon). Most fish species inhabiting these rivers have wide ranges in Europe or Central Europe, and are characteristic of standing or slowly running water. The typical rheophilic Danubian species, some of which occur in the neighboring Vardar River, are absent, as are most Danubian species of slowly flowing waters. It is also noticeable that the Struma River contains more Anatolian elements (*e.g.* presence of *Orthrias brandti* and a *Bicanestrinia* species) than the other more easterly rivers.

Only two fish species are endemic: *Barbus cyclolepis,* which is present in all three rivers, and *Cobitis* (*Bicanestrinia*) *punctilineata*, which occurs only in the Struma River basin. An endemic subspecies found in the Struma basin is the loach *Orthrias brandti bureschi*, which belongs to a polytypic species found from northern Iran and Anatolia to the Vardar River in Macedonia (Former Yugoslav Republic of Macedonia; FYROM). Other species occur only here and in the Macedonia-Thessaly region (see 4 below), such as *Chondrostoma vardarense* (which is closely related to the Central European *C. nasus*), and *Vimba melanops* (Drensky, 1951; Economidis, 1991).

A single species of hydrobioid snails is listed from this region, the often-mentioned eastern-Balkan *Grossuana serbica.* The Tethyan stenasellid genus

Balkanostenasellus occurs in this region, the southern area of the Danube Basin and the Vardar basin. Two species of *Bulgarocerberus* represent this endemic isopod genus with Tethyan ties, and other endemic isopod species occur, which belong to more widely-distributed genera. The hypogean watermite genus *Momonisia* is also endemic to this region.

4. The Macedonia-Thessaly region

This region encompasses the drainage network of the Vardar River and three more southerly rivers in Greek Macedonia and Thessaly.

The Macedonian – Thessaly fish fauna has marked Danubian ties, but is not strongly Central European in character. Three rheophilic and one non-rheophilic Danubian species are present as endemic subspecies (*Gobio uranoscopus elimeius, G. kessleri banarescui, Zingel streber balcanicus* and *Cobitis vardarensis vardarensis*; *C. taenia* from the Danube Basin is considered a subspecies of *vardarensis*). In the Vardar basin and Thessaly a rheophilic Danubian and Central European loach, *Orthrias barbatulus*, also occurs, and two recorded endemic *Barbus* species (*B. macedonicus* in the Vardar and Aliakmon, *B. thessalus* in the Pinios River) are more closely related to the Danubian (and Central European) *B. barbus* than the Thracian *B. cyclolepis*. The rheophilic crayfish *Astacus torrentiu* and the monospecific hydrobioid genus *Terranigra* (Radoman, 1983, 1985) also occur in both the Danube and Vardar basins.

None of these Danubian – Macedonian species is found in Thraki or the Adriatic watershed. Three species do occur both here and in Thraki (see 3 above), yet are absent from the Danube Basin: *Chondrostoma vardarense, Vimba melanops* and the prevalent Anatolian Caucasian *Orthrias brandti*, which has one subspecies in the Vardar (Sorić, 1999), and a second in the Struma. The endemic *Pachychilon macedonicum* has a single congener in the drainage area of lakes Skadar and Ohrid, linking this species to the western Balkan faunas (Vuković & Ivanović, 1971; Ivanović 1973).

Few hydrobioid snails are present, including the monospecific *Terranigra* in the headwaters of the Vardar and in a southern tributary of the Danube, *Horatia macedonica* (other *Horatia* species occur in the western Balkans) and *Grossuana serbica*, which is widely distributed in the eastern half of the Balkan Peninsula (Radoman, 1983).

Peracarid crustaceans of Tethyan origin present in the region are all hypogean, and include *Balkanostenasellus* (which has distinct subspecies in the southwestern part of the Danube Basin), *Microcerberus stygius* (which also occurs in Thraki), *Hadzia gjorgjevici* (a western Balkan species of *Hadzia* also occurs), and two *Ingolfiella*, one of which is closest to a South American species (Botoşăneanu, 1986). Thus, in contrast to the fish fauna, the hypogean fauna of Tethyan peracarids from Macedonia and Thessaly has closer affinities with the fauna of the western Balkans and Thraki than with the Danubian fauna.

5. The Dalmatian region

This region comprises the Adriatic watershed of the Balkan Peninsula, from Istria to the basin of Lake Skutari. It is drained by five relatively small but old rivers, including the Soča or Isonzo, the Zrmanja, the Krka, the Cetina and the Neretva, along with several lesser rivers, some without outlet to the sea.

The fish fauna is highly distinctive. Two monospecific genera are endemic, the salmonid *Salmothymus obtusurostris* (widely distributed in the region and related to *Acantholingua* from Lake Ohrid; Vuković & Ivanović, 1971; Kottelat, 1997) and the cyprinid *Aulopyge* (confined to a small area in the Krka and Cetina basins and a few inland rivers). The genus *Phoxinellus* is almost endemic, with eight species confined to distinct small areas in the region, and three others occurring in the north of the South Adriatic-Ionian region (see Section 6 below) (Karaman, 1972; Kottelat, 1977). The fauna also includes endemic species from a range of European and Eurasiatic genera (two *Chondrostoma*, six *Leuciscus* [subgenus *Squalius*], one *Telestes*, one *Rutilus*, three *Cobitis*; Kottelat, 1997, Bănărescu & Herzig Straschil, 1998). In the north of the region two Italian species occur, *Alburnus alburnus alborella* and *Barbus plebejus.* The fish fauna of the region lacks European genera which are present in other Balkan regions (*i.e. Gobio*, *Alburnoides*, *Barbus* other than the Italian *B. plebejus* and *Sabanejewia*).

The amphipods are richly represented, with six endemic genera (*Carinurella, Jugogammarus* and *Fontogammarus* of the Gammaridae and the hypogean *Typhlogammarus, Metohia* and *Accubogammarus* of the Tethyan Typhlogammaridae; Karaman, 1974; Bousfield, 1978), and endemic species from more wide-ranging genera. Other Tethyan malacostraceans endemic to the region are the atyid shrimp genus *Speleocaris*, four species of *Troglocaris*, the monospecific mysid *Troglomysis,* some species of *Monolistra* (an isopod genus which is also found in northern Italy and the southwestern tributaries of the Danube), *Limnosbaena*, and some species of the circum-Mediterranean *Monosbaena*.

A quite remarkable hypogean endemic is the mussel *Congeria kusceri,* single European survivor of a genus widely distributed in the continent from Eocene to Pliocene times, now restricted to the tropics. Other hypogean endemics include several triclad turbelarians, the sponge *Eunapius subterraneus,* and the snail *Theodoxus subrelictus* (Botoşăneanu, 1986).

The hydrobioids are the aquatic group represented by the highest number of endemic and non endemic species in the Dalmatian region. Six families are represented, including the Hydrobiidae (mainly euryhaline: two endemic genera; *Adriohydrobia* and *Adrioinsulana*), Emmericiidae (three endemic species of *Emmericia* occur here and a fourth is found in Italy), Lithogyphulidae (the monospecific *Lithoglyphulus* in the River Zrmanja; a second genus is found in the Danube Basin), Pyrgulidae (represented by *Pyrgula annulata,* which is widely distributed from Italy to Lake Skutari), and Bythynellidae (represented by a few species of the widely ranging *Bythinella*) (Figure 5). The sixth family, Orientallinidae, is by far the most varied family, with greatest abundance in the Dalmatian region and in Lake Ohrid). Nine genera of the family are endemic to the region, or extend south only into the South Adriatic-Ionian region (see below). Most of these are monospecific, although *Anagastina* contains six species. Four other genera of the Orientallinidae occur both here and in the southwestern tributaries of the Danube.

Thus the Dalmatian fauna is closest to that of the southwestern rivers of the Danube Basin, and has some affinities with the Thraki and Macedonia-Thessaly faunas, but none with that of the northern Danube. It is the only Balkan fauna which shows affinity to the Italian aquatic fauna: the genus *Emmericia* and the species

Pyrgula annuata are present only in Dalmatia and Italy, and two Italian fish species occur in the north of the region.

The aquatic fauna of the region is not uniform. Few species occur in all or most of the drainage networks, each of the five main rivers has its own endemic fish and hydrobioid snails species, and other species are present in only two or three neighbouring river basins.

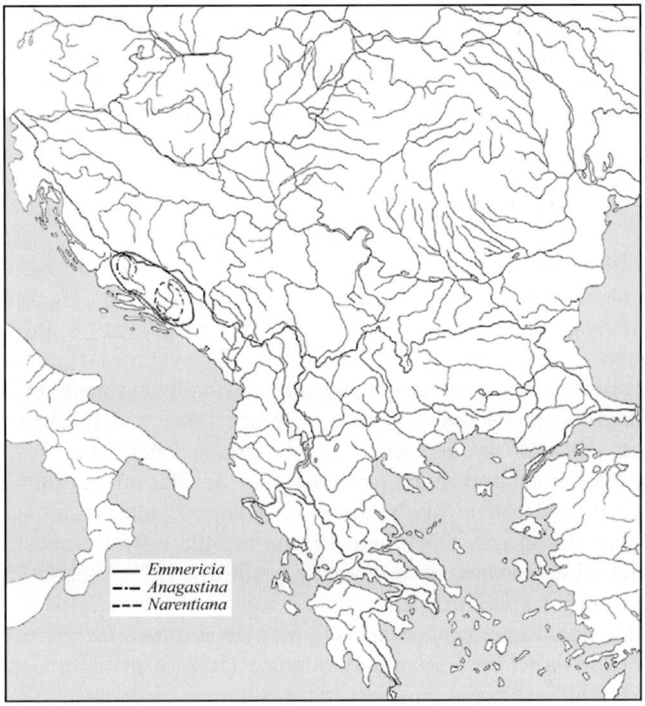

Figure 5. Map showing the distributional ranges of three genera of hydrobioid snails endemic to the western Balkan watershed.

6. The South Adriatic-Ionian region

This region encompasses the western watershed of the Balkan Peninsula, from Montenegro to the southern extremity of the Peloponnesos. It is drained by 16 small rivers and many lakes, including Skutari (= Skadar [Montenegro]), Ioannina, Trichonic, Stymphalia, Prespa and Ohrid. Prespa and Ohrid are treated separately here (sections 6a and 6b below). One fish genus is endemic to the region, *Tropidophoxinellus*, with two vicariant species in the south of the region. There are about 17 endemic fish species, the most remarkable being *Silurus* (*Parasilurus*) *aristotelis*, the only known Europen member of an East Asian subgenus. Other endemics include two species of the mainly Dalmatian *Phoxinellus*, three *Cobitis* species of the subgenus *Bicanestrinia* (the species from Dalmatia belong to *Cobitis* s.str.), two or three species of *Pseudophoxinus* (a genus occurring mostly in Syria and Anatolia), two *Barbus* species (including *B.* (*Luciobarbus*) *albanicus*, closely related to *B.* (*L.*) *graecus* in the eastern watershed of Greece), and *Orthrias pindus*. The Central Europen genera *Alburnoides* and *Gobio*, which are absent from the

Dalmatian region, are represented here by distinct subspecies. The occurrence of *Barbus peloponensis*, a species of the Danubian complex, is unexpected; the species occurs in in the Danube Basin, the rivers Nistru (Dniester) and Vistula, and in Macedonia-Thessaly, is absent from the eastern watershed of Greece south of Thessaly, but occurs again in the southwestern Adriatic region (Economidis, 1991; Economidis & Bănărescu, 1990).

The hydrobioid snails, which are very numerous in the Dalmatia region and in the lakes Ohrid and Prespa, are poorly represented in the South Adriatic-Ionian region; Radoman (1983, 1985) lists only one species of *Horatia* (a genus found in Dalmatia and the Vardar), a few Orientalidae and the endemic *Trichonia* in southwestern continental Greece. However, this poverty may be due to lack of study rather than actual absence of diversity.

6a. Lake Ohrid

Ohrid is the oldest and deepest lake in Europe, and has a high degree of endemism. The hydrobioid snails have the highest number of endemics among aquatic animals, including *Micropyrgula stankovici* (of Micrpyrgulidae), the subfamilies Chilopyrgulinae (five genera, six species), Ohridopyrguinae (three monospecific genera), four of the six genera of Pseudohoratiine (11 species), six genera of Orientalininae (16 species) and one subspecies of a more widely-distributed western Balkan species, *Orientalina cursa kicavica* (Radoman, 1983, 1985).

Other endemic prosobranchiates include four species of the Holarctic *Valvata* genus, and several pulmonate snails (seven *Gyraulus*, *Planorbis macedonicus*, three *Ancylus* and two *Acroloxus*; Ohrid and Prespa are the only European areas whose fauna have endemic species of pulmonates) (Illies, 1978; Radoman, 1985). An endemic fish genus, *Acantholingua*, occurs, which is closely related to *Salmothymus* (found in Dalmatia). Endemic fish species include *Rutilus ohridanus* (related to *R. basak* from Dalmatia), *Pseudophoxinus minutus* (from a genus generally found in Syria and Anatolia), and the problematic *Cobitis meridionalis*. Numerous other groups of animals have endemic species and genera in the lake and its tributaries, including turbelarians (one *Phagocata* and six *Dendrocoelum*), asellid isopoda (three *Proasellus*), and sponges (*Spongilla stankovici* and the remarkable monospecific *Ochridaspongia*, with uncertain family status, believed by Arndt (1937) to be related to a genus from Lake Tiberias, Israel and to another from Lake Baikal). A few species are shared only with Lake Prespa, while the fish *Pachychilon pictum* is shared with Lake Skutari and the Aoos River in northwestern Greece.

The lake and its tributaries also support a range of widely-distributed species including three prosobranchiates, fifteen pulmonate snails (Radoman, 1985), about ten fish (some represented by endemic subspecies), and a few species shared with western Balkan rivers or also with Italy.

6b. Lake Prespa

Located, like Lake Ohrid, in the south Adriatic-Ionian area, the smaller and younger Lake Prespa has far fewer endemics. Most are hydrobioid snails, including the genera *Prespolitorea, Prespiana* and *Malaprespia*, a total of four species of Orientalinidae, *Prespopyrgula*, and two species of *Parabythinella* (Bythinellidae). The most remarkable endemic fish is *Chalcalburnus belvica*, a member of a genus that includes three species in western Asia and one on the watershed of the Black, Caspian and Aral seas, with one subspecies in the Thraki region of the Balkan Peninsula (Economidis, 1986). Other endemic fish species include *Chondrostoma prespense* (related to the Central Europen *C. nasus*), *Barbus prespensis*, *Rutilus prespensis* (possibly a subspecies of *R. ohridanus*) and *Cobitis meridionalis* (Economidis, 1991; Kottelat, 1997).

7. The Attiko-Beotia region

This is a small region which incorporates the drainage basins of the rivers on the eastern watershed of continental Greece, south of the Sperchios River and of the Peloponnisos, and Euboea Island. The fauna is very poor. Only six fish species are present, all endemic and with restricted ranges within the region; *Rutilus ylikiensis, Pseudophoxinus beoticus, P. marathonicus, Scardinius graecus, Barbus graecus* (closely related to *B. albanicus* from the western watershed) and *B. euboicus* from the island of Euboea (apparently related to *B. cyclolepis* from the Thracian region). The genera to which these species belong, except for *Pseudophoxinus*, are widely distributed in Europe.

No hydrobioids have been recorded, although the absence of *Islamia* and *Graecanatolica* is remarkable since they occur both in other Balkan regions and in Anatolia, and Attiko-Beotia is the Balkan region geographically closest to Anatolia (Radoman, 1983, 1985).

The neritid snail *Theodoxus saulcyi* seems to be restricted to this region and to the island of Crete (Schütt, 1996).

ZOOGEOGRAPHY

The freshwater taxa present in the various Balkan regions belong, with few exceptions, to Holarctic or Palaearctic lineages (*i.e.* Euro-Siberian and East Asian). This generalisation applies to a wide range of groups including fish, crayfish, unionacean mussels, valvatid snails, hydrobioid snails, and Tethyan peracarids. Hydrobioid snails are very numerous in the western Balkan regions, and generally have ranges which extend to other European areas, and even North America. For example, Lithoglyphidae has several North American genera and a single European one, Orientalinidae, Bythinellidae, Hydrobiidae, and Pyrgulidae occur in many European and North American areas, Emmericidae in the western Balkans and in Italy, and the monogeneric Micropyrgulidae occurs in both Lake Ohrid and the Pyrenees; indeed, only the small Lithoglyphulidae is found only in the Balkan Peninsula.

The Tethyan families of peracarid crustaceans and Atyidae also have wide ranges from central America to southern Europe and beyond (East Africa, South-east Asia), but Balkan species from these families all belong to southern European or circum-Mediterranean genera.

It is widely accepted that the ancestors of the European or western Palaearctic genera and species have either an American origin (*e.g.* percid fishes, pleurocerid snails) or a Siberian or East Asian origin (most groups). American-origin groups entered Europe in the Palaeocene or early Eocene, whilst Asian groups entered in the Oligocene or even the Miocene. Since the last connection between America and

Europe involved Greenland and Scandinavia, and the route from East Asia to Europe was across Siberia, it can be argued that the both biogeographical groups initially colonised northern and Central Europe and only later spread into the southern peninsulas. The "classical" theory (Thienemann, 1950, Bănărescu 1960, 1989; Almaca, 1976) considered that the southern peninsulas were colonised independently, but Doadrio (1990) and Bianco (1990) advanced an alternative hypothesis. Analysis of the interrelations of *Barbus* (a cyprinid fish genus) enabled Doadrio to group them into two subgenera, *Barbus* s.str. and *Luciobarbus*. Both subgenera are present in western Asia, although the latter has more species. All Central European and Italian species and most Balkan species belong to *Barbus* s.str., whilst *Luciobarbus* comprises, besides many western Asian species, two Greek species with vicariant ranges, *B. graecus* and *B. albanicus*, and all Iberian and northwestern African species. Hsü (1974) demonstrated that during the Messinian Salinity Crisis the Mediterranean Sea was desiccated and replaced by a network of rivers and freshwater lakes, therefore Bianco (1990) argues that *Luciobarbus* species were enabled to disperse directly from western Asia to the Iberian Peninsula and northern Africa, whilst *Barbus* s.str. dispersed via Central Europe to Italy and the Balkan Peninsula.

A similar distribution pattern is found in two subgenera of *Cobitis*. *Cobitis* s.str. occurs in East Asia, Siberia, eastern and central Europe, Italy, and many rivers on the Aegean watershed of the Balkan Peninsula and the northern part of the western watershed (the Dalmatian region). The subgenus *Bicanestrinia* occurs in Iran, Anatolia, the Balkan watershed of the Black Sea, the river Struma (Thraki region) and the southern part of the western Balkan watershed (the South Adriatic-Ionian region). The dispersal of the latter subgenus took probably place, like that of *Luciobarbus*, through the former riverine network in the present Mediterranean Sea.

Some other genera present in the southern part of the Balkan Peninsula are shared with western Asia and are absent from the rest of Europe. These include the fish *Pseudophoxinus* (Lake Ohrid basin, the southern Adriatic-Ionian and Attiko-Beotia regions) and the hydrobioid snails *Islamia* (from Dalmatia and the southwestern tributaries of the Danube to Anatolia) and *Graecoanatolica* (Macedonia-Thessaly region and Anatolia).

The aquatic faunas of the various Balkan areas display multiple interrelationships. The northern Danube fauna has ties with Central Europe (Baltic and Atlantic watersheds), with the faunas of the southwestern tributaries of the river and of Macedonia-Thessaly, some connections with that of the Thraki region, and almost none with the western Balkan watershed. The fauna of the southwestern Danube tributaries has ties with the northern Danube and the western Balkans (especially the Dalmatian region). The fauna of Macedonia-Thessaly is most closely related to that of the Danube and the Thraki fauna has some links with western Balkan faunas and, curiously enough, with central-southern Anatolia. The Thraki aquatic fauna is closely linked with that of the central European, Danubian and Macedonian faunas, and also has ties with the western Balkan and even with the Anatolian faunas. The western Balkan fauna is most closely related to that of the southwestern Danube, and has ties with the Thraki and Macedonian regions. Italian affinities are seen only in the Dalmatian region, and central European and Danube affinities in the South–Adriatic Ionian region.

Several geological events have influenced the evolution of the Balkan Peninsula and of its aquatic fauna. Radoman (1985) summarises these as follows:

(1) the existence of the former Tethys Sea;
(2) the upheaval of the Carpatho-Balkans, Dinarides (Dinaric Alps) and Hellenides during the Late Oligocene and Miocene, separating the Tethys into a southern sea (the Mediterranean) and a northern one (the Paratethys, later to become brackish, then freshwater); as a consequence of this upheaval, wide areas of the Tethys became dry land;
(3) the existence of many lakes, derived from the Tethys and the Paratethys, which gradually became brackish, and then freshwater, and
(4) fluctuations of sea level in the Adriatic Sea, enabling faunal exchange between the western Balkans and Italy

Other factors which must be taken into account include:
(1) The former continental contact between the southern peninsula (mainly via Greece) and Anatolia, interrupted only in Pliocene or even in Pleistocene times;
(2) the desiccation of the Mediterranean Sea during the Messinian Salinity Crisis, which favored faunal dispersal between the Near East and the southern European peninsulas, and
(3) the continuous modification of the river network though repeated river capture.

Radoman (1985) considers that the rich Balkan hydrobioid fauna has a marine origin, being derived from a Tethyan, or Paratethyan fauna. Hydrobioids are known to include marine, brackish water and freshwater lineages, and a thorough comparison of the Balkan genera with those from other Holarctic or Palaearctic areas is necessary in order to determine the validity of this hypothesis. Radoman also explains the occurrence of many genera and even species on both sides of the Dinarides (*i.e.* both in Dinaric rivers and in the southwestern tributaries of the Danube) by assuming that they were already present before the uplifting of the mountains. This is considered unlikely, as species rarely if ever remain identical during long geological periods once interruption of the gene flow has taken place. In the author's opinion, the occurrence of the same species (and even genera) on both sides of the Dinarides is the result of relatively recent dispersal events. The Dinarides range represents an impenetrable barrier for fish and other groups of epigean freshwater animals, including the non-hydrobioid snails. Most hydrobioids are crenobiotic (able to live temporarily in underground waters), however, and it is possible that there are permanent or temporary contacts between the hypogean layers of water of the Adriatic and Danubian drainage systems which could enable dispersal to take place.

Radoman (1985) argues that Lake Ohrid originated from the Tethys or Paratethys, as with other Balkan lakes. The abundance of hydrobioid snails in the water bodies of the western Balkans and in Lake Ohrid actually suggests a common origin, but most Ohrid hydrobioids belong to other genera and even subfamilies than those of Dalmatia; this suggests a long evolution in isolation within the Lake Ohrid basin. However, the Ohrid fauna includes genera absent from other western Balkan water bodies, some of which show affinities with taxa inhabiting distant Palaearctic or Holarctic regions. An example is the hydrobioid *Micropyrgula stankovici*, considered by Radoman as representing a monospecific family endemic to Ohrid, although it is argued elsewhere that the genus has a second species in the Pyrenees (Illies, 1978). The lake also supports many Central European or more widely distributed species, *e.g.* the fish species, *Phoxinus phoxinus, Alburnoides bipunctatus*, *Gobio gobio*, and therefore the lake cannot have been totally isolated, since the presence of these species suggests that it was possible for new intruders to colonise the lake relatively recently.

As mentioned above, the Thraki fish fauna, and many of the aquatic invertebrates, are mainly Central European species. All these species are inhabitants of standing or very slowly flowing water, and are absent from other Balkan areas. Kosswig & Battalgil (1942, 1943) consider that these species arrived via dispersal from the Danube and other rivers into the Black Sea when it was a freshwater lake. At this point the Black Sea outlet was the Aegeopotamos, a large river flowing south and westwards, across a single landmass. The Thracian rivers were then tributaries of the Aegopotamos and the species were able to disperse into them. This hypothesis is adopted here.

In the Vardar, and the three more southerly rivers of Macedonia and Thessaly, four typical Danubian rheophilic fishes and a crayfish occur, all of which are absent from the Thracian and western Balkan drainage system, and one Central European species which is present in the Vardar and Danube and absent from the western Balkans and Thraki. These species would not have been able to disperse through the Black Sea and the Aegeopotamos, but must have come directly from the Danube Basin into the River Vardar by means of river capture: the upper Vardar is very close to the headwaters of Morava, a southern tributary of the Danube. The hydrobioid snail *Terranigra* is confined to the headwaters of the Vardar and Morava rivers, and its distribution supports the hypothesized river capture event.

CONCLUSIONS

The river network of the Balkan Peninsula includes rivers flowing in all directions, each having its own faunistic peculiarities and endemics. Seven regions, corresponding to groups of river drainages, are defined for the peninsula, together with the lakes of Ohrid and Prespa which are dealt with separately. The fish fauna of the Danube Basin is very unitary, although the hydrobioid snail fauna of the southwestern tributaries is quite distinct from that of the remaining part of the basin, being closely related to that of the western Balkan watershed. The degree of endemicity is very high, especially in Lake Ohrid and the rivers of Dalmatia (western watershed). Most fish and representatives of the aquatic groups are classed as continental. It is not clear whether the hydrobioid snails, the aquatic group including the highest number of species, should be considered as of continental or marine origin. Many aquatic animals of the peninsula are long-established, but there are some recent intruders, even in Lake Ohrid.

REFERENCES

Almaca, C. 1976. La spéciation chez les Cyprinidae de la peninsule ibérique. *Revue.des Travaux de l'Institute Pêche maritime* 40, 399-411.

Arndt, W. 1937. *Ochridaspongia rotunda* n.g.n.sp, ein neuer Süsswasserschwamm aus dem Ochridsee. Zugleich ein Überblick über die Spongillidenfauna der Balkanhalbinsel und ihrer Grenzländer. *Archiv für Hydrobiologie* 31, 636-677.

Bănărescu, P. 1960. Einige Fragen zur Herkunft und Verbreitung der Süsswasserfischfauna der europäisch-mediterranen Unterregion. *Archiv für Hydrobiologie* 57, 16-134.

Bănărescu, P. 1989. Zoogeography and history of the freshwater fish fauna of Europe. In: *The Freshwater Fishes of Europe*, Vol.1 (II) (J. Holcik, ed.), pp. 88-107. Aula Verlag, Wiesbaden.

Bănărescu, P. 1990. *Zoogeography of fresh waters, 1 General distribution and dispersal of freshwater animals*. Aula Verlag, Wiesbaden.

Bănărescu, P. 1992. *Zoogeography of Fresh Waters, 2 Distribution and dispersal of freshwater animals in North America and Eurasia*. Aula Verlag, Wiesbaden.

Bănărescu, P. & Herzig Straschil, B. 1998. Beitrag zur Kenntnis der *Leuciscus*-Untergattung *Telestes* Bonaparte (Pisces: Cyprinidae). *Annalen des Naturhistorisches Museum Wien* 100-B, 405-424.

Bianco, P.G. 1990. Potential role of the paleohistory of the Mediterranean and Paratethys basins on the early dispersal of Europe-Mediterranean freshwater fishes. *Ichthyological Exploration of Freshwaters* 1, 167-184.

Botoşăneanu, B. (ed.). 1986. *Stygofauna Mundi: a faunistic, distributional and ecological synthesis of the world fauna inhabiting subterranean waters* Dr.W.Backhuys, Leiden.

Bousfield, E.L. 1978. A revised classification and phylogeny of amphipod crustaceans. *Transactions of the Royal Society of Canada* 16, 343-390.

Dancău, D. 1976. Sur un nouveau amphipode souterrain de Roumanie: *Pontoniphargus racovitzai* n. g. n. sp. Livre du centainaire Emil Racovitza (1868 – 1968) *Acad. Roum.* 275-285.

Doadrio, I. 1990. Phylogenetic relationships and classification of western palaearctic species of the genus *Barbus* (Osteichthyes). *Aquatic Living Resources* 3, 265-282.

Drensky, P. 1951. [*Ribite v B'lgarja.*] Izdanie na B'lgarskata Akademija na naukite, Sofia. [in Bulgarian.]

Economidis, P. 1986. *Chalcalburnus belvica* (Karaman, 1924), (Pisces, Cyprinidae), nouvelle combinaison taxinomique pour la population provenant du lac Petit Prespa (Macédoine, Grèce). *Cybium* 10, 85-90.

Economidis, P. 1991. *Check list of freshwater fishes of Greece (recent status of threats and protection)*. Hellenic Society for the Protection of Nature, special publication, Athens.

Economidis, P. & Bănărescu, P. 1990. The distribution and origins of freshwater fishes in the Balkan Peninsula, especially in Greece. *Intern. Rev. Ges. Hydrobiol.* 79, 257-283.

Hsü , K.J. 1974. The Miocene desiccation of the Mediterranean and its climatic and zoogeographical implications. *Naturwissenschaften*, 137-142.

Illies, J. (ed.). 1978. *Limnofauna Europaea*. 2nd edition. Gustav Fischer Verlag, Stuttgart.

Ivanović, B.M. 1973. *Ichthyofauna of Skadar Lake*. Institute of Biological Sciences, Titograd.

Karaman, G. 1974. *Crustacea, Amphipoda. Catalogus Fauna Jugöslaviae* III/3. Acad. Scient et Artium Sloven., Ljubljana.

Karaman, M. 1972. Süsswasserfische der Türkei, 9. Teil. Revision einiger kleinwüchsiger Cyprinidengattungen, *Phoxinellus*, *Leucaspius*, *Acanthobrama*_usw aus Südeuropa, Kleinasien, Vorderasien und Nordafrika. *Mitteilungen des Hamburg Zoologisches Museum und Institut* 69, 115-155.

Kosswig, C. & Battalgil, F. 1942. Zoogeographie der türkischen Süsswasserfische. *Revue Faculté des Sciences, Université d'Istanbul, B. Sciences Naturelles* 7, 145-165.

Kosswig, C. & Battalgil, F. 1943. Beiträge zur türkischen Faunennegeschichte. 1. Süsswasserfische. *Comptes Rendus annuels et Archives de la Société Turque des Sciences Physiques et Naturelles* 8, 29-63.

Kottelat, M. 1997. European freshwater fishes. *Biologia, Bratislava* 52 Supplement 5, 1-271.

Radoman, P. 1983. Hydrobioidea, a superfamily of Prosobranchia (Gastropoda). I. Systematics. *Serbian Academy of Sciences and Arts Monographs* 547, 1-256.

Radoman, P. 1985. Hydrobioidea, a superfamily of Prosobranchia (Gastropoda). II. Origin, zoogeography, evolution in the Balkans and Asia Minor. *Monographs of the Beograd Faculty of Sciences Department Biol.* 1, 1-173.

Schütt, M. 1996. The dissemination of Theodoxus species in Greece, including the islands (Moll. Prosobr.). *Biologia Gallo-Helenica* 12, 283-290.

Sorić, V.M. 1999. First finding of *Orthrias brandti* (Cobitidae) in the drainage area of the Vardar River (Fyrom) and description of the subspecies *Orthrias brandti macedonicus* ssp.nov. *Ichthylogia (Beograd)* 31, 81-87.

Thienemann, A. 1950. *Verbreitungsgeschichte der Süsswassertierwelt Europas*. Schweizerbartische Verlag, Stuttgart.

Vuković, T., & Ivanović, B. 1971. *Slatkovodne ribe Jugoslavije*. Zemaljski Muzej Bosne i Hercegovine, Sarajevo.

13. THE BIVALVE MOLLUSC FAUNA OF ANCIENT LAKES IN THE CONTEXT OF THE HISTORICAL BIOGEOGRAPHY OF THE BALKAN REGION

ALEXEI V. KORNIUSHIN*†

INTRODUCTION

One of the most remarkable features of the Balkan freshwater environment is the presence of so called ancient or long-lived lakes, Ohrid and Prespa being the most famous among them. Existing continuously for several millions of years, such lakes are distinguished by their rich and unique faunas (Stanković, 1960; Starobogatov, 1970; Taylor, 1988; Wesselingh & Michel, 2001), and undoubtedly constitute "hot spots" of biodiversity. Investigations of these peculiar habitats, and, in particular, their rich mollusc fauna, are important for biogeographical regionalisation, reconstructing the history of the fauna and understanding the mechanisms of speciation (Starobogatov, 1970; Taylor, 1988; Wesselingh & Michel, 2001). To date, such investigations have focused on the extremely diverse gastropod faunas of Tanganyika and Baikal, and the ancient lakes of Sulawesi, where endemism may reach generic or even family level. Modern methods of cladistic phylogenetic analysis on phenetic and molecular data have been applied to clarify within- and between-group relationships of endemic taxa (Glaubrecht, 2001; Sitnikova, 2001; Todd & Michel, 2001; West & Michel, 2001). In contrast, the molluscs of Balkan lakes have been reviewed in several classical works (Stanković, 1960; Maassen, 1980; Radoman, 1983, 1985; Meier-Brook, 1983; Dhora & Welter-Schultes, 1996) but have not yet been the subject of up-to-date multidisciplinary research.

The major freshwater bivalve mollusc families - Unionidae, Corbiculidae, Sphaeriidae and Dreissenidae - have not received much attention in the aforementioned works. However, bivalves can also be good models for biogeographic and evolutionary studies because of their worldwide distribution and ancient age (see below). Endemic bivalve taxa from ancient lakes are apparently not so numerous as gastropods, but a number of endemic species have been described from Ohrid and Prespa (Kuiper, 1960, 1987; Lvova & Starobogatov, 1982), as well as from Baikal, Biwa, the lakes of Sulawesi, the African Rift lakes and Titicaca (see Mori, 1938; Kuiper, 1966; Djajasasmita, 1975; Kuiper & Hinz, 1984; Slugina & Starobogatov, 1999 for reviews). Only one endemic genus is currently recognised - *Possostrea* (Corbiculidae) from Lake Poso, Sulawesi (Bogan & Bouchet, 1998). Several recent papers discuss the relationships of endemic taxa to widely-distributed species (Starobogatov & Sitnikova, 1990; Korniushin, 1997; Korniushin *et al.*, 2000; Bogan & Bouchet, 1998), but molecular data are available for only a few taxa (Glaubrecht *et al.*, in press).

Comparison of the fauna of ancient lakes with that of adjacent areas is also important for understanding the history of each lake. The fauna of Ohrid was first analysed in this respect by Kuiper (1960). Several papers on the bivalve mollusc fauna of the Balkans were published afterwards (Beron & Georgiev, 1967; Angelov,

H.I. Griffiths et al. (eds.), Balkan Biodiversity, 219–241.

1971, 1976, 1984, 2000; Kuiper, 1981; Schütt, 1985; Kinzelbach, 1986; Sattmann & Reischütz, 1988; Frank, 1988; Reischütz & Stummer, 1989; Dyduch-Falniowska, 1989; Dhora & Welter-Schultes, 1996); important fossil records from the Mediterranean region are provided by Becker-Platen & Kuiper (1979). However, most of these works only record species; the status of many taxa is still controversial and needs to be revised in the light of recent developments in freshwater bivalve mollusc systematics (Starobogatov, 1994; Nagel *et al.*, 1998; Korniushin, 1999, 2001; Stepien *et al.*, 2001).

Thus, our knowledge of freshwater bivalve molluscs in the Balkan region is still insufficient. The aim of this paper is to review the existing literature, with a special emphasis on the ancient lakes of Ohrid and Prespa, and those of adjacent countries - Macedonia (Former Yugoslav Republic of Macedonia; FYROM), Albania, Bulgaria and Greece. Faunistic data on the other Balkan countries are included for purposes of comparison. Some new records are provided in the framework of this study, the status of difficult taxa is discussed, and in some cases taxonomic changes are suggested. The biogeographical composition of the fauna is analysed, and possible modes of origin and measures for its conservation are discussed.

REVIEW OF THE MAJOR FRESHWATER BIVALVE FAMILIES AND THEIR REPRESENTATION IN THE BALKANS

Freshwater bivalve molluscs are represented in the study region by three families: Unionidae, Sphaeriidae and Dreissenidae. Neither Margaritiferidae, which is distributed in North and Central Europe, nor extant Corbiculidae, known in the Palaearctic region from the eastern Mediterranean and introduced to North America and some European countries (Araujo *et al.*, 1993; Glöer & Meier-Brook, 1998; Gittenberger *et al.*, 1998), have been recorded in the Balkans.

Unionidae (najades)

This family comprises the largest freshwater bivalves (shell length in European species up to 200 mm) and occurs mostly in the larger water bodies - rivers, ponds and lakes. Adult animals are partly embedded in sediments (Fechter & Falkner, 1990). The offspring are brooded in water tubes of gills and released as peculiar larvae (glochidia) which parasitise fish (Nagel *et al.*, 1998). The general range of the family includes Eurasia, Africa and North America, and its geological age is estimated as Jurassic (Kolesnikov, 1977; Taylor, 1988). Within Europe, most specialists recognise nine native species; one more (*Sinanodonta woodiana* Lea, 1834) was introduced from East Asia (Nagel *et al.*, 1998; Falkner *et al.*, 2001). A taxonomic system with many more splits was developed in the former USSR (see Korniushin, 1998b for review), but it is not used in any other country. All European native species are recorded in the Balkans (Sattmann & Reischütz, 1988; Frank, 1988; Reischütz & Stummer, 1989; Fechter & Falkner, 1990; Dhora & Welter-Schultes, 1996). The list of these species arranged according to the system suggested by Falkner *et al.* (2001) is provided below:

Potomida littoralis (Cuvier, 1798)
Microcondylaea bonellii (A. Férussac, 1827) (= *M. compressa* (Menke, 1828)
Unio (Unio) pictorum (Linnaeus, 1758)
U. (U.) tumidus Philippson, 1788

U. (U.) mancus (Lamarck, 1819) (=*U. elongatulus* C. Pfeiffer, 1825)
U. (Crassiana) crassus Philippson, 1788
Anodonta (Anodonta) cygnea (Linnaeus, 1758)
A. (A.) anatina (Linnaeus, 1758)
Pseudanodonta complanata (Rossmässler, 1835).

Most of the listed species are widely distributed (European or Euro-Siberian). The exceptions are *Potomida littoralis* and *Unio mancus,* which are restricted to the Mediterranean region, and *Microcondylaea bonellii*, which is characterised by a disjunct range with one area including the basin of the River Po, outflows of some Alpine lakes (Lugano, Como and Garda) and the north coast of the Adriatic sea, and another area stretching from Montenegro (inflows to Lake Skutari [= Shkodra]) across Albania to Lake Ohrid (Fechter & Falkner, 1990). Jaeckel *et al.* (1957) reported the presence of the latter species in Greece, but these records were not confirmed by later investigations; recently, it has been recorded in Croatia (Bössneck, 2002).

Subspecies are distinguished in most of the European species of najades (Glöer & Meier-Brook, 1998; Falkner *et al.*, 2001). The subspecific status of the Balkan forms is defined in several papers, but there are some discrepancies between them. Maassen (1980) and Sattmann & Reischütz (1988), for example, identified *Unio crassus* from lakes Ohrid and Kastoriás as *U. c. carneus* Küster, 1878, but Frank (1988) assigned this form to *U. c. bruguierianus* Bourguignat, 1853, which she considered to be distributed from the Aegean region across Asia Minor to the Transcaucasus, the River Tigris, Syria and Oront. Falkner (1994) restricted the range of the latter subspecies to eastern Greece and western Asia Minor. *Anodonta cygnaea deplanata* Gallenstein, 1852 was reported from the area of Ohrid and Prespa (Maassen, 1980; Reischütz & Stummer, 1989), while Falkner (1994) applied the name *A. c. gravida* Drouet, 1879 to the form distributed from central Greece to Thrakia. *Unio pictorum gaudioni* Drouet, 1881 and *U. mancus kruperi* Drouet, 1879 were also listed for the Balkan fauna by Falkner (1994). However, the latter author noted that the status of many forms from the eastern Mediterranean is still not clear, and their distribution ranges are not yet delimited precisely. Furthermore, genetic research by Nagel *et al.* (1998) revealed rather intricate relationships between unionid populations in Mediterranean countries, which do not agree closely with morphologically-based classifications; subspecies were not defined in this study. More recent genetic research by Nagel & Badino (2001), which was also based on allozyme frequencies, suggested the subspecific status of *U. mancus* within *U. pictorum*, but did not resolve the taxonomy of the Mediterranean *Anodonta* and *U. crassus* forms. Thus, a lot of taxonomic work still needs to be done with this group before it can be of value in biogeographic research, especially in such a complicated region as the Balkan Peninsula.

Sphaeriidae

These molluscs are distinguished by their small size (adult shells of some species are only 2 mm long, with the largest species up to 25 mm long). They inhabit a great variety of freshwater habitats (from springs, temporary pools and peat bogs to large rivers and lakes), and are usually completely embedded in the substrate (Korniushin, 1996; Glöer & Meier-Brook, 1998; Gittenberger *et al.*, 1998). In contrast to the previous group, they are hermaphroditic and reproduce by brooding in special pouches within gills; their development is direct, without any swimming larvae (Korniushin, 1996). The family is cosmopolitan and its geological age is thought to be Jurassic (Kolesnikov, 1977) or Cretaceous (Taylor, 1988). The taxonomy of the group is still in debate (Starobogatov, 1970; Korniushin, 1996, 1998a, 1999, 2001). Here we use the system recognising three genera (*Sphaerium, Musculium* and *Pisidium*), which was suggested for the European database system by Falkner *et al.* (2001). Korniushin (1999, 2001) argued the generic status of *Amesoda* and several subgroupings of *Pisidium* s.l. - *Lacustrina, Euglesa* and *Neopisidium*. These suggestions are partly supported by recent molecular studies (Cooley & Ó Foighil, 2000), but the final generic classification cannot be confirmed as yet due to some contradictions between molecular and morphological phylogenies

The European sphaeriid fauna comprises 32 species (Korniushin, 1999, 2001; Korniushin *et al.*, 2000; Falkner *et al.*, 2001), 19 of them being recorded in the study area (Table 1). Some of the taxa are worthy of special comment.

Sphaerium rivicola (Lamarck, 1818). Gambetta (1930) described a new species from the Albanian waters of lake Ohrid - *Pisidium parenzani*. No material of this species is currently available, and its status remains dubious, but its relatively large size (length 13 mm, height 10 mm, breadth 7 mm), central umbones, elevated ligament and strongly curved cardinal teeth, which are clearly visible on the published figures (Gambetta, 1930: pl. 1, figs 7, 8), suggest that *P. parenzani* belongs to the genus *Sphaerium* (see also Kuiper, 1960) and is most probably conspecific with *Sphaerium rivicola* (Lamarck, 1818). Until now, the southernmost record of the latter species in the Balkans was from the Maritsa river in Bulgaria (Angelov, 2000). If confirmed, the record of Gambetta (1930) will considerably extend this range.

Sphaerium nucleus (Studer, 1820). This taxon is usually treated as a variety of *S. corneum* (Linnaeus, 1758); as such, it was included in the list of Bulgarian species by Angelov (1976). However, Glöer & Meier-Brook (1998) and Falkner (2000) consider it a distinct species. Characters supporting the species status of this taxon and distribution data (including one record from Slovenia) are summarised by Korniushin (2001). The presence of this species in Greece is confirmed by the as yet unpublished material in Senckenbergsmuseum, Frankfurt-am-Main (SMF 86927), labelled „Griechenland, See vor Vretori“ (Greece, the lake at Vretori), which was identified by A. Korniushin in 1995.

Pisidium (Euglesa) personatum Malm, 1853. Korniushin (1996) has shown that specimens of *P. personatum* from southern Ukraine and the Aegean Islands differ from those of Central European populations in having a very small outer demibranch, which in young specimens is not visible at all. This character was also observed in specimens from a small stream flowing into Lake Ohrid near Peštani, Macedonia (FYROM) collected by the author in 1998. It seems probable that the South European form of *P. personatum* is a distinct subspecies or even a species, and that a valid name for it might be *Pisidium bulgaricum* Odhner, 1929, based on Bulgarian specimens without the outer demibranch. However, revision of the type material of *P. bulgaricum* and a study of more numerous collections from different countries are necessary, before any taxonomic decision on this matter is possible. Specimens collected by A. Angelov in the Vitosha Mountains, for example, near the supposed type locality of *P. bulgaricum*, and revised by this study, had a well-developed outer demibranch and therefore were defined as *P. casertanum*.

Table 1. Species diversity of the family Sphaeriidae in Balkan lakes compared to adjacent countries. Sources: Ohrid - Kuiper (1960), Maassen (1980), Dhora & Welter-Schultes (1996), Korniushin et al. *(2000); Prespa and Mikra Prespa - Angelov (1971), Kuiper (1987), Reischütz & Stummer (1989), Korniushin* et al. *(2000); Macedonia - Angelov (1971), Kuiper (1981), original data (*P. personatum*); Albania - Dhora & Welter-Schultes (1996); Greece - Schütt (1985), Dyduch-Falniowska, 1989, original data (*S. nucleus*); Bulgaria - Angelov (1976, 1984, 2000). * - taxa alluded to in text; endemic taxa are underlined.*

Species	Ohrid	Prespa +Mikra Prespa	Macedonia except lakes	Albania	Greece	Bulgaria
Sphaerium rivicola (Lamarck, 1818)*	?					+
S. corneum (Linnaeus, 1758)					+	+
S. nucleus (Studer,1820)*					+	+
Musculium lacustre (Müller, 1774)			+	+	+	+
Pisidium amnicum (Müller, 1774)	+				+	+
P. casertanum (Poli, 1791)			+	+	+	+
P. c. ponderosum Stelfox, 1918	+	+				
P. personatum Malm, 1855*	+	+	+		+	+
P. obtusale (Lamarck, 1818)						+
P. henslowanum (Sheppard, 1823)	+	+				+
P. supinum A. Schmidt, 1851						+
P. s. subtruncatum Malm, 1855	?	+	+		+	+
P. s. recalvum Kuiper, 1960	+					
P. nitidum Jenyns, 1832					+	+
P. edlaueri Kuiper, 1960	+					
P. maasseni Kuiper, 1987		+				
P. milium Held, 1836			+		+	+
P. pseudosphaerium Favre, 1927						+
P. moitessierianum Paladilhe, 1866	+	+			+	+
P. tenuilineatum Stelfox, 1918	+		+		+	
P. sogdianum (Izzatullaev & Starobogatov, 1986)*			+		+	
Total no. of species	**9**	**6**	**7**	**2**	**12**	**15**

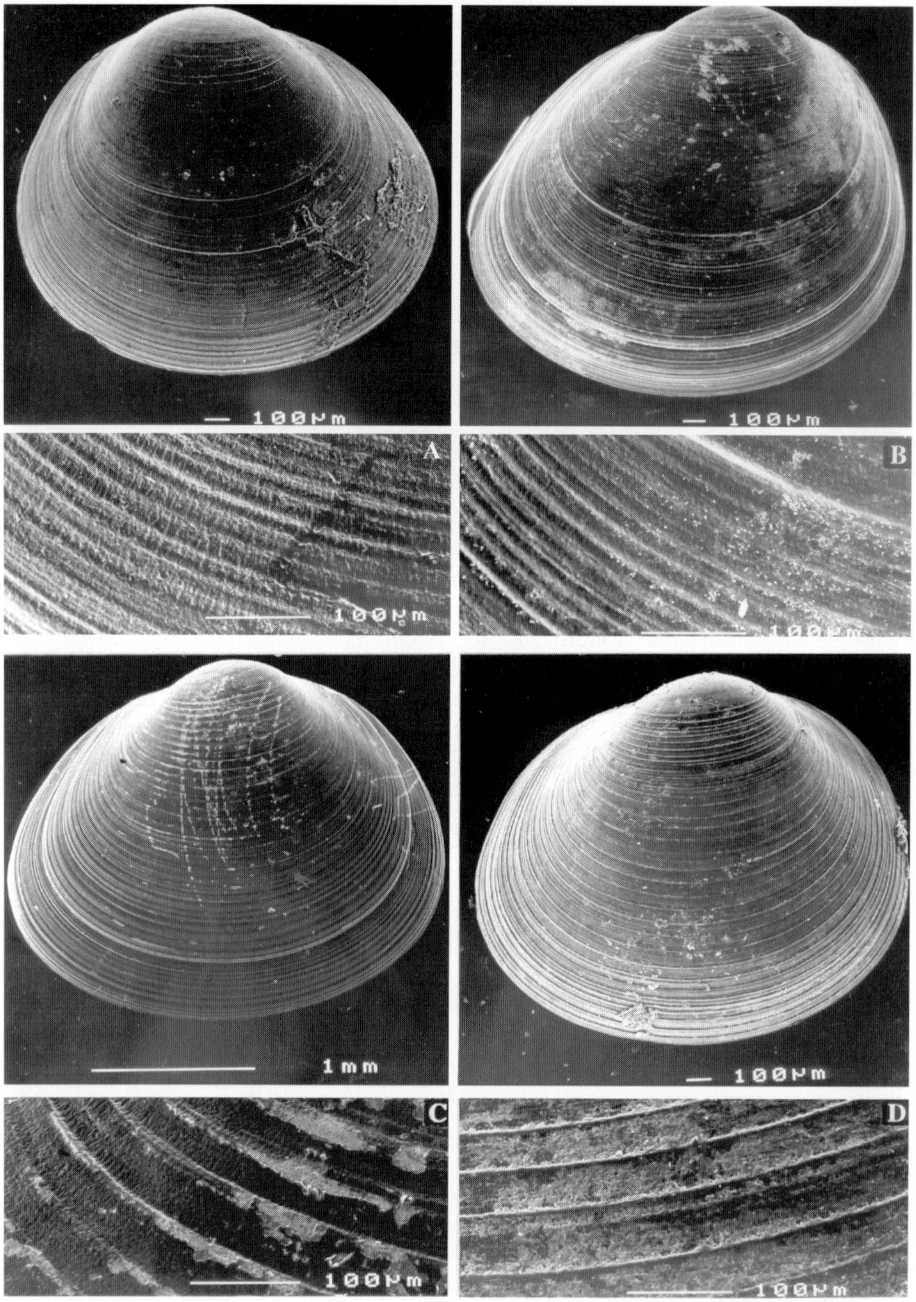

Figure 1. Pisidium sogdianum *and* P. annandalei, *external view of shell and surface sculpture: A* - P. sogdianum *from Greece, leg. E. Gittenberger, collection of the Nationaal Naturhistorisch Museum, Leiden); B* - P. sogdianum *from Kyrgyzia, collection of U. Bößneck, Erfurt); C* - P. annandalei *from Nepal, leg. H. Nesemann, collection of A.V. Korniushin, Kiev); D* - P. annandalei *from Thailand, leg. R. Brandt, collection of the Nationaal Naturhistorisch Museum, Leiden).*

Pisidium (Odhneripisidium) sogdianum (Izzatullaev & Starobogatov, 1986). Here we refer to the taxon earlier recorded in the Balkan region as *Pisidium vincentianum* Woodward, 1913 (Kuiper, 1960) and than re-identified as *P. annandalei* Prashad, 1925 (Kuiper, 1962, 1981). The heterogeneity of *P. annandalei* sensu Kuiper was suggested by Izzatullaev & Starobogatov (1986); Korniushin (1999) distinguished two subspecies differing in size and patterns of sculpture. As shown by SEM study (Figure 1), the differences in sculpture are even more pronounced than suggested in the aforementioned papers. The shell surface of *P. annandalei* specimens from Mediterranean countries and Central Asia (within the borders of the Palearctic region) is characterised by fine, dense striations (50-55 per 1 mm) (Figure 1A-D). In contrast, shells from Nepal, which is near to the type locality (Bihar, India), as well as those from other localities in tropical Asia, are covered by widely spaced (up to 25 per 1 mm) fine ribs (Figure 1E-H). The first form also differs in its smaller size (up to 2.5 mm long, compared to up to 3.8 mm in *P. annandalei* s. str.) and narrow, somewhat protruding umbones. Thus, there are sufficient grounds to treat this form as a distinct species. Revision of the type materials from the Zoological Institute of the Academy of Sciences, Saint Petersburg (Korniushin, 1999) confirmed that *Odhneripisidium sogdianum* Izzatullaev & Starobogatov, 1986 is a valid name for it.

In our opinion, all Palaearctic localities of *Pisidium annandalei* provided by Kuiper (1981) should be referred to as *P. sogdianum*. Thus, the recent range of this species comprises the Mediterranean countries (southern Italy, Macedonia [FYROM] , Greece, Turkey, Syria, Jordan and Israel) and Central Asia (Iran and the former Central Asian republics of the USSR) (Kuiper, 1981; Izzatullaev & Starobogatov, 1986). It is noteworthy that in the Neogene it has also been reported from Central Europe (southern Bavaria) (see Becker-Platen & Kuiper, 1979). Its recent distribution can therefore be treated as relict.

The Sphaeriid fauna of the Balkan Peninsula is less diverse than that of North and Central Europe (Falkner *et al.*, 2001), but is rather rich in comparison with other southern European countries, such as Romania, where 14 species have been recorded (Grossu, 1962, with additions by Korniushin, 2001). All the species, with the exception of lakes endemics, are characterised by wide ranges.

Dreissenidae

This family comprises medium-sized molluscs (shell length of adults about 20-30 mm) with an attached mode of life and preference for hard substrates. *Dreissena* species live in rivers, estuaries, and lakes, including brackish waters such as the Caspian Sea; the only extant species of the genus *Congeria* - *C. kusceri* Bole, 1962, is troglobiont (Morton *et al.*, 1998). In contrast to other freshwater bivalves, most representatives of this group have planktonic veliger larvae, but direct development and brooding in gills was suggested for *C. kusceri* (Morton *et al.*, 1998). The modern distribution of one species - *Dreissena polymorpha* (Pallas, 1771) - is European, while other species of *Dreissena* are restricted to the Ponto-Caspian region (Starobogatov, 1970, 1994). The extant *Congeria* is endemic to the Dinaric karst region. Finally, two species of *Dreissena* were introduced to North America (Starobogatov, 1994; Morton *et al.*, 1998; Stepien *et al.*, 2001). The geological age of this group is Caenozoic and its evolution took place mainly in the Paratethian basin (Starobogatov, 1994; Morton *et al.*, 1998).

According to Kinzelbach (1986), *Dreissena polymorpha polymorpha* is basically restricted in the Balkan Peninsula to the watercourses of the Danube basin. More southern populations are morphologically distinct, and species status has been attributed to the forms discussed below.

Dreissena stankovici Lvova & Starobogatov, 1982. According to the original description, this endemic species of Lake Ohrid is distinguished by the peculiar form of its shell (in contrast to *D. polymorpha*, the keel is placed rather high, the ventral surface is not flattened and the dorsal edge forms a high wing), the form of muscle scars (the anterior adductor scar is equilateral, the pedal retractor scar broad), the presence of large papillae between siphons and numerous papillae of inhalant siphon. It was also mentioned in the original description, that in shell form the new species is more similar to *D. (Pontodreissena) bugensis* (Andrussov, 1897) than to *D. (Dreissena) polymorpha* (Pallas, 1771), and a new subgenus *Carinodreissena* Lvova & Starobogatov, 1982 was established. However, some subsequent authors still treated Ohridan *Dreissena* as *D. polymorpha* ssp. or simply *D. polymorpha* (Kinzelbach, 1986; Dhora & Welter-Schultes, 1996) without any comment on the species described by Lvova & Starobogatov (1982); it seems that the Russian work has remained unnoticed by West European malacologists.

Recently we had an opportunity to assess the diagnostic characters of *D. stankovici* on material collected in 1998 near the town of Ohrid, Macedonia [FYROM] by comparison with samples of *D. polymorpha* and *D. bugensis* collected in the Ukraine. The differences between these species in shell profile, as well as in the form of anterior adductor and its scar (Lvova & Starobogatov, 1992), are confirmed (Figures 2, 3 & 4). The most remarkable anatomical distinction of *D. stankovici* is the presence of two pairs of large papillae between the siphons; in *D. polymorpha* smaller intersiphonal papillae are organized in two rows (Figure 4), as is also clearly visible on the published photographs of living specimens (Fechter & Falkner, 1990; Gittenberger *et al.*, 1998). In having numerous papillae of inhalant siphon, *D. stankovici* is similar to *D. bugensis*, but the papillae of the latter species are smaller. The Ohridan *Dreissena* is also distinguished by its relatively large posterior adductor, which is clearly divided into two parts ("slow" and "fast"), while the size of the pedal retractor proved to be rather variable. Thus, at least some of the characters reported by Lvova & Starobogatov (1982) are constant, and therefore *D. stankovici* is recognised here as a distinct species, endemic to Lake Ohrid.

The status of *Dreissena* from Lake Prespa is still not clear; it is usually recorded as *D. polymorpha* (Frank, 1988; Reischütz & Stummer, 1989), but its characters need to be tested in the light of the data presented above.

D. blanci (Locard, 1893). In western Greece and Lake Skutari in Albania, this taxon is treated as a subspecies of *D. polymorpha* by Kinzelbach (1986), and as a distinct species by Fechter & Falkner (1990) and Dhora & Welter-Schultes (1996). Starobogatov (1994) considered the taxon to be closely related to *D. polymorpha*. It is distinguished first of all by its sharp keel (Fechter & Falkner, 1990). Specimens from Lake Dojran in Macedonia (FYROM) collected by R. Kilias in 1977 and deposited in the collection of the Museum für Naturkunde Berlin (ZMB) also show this character (Figure 3) and thus may represent *D. blanci*. Anatomical characteristics of this species are not known in detail, but the published photographs

(Fechter & Falkner, 1990) show more similarity to *D. polymorpha* than to *D. stankovici*.

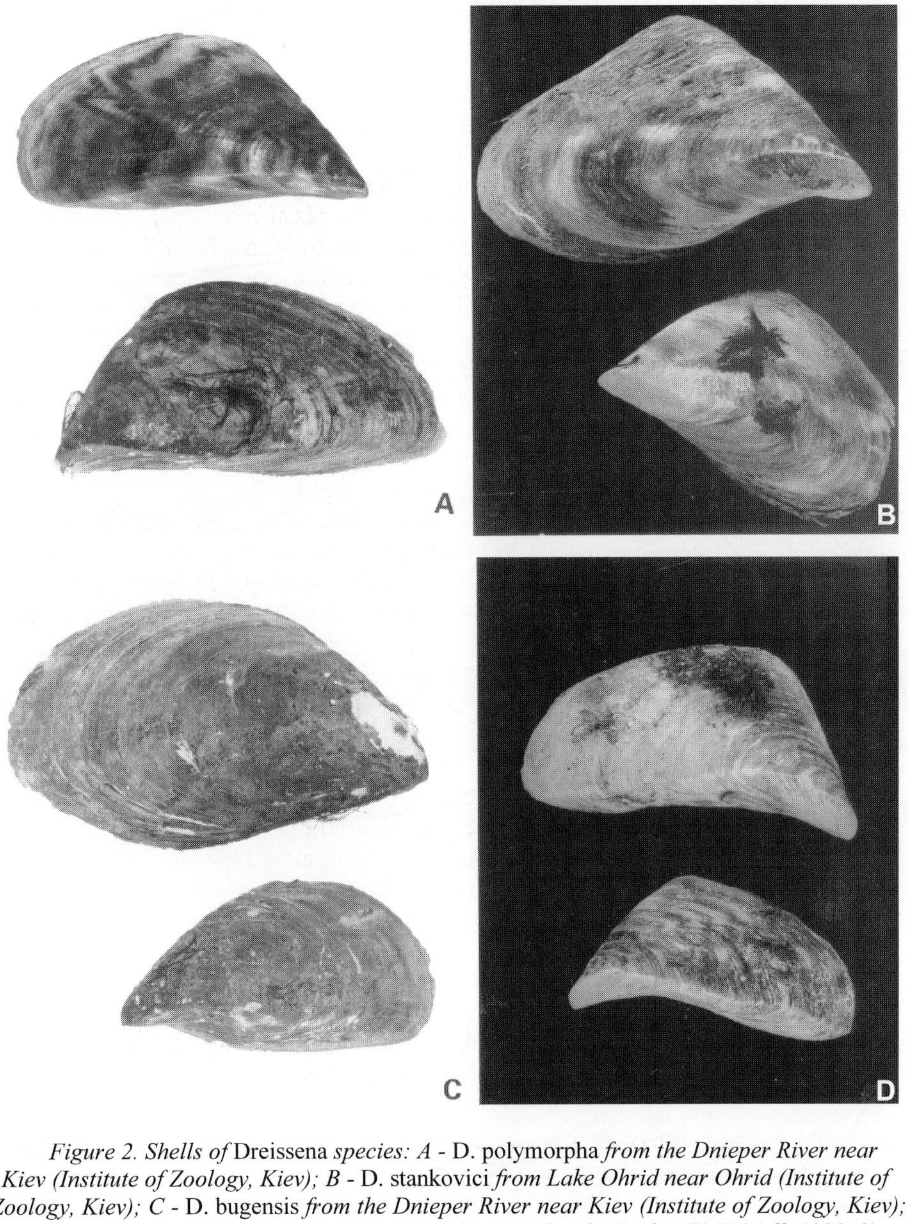

Figure 2. Shells of Dreissena *species: A -* D. polymorpha *from the Dnieper River near Kiev (Institute of Zoology, Kiev); B -* D. stankovici *from Lake Ohrid near Ohrid (Institute of Zoology, Kiev); C -* D. bugensis *from the Dnieper River near Kiev (Institute of Zoology, Kiev); D –* D. blanci *(?) from Lake Dojran (Museum für Naturkunde, Berlin, ZMB Moll. 103048).*

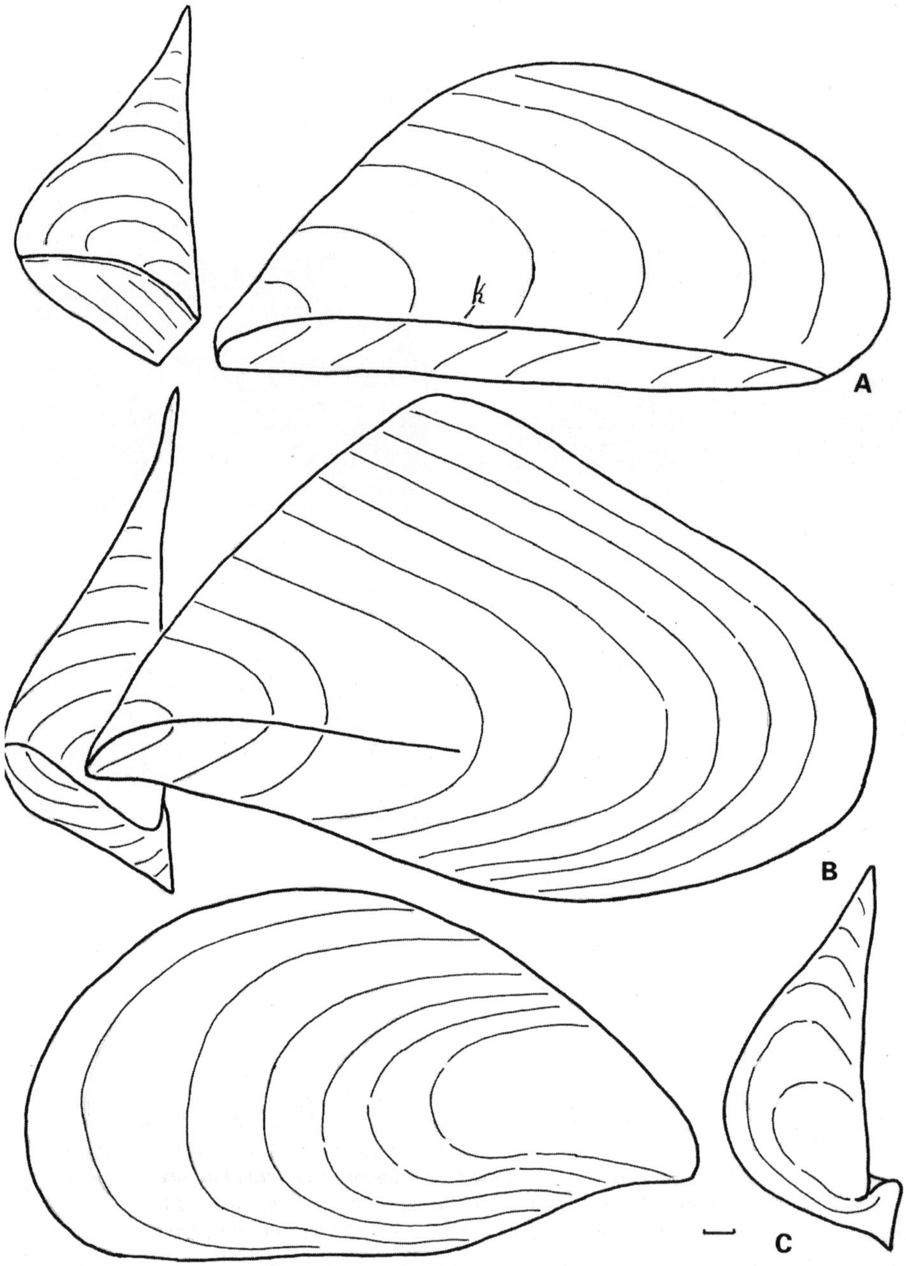

Figure 3. Shells of Dreissena *species (the same specimens as shown in Figure 2 A - C, respectively): lateral and frontal views. k - keel. Scale bar = 1 mm.*

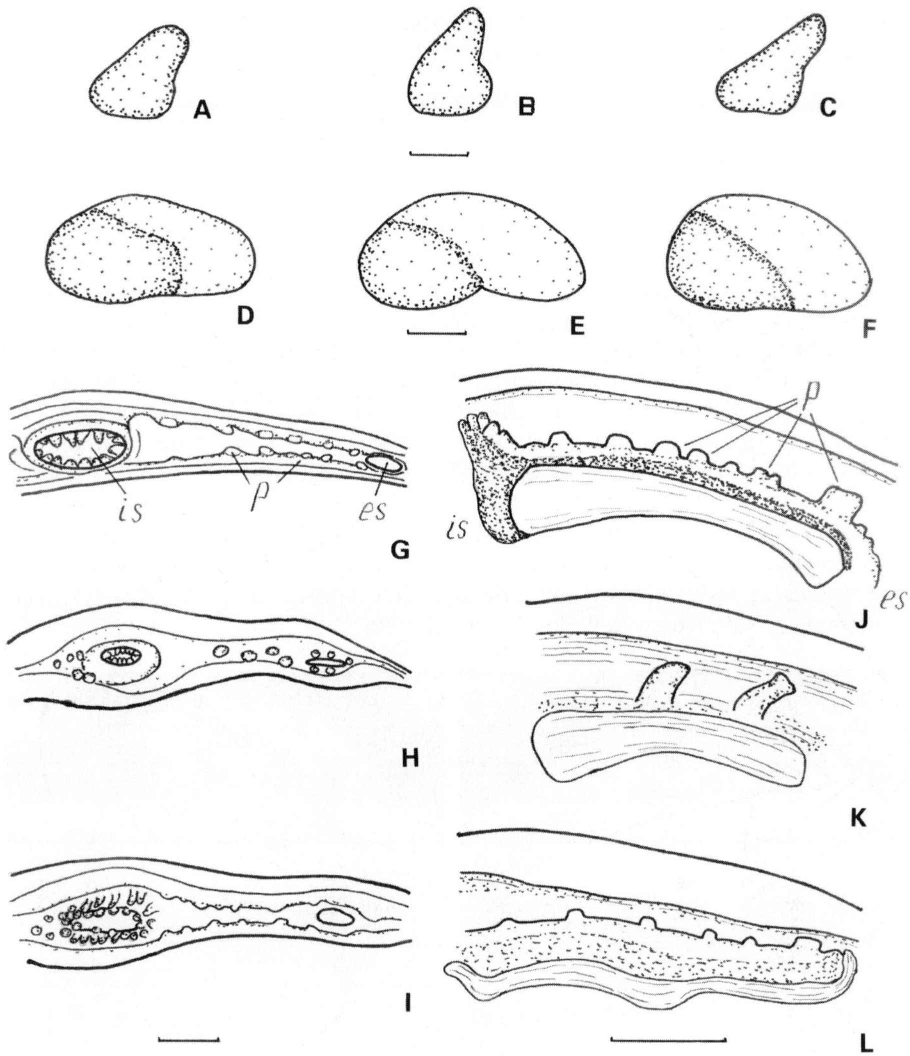

Figure 4. Dreissena *species (the same specimens as shown in Figure 2): anatomical characteristics. A - anterior adductor of* D. polymorpha *from left; B - the same of* D. stankovici*; C - the same of* D. bugensis*; D - posterior adductor of* D. polymorpha *from left; E - the same of* D. stankovici*; F - the same of* D. bugensis*; G - siphonal area of mantle exteriorly in* D. polymorpha*; H - the same in* D. stankovici*; I - the same of* D. bugensis*; J - mantle lobes between siphons in* D. polymorpha*, right side interior view, inner lobe of mantle dissected; K - the same of* D. stankovici*; L - the same of* D. bugensis*; is - inhalant siphon, es - exhalant siphon, p - papillae. Scale bar = 1 mm.*

THE BIVALVE MOLLUSC FAUNA AND BIOGEOGRAPHY OF LAKES OHRID AND PRESPA

According to the literature (Kuiper, 1960; Angelov, 1971; Maassen, 1980; Dhora & Welter-Schultes, 1996) and our own unpublished data, the bivalve fauna of Lake Ohrid includes 13 species: three species of Unionidae - *Unio crassus bruguierianus* (Figure 5A), *Anodonta cygnaea deplanata* and *Microcondylaea bonellii* (Figure 5B), one species of Dreissenidae - *Dreissena stankovici* (Figures 3 & 4), and nine species of Sphaeriidae (Table 1, Figures 6 & 7). Nine species of bivalves including *Anodonta cygnaea, A. anatina, Dreissena cf. polymorpha* and six sphaeriids (Table 1) are recorded in Prespa (Angelov, 1971; Kuiper, 1987; Sattmann & Reischütz, 1988; Reischütz & Stummer, 1989). The fauna of these lakes, and Ohrid in particular, is rather rich in comparison with the adjacent countries of Macedonia (FYROM) and Albania (Table 1). One should take into account, however, that the bivalves of Albania are still poorly known. The richest bivalve fauna among the Balkan countries is reported for Bulgaria (Angelov, 2000), but several species of Bulgarian sphaeriids are restricted to the Danube or its tributaries (*Pisidium henslowanum, P. pseudosphaerium* and *P. moitessierianum*), while the only record of *P. supinum* needs confirmation. It is noteworthy that bivalves are also well represented in lakes Skutari and Trichonis (ten species in each lake) (Dyduch-Falniowska, 1989; Dhora & Welter-Schultes, 1996).

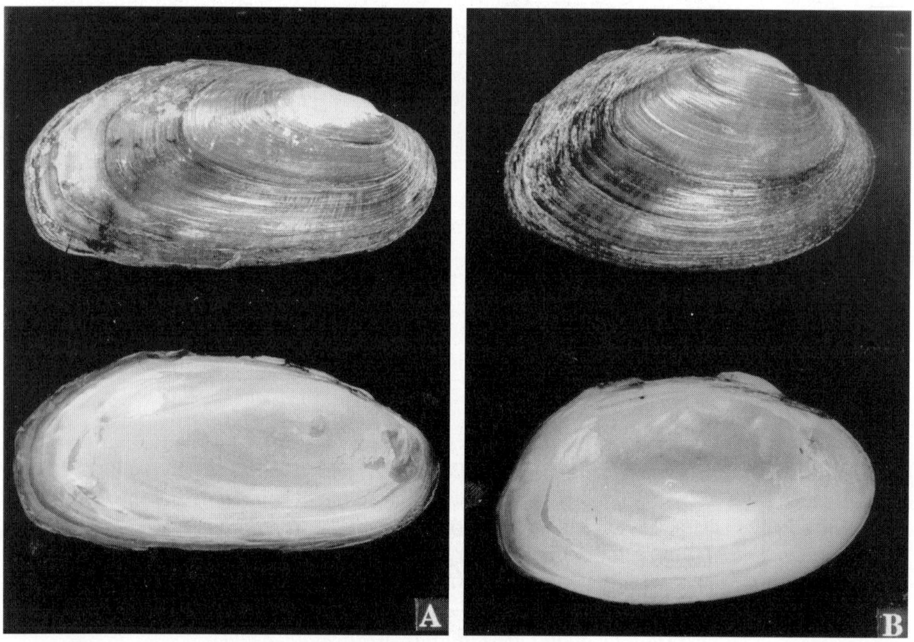

Figure 5. Unionidae from Lake Ohrid (all from Museum für Naturkunde, Berlin). A – Microcondylaea bonelii *from an unknown locality (ZMB Moll. 103044); B –* Unio crassus, *2 km west of Struga (ZMB Moll. 103046).*

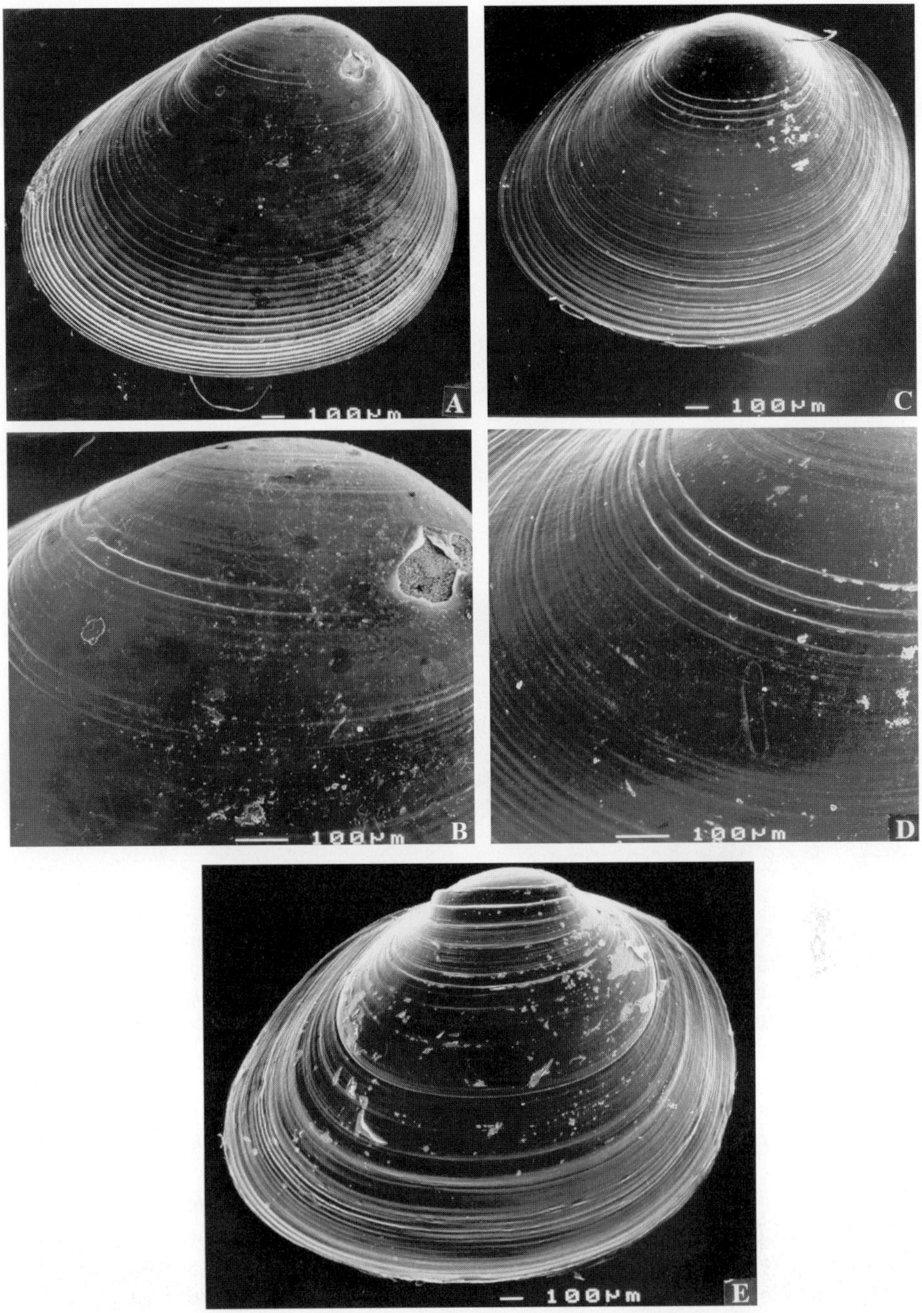

Figure 6. Pisidium (Cingulipisidium) edlaueri *from Lake Ohrid (A - lateral view, B - umbonal area) in comparison with* P. (Cingulipisidium) nitidum *(C lateral view, D - umbonal area) and* P. (Cingulipisidium) crassum *(E) from Brandenburg, Germany.*

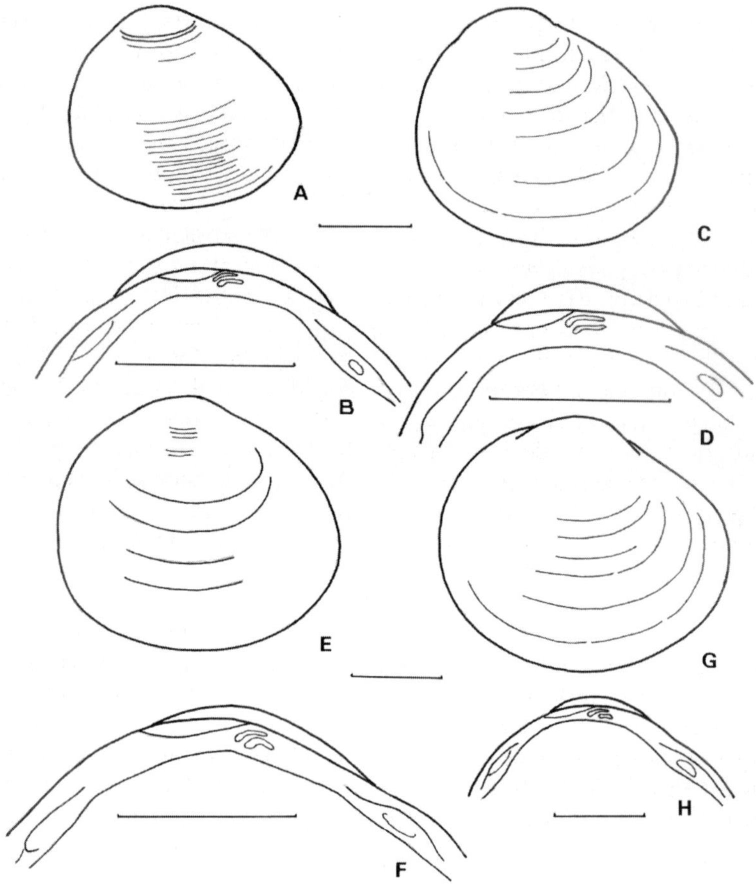

Figure 7. Examples of parallel evolution among sphaeriids from ancient lakes: A, B - P. (Cingulipisidium) edlaueri *from Lake Ohrid (lateral view of shell and hinge); C, D -* P. (Pseudeupera) subtruncatum recalvum *from Lake Ohrid (lateral view of shell and hinge); E, F -* P. (Neopisidium) raddei *from Baikal (lateral view of shell and hinge); G, H -* Musculium titicacense *from Lake Titicaca (lateral view of shell and hinge). Scale bar = 1 mm.*

Most of the species inhabiting Ohrid and Prespa have broad ranges: Holarctic (*Pisidium casertanum*), Palaearctic (*Pisidium amnicum* and *P. henslowanum*) or European or Euro-Siberian (*Unio crassus, Anodonta cygnea, Pisidium moitessierianum, P. tenuilineatum* and *P. personatum*). It is noteworthy that the peculiar Mediterranean fauna is represented by only one species of Unionidae (*Microcondylaea bonellii*); other species with a Mediterranean distribution known from neighbouring countries (*Unio mancus, Potomida littoralis* or *Pisidium sogdianum)* do not occur in the lakes analysed here.

Three endemic bivalve taxa are known from Ohrid and another one from Prespa; no endemics are known from other lakes. Thus, the fauna of bivalves is much less diverse than that of gastropods (several endemic genera and dozens of species in Ohrid alone) (Dhora & Welter-Schultes, 1996).

As far as we can judge from morphological characters, endemic species of sphaeriids from Ohrid and Prespa are closely related to the Holarctic *P. nitidum*; the presence of concentric ribs around the umbones, somewhat elongated presiphonal mantle suture, weakened mantle muscle bundles and open nephridia with a broadened dorsal lobe can be regarded as synapomorphies for *P. nitidum* and the two endemic species (Korniushin *et al.*, 2000; see also Figure 6). The principal distinctive characters of *P. edlaueri* are its relatively high triangular shell and broad hinge plate (Figures 6 & 7); *P. maasseni* also has a broadened hinge plate, but its shell is considerably lower and less convex. A somewhat similar high, thick-walled and broad-hinged form is recorded in western and Central Europe as *P. nitidum* f. *crassa, P. crassum* Stelfox, 1918, or *Cingulipisidium crassum* (Korniushin, 1996; Glöer & Meier-Brook, 1998; Gittenberger *et al.*, 1998) (Figure 6D). This form is also known from the Lower Pleistocene of Turkey, and some shells figured in Becker-Platen & Kuiper (1979) are superficially rather similar to *P. edlaueri* and *P. maasseni*. It is hard to decide whether this similarity is due to phylogenetic relationship or convergency; molecular studies may be helpful in elucidating the relationships between *P. edlaueri*, *P. maasseni* and *P. crassum*. The general distribution range of *Pisidium subtruncatum* represented in Lake Ohrid by the endemic subspecies is also Holarctic.

In general, the composition of the bivalve mollusc fauna in Ohrid and Prespa and the affinities of the endemic species are in accord with the biogeographical hypothesis of Stanković (1960), which suggests that the origin of the lakes' fauna is from freshwater taxa distributed across Eurasia in the Tertiary. However, only one species - *Microcondylaea bonellii* - can be treated as a relict; the recent distribution of all other species is still wide, or they are related to such widely-distributed species. Analysing the particularly strong affinities across a range of animal groups, between endemic Lake Ohrid species and those inhabiting watercourses of the Dinaric region, Stanković (1960) suggested that the Dinaric fauna was an important source for colonisation of the lake. Some sphaeriids really would be able to penetrate the lake in this way, since *Pisidium amnicum*, *P. casertanum*, *P. personatum*, *P. subtruncatum*, *P. henslowanum, P. milium, P. nitidum,* and *P. tenuilineatum* have all been recorded in Slovenia, and the first four also in Dalmatia and Bosnia (Kuiper, 1981). However, other modes of colonisation cannot be excluded, since most of these species also occur in the Danube basin (Grossu, 1962; Angelov, 1976) and in some lakes in Greece (Dyduch-Falniowska, 1989).

On the other hand, the Paratethian affinities of the Balkan lacustrine fauna are rather clear, demonstrated by presence of *Dreissena*. Starobogatov (1994) suggested a Pannonian origin for the Ohridan endemic *D. stankovici*, since a Pannonian Pliocene species *D. cristellata* (Roth, 1881) was considered ancestral for the subgenus *Carinodreissena.* Another fossil species - *D. filifera* (Andrusov, 1897) from the west coast of the Aegean Sea was included in the same subgenus; both fossil species have a keel similar to that described in *D. stankovici*. It is noteworthy that a Pannonian origin was also suggested for the subgenus *Pontodreissena*, represented now in the Caspian Sea by *D. rostriformis* (Deshayes, 1838) and in the rivers and estuaries of the Black Sea basin by *D. bugensis* (Andrusov, 1897) (Starobogatov, 1994). *Pontodreissena* and *Carinodreissena* were considered by Starobogatov (1994) to be sister taxa, forming a terminal clade alongside *Dreissena* s.str. *Dreissena blanci* was included by Starobogatov (1994) in the latter subgenus, but its affinities were not discussed. The above scenario is based both on the morphology of the extant taxa and on fossil records. It can be tested in part by checking the suggested close relationship between *D. stankovici* and *D. bugensis* on the one hand, and between *D. blanci* and *D. polymorpha* on the other.

A Paratethian origin can be also hypothesised for the other endemic Balkan dreissenid - the troglodytic *Congeria kusceri* known from several caves in the Dinaric karst region (Morton *et al.*, 1998). The close relationship of this endemic bivalve to *Mytilopsis* was recently shown by sequencing COI and 16S DNA genes (Stepien *et al.*, 2001). In contrast to *D. stankovici*, which is similar to its congeners in basic morphology, the only extant *Congeria* differs from other dreissenids in many respects - the absence of pigmentation, reduction of the foot and some pedal retractors, large gills, small labial palps, a stomach with simplified structure, and the absence of planktonic larvae and brooding of young in gills. All these evolutionary changes can be considered as adaptations to a subterranean mode of life (Morton *et al.*, 1998), but one cannot ignore the fact that at least some of them were shared by the extinct *Congeria* species distributed in the Paratethys Sea, and the Pannonian basin in particular (Starobogatov, 1994).

Thus, the bivalve fauna of Lake Ohrid is apparently heterogeneic and includes a variety of taxa of different origins and biogeographic status. Therefore, connections with different ancient basins, such as the Dinaric and Pannonian basins, can be hypothesised. New investigations involving molecular techniques are needed in order to test these hypotheses.

It is noteworthy that lacustrine populations of the widely distributed species of sphaeriids (*e.g. Pisidium henslowanum*, *P. moitessierianum* and *P. nitidum*) mentioned by Kuiper (1960, 1987) and Dyduch-Falniowska (1989) are isolated from each other and from the main distribution range. Therefore, the genetic distinctness of these populations may be also suggested, and this seems to be another interesting problem to study.

BIVALVE MOLLUSCS AND BIOGEOGRAPHICAL REGIONALISATION OF THE BALKANS

To date, data on bivalve mollusc distribution do not contribute much to the biogeographical regionalisation of the Balkan Peninsula, since local endemism at the species level is rare, and intraspecific taxonomy is still in debate. Nevertheless, these data provide additional support to the high rank of Lake Ohrid, which has been suggested in the literature (Starobogatov, 1970). Bivalve molluscs show no evidence of there being a relationship between the faunas of Ohrid and Prespa. As will be shown below, some similarities between endemic taxa from these two lakes have most probably developed due to parallel evolution.

The southern Balkan countries (Greece, Thraki and, in part, Macedonia [FYROM]) are marked by the presence of several species with a Mediterranean distribution, including *Pisidium sogdianum* with its northernmost record in the Vardar basin, Macedonia (Kuiper, 1981), and *Potomida littoralis*. Mediterranean species are lacking in the northern part of the Balkans, which is under stronger influence from the Danube basin. Further delimitation of subspecies in unionids, and clarification of the status and affinities of the Balkan *Dreissena*, may provide more detailed information about this intricate region.

SOME GENERAL TRENDS IN THE FORMATION OF THE FAUNA AND EVOLUTION OF SPHAERIIDS IN ANCIENT LAKES

The family Sphaeriidae is selected for further analysis, since it is the most diverse and widely distributed of lacustrine bivalves. A comparison of the faunas of different ancient lakes (Table 2) shows some common trends, which are detailed below.

(1) Widely-distributed taxa are present in many lakes alongside endemic species; thus the lakes' fauna is not sharply delimited from that of adjacent areas.

(2) No high ranking lacustrine endemics are known in this family.

(3) The number and degree of morphological distinctness of endemic sphaeriid taxa in Ohrid and Prespa are approximately the same as in the oldest lakes Baikal and Tanganyika (see Glaubrecht, 2001 and Sitnikova, 2001 for age estimations). A particularly rich endemic fauna is also found in Lake Victoria, a lake which is considered to be comparatively young (see Wesselingh & Michel, 2001).

(4) All endemic species are closely related to the widely-distributed congeners. They differ mainly in shell characters, such as proportions and pattern of sculpture and hinge. In most cases possible ancestors of endemic species are known from the adjacent area or at least from the same biogeographic region.

(5) No extensive speciation has taken place amongst the sphaeriids. In some lakes pairs of closely related species occur which exhibit different preferences for substrate or depth (*e.g. Musculium stuhlmanni* and *M. nynzae* in Victoria); a widely-distributed species may also occur together with its endemic relative showing different ecological preferences (*e.g. Pisidium pirothi* and *P. fistulosum* in Victoria, *P. lauricochae* and *P. titicacense* in Titicaca), but in all these cases the distinction between species is not entirely clear (Mandahl-Barth, 1988; Korniushin, 1998a). An extensive sampling programme across the range of different habitats will be necessary in order to demonstrate this rigorously.

In addition to sphaeriids, some other bivalve families are represented in ancient lakes by endemic taxa. A rather rich endemic fauna of Unionidae is known from the African Rift Lakes (see Mandahl-Barth, 1988 for review). Several endemic species and even a monotypic genus of Corbiculidae with a very peculiar mode of life have been reported from the lakes of Sulawesi, Indonesia (Bogan & Boucheet, 1998). The endemic species of *Dreissena* (Dreissenidae) from Lake Ohrid was alluded to above. The co-existence of several (two to four) congeneric unionid species has been reported for some African lakes (Mandahl-Barth, 1988), but the possibility of more extensive sympatric speciation has not yet been demonstrated, either for this or for any other freshwater bivalve group. In contrast to the sphaeriids, which differ in ancient lakes in conchological characters alone, the corbiculids of Sulawesi also exhibit anatomical differences and peculiar modes of reproduction (Bogan & Bouchet, 1998; Korniushin & Glaubrecht, 2001). Thus, some of the above evolutionary trends can be attributed to all freshwater bivalve groups entering ancient lakes, but sphaeriids seem to be the most conservative amongst them.

On the other hand, the evolutionary trends characterising bivalves are apparently in contrast to those observed among gastropods. Snail faunas are much more diverse and differ between lakes in the number and rank of endemic taxa (Dhora & Welter-Schultes, 1996; West & Michel, 2001; Sitnikova, 2001). Endemic clades comprising dozens of closely-related species - so called *species flocks* - have been found in some lakes (Glaubrecht, 2001; Todd & Michel, 2001), and endemic and widely-distributed faunas do not usually mix (Stanković, 1960; Slugina & Starobogatov, 1999). In our opinion, these discrepancies can be explained by the general evolutionary conservativeness of freshwater bivalves in comparison with gastropods

*Table 2. Endemic taxa of small bivalves (Sphaeriidae) in ancient lakes and their possible affinities. * - the data on Lake Tanganyika are fragmentary, and the presence of more than one endemic species is possible.*

Lake	Total # sphaeriid species	Endemic taxa	Related widely-distributed taxa	Source
Baikal	6	*Sphaerium baicalense* Dybowski, 1902;	*S. corneum*	Slugina & Starobogatov, 1999 Korniushin, 1996
		Pisidium raddei Dybowski, 1902	*P. conventus* Clessin, 1877	
Ohrid	8	*Pisidium edlaueri*	*P. nitidum*	Kuiper, 1960; Korniushin *et al.*, 2000
		P. subtruncatum recalvum	*P. s. subtruncatum*	
Prespa	5	*P. maasseni* Kuiper, 1987	*P. nitidum*	Kuiper, 1987; Korniushin *et al.*, 2000
Biwa	3	*Pisidium casertanum lacustre* Woodward in Mori, 1938	*P. c. ponderosum*	Mori, 1938; Korniushin, 1997
		P. kawamurai Mori, 1938	*P. dilatatum* Westerlund, 1897	
Titicaca	4	*Musculium titicacense* (Pilsbry, 1924)	*M. lauricochae* (Philippi, 1869)	Kuiper & Hinz, 1984; Korniushin, 1998a
Tanganyika *	2	*Pisidium giraudi* Bourguignat, 1885	*P. pirothi* Jickeli, 1881	Kuiper, 1966
Victoria-Albert system	4	*Musculium stuhlmanni* (Martens, 1897) *M. nyanzae* (Smith, 1892)	*M. hartmanni* Jickeli, 1874	Mandahl-Barth, 1988; Kuiper, 1966; Korniushin, 1998a; Korniushin, unpbl.
		Pisidium fistulosum Mandahl-Barth, 1954	*P. pirothi*	

Endemic sphaeriid taxa inhabiting different lakes and belonging to different genera and subgenera (as shown by their different anatomies) often show remarkable similarity in some shell characters, in particular in their relatively high rounded or triangular outline and broad hinge (Figure 7). These characters may have evolved independently as specific adaptations to the lacustrine environment, such as greater water depth, or coarse sediment. The convergent origin of similar conchological characters in endemic *Gyraulus* (Gastropoda, Planorbidae) species from ancient lakes was shown by Meier-Brook (1983), while Mandahl-Barth (1988) suggested that the smaller size and more well-developed sculpture which distinguishes African lacustrine unionids from related species from rivers are adaptations to life in big lakes.

Therefore, we see no evidence for a common origin for the bivalve taxa of remote lakes, such as Baikal, Ohrid and Biwa, from a common ancestor distributed across Eurasia in the Tertiary, as has been suggested by some earlier authors (see Stanković, 1960; Slugina & Starobogatov, 1999 for reviews). The same could be said of the similarity of the Balkan endemics *Pisidium edlaueri* and *P. maasseni* to the central European *P. crassum*, and to the similarity between some forms inhabiting Ohrid and Prespa - *P. edlaueri* and *P. maasseni*, *P. subtruncatum recalvum* and *P. subtruncatum* ssp. (Korniushin *et al.*, 2000). All these similarities concern shell and hinge proportions and might have developed via parallel evolution. However, molecular data are needed to confirm or reject either hypothesis.

CONSERVATION ASPECTS

Three species of unionids represented in the Balkans have an international protection status: *Unio crassus* is included in the IUCN list, while *U. elongatulus* (=*U. mancus*) and *Microcondylaea compressa* (= *M. bonellii*) are protected by the Bern Convention. Fechter & Falkner (1990) reported a drastic decline of the latter species in the Italian part of its range, however, so conservation of the Balkan populations may be crucial for its survival. With their limited ranges, some other Balkan bivalves may be in urgent need of protection, but this group is apparently neglected in the international conservation lists (Bouchet *et al.*, 1999). The endemics of Ohrid and Prespa, as well as *Congeria kusceri* are of particular concern. The conservation status of *Pisidium sogdianum* might be also questioned at least in the Mediterranean region, where its occurrence is only sporadic (Kuiper, 1981); it may be significant that the habitats of this species (springs and streams) are rather sensitive to pollution.

Bouchet *et al.* (1999) stressed the necessity for protection of rare subspecies. The Balkan fauna of najades needs to be investigated in this respect. Apparently, some subspecies represented here might need conservation measures because of their restricted range, low abundance and/or sensitivity to pollution. Furthermore, the species common in other areas, such as *P. amnicum*, *P. henslowanum*, and *P. moitessierianum* (all found in Ohrid), as well as *P. nitidum* Jenyns, 1832 are now represented in the Balkans by scarce isolated populations (Angelov, 1971; Diduch-Falniowska, 1989). These populations may be genetically different from those living on the "mainland" and may therefore be as much in need of protection as endemic species and subspecies.

CONCLUSIONS

The Balkan lakes Ohrid and Prespa are rather peculiar in regard to their bivalve fauna, even when compared to more ancient lakes with a generally higher degree of endemism, such as Baikal and Tanganyika. The principal peculiarity is the presence of endemic species and subspecies of *Pisidium* and *Dreissena*. Anatomical investigation of the endemic *Pisidium* species confirms their relationship to widely-distributed taxa. The prevalence of the widely-distributed species in the fauna, and the Holarctic affinities of the endemic taxa, make it difficult to trace the sources or modes of colonisation of ancient Ohrid and Prespa by strictly freshwater groups. On the other hand, the presence of *Dreissena* and similarity of endemic Ohridan forms to fossil taxa from the Pannonian and Aegean basins may indicate certain Paratethian affinities of the lake fauna. Similarities between endemic taxa from remote lakes in shell and hinge proportions have apparently developed due to parallel evolution. Application of modern methods of phylogenetic reconstruction, including molecular phylogenetics, is needed for verification of hypotheses concerning ways of formation of the Balkan fauna of bivalve molluscs, as well as the origin and evolution of endemic taxa. It is also worth noting that the extremely diverse gastropod fauna of Ohrid and some other Balkan lakes is also awaiting such a revision.

ACKNOWLEDGEMENTS

This paper was presented at an international meeting in Koper, Slovenia due to the generous support of the European Science Foundation. The travel to Ohrid, Macedonia in 1998 was partly supported by the Ralf Brown Foundation (UK) and the International Solomon University (Ukraine). The author is grateful to Prof. Z. Krstanovski and Dr. G. Kostoski for their hospitality in Ohrid, and to the curators of malacological collections in Senckenberg Museum, Frankfurt/ Main (Dr. R. Janssen), Museum für Naturkunde, Berlin (Dr. M. Glaubrecht) and Nationaal Naturhistorisch Museum, Leiden (Prof. E. Gittenberger) for the kind permission to study museum materials relevant to this review. The final stage of the study was supported by the A. von Humboldt Foundation. SEM study was made possible by Dr. E. Wäsch, and all photographs were processed by Mrs. V. Heinrich (Berlin).

REFERENCES

Angelov, A. 1971. Beitrag zur Kenntnis der Pisidienfauna (Molluska - Sphaeriidae) Mazedoniens. *Fragmenta Balkanica* 8, 53-55

Angelov, A. 1976. [Revision der Fam. Pisidiidae Gray, 1857 (Bivalvia - Mollusca) in Bulgarien]. *Annuaire de l'Universitè de Sofia Facultè de Biologie* 69, 109-119 (in Bulgarian, with German summary).

Angelov, A. 1984. [*Pisidium supinum* Schmidt, 1850 has been found in Bulgaria]. *Acta Zoologica Bulgarica* 24, 80-82 (in Bulgarian, with English summary).

Angelov, A. 2000. *Mollusca (Gastropoda et Bivalvia) aquae dulcis. Catalogus Faunae Bulgaricae* 4. Pensoft & Buckhuys, Sofia & Leiden.

Araujo, R., Moreno, D. & Ramos, M.A. 1993. The Asiatic clam *Corbicula fluminea* (Müller, 1774) in Europe. *American Malacological Bulletin* 10, 39-49.

Becker-Platen, J.D. & Kuiper, J.G.J. 1979. Sphaeriiden (Mollusca, Lamellibranchia) aus dem Känozoikum der Türkei (Känozoikum und Braunkohlen der Türkei, 13.). *Geologischer Jahrbücher* B33, 159-185.

Beron, P. & Georgiev, V. 1967. Essai sur la faune cavernicole de Bulgarie. II. Resultats biospeleologiques de 1961 a 1965. *Bulletin de l'Institute de Zoologique et Museé* 4,151-212.

Bogan, A., & Bouchet, P. 1998. Cementation in the freshwater bivalve family Corbiculidae (Mollusca: Bivalvia): a new genus and species from Lake Poso, Indonesia. *Hydrobiologia* 389, 131-139.

Bössneck, U. 2002. Lebendnachweise von *Microcondylaea bonellii* (A. Férussac 1827) und *Pisidium tenuilineatum* Stelfox 1918 aus Istrien (Kroatien) (Bivalvia: Unionidae and Sphaeriidae). *Malakologische Abhandlungen Staatliches Museum für Tierkunde Dresden* 20, 313-317.

Bouchet, P., Falkner, G. & Seddon, M. B. 1999. Lists of protected land and freshwater molluscs in the Bern Convention and European Habitats Directive: are they relevant to conservation? *Biological Conservation* 90, 21-31.

Cooley, L.R. & Ó Foighil, D. 2000. Phylogenetic analysis of the Sphaeriidae (Mollusca: Bivalvia) based on partial mitochondrial 16S rDNA gene sequences. *Invertebrate Biology* 119, 299-308.

Dhora, Dh. & Welter-Schultes, F.W. 1996. List of species and atlas of the non-marine molluscs of Albania. *Schriften zur Malakozoologie* 9, 90-197.

Djajasasmita, M. 1975. On the species of the genus *Corbicula* from Celebes, Indonesia (Mollusca, Corbiculidae). *Bulletin Zoologisch Museum Universiteit van Amsterdam* 4, 83-87.

Dyduch-Falniowska, A. 1989. Remarks on the Sphaeriidae of Greece. *Basteria* 53, 15-18.

Falkner, G. 1994. Systematik vorderorientalischer Najaden als Vorstudie zur Bearbeitung archäologischer Funde. *Forschungen und Berichte zur Vor- und Frühgeschichte in Baden-Würtemberg* 53, 135-162 (*Beiträge zur Archäozoologie und Prähistorischen Anthropologie. 8. Arbeitstreffen der Osteologen, Konstanz 1993.* M. Kokabi & J. Wahl, eds.)

Falkner, G. 2000. *Sphaerium (Nucleocyclas) nucleus* (S. Studer 1820) in Bayern (Bivalvia: Sphaerioidea). *Heldia* 3, 11-18.

Falkner, G., Bank, R. A. & von Proschwitz, T. 2001. CLECOM-project. Check-list of the non-marine Molluscan species-group taxa of the states of Northern, Atlantic and Central Europe (CLECOM I). *Heldia* 4, 1-76.

Fechter, R. & Falkner, G. 1990. *Weichtiere. Europäische Meeres- und Binnenmollusken.* Mosaik Verlag, München.

Frank, Ch. 1988. Beitrag zur Kenntnis der Molluskenfauna der östlichen Mittelmeerländer. Teil III (2): Zusammenfassung der Sammelergebnisse der Jahre 1982-1985 vom kontinentalen Griechenland, dem Peloponnes, den Nördlichen Sporaden sowie einigen Inseln des Ionischen und Ägäischen Meeres. *Malakologische Abhandlungen Staatliches Museum für Tierkunde Dresden* 13, 1-22.

Gambetta, L. 1930. Molluschi terrestri e lacustri d'Albania. *Atti della Accademia Scientifica Veneto - Trentino - Istriana, Padova* 21, 51-74.

Gittenberger, E., Janssen, A. W., Kuijper, W. J., Kuiper, J. G. J., Meijer, T., van der Velde, G. & de Vries, J. N. 1998. *De nederlandse Zoetwatermollusken. Recente en fossiele Weekdieren uit zoet en brak Water*. Naturalis, Leiden.

Glaubrecht, M. 2001. An evolutionary ecology approach to the presumptive adaptive and endemic radiations of freshwater Cerithioidea (Mollusca, Gastropoda) in East Africa and on Sulawesi. In: *Abstracts, World Congress of Malacology 2001, Vienna, Austria* (L. Salwini-Plawen, J. Voltzow, H. Sattmann & G. Steiner, eds.), p. 122. Unitas Malacologica, Vienna.

Glaubrecht, M., Rintelen, T. v. & Korniushin, A.V. in press. Toward a systematic revision of brooding freshwater Corbiculidae in Southeast Asia (Bivalvia, Veneroida): on shell morphology, anatomy and molecular genetics of endemic taxa from islands in Indonesia. *Malacologia.*

Glöer, P. & Meier-Brook, C. 1998. *Süsswassermollusken.* Deutscher Jugendbund für Naturbeobachtung, Hamburg.

Grossu, A.V. 1962. Bivalvia (Scoici). Fauna Republicii Populare Romîne. Vol. III, fasc. 3. Editura Academiei Republicii Populare Romîne, Bucureşti.

Izzatullaev, Z. & Starobogatov, Ya.I. 1986. [*Odhneripisidium* species from the fauna of the USSR]. [*Trudy Zoologicheskogo Instituta Akademii Nauk SSSR*] 148: 54-70 (in Russian with English summary).

Jaeckel, S.G., Klemm, W. & Meise, W. 1957. Die Land- und Süsswassermollusken der nördlichen Balkanhalbinsel. *Abhandlungen und Berichte aus dem Staatlichen Museum für Tierkunde Dresden* 23, 141-205.

Kinzelbach, R. 1986. The recent distribution of the Zebra Mussel, *Dreissena polymorpha*, in the Aegean Region and in Anatolia. *Zoology in the Middle East* 1, 132-138.

Kolesnikov, Ch.M. 1977. [Sistema i proiskhozhdenije limnicheskikh dvustvorok mezozoja]. Paleontologicheskij Zhurnal No.3, 42-54 (in Russian).

Korniushin, A.V. 1996. [*Bivalve molluscs of the superfamily Pisidioidea in the Palaearctic Region. Fauna, systematics, phylogeny*]. Kiev: Schmalhausen Institute of Zoology (in Russian, with English summary).

Korniushin, A.V. 1997. Morphometrical and anatomical characteristics of *Pisidium casertanum* (Poli) from the Lake Biwa (Japan) (Bivalvia: Eulamellibranchiata: Pisidiidae). *Malakologische Abhandlungen Staatliches Museum für Tierkunde Dresden* 18, 53-57.

Korniushin, A.V. 1998a. Notes on the anatomy of some species of *Sphaerium* s.l. (Mollusca, Bivalvia) from the tropical regions with revision of their taxonomic status. *Vestnik Zoologii* 32, 3-12.

Korniushin, A.V. 1998b. Review of the studies on freshwater bivalve mollusc systematics carried out by the Russian taxonomic school. *Malacological Review* Suppl. 7 (Bivalvia I), 65-82.

Korniushin, A.V. 1999. Anatomical investigation and taxonomic revision of pill clams of the genus *Pisidium* s.l. (Bivalvia: Sphaeriidae) in the Palearctic region. *Malacological Review* Suppl. 8 (Freshwater Mollusca I), 31-36.

Korniushin, A.V. 2001. Taxonomic revision of the genus *Sphaerium* sensu lato (Bivalvia Sphaeriidae) in the Palaearctic Region, with some notes on the North American species. *Archiv für Molluskenkunde* 129, 77-122.

Korniushin, A.V. & Glaubrecht, M. 2001. Review of endemic species of heterodont bivalves from the ancient lakes. In: *Abstracts, World Congress of Malacology 2001, Vienna, Austria* (L. Salwini-Plawen, J. Voltzow, H. Sattmann & G. Steiner, eds.), p. 179. Unitas Malacologica, Vienna.

Korniushin, A.V., Krstanovski, Z. & Kostoski, G. 2000. Anatomical evidence of close affinity between endemic species of *Pisidium* (Bivalvia, Sphaeriidae) from some ancient lakes and the widely distributed taxa. *Journal of zoological systematics and evolutionary research* 38, 81-86.

Kuiper, J.G.J. 1960. Die Pisidien des Ochridsees, Mazedonien, nebst Bemerkungen über die Verbreitung der Pisidien in der Balkanhalbinsel und den Donauländern. *Beaufortia* 7, 219-231.

Kuiper, J.G.J. 1962. Etude critique de *Pisidium vincentianum. Bulletin Institut royal des Sciences naturelles de Belgique* 38, 1-19.

Kuiper, J.G.J. 1966. Les espèces africaines du genre *Pisidium*, leur synonymie et leur distribution (Mollusca, Lamellibranchiata, Sphaeriidae). *Annales du Musée Royal de l'Afrique Central, Sci. Zool.* 151, 1-78.

Kuiper J.G.J. 1981. The distribution of *Pisidium tenuilineatum* Stelfox and *Pisidium annandalei* Prashad in the Mediterranean area. *Basteria* 45, 79-84.

Kuiper, J.G.J. 1987. *Pisidium maasseni* n.sp., a new species from Lake Prespa,Yugoslavia (Bivalvia, Sphaeriidae). *Basteria* 51, 163-165

Kuiper, J.G.J. & Hinz, W. 1984. Zur Fauna der Kleinmuscheln in den Anden (Bivalvia: Sphaeriidae). *Archiv für Molluskenkunde* 114, 137-156.

Lvova, A.A. & Starobogatov, Y.I. 1982. New species of *Dreissena* (Bivalvia: Dreissenidae) from Lake Ochrid. *Zoologicheskij Zhurnal* 61, 1749-1752.

Maassen, W.J.M. 1980. De zoetwattermollusken van het Ohriden Prespameer en hun omgeving. *De Kreukel* 16, 81-98.

Mandahl-Barth, G. 1988. *Studies on African freshwater bivalves*. Danish Bilharziasis Laboratory, Charlottenlund.

Meier-Brook, C. 1983. Taxonomic studies of *Gyraulus* (Gastropoda: Planorbidae). Malacologia 24, 1-113.

Mori, S. 1938. Classification of Japanese *Pisidium*. Mem. Coll. sci. Kyoto Imp. Unv., ser.B, 14, 255-278.

Morton, B., Velkovrh, F. & Sket, B. 1998. Biology and anatomy of the "living fossil" *Congeria cusceri* (Bivalvia: Dreissenidae) from subterranean rivers and caves in the Dinaric karst of the former Yugoslavia. *Journal of Zoology London* 245, 147-174.

Nagel, K.-O. & Badino, G. 2001. Population genetics and systematics of European Unionoidea. In: *Ecological Studies, vol. 145. Ecology and Evolution of the Freshwater Mussels Unionoida* (G. Bauer & K. Wächtler, eds.), pp. 51-80. Springer-Verlag, Berlin & Heidelberg.

Nagel, K.-O., Badino, G. & Celebrano, G. 1998. Systematics of the European najades (Bivalvia: Margaritiferidae and Unionidae): a review and some new aspects. *Malacological Review* Suppl. 7 (Bivalvia I), 83-104.

Radoman, P. 1983. Hydrobioidea, a superfamily of Prosobranchia (Gastropoda). I. Systematics. *Serbian Academy of Sciences and Arts Monographs* 547, 1-256.

Radoman, P. 1985. Hydrobioidea, a superfamily of Prosobranchia (Gastropoda). II. Origin, zoogeography, evolution in the Balkans and Asia Minor. *Monographs of the Beograd Faculty of Sciences Department Biol.* 1, 1-173.

Reischütz, P.L. & Stummer, B. 1989. Ein Beitrag zur Molluskenfauna der dessaretischen Seen. *Malakologische Abhandlungen Staatliches Museum für Tierkunde Dresden* 14, 105-109.

Sattmann, H. & Reischütz, P.L. 1988. Beiträge 2. Molluskenfauna Nordgriechenlands. *Annalen Naturhistorisches Museum Wien* 90B, 204-214.

Schütt, H. 1985. Malakozoologische Kurzberichte (10). Die Mollusken des Vegoritis-Sees in Makedonien. *Mitteilungen Zoologischer Gesellschaft Braunau* 4, 301-302.

Sitnikova, T.Ya. 2001. Baikalian gastropodes: species diversity and evolution. In: *Abstracts, World Congress of Malacology 2001, Vienna, Austria* (L. Salwini-Plawen, J. Voltzow, H. Sattmann & G. Steiner, eds.), p. 330. Unitas Malacologica, Vienna.

Slugina, Z.V., & Starobogatov, Ya.I. 1999. *Guide and key to the Bivalvia of Lake Baikal*. Izdalelstvo SO RAN, Novosibirsk (in Russian with English summary).

Stanković, S. 1960. *The Balkan Lake Ochrid and its living world. Monographiae Biologicae, vol. 9.* Uitgeverij Dr. W.Junk, den Haag.

Starobogatov, Ya.I. 1970. [*Fauna Molluskov i Zoogeograficheskoye Rajonirovaniye Kontinnentalnykh Vodoemov Zemnogo Shara*]. Nauka, Leningrad (in Russian).

Starobogatov, Ya.I. 1994. [Sistematika i paleontologija]. In: [*Sistematika, evolyutsiya i prakticheskoe znachenie Dreissena polymorpha (Pall.) (Bivalvia, Dreissenidae)*] (Ya. I. Starobogatov, ed.), pp.18-46. Nauka, Moscow (in Russian).

Starobogatov, Ya.I. & Sitnikova, T.Y. 1990. The pathway of species formation of the Lake Baikal molluscs. *Zhurnal Obsshchej Biologii* 51, 499-512 (in Russian with English summary).

Stepien, C.A., Morton, B., Dabrowska, K.A., Guarnera, R.A., Radja, T. & Radja, B. 2001. Genetic diversity and evolutionary relationships of the troglodytic "living fossil" *Congeria kusceri* (Bivalvia: Dreissenidae). *Molecular Ecology* 10, 1873-1879.

Taylor, D.W. 1988. Aspects of freshwater mollusc ecological biogeography. *Palaeogeography, Palaeoclimatology, Palaeoecology* 62, 511-576.

Todd, J.A. & Michel, E. 2001. Species diversity of the *Lavigeria* species flock (Gastropoda: Thiaridae) of Lake Tanganyika - a re-evaluation using multiple datasets. In: *Abstracts, World Congress of Malacology 2001, Vienna, Austria* (L. Salwini-Plawen, J. Voltzow, H. Sattmann & G. Steiner, eds.), p. 355. Unitas Malacologica, Vienna.

Wesselingh, F.P. & Michel, E. 2001. Long-lived lakes as laboratories and archives for molluscan evolution. In: *Abstracts, World Congress of Malacology 2001, Vienna, Austria* (L. Salwini-Plawen, J. Voltzow, H. Sattmann & G. Steiner, eds.), p. 388. Unitas Malacologica, Vienna.

West, K. & Michel, E. 2001. Diversification patterns in the Tanganyican thiarid gastropod superflock. In: *Abstracts, World Congress of Malacology 2001, Vienna, Austria* (L. Salwini-Plawen, J. Voltzow, H. Sattmann & G. Steiner, eds.), p. 390. Unitas Malacologica, Vienna.

14. A FAUNISTIC REVIEW OF THE MODERN AND FOSSIL MOLLUSCAN FAUNA FROM LAKE PAMVOTIS, IOANNINA, AN ANCIENT LAKE IN NW GREECE: IMPLICATIONS FOR ENDEMISM IN THE BALKANS

MICHAEL R. FROGLEY* AND RICHARD C. PREECE

INTRODUCTION

The Balkans are justifiably famous for being natural 'hotspots' of floral and faunal biodiversity and endemism. Much of this biological diversity is witnessed in the ancient lakes that exist (or previously existed) in the region, which have provided stable, long-lived arenas within which evolutionary processes have been able to operate. One group that ably demonstrates such diversity is that of the gastropod molluscs. Their wide dispersal, morphological variability and relative ease of identification make them ideal candidates for addressing questions concerning patterns and processes of biodiversity and evolution over time (*e.g.* Michel, 1996). Furthermore, the fact that they are often well preserved in fossil sequences means that they can provide a temporal dimension to evolutionary investigation, something that is often difficult to resolve when using groups more traditionally associated with this kind of work, such as the cichlid fishes.

The molluscan faunas of the Balkans (and Greece in particular) have been the subject of study for at least the last two centuries (*e.g.* Butot & Welter-Schultes, 1994). Isolated ancient lakes from the region, and in particular Ohrid, have often been a focus for detailed investigation, because of their remarkable endemic faunas (*e.g.* Hadžišče, 1956; Hubendick & Radoman, 1959; Stanković, 1960; Meier-Brook, 1983; Stanković, 1985; Schütt, 1987; Sattmann & Reischütz, 1988; Dhora & Welter-Schultes, 1996). Despite this, the malacology of many other key localities from the area still remains poorly documented.

In this paper we first briefly review the main controls thought to influence the radiation of gastropods in ancient lakes and then present the modern and fossil aquatic molluscan fauna from a 'typical' but previously poorly known site: Lake Pamvotis, located in the Ioannina Basin in NW Greece. Comparisons are then drawn with the gastropod fauna derived from other extant and fossil ancient lakes from across the Balkan region (bivalves are considered by Korniushin elsewhere in this volume), before highlighting some common patterns and providing some thoughts on the implications for speciation in ancient Balkan lake systems.

Gastropod dispersal

The dominant controls influencing gastropod radiation in ancient lakes is a widely debated topic (see, for example, the excellent review by Michel, 1994). Whilst it is beyond the scope of this paper to reproduce these discussions in detail, it is

H.I. Griffiths et al. (eds.), Balkan Biodiversity, 243–260.

nevertheless useful to summarise some of the main concepts prior to presenting and discussing data from specific Balkan lake sites.

High gastropod diversity in a lake setting rarely results simply from multiple invasions of unrelated taxa (Michel, 1994); rather, it is likely to be due to a combination of several inter-related factors. These include: (a) the presence and ecological preferences of certain gastropod lineages; (b) the physical characteristics of the lake basin (*e.g.* depth, areal extent, bathymetry, longevity); (c) external factors (*e.g.* climate); and (d) local biotic and abiotic conditions. The first two of these factors are usually thought to be the most important and hence warrant further discussion.

It is well known that in considering the two main sub-classes of freshwater gastropods, diversity amongst prosobranchs is usually higher than that amongst pulmonates (*e.g.* Whittaker, 1960; Boss, 1978). This characteristic is probably due to a combination of their physiology, dispersal and reproductive/sexual strategies (summarised in Table 1). Prosobranchs breathe using gills, which allows them to occupy not only shallow water environments, but also deeper habitats which are likely to be more stable with regard to environmental variables such as temperature and salinity. The highest expression of prosobranch endemism is seen amongst families that are gonochoristic (*i.e.* possess two separate sexes) and brood their eggs, such as the Hydrobiidae, Baicallidae, Benedictiidae and Thiaridae (Michel, 1994). On the other hand, pulmonates use a simple 'lung' mechanism to breathe, tying them to shallower (often littoral or marsh) habitats. This physiological restriction, combined with their vulnerability for dispersal (by living predominantly in marginal environments they are more likely to be dispersed passively by animal vectors than the deeper-living prosobranchs), means that pulmonates have to cope regularly with often highly unstable environmental conditions. This requirement is also reflected in their reproductive and sexual strategies. Pulmonates are not brooders (such a tactic requires stable environments), but instead produce eggs which are often tough and resistant; they are also all hermaphrodite, with some of the most diverse families also capable of self-fertilisation (*e.g.* the Planorbidae and Ancylidae). Consequently, it is the poorly-dispersing prosobranchs that enjoy relatively higher habitat stability than the pulmonates. This situation encourages the development of ecological and environmental specialisation, ultimately leading to lineage divergence, speciation and endemism.

It is interesting to note that certain pulmonate groups which are small in size have developed vascular haemoglobin and secondary gill folds (such as members of the Planorbidae), or pseudo-gill systems (such as members of the Ancylidae and Acroloxidae), that enables them to dive deeper and for longer. Significantly, these pulmonate groups have all undergone more extensive endemic radiation in long-lived lakes than their larger counterparts, such as the Lymnaeidae and Physidae (Boss, 1978). A similar propensity to radiate is also seen amongst some of the prosobranch families (*e.g.* the Hydrobiidae, Thiaridae, Valvatidae and Viviparidae), although the reasons why these groups should radiate more readily than, say, the Bithyniidae or Ampullariidae, are not so well defined (Michel, 1994). It is also worth noting that amongst ancient lakes globally, an almost exclusive radiative pattern is witnessed between hydrobiids and thiarid (i.e. either one group or the other radiates, never both) (Davis, 1982). Whilst it is possible that this is related to ecological requirements, as Michel (1994) points out, it is not yet clear what those requirements may be.

Table 1. Comparison of main diversification processes amongst prosobranch and pulmonate gastropods. [1] Valvatidae are probably the most notable exception amongst the prosobranchs (they are hermaphrodite).

PROSOBRANCHS	PULMONATES
	Physiology
Breathe using gills	Breathe using a 'lung'
Have wide depth tolerance	Often tied to surface/littoral waters
Often found in main (deeper) parts of lakes	Most commonly found in shallow, O_2-depleted waters
More likely to be stenotopic	More likely to be eurytopic
Widespread in stable environments	Widespread in unstable environments
	Dispersal Strategies
Passive	Passive
Less likely to be dispersed (deeper habitats)	More likely to be dispersed (nearer surface)
	Reproduction
Oviparous and/or ovoviviparous	Oviparous
Gonochoristic[1]	Hermaphrodite

Clearly, the process of radiation cannot operate unless the physical basin characteristics provide the range of habitats required. Depth, geological origin, steepness of basin sides, substrate variability and longevity are some of the factors that are thought to influence molluscan radiation. For example, lakes formed by tectonic processes are often host to significant faunal radiations (*e.g.* Michel, 1994), a characteristic which is probably related to a combination of increased ecological space (Davis, 1982) and variable littoral substrates and habitats (*e.g.* Fryer & Iles, 1972; Michel *et al.*, 1992; Michel, 1994).

Lake Pamvotis

Lake Pamvotis (39°45'N, 20°51'E) is situated 470 m above sea-level (a.s.l.) in the Ioannina basin on the western flank of the Pindus Mountain Range, in the region of Epirus of NW Greece (Figure 1). It is thought that solution of the Mesozoic limestone basement in the Plio-Pleistocene created a depression known as a *polje* which, coupled with tectonic subsidence along a series of boundary faults, led to the development of the lake basin. Extensive coring by the Greek Institute of Geology and Mineral Exploration (IGME) has revealed that the floor of the basin was sealed by braided river deposits that were quickly succeeded by lacustrine sediments (IGME, *unpublished data*). Evidence from lignite horizons suggests that several sub-basins developed, each of which was subject to a variable tectonic history.

The modern lake currently has a maximum depth of around 10 m and an areal extent of around 23 km^2. It is approximately 8 km in length (along its longest axis) and has a maximum width of approximately 5 km. Throughout historical times, the lake was more extensive than it is today. At the end of the nineteenth century, part of the lake and the marsh around Kato Lapsista in the northern part of the basin (approximately 10 km NW of the present extent of the lake) were artificially drained for agricultural purposes (Figure 1).

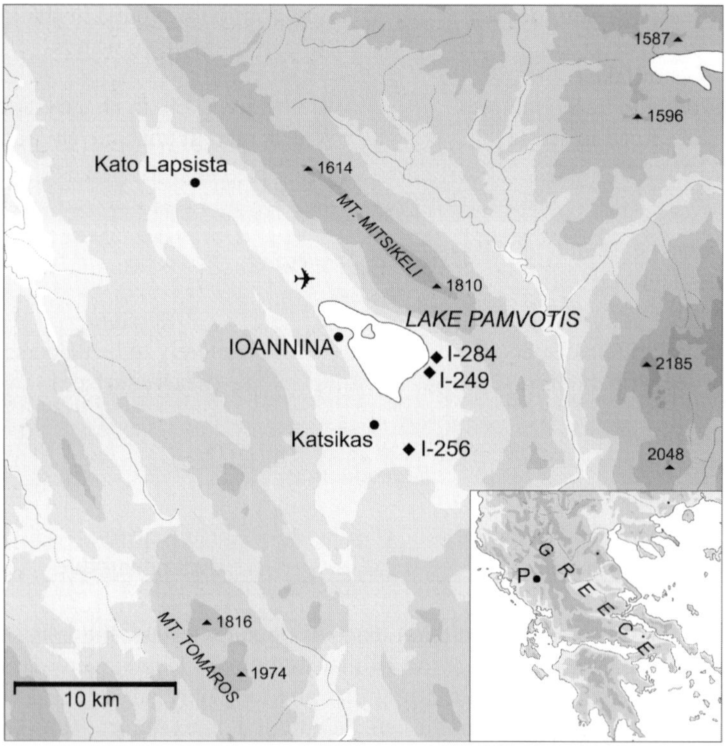

Figure 1. Map showing location of Lake Pamvotis (P), NW Greece. Core sites (filled diamonds) and places mentioned in the text (including Ioannina airport) are also shown. Map amended from Tzedakis et al. *(2002b).*

Lake Pamvotis is a typical temperate, fresh-water, eutrophic lake (average pH ~7.6, average salinity ~0.35 ‰), with a water chemistry dominated by calcium (Ca^{2+}) and bicarbonate (HCO_3^-) ions (Overbeck, in Anagnostidis & Economou-Amilli, 1980; Frogley, 1997). Abundant phytoplankton in the surface waters thrive as a result of the pollutants introduced from various human and industrial sources, as well as from agricultural pesticides. Falling fish stocks and the constant presence of unpleasant dark green algal blooms mean that pollution of the lake has become a serious ecological concern (*e.g.* Albanis *et al*., 1986; Kalogeropoulos *et al*., 1994).

Table 2. Summary of fossil molluscan fauna recovered from the sediments of Lake Pamvotis. E denotes species endemic to Lake Pamvotis; T denotes taxa having their type locality at Lake Pamvotis or its immediate environs; Pulm. = pulmonates. [1] Viviparus megarensis *may be a synonym of* Viviparus brusinai. [2] *Description of* Pyrgula *sp. by Guernet* et al. *(1977) matches that of* Pyrgula brusinai. *Their material must be examined before a critical assessment can be made.* [3] *Description of* Planorbis *sp. by Dollfus (1922) matches that of* Gyraulus janinensis.

			PLIO-PLEISTOCENE					HOL.
			Dollfus (1922)	Aubouin (1959)	Gillet (1963)	Guernet *et al.* (1977)	**I-256**	**I-284**
	Gastropods							
	Theodoxus micans (GAUDRY AND FISCHER)		+	-	+	+	+	-
	Theodoxus licherdopoli STEFANESCU		-	+	-	-	-	-
	Theodoxus quadrifasciatus (BIELZ)		-	+	-	-	-	-
	Bithynia graeca (WESTERLUND)	T	+	+	+		+	+
	Bithynia cf. *meridionalis* (FRAUENFELD)		-	-	-	+	-	-
	Neumayria janinensis DOLLFUS	E?	+	+	+		+	-
	Viviparus clathrata (DESHAYES)		+	-	+	-	+	-
	Viviparus brusinai (NEUMAYR)		+	+	-	+	-	-
Prosobranchs	*Viviparus megarensis* (FUCHS)[1]		-	+	-	-	-	-
	Viviparus mamillatus janinensis (MOUSSON)	E	-	-	-	-	-	+
	Valvata sp.		-	-	-	+	-	-
	Valvata cristata MÜLLER		-	-	-	-	-	+
	Valvata piscinalis (MÜLLER)		-	-	-	-	-	+
	Pyrgula brusinai TOURNOUER		+	+	+	+[2]	+	-
	Pyrgula eugeniae (BERNADI)		-	-	-	+	-	-
	Melanopsis costata (OLIVIER)		+	+	+	+	+	-
	Melania curvicosta DESHAYES		+	+	+	+	-	-
	Melania rhodensis BUKOWSKI		-	+	-	-	-	-
	Hydrobia janinensis (DOLLFUS)	E?	-	-	-	+	-	-
	Ancylus sp.		-	-	-	+	-	-
Pulm.	*Planorbis planorbis* (LINNAEUS)		-	-	-	-	-	+
	Gyraulus janinensis (MOUSSON)	E	+[3]	-	-	-	-	+
	Bivalves							
	Pisidium nitidum JENYNS		-	-	-	-	-	+
	Dreissena (*Carinodreissena*) sp.	E?	-	-	-	-	-	+

Table 3. Comparison of records of modern aquatic molluscan fauna collected from Lake Pamvotis. Taxa in brackets (+) denote occurrences that were originally either misidentified or unrecognised. E denotes species endemic to Lake Pamvotis; T denotes taxa having their type locality at Lake Pamvotis or its immediate environs. M: Mousson (1859); D: Dollfus (1922); S: Schütt (1962, 1980); R: Reischütz & Sattman (1990); F: Frogley (1997), Frogley & Preece (in prep.).

			M	D	S	R	F
	Gastropods						
Prosobranchs	*Bithynia tentaculata* (LINNAEUS 1758)		(+)	-	-	-	-
Prosobranchs	*Bithynia graeca* (WESTERLUND 1879)	T	(+)	(+)	-	+	+
Prosobranchs	*Bithynia* sp.		(+)	-	-	-	-
Prosobranchs	*Viviparus mamillatus janinensis* (MOUSSON 1859)	E	+	+	-	+	+
Prosobranchs	*Valvata cristata* MÜLLER 1774		-	-	-	+	+
Prosobranchs	*Valvata piscinalis* (MÜLLER 1774)		+	-	-	-	+
Prosobranchs	*Belgrandiella* (*Belgrandiella*) *haesitans* (WESTERLUND 1881)		-	-	+	+	-
Prosobranchs	*Bythinella charpentieri* (ROTH 1855)		(+)	-	-	-	-
Prosobranchs	*Horatia* (*Neohoratia*) *epirana* SCHÜTT 1962	T	-	-	+	+	+
Prosobranchs	*Paladilhiopsis* (*Paladilhiopsis*) *janinensis* (SCHÜTT 1962)	E	-	-	+	+	-
Prosobranchs	*Trichonia trichonica* RADOMAN 1973		(+)	-	-	-	+
Prosobranchs	*Semisalsa steindachneri* (WESTERLUND 1902)		-	-	+	+	-
Pulmonates	*Lymnaea stagnalis* (LINNAEUS, 1758)		+	+	-	-	+
Pulmonates	*Radix auricularia* (LINNAEUS 1758)		(+)	+	-	-	+
Pulmonates	*Radix peregra* (MÜLLER 1774)		+	-	-	+	-
Pulmonates	*Stagnicola palustris* (MÜLLER 1774)		-	-	-	-	+
Pulmonates	*Physella acuta* (DRAPARNAUD 1805)		-	-	-	-	+
Pulmonates	*Planorbis carinatus* MÜLLER 1774		+	+	-	-	+
Pulmonates	*Planorbis planorbis* (LINNAEUS 1758)		+	-	-	+	+
Pulmonates	*Gyraulus* cf. *piscinarum* (BOURGUIGNAT 1852)		(+)	-	-	-	+
Pulmonates	*Gyraulus janinensis* (MOUSSON 1859)	E	+	-	-	-	+
Pulmonates	*Gyraulus crista* (LINNAEUS 1758)		-	-	-	-	+
Pulmonates	*Segmentina nitida* (MÜLLER 1774)		-	-	-	-	+
Pulmonates	*Hippeutis complanatus* (LINNAEUS 1758)		-	-	-	-	+
Pulmonates	*Planorbarius corneus* (LINNAEUS 1758)		+	+	-	-	+
Pulmonates	*Ancylus fluviatilis* MÜLLER 1774		+	-	-	+	-
Pulmonates	*Ferrissia* sp.		-	-	-	-	+
	Bivalves						
	Anodonta anatina (LINNAEUS 1758)		-	+	-	-	+

Anodonta cygnea (LINNAEUS 1758)		+	-	-	-	+
Corbicula sp.		(+)	-	-	-	-
Musculium lacustre (MÜLLER 1774)		-	-	-	-	+
Sphaerium corneum (LINNAEUS 1758)		+	-	-	-	-
Pisidium casertanum (POLI 1791)		-	-	-	+	-
Pisidium obtusale (LAMARCK 1818)		-	-	-	-	+
Pisidium milium HELD 1836		-	-	-	-	+
Pisidium nitidum JENYNS 1832		-	-	-	-	+
Dreissena (*Carinodreissena*) sp.	E?	(+)	(+)	-	-	+

MOLLUSCAN FAUNA

Fossil Fauna

There have been few studies on the fossil molluscan fauna of the lake (Table 2). Dollfus (1922) described nine species from Upper Pliocene lacustrine sediments from a site south-west of the village of Katsikas, in the south of the basin (Figure 1). This site was subsequently re-sampled by Aubouin (1959), Gillet (1962) and Guernet *et al.* (1977), bringing the total number of fossil molluscs recorded from this location to eighteen. More recently, a pilot study was carried out on the basal lake sediments from a 130 m long borehole taken from the margin of the basin south-east of Katsikas (core I-256, Figure 1). The basal section of the core reached bedrock and, on the basis of palaeomagnetic and faunal evidence, was thought to be Plio-Pleistocene in age (Frogley, 1997). The sediments yielded six aquatic molluscan taxa, all of which had previously been recorded from the Katsikas sediments (Table 2).

Specimens were also recovered from the Holocene interval of a 319 m core (I-284), drilled by IGME in 1989 on the eastern shore of the existing lake (Figure 1). An age model was established for this sequence (Frogley, 1997), based on correlation with the pollen stratigraphy from an adjacent core (I-249) situated in the same sub-basin (Tzedakis, 1994). This was coupled with twelve AMS radiocarbon determinations from the upper part of the core, a set of U-series dates from the last interglacial interval and detailed palaeomagnetic data. Improvements to this age model were achieved by subsequent palynological work (Lawson, 2001; Tzedakis *et al.*, 2002a, b) and on the basis of this revised chronology, the I-284 record appears to extend back approximately 450,000 years. Only the top ~145 m contains molluscan remains, most of which are too poorly preserved for confident identification; better material occurs throughout the Holocene interval of the core (approximately the top 15 m) and these data are incorporated into Table 2.

Modern Fauna

Previous studies of the modern fauna are also limited (Table 3). The earliest and most comprehensive was published by Mousson (1859), who described a total of thirty-eight molluscan species, of which thirteen were aquatic. In 1918, Marius Dalloni collected eight species from the lake, a list of which was subsequently published by Dollfus (1922) as part of a wider discussion of the fossil taxa known from the Katsikas site. Schütt (1962, 1980) described four species of hydrobiid collected from the northern shore of the lake, Pamvotis being the type site for three of these. Finally, whilst on a field tour of the Epirus region of north-western Greece, Reischütz & Sattmann (1990) visited Ioannina and collected thirty molluscan taxa, of which eleven were aquatic. The most recent study was carried out during September, 1994 and April, 1998, when twenty-six aquatic taxa were recovered from thirteen sites located around the edge of the lake and on the island in its centre; one additional sample was taken from the former lake bed soils around Kato Lapsista (Frogley, 1997; Frogley & Preece, in preparation).

Lake Pamvotis: Discussion

Taxa from the early (Plio-Pleistocene) history of Lake Pamvotis are significantly different from those found more recently. Indeed, the Holocene fauna bears considerable similarity to the modern fauna, even allowing for the inevitable biasing of the fossil record. Thus we can consider at least two phases in the lake's history: an 'early' phase and a 'later' one.

Sixteen out of eighteen taxa recovered from the early phase are prosobranchs (89%), several of which (*e.g. Pyrgula brusinai*, *Melanopsis costatus*, *Viviparus clathrata*, *Melania tourneri*) possess ornate shell morphologies. It is well established that many ancient lake endemics (including gastropods) often develop these so-called 'exotic' shell morphologies over time. Well-documented examples include the *Gyraulus* clades from Lake Steinheim (*e.g.* Gorthner & Meier-Brook, 1985; Reif, 1985) and the *Pyrgula* clades from Lake Ohrid (*e.g.* Radoman, 1959). There is some debate whether morphological change arises as a result of coevolution (*e.g.* West *et al.*, 1992; West & Cohen 1994) or from other, genetically-related pressures (see discussion in Michel, 1994). Whatever the process(es) involved, the longevity and stability of ancient lake systems undoubtedly provides an appropriate arena for their uninterrupted operation.

It is not easy to assess the percentage of endemic gastropod taxa in the early lake, since the fossil record is likely to be biased (and therefore almost certainly incomplete) and the pedigree of the many of the taxa found is difficult to ascertain confidently. However, if we assume that the *Gyraulus* described by Dollfus (1922) is *G. janinensis* (see Table 2), then 17% of the gastropods from the early phase of the lake are endemic (Figure 2). One further point to note is that the early phase of Lake Pamvotis is also characterised by the presence of thiariids (and the corresponding absence of hydrobiids). Whilst this may be a function of a fossil record biased towards larger taxa, it is more likely to be a reflection of the exclusive thiarid/hydrobiid distribution in ancient lakes mentioned previously.

The gastropod assemblage from the later phase of Lake Pamvotis (Holocene and modern) is of a completely different character (Figure 2). As might be expected given the opportunity to sample from the current lake, the faunal count is higher (27 taxa), though significantly, the fauna is now dominated by pulmonates (55%). In addition, none of the prosobranchs in this later phase are thiarids (though hydrobiids are now present) and none of the taxa can be described as possessing exotic shell morphologies. Despite the more representative faunal count the degree of gastropod endemism is nevertheless low (12%) and all the other changes represent a marked shift in faunal composition.

The most plausible reason for some of these dramatic changes is that of changing depth, brought about by conditions of increased aridity and decreased temperatures during much of the Quaternary. The physiography of the Ioannina basin shows that its floor is inclined only gently, but that the sides are relatively steep. If the Plio-Pleistocene levels of the lake were significantly higher than they are today, then there would be a distinct lack of shallow, littoral habitats, thereby favouring prosobranch taxa. However, any reduction in lake-levels would swiftly ensure the proliferation of shallow environments, which would in turn favour pulmonate taxa. It is also possible that a radical and sustained change in lake-level might also be responsible for a major reorganisation of the prosobranchs, as competition for increasingly limited habitats became intense. The more stenotopic taxa in particular would be susceptible to extinction (especially if the changes happened rapidly), whereas more cosmopolitan taxa, capable of tolerating or adapting to the new conditions, would stand a better chance of survival. Generally speaking, those gastropod taxa with more ornate shell morphologies tend to be the most specialised in terms of habitat requirements; this may explain their absence from the later phase of the Pamvotis record. Whether or not lake-level changes (and associated habitat changes) might also be responsible for the disappearance of the thiarids (and subsequent appearance of the hydrobiids), it is not possible to say.

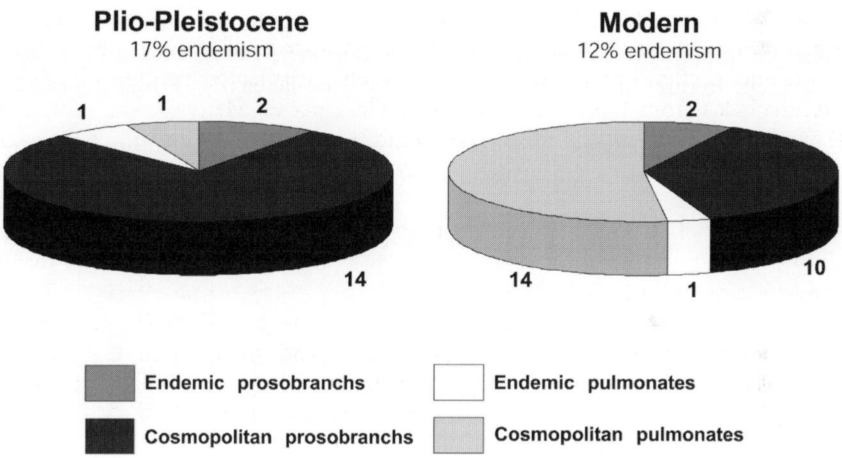

Figure 2. Composition of the gastropod fauna of Lake Pamvotis. Modern = modern plus Holocene records. See text for details.

There is limited evidence in the Ioannina basin for the lake having been significantly deeper in the past. Prior to major industrial and residential development of the area, Higgs *et al.* (1967) claim to have observed a series of former beach terraces and deposits around the periphery of the basin. Situated at approximately 3.2 m above the current lake-level, the terraces themselves were undated, but Higgs *et al.* (1967) made the assumption that they were of Last Glacial Maximum age. Whilst the argument for a lake high-stand at the Last Glacial Maximum has recently been discounted (Galanidou *et al.*, 2000), a significantly older age for the terraces might provide evidence for former high lake levels. Support for a much deeper lake comes from alleged (undated) beach deposits identified to the west of Ioannina airport and situated at least 12 m above the present-day level of the lake (Higgs & Vita-Finzi, 1966). In addition, Waltham (1970) maintains that concretions occurring in the roof of the Perama Cave system are phreatic in origin, suggesting that the lake must have been another 10-15 m deeper than it is today for these features to form. Furthermore, the mollusc-bearing Plio-Pleistocene sediments derived from the Katsikas site in the south of the basin were located several metres above the current lake level (Guernet *et al.*, 1977). Whilst these deposits might also provide support for the lake being significantly deeper in the past than it is today, it must be remembered that the basin is located within an active tectonic region that has been under compression since the beginning of the Pleistocene (Clews, 1989). Evidence exists for tectonically-controlled uplift from up-arched fluvial deposits in the south of the basin (King & Bailey, 1985), as well as from stratigraphical data derived from borehole studies (Tzedakis, 1994; IGME, unpublished data), suggesting that any indications for former lake high-stands need to be considered critically, in conjunction with robust, independent dating information.

COMPARISON WITH OTHER ANCIENT BALKAN LAKES

It is useful to compare the patterns of endemism and gastropod radiation from the Pamvotis record with other fossil and extant ancient lakes from the Balkans (Figures 3 & 4; Table 4). In view of the difficulties of defining the term 'ancient lake' to the satisfaction of both geologists and biologists (Gorthner, 1994), we shall here use the term 'ancient' to refer to long-lived modern or palaeo-lakes (*e.g.* Frogley *et al.*, 2002).

Table 4. Comparison of faunal and basin characteristics of the Balkan lake sites discussed in the text.

		FAUNAL CHARACTERISTICS				Basin
Lake	Fossil/Modern	Pulmonate or Prosobranch dominated?	Radiation in thiarids or hydrobiids?	Endemism: high or low?	Exotic shell morphologies present?	Deep/shallow
Pannon	F	Proso.	T + H	H	Yes	D
Kos	F	Proso.	T	H	Yes	D
Pamvotis (early)	F	Proso.	T	L	Yes	D?
Ohrid	M	Proso.	H	H	Yes	D
Skutari	M	Pulm.	H	L	No	S
Prespa	M	Pulm.	H	L	No	S
Dojran	M	Pulm.	H	L	No	S
Pamvotis (late)	M	Pulm.	H	L	No	S

Ohrid

Arguably the most famous ancient lake from the region is Ohrid, located on the border between the Former Yugoslavian Republic (FYR) of Macedonia and Albania. Thought to be Late Pliocene in age, Lake Ohrid is situated approximately 695 m a.s.l., is 295 m at its maximum depth and has an area of *ca.* 350 km^2. The lake experiences minimal run-off from the surrounding catchment (dominantly karstic in nature) and is instead fed almost exclusively by springs on the eastern shore

(Salemaa, 1994). The lake is linked to Lake Prespa through underground karstic channels and is drained by the Black Drin River to the north. In contrast to Pamvotis, not only is the entire physiography of Ohrid different, but the fauna is of a completely different character. The lake is justifiably famous for the fabulous degree of endemism of its biota: 64% in oligochaete worms, 66% in ostracods, 60% in fish and over 70% in isopod and amphipod crustaceans (Stanković, 1960; Salemaa, 1994). Amongst the gastropods, fifty-six of the seventy-two prosobranch-dominated taxa are endemic (78%), many of which display exotic morphologies (*e.g.* Snegarova, 1954; Stanković, 1960); there is also significantly more endemism amongst the prosobranchs than the pulmonates (Figure 4). The only similarities the gastropod fauna of Pamvotis has with Ohrid is seen in the earlier (Plio-Pleistocene) phase of the lake's history (Table 4), when both lakes are dominated by prosobranchs (although the difference in thiarid/hydrobiid dominance of the fauna remains different).

Figure 3. Map showing location of the ancient lake basins discussed in the text. Slov. = Slovenia; Mont. = Montenegro; Alb. = Albania; FYR Maced. = Former Yugoslavian Republic of Macedonia.

Lake Pannon

In central and eastern Europe, many of the longest-lived lake basins (some of which are still extant) owe their origins to the Paratethys, an aquatic bio-province consisting of a series of tectonic basins that formed in the hinterland of the mountain chains being uplifted by Alpine orogenic activity in the mid-Tertiary. Perhaps the most significant of these basins was the Pannonian, which became separated from the rest of the Paratethys sometime towards the end of the Middle Miocene, around 12 million years ago (Magyar *et al*., 1999). Uplift of the southern and eastern Carpathians allowed the formation of a vast, isolated water body (Lake Pannon) that subsequently underwent significant desiccation before the later incursion of fresh water and the establishment of a water body that survived for the next 7 million years, finally disappearing in the Early Pliocene as a result of sedimentary infilling (Magyar *et al*., 1999). Fluvial inputs led the initially brackish lake to become progressively less saline, triggering a massive radiation in aquatic faunas, including the development of many endemic taxa, including cardiid bivalves (cockles) and dreissenid mussels.

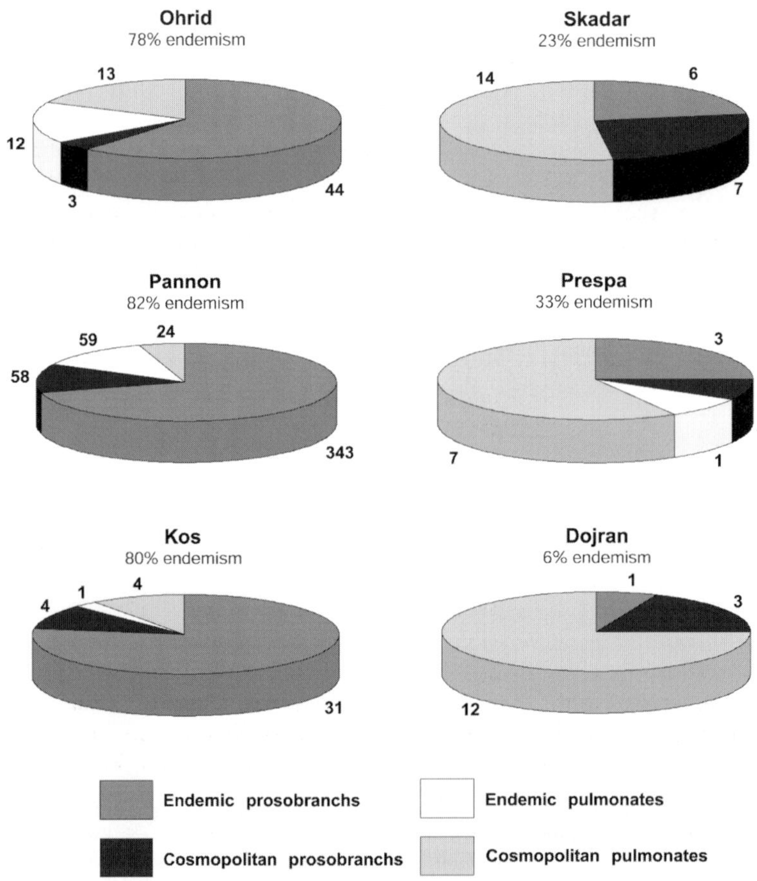

Figure 4. Gastropod faunal composition of several Balkan ancient lakes.

At its most extensive *ca.* 9.5 million years ago, Lake Pannon covered ~150,000 km^2 and was in excess of 1 km in depth (Magyar *et al.*, 1999). Although the status of some gastropod taxa is uncertain, 524 have been described, of which 82% are thought to be endemic; many of these, such as members of the *Melanopsis* clade, possess exotic shell morphologies (Geary, 1990; Müller *et al.*, 1999). Most endemism is represented in the prosobranchs (343 taxa) when compared with the pulmonates (59 taxa) (Figure 4) and, unusually, there appears to be significant radiation in both the hydrobiids and the thiarids (Müller *et al.*, 1999). This may be related to the uncertainty with which the limits of such a vast area as Lake Pannon are defined (for example, what might constitute the main body of the lake and what might constitute a ponded-out or geographically close, but isolated, water body?). These difficulties aside, it would appear that the faunal composition of Pannon is more comparable with those of Lake Ohrid than of modern-day Pamvotis (Table 4).

Kos

Another well-known fossil ancient lake is that which formerly occupied part of what is now the Greek island of Kos (Figure 3) during the Plio-Pleistocene, when the island was still part of the Anatolian landmass. Several large lakes were created in the area by tectonism associated with the formation of a volcanic arc, though frequent marine transgressions from the south left the lake area in the east largely untouched (Willmann, 1985). Over 400 m of freshwater sedimentary deposits from this part of the island indicates that lake-levels fluctuated markedly over a period of 4-5 million years and, although depths are difficult to estimate, Willmann (1985; p.303) argues that they were, at times, "considerable". Again, the faunal composition of the Kos lake bears more similarity with Ohrid and Pannon than with modern-day Pamvotis (Table 4). Twenty-one species of gastropod have been described, with associated sub-species pushing the total to around forty taxa. Of these, thirty-two (80%) are endemic (31 prosobranchs, 1 pulmonate), many taxa possess exotic shell morphologies (including such species as *Viviparus brevis*, *Melanopsis gorceixi* and *Theodoxus doricus*), and the thiarid taxa outnumber the hydrobiids by a factor of 9:1 (Figure 4).

Skutari

An extant lake site that is perhaps more comparable with Pamvotis is Lake Skutari (Scutari, Shkodra), located at 121 m a.s.l. on the border between Montenegro and Albania (Figure 3). The largest lake of the Balkan Peninsula, Skutari is *ca.* 40 km long and between 6.4 and 12.9 km wide. Once an inlet of the Adriatic, the lake is now separated from the sea by an alluvial isthmus; it is fed by the Moraca River and drains southeastward to the Drin River. Like Pamvotis, Skutari is relatively shallow and experiences no true thermocline, though some depressions in the lake exist that are up to 60 m deep and are known to be sources of groundwater. The origin and age of the lake are unclear, though most authorities consider it to be a polje of at least Plio-Pleistocene (and possibly Tertiary) age. In terms of its gastropod fauna, twenty-seven taxa have been described from the modern lake (Dhora & Welter-Schultes, 1996), six of which (all species of the hydrobiid *Anagastina*) are endemic (Figure 4). Like Pamvotis, none of the taxa possess exotic shell morphologies and thiarids are absent from the fauna (Table 4).

Prespa

Another lake whose gastropod fauna bears similarity to modern-day Pamvotis is Prespa, a 290 km^2 body of water that straddles the borders of SW Macedonia, NW Greece, and eastern Albania (Figure 3). At 853 m a.s.l. it is the highest major lake in the Balkans and is thought to be connected to Lake Ohrid by underground karstic channels. The malacology of Prespa (and the associated lake to which it is connected, Mikri Prespa) has been poorly studied and so the current list of twelve gastropod taxa is almost certainly incomplete (Maassen, 1980; Sattmann & Reischutz, 1988; Petridis & Sinis, 1994). Whilst the fauna is dominated by pulmonates, most endemism is seen in the prosobranchs (3 prosobranchs, 1 pulmonate) and no taxa display exotic shell morphologies (Figure 4). Amazingly, despite the fact that Prespa is comparable to Ohrid in terms of age, origin and surface area, the gastropod faunas of the two lakes (so far as they have been described) are significantly different (Table 4). This raises questions as to why, in two similar lakes that are in close geographical proximity, one now possesses an incredibly diverse gastropod fauna, whereas the other does not. Part of the answer may lie in the physical differences between the two basins: Prespa is over 150 m higher than Ohrid and, perhaps more importantly, is considerably shallower (maximum depth of Prespa = 54 m; maximum depth of Ohrid = 295 m). In addition, Ohrid is largely steep-sided, whereas Prespa has a much gentler bathymetry which provides a greater area of shallow, littoral environments, thereby accounting for the dominance of pulmonates in Prespa and prosobranchs in Ohrid.

Dojran

Finally, we consider Lake Dojran, situated at *ca.* 150 m a.s.l. on the borders of FYR Macedonia and Greece (Figure 3). Like Pamvotis, Dojran is thought to be Plio-Pleistocene in age and also to have formed as a polje within a karstic basin. Currently around 40 km^2 in area, Lake Dojran has been drained artificially over the past 30 years for water extraction purposes, reducing its average depth by a metre every decade to a level of *ca.* 6 m by 1997 (Ryan & Griffiths, 2001). The relatively poorly-studied gastropod fauna is also similar to modern-day Pamvotis (Figure 4; Table 4): sixteen taxa are known (dominated by pulmonates), there are no thiarids, and none of the taxa display exotic shell morphologies (Stanković, 1985). Endemism amongst the gastropods is low (6%) and represented by the single hydrobiid species *Graecoanatolica macedonica*, which may have recently become extinct (Ryan & Griffiths, 2001). Interestingly, Dojran is the relict of former Lake Peonic, a significant body of water thought to have been three times larger and ten times deeper than the present lake (Cvijić, 1911; Stanković, 1931). If this was the case, then it would be useful to test if the fossil gastropod material from that older lake reflected the modern day Dojran fauna, or whether it bore more similarity to the early (Plio-Pleistocene) phase of Lake Pamvotis.

CONCLUSIONS

In comparing the Pamvotis fauna with other Balkan ancient lakes, several marked patterns have emerged. (1). Gastropod diversity is higher in fossil lakes than in extant lakes (with the exception of Ohrid). This is clearly seen amongst the examples discussed here, where endemic fauna from fossil lakes is around (or even exceeds) 80%. Amongst the extant lakes, only Ohrid witnesses such a high degree of endemism; the more usual value is less than 30%. (2). Gastropod clades that have developed exotic shell morphologies are not seen in extant ancient lakes (again, Ohrid is the exception); the fossil lakes display an abundance of exotic forms. (3). Excluding Lake Pannon (discussed above), the pattern of endemic exclusivity between thiarids and hydrobiids seen elsewhere in the world appears to hold true in the Balkans. (4). The fossil records are dominated by prosobranchs; the modern records are mostly dominated by pulmonates.

It is perhaps this last point that provides the key in understanding some of these patterns, and to why Ohrid appears to be unique amongst the extant ancient lakes of the region. Pannon aside, all of the other Balkan lakes considered here are of broadly comparable age and would therefore have experienced broadly similar climatic variations in moisture availability during their history. Why then is the fauna of Ohrid so different? The reason is unlikely to be due solely to sheer areal extent, since Prespa and Skutari are comparable in this respect and yet neither display any of the faunal characteristics of Ohrid. More likely, is the relationship between overall depth and basin physiography. Most of the modern lakes are relatively shallow, having broad, flat basin floors that provide ample littoral environments. Whilst these favour (generally eurytopic) pulmonates, these are not environments conducive to encouraging extreme gastropod radiation. On the other hand, the fossil lakes were often extremely deep, which would instead have favoured the proliferation of (largely stenotopic) prosobranch taxa. Of all the extant lakes, only Ohrid fits comfortably into this latter category, its 295 m depth and relatively steep sides providing a stable environment for evolutionary processes to operate, buffered from the extremes of external (climatic) factors.

Nevertheless, as discussed above, there is limited evidence for at least some of these extant lakes (Pamvotis, Dojran) having been significantly deeper in the past. In the case of Pamvotis, this supposition is supported by the fossil fauna, which (in its early phase) displays some of the characteristics comparable with other, deeper ancient lakes (e.g. exotic shell morphologies, the presence of thiarids). It seems plausible that at some stage in their history, these lakes may have undergone a shallowing that, because of the physical characteristics of their basins (steep sides but broad, flat floors), would have opened up different ecological space, precipitating fierce competition for the extensive new littoral habitats. Whether this shallowing took place gradually, or whether it represents a crisis of some kind (possibly tectonic or climatic) that effectively 'reset' the evolutionary clock in the lake, is difficult to say from the limited evidence so far available. Even if the shallowing process occurred relatively suddenly, it would be difficult to establish if this was truly a response to a (Balkan-wide?) externally-forced crisis or merely the crossing of a critical depth threshold as lake-levels reduced. One possible strategy might be to examine genetic data from taxa derived from lakes known to have been deeper in the past and compare it with data derived from taxa present in a 'control' lake, Ohrid being the obvious candidate. Alternatively, high-resolution biostratigraphical studies on the faunas from the lesser studied sites, supported by robust dating (possibly using tephras or uranium-series methods), might indicate the way forward. Either way, it is clear that the ancient lakes of the Balkans demonstrate real potential in helping us to understand the rates, patterns and processes of faunal radiation and endemism. Nevertheless, it is also clear that to realise this potential a significant amount of work remains to be done on the modern and, particularly, fossil sequences of these remarkable biological archives.

ACKNOWLEDGEMENTS

We would like to thank D.C. Aldridge, G. Faulkner, M.P. Kerney, T. Meijer, C. Meier-Brook and H. Schütt for helping to identify certain specimens, T. Meier (Zürich Museum of Zoology) for allowing loan of the material from the Mousson Collection, and A. Lauriat-Rage and P. Lozouet (Muséum National d'Histoire Naturelle, Paris) for arranging access to the Dollfus specimens. We are also grateful to H.I. Griffiths (University of Hull) and F. Welter-Schultes (Zoological Institute, Göttingen), who helped track down some of the more obscure literature, and to Ian Lawson (University of Aberdeen) for help in the field and for permission to use and amend the map used in Figure 1. The Director General and staff of IGME are acknowledged for their co-operation and for making the Ioannina cores available for study. We would also like to extend particular thanks to Chronis Tzedakis (University of Leeds) and Frank Wesselingh (National Museum of Natural History, Leiden), who provided measured and constructive criticism on an earlier draft of this manuscript, and to Huw Griffiths and Boris Kryštufek, whose invitation to participate in the ESF biodiversity workshop in Slovenia precipitated this review.

NOTE ADDED IN PROOF

Our attention has been drawn to a recent paper of which we were previously unaware (Reischütz & Reischütz, 2002). The authors describe three new species of pyrgulid (*Pyrgula falkneri*, *P. acicula* and *P. pambotis*) recovered from a chalky and peat-like deposit of unknown age from the floor of the Ioannina basin which bear no resemblance to the Upper Pliocene pyrgulids described by Dollfus (1922) (see also Table 2). Reischütz & Reischütz (2002) are uncertain as to the status of these taxa and, since no living representatives have been found in the modern lake, it seems plausible that they are fossil shells, possibly from the 'early' phase of the lake's history. Further work is clearly required to ascertain not only the age of the deposits, but also whether these taxa may be endemic to the lake.

Reischütz & Reischütz (2002) also provide a list of aquatic molluscan taxa collected from the modern lake. This list includes many of the species detailed in Table 3; in addition they include *Theodoxus varius varius* (Rossmässler 1835), *Viviparus hellenicus* (Clessin 1879) and *Dreissena blanci* Westerlund 1890. They are also able to identify the *Ferrissia* as *F. clessiniana* (Jickeli 1882). Critical comparison of their material would be desirable, particularly with regard to the *Dreissena* and *Viviparus* taxa.

REFERENCES

Albanis, T.A., Pomonis, P.J. & Sdoukos, A.T. 1986. Seasonal fluctuations of organochlorine and triazine pesticides in the aquatic system of the Ioannina Basin (Greece). *The Science of the Total Environment* 58, 243-253.

Anagnostidis K. & Economou-Amilli A. 1980. Limnological studies on Lake Pamvotis (Ioannina), Greece. *Archiv für Hydrobiologie* 89, 313-342.

Aubouin, J. 1959. Contribution à l'étude géologique de la Grèce septentrionale: les confins de l'Épire et de la Thessalie. *Annales Géologiques des Pays Helléniques* 10, 1-525.

Boss, K.J. (1978). On the evolution of gastropods in ancient lakes. In: *Pulmonates, vol. 2A: systematics, evolution and ecology* (V. Fretter & J. Peake, eds.), pp. 385-428. Academic Press, San Diego.
Butot, L.J.M. & Welter-Schultes, F.W. 1994. Bibliography of the mollusc fauna of Greece. *Schriften zur Malakozoologie* 7, 1-160.
Cvijić, J. 1911. L'ancien Lac Egéen. *Annales de Géographie* 20, 233-259.
Clews, J.E. 1989. Structural controls on basin evolution: Neogene to Quaternary of the Ionian zone, Western Greece. *Journal of the Geological Society* 146, 447-457.
Davis, G.M. 1982. Historical and ecological factors in the evolution, adaptive radiation and biogeography of freshwater molluscs. *American Zoologist* 22, 375-395.
Dhora, D.H. & Welter-Schultes, F.W. 1996. List of species and atlas of the non-marine molluscs of Albania. *Schriften zur Malacozoologie* 9, 90-197.
Dollfus, G.F. 1922. Faune malacologique du Miocène supérieur de Janina en Epire. *Bulletin de la Société Gèologique Française* 22, 101-123.
Frogley, M.R. 1997. The biostratigraphy, palaeoecology and geochemistry of a long lacustrine sequence from NW Greece. PhD thesis. University of Cambridge, Cambridge.
Frogley, M.R., Griffiths, H.I. & Martens, K. 2002. Modern and fossil ostracods from ancient lakes. In: *The Ostracoda: Applications in Quaternary Research* (J.A. Holmes & A.R. Chivas, eds.). *Geophysical Monograph* 131, American Geophysical Union, 167-184.
Fryer, G. & Iles, T.D. 1972. *The cichlid fishes of the great lakes of Africa*. Oliver and Boyd, Edinburgh.
Galanidou, N., Tzedakis, P.C., Lawson, I.T. & Frogley, M.R. 2000. A revised chronological and palaeoenvironmental framework for the Kastritsa rockshelter, northwest Greece. *Antiquity* 74, 349-355.
Geary, D.H. 1990. Patterns of evolutionary tempo and mode in the radiation of *Melanopsis* (Gastropoda; Melanopsidae). *Paleobiology* 16, 492-511.
Gillet, S. 1962. Remarques sur des Gastropodes de quelques gisements du Pliocène et du Quaternaire d'Épire. *Praktika Akadamiae Athens* 37, 260-273.
Gorthner, A. 1994. What is an ancient lake? In: *Speciation in Ancient Lakes* (K. Martens, B. Goddeeris & G. Coulter, eds.). *Advances in Limnology* 44, 97-100.
Gorthner, A. & Meier-Brook, C. 1985. The Steinheim Basin as a palaeo-ancient lake. In: *Sedimentary and Evolutionary Cycles. Lecture Notes in Earth Sciences 1* (U. Bayer & A. Seilacher, eds.), pp. 322-334. Springer-Verlag, Berlin.
Guernet, C., Sauvage, J. & Soulié-Marsche, I. 1977. Le Levantin de Joanina (Épire, Grèce): observations stratigraphiques et paléontologiques. *Géobios* 10, 297-309.
Hadžišče, S. 1956. Beitrag zur Kenntnis der Gastropodenfauna des Ohridsees. *Receuil des Travaux, Skopje*, 4 (1), 57-107.
Higgs, E. S. Vita-Finzi, C., Harris, D. R. & Fagg, A. E. 1967. The climate, environment and industries of Stone Age Greece: Part III. *Proceedings of the Prehistoric Society* 33, 1-29.
Hubendick, B. & Radoman, P. 1959. Studies on the *Gyraulus* species of Lake Ochrid. Morphology. *Arkiv för Zoologi, Series 2* 12 (16), 223-243.
Kalogeropoulos, N., Karayannis, M. I., Vassilaki-Grimani, M. & Grimanis, A. P. 1994. Application of trace element determination and multivariate statistics to a pollution study of Lake Pamvotis, NW Greece. *Fresnius Environmental Bulletin* 3, 187-192.
King, G. & Bailey, G. 1985. The palaeoenvironment of some archaeological sites in Greece: the influence of accumulated uplift in a seismically active region. *Proceedings of the Prehistoric Society* 51, 273-282.
Lawson, I.T. 2001. The Late Glacial and Holocene environmental history of Greece. PhD thesis. University of Cambridge, Cambridge.
Maassen, W.J.M. 1980. De zoetwatermollusken van het Ohrid en Prespamer en hum omgeving. *De Kreukel* 16 (7/8), 81-98.
Magyar, I., Geary, D.H. & Müller, P. 1999. Paleogeographic evolution of the Late Miocene Lake Pannon in Central Europe. *Palaeogeography, Palaeoclimatology, Palaeoecology* 147, 151-167.
Meier-Brook, C. 1983. Taxonomic studies on *Gyraulus* (Gastropoda: Planorbidae). *Malacologia* 24, 1-113.
Michel, E. 1994. Why snails radiate: a review of gastropod evolution in long-lived lakes, both recent and fossil. In: *Speciation in Ancient Lakes* (K. Martens, B. Goddeeris & G. Coulter, eds.). *Advances in Limnology* 44, 75-96.
Michel, E., Cohen, A.S., West, K., Johnston, M.R. and Kat, P.W. 1992. Large African lakes as natural laboratories for evolution: examples from the endemic gastropod fauna of Lake Tanganyika. *Mitteilungen der Internationalen Vereinigung für Theoretische und Angewandte Limnologie* 23, 85-99.

Mousson, A. 1859. Coquilles Terrestres et Fluviatiles. Recueilles dans l'Orient par M. le Dr Alexandre Schlaefli. *Vierteljahrsschrift der Naturforschenden Gesellschaft in Zürich* 4, 37-71.

Müller, P., Geary, D.H. & Magyar, I. 1999. The endemic molluscs of the Late Miocene Lake Pannon: their origin, evolution and family-level taxonomy. *Lethaia* 32, 47-60.

Petridis, D. & Sinis, A. 1994. Benthos of Lake Mikri Prespa (North Greece). *Hydrobiologia* 304, 185-196.

Radoman, P. 1959. On speciation in the Ohrid relict species *Pygula sturanyi* (Gastropoda). *Proceedings of the Malacological Society* 33, 148-158.

Reif, W.-E. 1985. Endemic evolution of *Gyraulus kleini* in the Steinheim Basin (planorbid snails, Miocene, southern Germany). In: *Sedimentary and Evolutionary Cycles. Lecture Notes in Earth Sciences 1*. (U. Bayer & A. Seilacher, eds.), pp. 256-294. Springer-Verlag, Berlin.

Reischütz, A. & Reischütz, P.L. 2002. Helleniká pantoía 1: Pygulidae (Gastropoda: Prosobranchia) aus dem Limni Pamvotis (Epirus, Greichenland). *Nachrichtenblatt der Ersten Vorarlberger Malakologischen Gesellschaft* 10, 1-4.

Reischütz, P.L. & Sattmann, H. 1990. Beiträge zur Molluskenfauna des Epirus, II. *Annalen des Naturhistorischen Museums in Wien* B91, 253-272.

Ryan, S. & Griffiths, H.I. 2001. The decline and probable extinction of *Graecoanatolica macedonica* (Gastropoda: Orientallinidae) in Balkan Lake Dojran. *Journal of Conchology* 37, 261-265.

Salemaa, H. 1994. Lake Ohrid. In: *Speciation in Ancient Lakes* (K. Martens, B. Goddeeris & G. Coulter, eds.). *Advances in Limnology* 44, 55-64.

Sattmann, H. & Reischütz, P.L. 1988. Beiträge zur Molluskenfauna Nordgreichenlands. *Annalen des Naturhistorischen Museums in Wien* B90, 203-214.

Schütt, H. 1962. Neue Süßwasser-Prosobranchier Greichenlands. *Archiv für Molluskenkunde* 91, 157-166.

Schütt, H. 1980. Zur Kenntnis greichischer Hydrobiiden. *Archiv für Molluskenkunde*, 110, 115-149.

Schütt, H. 1987. Limnische Mollusken aus älterem Quartär Makedoniens. *Zoologische Mededelingen* 61 (8), 113-121.

Snegarova, L. 1954. La faune des gastropodes du Lac d'Ohrid. *Acta Musei Macedonici Scientiarum Naturalium* II, 55-64.

Stanković, S. 1931. Sur les particularités limnologiques des lacs Égéens. *Verhandlungen der International Vereinigung für Theoretische und Angewandte Limnologie* 5, 158-190.

Stanković, S. 1960. The Balkan Lake Ohrid and its living world. *Monographiae Biologicae* 9, 1-357.

Stanković, S.V. 1985. A contribution to the knowledge of Gastropods fauna of Dojran Lake and the surrounding waters. *Fragmenta Balcanica* 12, 141-152.

Tzedakis, P.C. 1994. Vegetation change through glacial - interglacial cycles: a long pollen sequence perspective. *Philosophical Transactions of the Royal Society* B345, 403-432.

Tzedakis, P.C., Frogley, M.R. and Heaton, T.H.E. 2002a. Duration of Last Interglacial conditions in northwestern Greece. *Quaternary Research* 58, 53-55.

Tzedakis, P.C., Lawson, I.T., Frogley, M.R., Hewitt, G.M. and Preece, R.C. 2002b. Buffered tree population changes in a Quaternary refugium: evolutionary implications. *Science* 297, 2044-2047.

Waltham, A.C. 1970. The karstlands of the Ioannina region, NW. Greece. *Journal of the British Speleological Association* 6, 1-11.

West, K. & Cohen, A. 1994. Predator-prey coevolution as a model for the unusual morphologies of the crabas and gastropods of Lake Tanganyika. In: *Speciation in Ancient Lakes* (K. Martens, B. Goddeeris & G. Coulter, eds.). *Advances in Limnology* 44, 267-283.

West, K., Cohen, A. & Baron, M. 1991. Morphology and behaviour of crabs and gastropods from Lake Tanganyika, Africa: implications for lacustrine predator-prey coevolution. *Evolution* 45 (3), 589-607.

Whittaker, R.H. 1960. Community diversity. *Ecological Monograph* 30, 306-335.

Willmann, R. 1985. Responses of the Plio-Pleistocene freshwater gastropods of Kos (Greece, Aegean Sea) to environmental changes. In: *Sedimentary and Evolutionary Cycles. Lecture Notes in Earth Sciences 1*. (U. Bayer & A. Seilacher, eds.), pp. 295-321. Springer-Verlag, Berlin.

15. FOSSIL OSTRACODS, FAUNISTICS AND THE EVOLUTION OF REGIONAL BIODIVERSITY

HUW I. GRIFFITHS AND MICHAEL R. FROGLEY*

INTRODUCTION

The remarkable biodiversity of the Balkan region is well established (*e.g.* Gaston & David, 1994; Blondel & Aronson, 1999). The long and complex geological history of the region, coupled with its topographic variability, has ensured that a wide variety of ecological niches have been maintained even over the climatic extremes of the Quaternary, when the peninsula acted as a southern refugium for many north-temperate species (*e.g.* Tzedakis, 1993; Tzedakis *et al.*, 2002). These niches, of course, extend to the aquatic environment, where high levels of both regional and local endemism are reflected in many diverse groups. However, whilst it would seem plausible that modern aquatic faunal distribution patterns should reflect the biogeographical evolution of the region, the extreme complexity of Balkan physiographical history means that attempts to document this evolution remain difficult without large biological datasets, which are often unavailable.

Nevertheless, some progress has recently been made using modern phylogenetic and taxonomic techniques on freshwater fish populations (Figure 1). For example, Economidis & Banarescu (1991) examined taxa from 55 lake, river and stream sites across the region and, on a taxonomic basis, proposed five main biogeographical provinces linked, in part, to major drainage basin structure: Danubian, Ponto-Aegean, Attiko-Boetian, Dalmatian and South Adriatic-Ionian. In addition, they proposed that the endemic species seen originally descended from either European, Euro-Siberian or Palaearctic stock (Economidis & Banarescu, 1991). Although genetic work in part supports these conclusions and in part suggests that the boundaries between the provinces may be more complicated (*e.g.* Bianco, 1990; Doadrio & Carmona, 1998; Durand *et al.*, 1999), researchers agree that to have full confidence in the results of such investigations, both genetic and taxonomic data must be validated by means of comprehensive fossil evidence. Whilst this presents certain obvious difficulties when dealing with fish populations, other groups are perhaps more suited to such analyses.

The Ostracoda have been a subject of study in the Balkans for almost the last two centuries. These tiny, millimetre-scale bivalved crustaceans are ubiquitous in a variety of aquatic habitats from fully freshwater to fully marine, including marshes, estuaries, lagoons, rivers and lakes. Their calcareous shells often preserve well in the sediments of such environments and, owing to the stenotopic nature of many species, are able to contribute usefully to palaeoenvironmental and palaeoecological reconstructions. These factors, coupled with their passive dispersal mechanisms, make them ideal tools for studying the biogeographical evolution of a region over geological timescales (*e.g.* Frogley *et al.*, 2002). Indeed, modern faunas are relatively well described in the Balkans (particularly from lacustrine environments) and fossil records are often correspondingly numerous and detailed (*e.g.* Griffiths,

H.I. Griffiths et al. (eds.), Balkan Biodiversity, 261–272.

1995). Consequently, this paper aims to highlight their suitability in addressing biogeographical and evolutionary issues from this region.

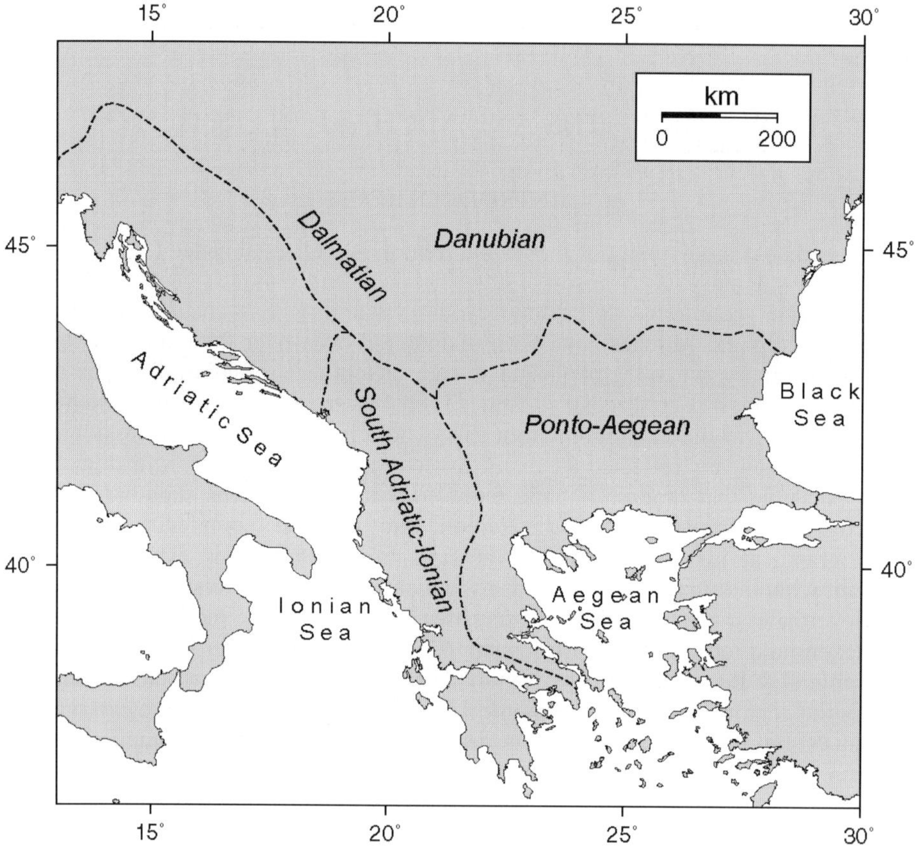

Figure 1. Map showing the Balkan region and principal biogeographical provinces based on freshwater fish fauna. Further subdivisions of the Ponto-Aegean region (Economidis & Banarescu, 1991), the South Adriatic-Ionian region (Bianco, 1990) and the existence of a central Greek region (Durand et al.*, 1999) have also been suggested (not shown).*

BALKAN OSTRACODS: MODERN DISTRIBUTIONS

Although there is only limited knowledge of groundwater ostracod distributions in the Balkans (*e.g.* Danielopol, 1992), occurrences of freshwater taxa are generally well documented (Figure 2), albeit mostly from a qualitative perspective. The most studied of these lacustrine sites are undoubtedly Lake Ohrid on the Albanian/Macedonian border (*e.g.* Stanković, 1960; Novevska *et al.*, 1985; Serafimova-Hadžišče, 1985), the Prespa lakes (Prespa and Mikri Prespa) shared between Macedonia, Greece and Albania (*e.g.* Petkovski, 1960b; Crivelli & Catsadorakis, 1997), Lake Pamvotis in north-western Greece (Frogley *et al.*, 2001; 2002), Lake Vegoritis in northern Greece (e.g Petkovski & Keyser, 1992) and Lake Dojran on the Greek/Macedonian border (Griffiths *et al.*, 2002). Typically (though not exclusively), ostracod endemnicity in these lakes is comprised of candonids (Candoninae) and cytheroids (Cytheroidea), groups that also often characterise geologically long-lived lake basins (Forester, 1991). For example, at an age of around 2-3 Ma, Lake Ohrid is probably the oldest lake in the region and, along with its surrounding wetlands, is host to 52 ostracod species, of which 33 are endemic. These include three species of *Leptocythere*, four of *Paralimnocythere* and twenty-five candonids, twenty-three of which are ascribed to *Candona* sensu stricto and probably represent an endemic species flock (Holmes, 1937; Klie, 1934, 1939a,b, 1941; Mikulič, 1961; Petkovski, 1960a, 1969a,b).

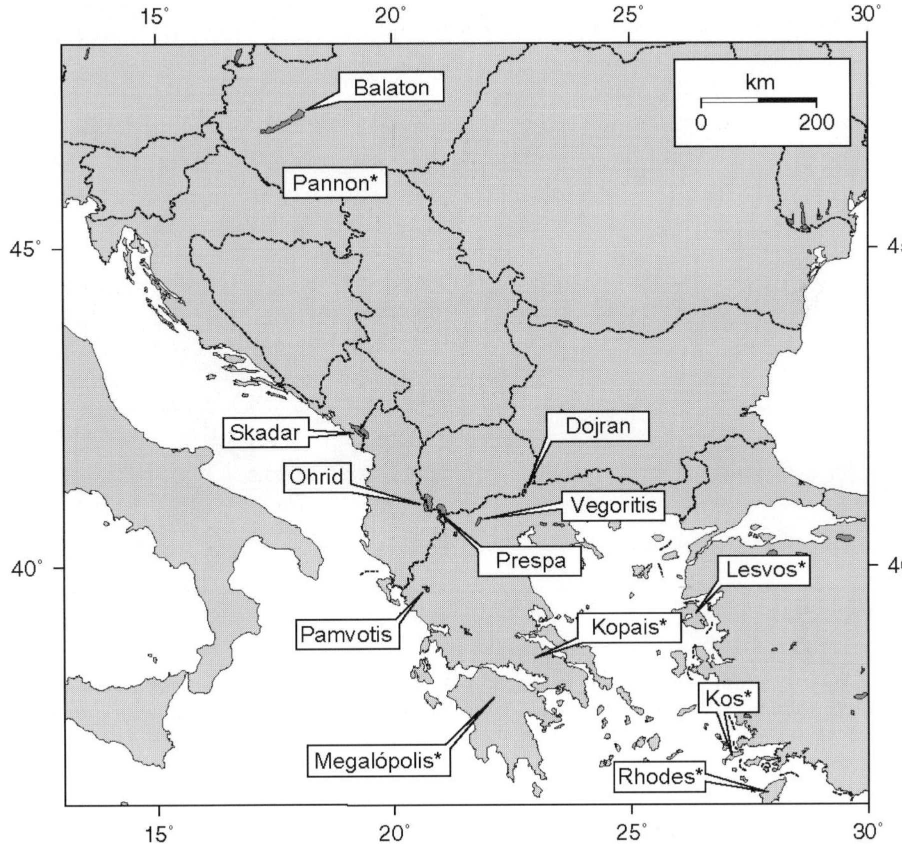

Figure 2. Map showing the principal modern and fossil (starred) lake sites across the Balkans.

BALKAN OSTRACODS: FOSSIL DISTRIBUTIONS

In spatial and numerical terms, records of fossil freshwater ostracod assemblages in the Balkans are uneven, with coverage in Croatia, Serbia and Greece being relatively good in comparison with other countries from the region (*e.g.* Griffiths, 1995; Griffiths *et al.*, 2002). Endemic taxa are particularly well represented in Neogene sediments; for example, in the Megalópolis Basin of Greece (Lüttig, 1968; Hiltermann & Lüttig, 1969), as well as in Pannonian Basin deposits seen at Metohija and Vojvodina in Serbia (*e.g.* Krstić, 1975, 1985, 1988) and at Slavonia and Posavina in Croatia (*e.g.* Sokač, 1976; 1978) (Figure 3).

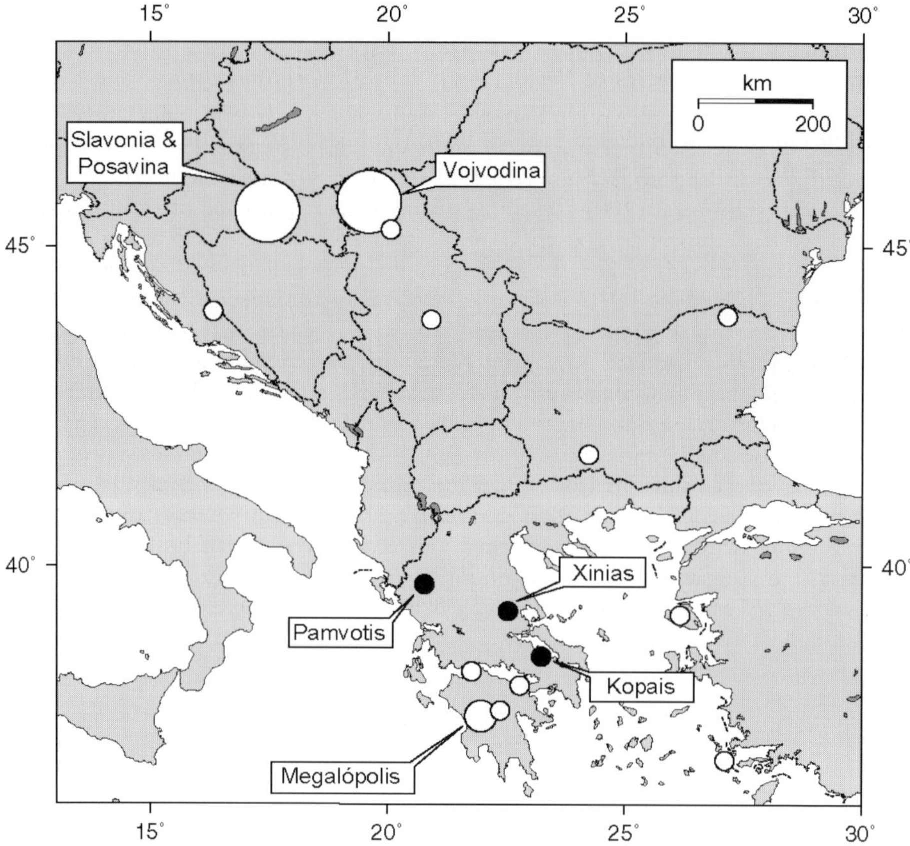

Figure 3. Map showing the principal fossil freshwater ostracod records across the Balkans south of 47° latitude. White circles denote that dating of the sequence is relatively poor; black circles indicate that dating is relatively robust; large circles indicate multiple studies. Data taken from Griffiths (1995). Lake Xinias (Digerfeldt et al., 2000) reportedly possesses an typical European Late Holocene ostracod fauna but it has not yet been formally described (H.I. Griffiths, unpublished information).

Undoubtedly the biggest problem facing the study of such records (particularly those extending back beyond the Holocene) is establishing a robust chronology, since stratigraphical control for most existing sequences is often poor by modern standards. Only two of the temporally long continuous records from the region that possess good dating control have published ostracod data, namely Lake Pamvotis from north-western Greece (Frogley *et al.*, 2001, 2002) and Lake Kopais from eastern Greece (Griffiths *et al.*, 2002). While the Kopais ostracod record probably only extends into the last glacial (P.C. Tzedakis, personal communication), that from the 319 m long I-284 borehole at Pamvotis stretches back around 420 ka. Another core from the same basin (I-256) is currently undated, but molluscan biostratigraphical analysis suggests that it is of probable Plio-Pleistocene age:

interestingly, the ostracod and molluscan assemblages from the bottom section of this core are significantly different to those found in the younger deposits (see also Frogley & Preece, this volume).

OSTRACOD FAUNISTICS

By means of a comprehensive literature review it is possible to begin identifying Balkan biogeographical patterns based on the occurrence of selected zoogeographically significant modern and fossil freshwater ostracod taxa. For example, Griffiths *et al.* (2002) demonstrated that the occurrence of several taxa in last glacial and Holocene deposits from the Lake Kopais Basin (eastern Greece) could be ascribed distinct Ponto-Aegean affinities. One of these, *Tyrrhenocythere donetziensis* (Dubowsky, 1926), is an extant euryhaline species known from the Black, Caspian and Aral Seas (e.g Boomer *et al.*, 1996) and from sites across Anatolia (Figure 4). Modern Balkan populations are unknown and it has only been reported from one other fossil locality apart from Kopais, namely the basal Plio-Pleistocene deposits of Lake Pamvotis (Griffiths *et al.*, 2002). However, a note of caution must be sounded in ascribing the Pamvotis material to *T. donetziensis*, since all valves recovered thus far from these levels have been juvenile and it is possible that they should instead be ascribed to the more common *T. amnicola* (Sars, 1888), which is found widely in the modern and Quaternary deposits of Italy, Greece, the islands of the Aegean and in south-central Anatolia (*e.g.* Devoto, 1965; Maness & Kaesler, 1985; Fernando-Gonzalez *et al.*, 1994, Galoukas *et al.*,1995; H.I. Griffiths, unpublished data).

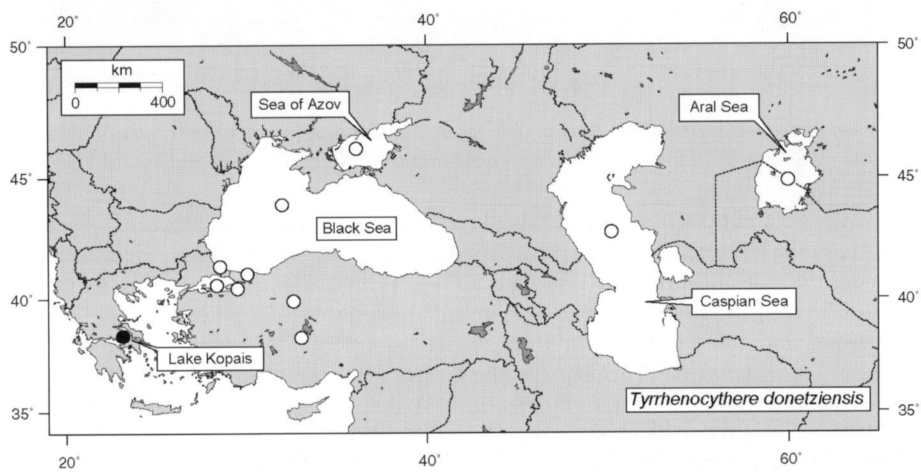

Figure 4. Map showing the distribution of Tyrrhenocythere donetziensis (after Griffiths et al.*, 2002). White circle denotes modern locality; black circle denotes fossil locality.*

Another taxon found in the deposits of Lake Kopais is *Candona angulata meridionalis* Petkovski, 1958, which would also appear to have a Ponto-Aegean distribution (Griffiths *et al.*, 2002). Extant populations of this oligohaline taxon are also known from the Black Sea and from Lake Dojran (Petkovski, 1958) (Figure 5).

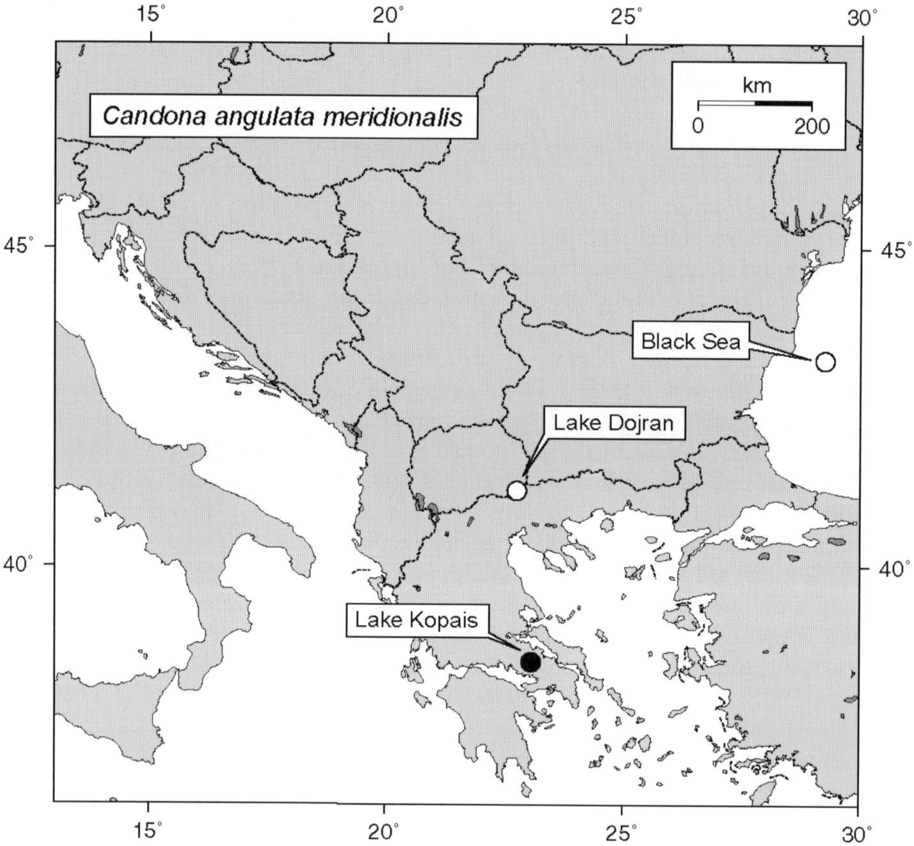

Figure 5. Map showing the Balkan distribution of Candona angulata meridionalis. *White circles denote modern localities; black circle denotes fossil locality. See text for references.*

Quite different distributions are displayed by freshwater leptocytherids, of which only seven extant species are presently known from the Balkans (Petkovski & Keyser, 1992; Karanović & Petkovski, 1999). With the exception of *Leptocythere fluviatilis* Klie, 1939 and *L. pseudoproboscidea* Karanović & Petkovski, 1999 all are lacustrine, and most are known from single lake basins (Figure 6). Extinct taxa include *L. karamani* Klie, 1939 from the Upper Pliocene of the Megalópolis Basin (Hiltermann & Lüttig, 1969), *L. picturata* (Livental, 1929) and *L. saljanica* (Livental, 1929) from the Lower Pleistocene Prevlaka OS-1 borehole, and what is probably an undescribed endemic (designated *Leptocythere* sp. A) from Lake Pamvotis that survived into the Holocene (Frogley *et al.*, 2001). In terms of its fossil record, this genus displays a distinct Southern Adriatic-Ionian and Dalmatian distribution; however, modern populations have also been described from lakes in western Anatolia (Altınsaçlı and Yılmam, 1995) (Figure 6).

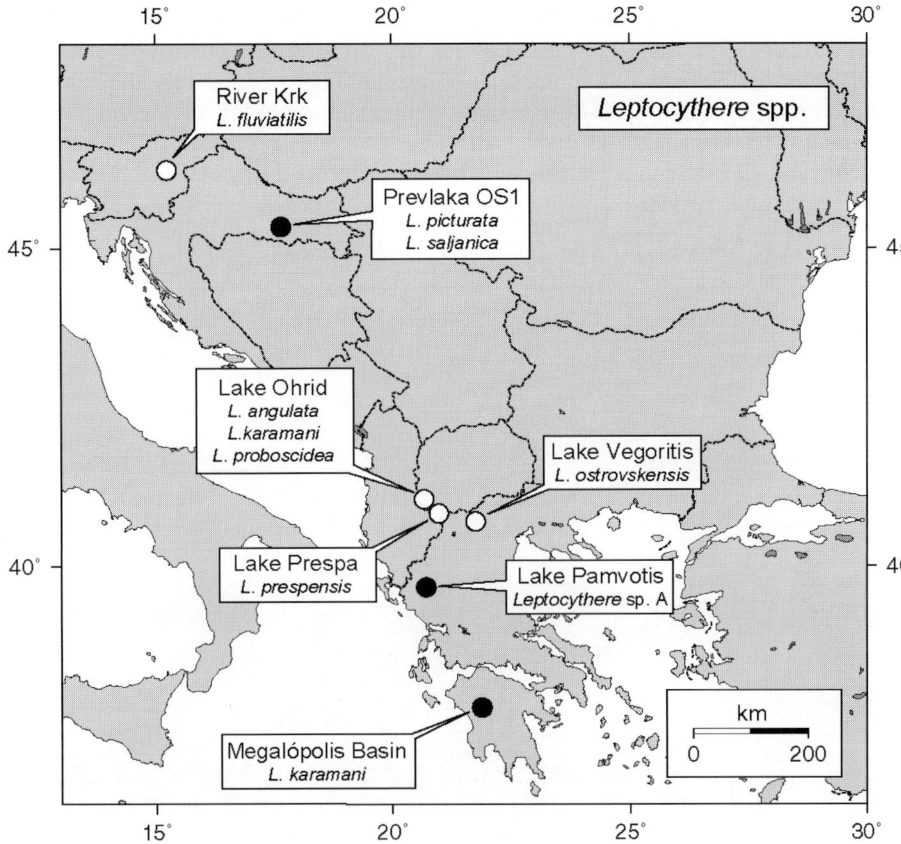

Figure 6. Map showing the Balkan distribution of freshwater leptocytherid taxa. White circle denotes modern locality; black circle denotes fossil locality. See text for references.

Although the data set is somewhat more restricted, a similar pattern can also be seen in the occurrence of *Candona parvula* Mikulić, 1961 (Figure 7). Originally described as an extant species inhabiting the sublittoral zone of Lake Ohrid (Stanković, 1960), *C. parvula* is also known from the last interglacial and last glacial deposits of Lake Pamvotis (Frogley *et al.*, 2001). Here, it disappears from the sequence at around 23 k yr. BP, just prior to the most extreme phase of the last glaciation (Galanidou *et al.*, 2000). It has been suggested by Frogley *et al.* (2001) that *C. parvula* is probably conspecific with *C. pančići* Krstić, 1998 described from the Pliocene deposits of Metohija in Kosovo (Krstić, 1998), although this has not yet been formally verified.

Despite the fact that the few data sets presented here are limited in temporal and spatial extent, they are sufficient to identify clear biogeographical differences between Balkan lake sites. The occurrences of *Leptocythere* spp. and *C. parvula* are suggestive of South Adriatic-Ionian and Dalmatian distributions, whereas records of *T. donetziensis* and *C. angulata meridionalis* are much more noticeably Ponto-Aegean in distribution. In addition, the zoogeographical difference between eastern and western Greece that has been noted by workers in other fields (*e.g.* Bianco, 1990;

Economidis & Banarescu,1991; Durand *et al.*, 1999) is clearly apparent in the ostracod record. The underlying reasons for this division (along with others noted elsewhere in the Balkans) are complex and are unlikely to be purely the result of migration and/or evolution. The elaborate geological evolution of the region, the modern-day tectonic and climatic regimes, and the resulting differences in hydrology are all likely to be factors which have influenced faunal distributions (*e.g.* Blondel & Aronson, 1999).

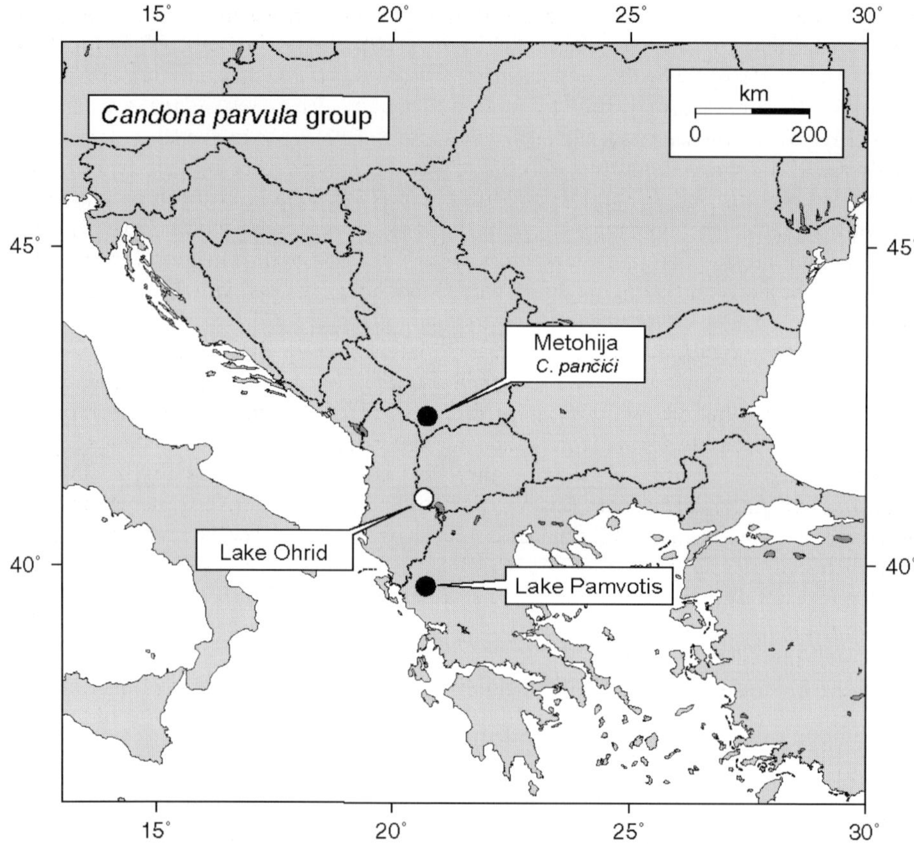

Figure 7. Map showing the Balkan distribution of Candona parvula *group taxa. Occurrence of* C. pančići *at Metohija may represent a conspecific population (see text). White circle denotes modern locality; black circle denotes fossil locality. See text for references.*

BALKAN AQUATIC BIOGEOGRAPHY

The lacustrine deposits in the northern part of the Balkans owe their origins largely to the Paratethys, a shallow sea of considerable extent that consisted of a series of tectonic basins that formed in the foreland of the mountain chains being uplifted by Alpine orogenic activity during the mid-Tertiary. These included the Alps, Dinarides, Carpathians, Pontides, Caucasus and, in the east, the Elburz Mountains. The intervening Paratethyan basins, which included the Pannonian (originally stretching across the northern Balkans and present-day Hungary), the Pontian (currently occupied by the Black Sea) and the Caspian, were repeatedly isolated from (and subsequently reconnected to) each other and the Tethys Ocean (essentially the fore-runner to the present-day Mediterranean). As a result, salinities in these basins fluctuated markedly before final fragmentation and isolation in the late Miocene and early Pliocene as many of the mountain chains completed their uplift and more stable drainage regimes became established.

Current knowledge concerning the origins of the lacustrine faunas in the southern part of the region is less clear, despite a significant body of research based around late 19th and early 20th century fresh-water molluscan studies. Radoman (1985) summarises one viewpoint, beginning with the early work of Cvijić (*e.g.* 1900, 1911, 1924), who argued for the presence of an Aegean 'superlake' of probable Miocene and/or Lower Pliocene age in what is known as the Vardar area (stretching from present-day Macedonia to western Turkey in the east and as far south as Rhodes and, possibly, Crete). Cvijić suggested that this body of water gradually 'ponded out' in the Upper Pliocene and Pleistocene to become a zone of individual lakes. An opposing viewpoint is put forward by Keraudren (1975), who suggests that the Aegean lakes owe their origins to a series of transgressions into the region by the Aegean Sea at the end of the Pliocene by waters with 'Caspi-brackish' affinities.

In fact, the situation is probably rather more complex than either of these two scenarios. As previously discussed elsewhere (Frogley *et al.*, 2001), some of the ostracod genera that are widespread in the Aegean lakes (and particularly those of Lake Ohrid) resemble forms from other ancient relict faunas, such as that of Lake Baikal (Stanković, 1960). However, others seem to have more localised Balkano-Anatolian affinities and are present throughout the area in the Neogene. A possible analogue is provided by ongoing molecular work at Lake Baikal, which suggests that different faunal groups within the lake are of different ages: some benthic groups (Gammaridae, Baicallidae) appear to have speciated in Miocene lake basins that predate the present lake, whilst cotteid fish appear to have speciated more recently (Sherbakov, 1999).

In the broadest terms, it would seem that the faunas of the western Balkans (and especially of tectonic lakes such as Ohrid) are probably phylogenetically older than those of the eastern and central Balkans. Nevertheless, the lack of robust chronological data for most fossil sites highlighted earlier means that it is not yet possible to resolve the issue concerning the 'true' origin of the Aegean lakes using ostracod faunas alone. However, it does seem likely that ostracod speciation in the Balkan Neogene has been largely by vicariance, driven by progressive hydrological isolation.

CONCLUSIONS

What emerges from this study, therefore, is that the major hindrance to advancing Balkan palaeobiogeographical studies is not necessarily the number of sites, but rather the confused stratigraphies of the region's fossil deposits. In addition, there is also increasing recognition that large, quantitative modern data sets are required if fossil assemblages are to be properly interpreted. Even relatively small water bodies can exhibit a wide range of habitats and this often leads to associated diversity in microscopic faunal assemblage composition, both within and between different lakes and wetlands. It is encouraging, therefore, that the ongoing coring of extant and recently extinct lakes in the region should go some way to addressing these problems, particularly as many of these studies are multi-proxy in nature. By adopting such a strategy, a range of complementary evidence can be used to resolve specific issues related to, for example, whether tectonics or climatic variability are driving the biological and environmental changes witnessed in the fossil record over certain time intervals. Ostracods are an ideal faunal group for tackling biogeographical questions; with the application of modern taxonomic procedures coupled with genetic techniques (particularly mitochondrial-DNA analysis), they potentially offer a means by which we can better understand the evolution of this complex region.

ACKNOWLEDGEMENTS

We would like to thank Jane Reed (University of Hull) for patience, encouragement and useful discussion; Sue Griffiths (University of Swansea) for making available the Kopais ostracod dataset; and Becky Turner (University of Plymouth) for somehow recovering the original figures.

CO-AUTHORS NOTE

This paper was originally presented at the Balkan Biodiversity Workshop by Huw in typically flamboyant style (timetabled – by Huw – immediately before coffee, as I recall, thereby encouraging "succinct questioning and even more succinct answers", as he put it). Alas, the paper presented here is a poor shadow of what it should have been: Huw was renowned for keeping many of his ideas in his head and, tragically, he passed away suddenly before anything could be properly committed to paper. Consequently, the resulting distillation of a seemingly random collection of figures and a few scribbled notes that I had made following our many telephone conversations can only serve as a faint echo of what would undoubtedly have been a manuscript brimming with enthusiasm and casually provocative concepts. I therefore take full responsibility for any errors or omissions contained herein: you can be sure that Huw would *never* have let me get away with even half of it. He was an inspirational colleague and, more importantly, a much-loved friend who is already enormously missed. MF

REFERENCES

Altınsaçlı, S. & Yılmam, S. 1995. [The Ostracoda (Crustacea) fauna of Lake Terkos (Lake Durusu).] Terkos Gölü (Durusu Gölü) Ostrakod (Crustacea) Faunasi. *Turkish Journal of Zoology* 19, 207-212 (in Turkish with English abstract).

Bianco, P.G. 1990. Potential role of the palaeohistory of the Mediterranean and Paratethys basins on the early dispersal of Euro-Mediterranean freshwater fishes. *Ichthyological Exploration of Freshwaters* 1, 167-184.

Blondel, J. & Aronson, J. 1999. *Biology and wildlife of the Mediterranean region.* Oxford University Press, Oxford.

Boomer, I., Whatley, R.C. & Aladin, N.V. 1996. Aral Sea Ostracoda as palaeoenvironmental indicators. *Lethaia* 29, 77-85.

Crivelli, A.J. & Catsadorakis, G. (eds.) 1997. Lake Prespa, Northwestern Greece. *Hydrobiologia*, 351, 1-196.

Cvijić, J. 1900. Karsna polja zapadne Bosne i Hercegovine. *Glas Srpske Kraljevske Akademije* 59, 59-182.

Cvijić, J. 1911. *Osnovi za geografiju i geologiju Makedonije i stare Srbije III. Glasnik* Srpske Kraljevske Akademije, Beograd, 689-1272.

Cvijić, J. 1924. *Geomorfologija*. Državna štamparija, Beograd, 1-588.

Danielopol, D.L. 1981. Distribution of ostracods in the groundwater of the North Western Coast of Euboea (Greece). *International Journal of Speleology* 11, 91-103.

Devoto, G. 1965. Lacustrine Pleistocene in the lower Liri Valley. *Geologica Romana* 4, 291-368.

Digerfeldt, G., Olsson, S. & Sandgren, P. 2000. Reconstruction of lake-level changes in Lake Xinias, central Greece, during the last 40,000 years.

Doadrio, I. & Carmona, J.A. 1998. Genetic divergence in Greek populations of the genus *Leuciscus* and its evolutionary and biogeographical implications. *Journal of Fish Biology* 53, 591-613.

Durand, J.-D., Guinard, B. & Bouvet, Y. 1999. Local and global multivariate analysis of geographical mitochondrial DNA variation in *Leuciscus cephalus* L. 1758 (Pisces: Cyprinidae) in the Balkan Peninsula. *Biological Journal of the Linnean Society* 67, 19-42.

Economidis, P.S. & Banarescu, P. 1991. The distribution and origin of freshwater fishes in the Balkan Peninsula, especially Greece. *Internationale Revue der Gesamten Hydrobiologie* 76, 257-283.

Fernandez-Gonzalez, M., Frydas, D., Guernet, C. & Mathieu, R. 1994. Foraminifères et ostracodes du Pliopleistocène de la région de Patras (Grèce). Intérêt stratigraphique et paléogéographique. *Revista Española de Micropaleontologia* 26, 89-108.

Forester, R.M. 1991. Pliocene-climate history of the western United States derived from lacustrine ostracodes. *Quaternary Science Reviews* 10, 133-146.

Frogley, M.R. 1997. *The Biostratigraphy, Palaeoecology and Geochemistry of a Long Lacustrine Sequence from NW Greece.* Ph.D. thesis, Univ. of Cambridge.

Frogley, M.R. & Preece, R.C. (this volume). A faunistic review of the modern and fossil molluscan fauna from Lake Pamvotis, Ioannina, an ancient lake in NW Greece: implications for endemism in the Balkans. *In*: Griffiths, H.I., Kryštufek, B. & Reed, J.M. (eds.) *Balkan Biodiversity*. Kluwer, Dordrecht.

Frogley, M.R., Griffiths, H.I. & Heaton, T.H.E. 2001. Historical biogeography and Late Quaternary environmental change of Lake Pamvotis, Ioannina (north-western Greece): evidence from ostracods. *Journal of Biogeog*raphy 28, 745-756.

Frogley, M.R., Griffiths, H.I. & Martens, K. 2002. Recent and fossil ostracods from ancient lakes. In: *The Ostracoda, Applications in Quaternary Research* (J.A. Holmes, & A. Chivas, eds.), pp. 167-184. Geophysical Monograph Series 131, American Geophysical Union, Washingtion DC.

Galanidou, N., Tzedakis, P.C., Lawson, I.T. & Frogley, M.R. 2000. A revised chronological and palaeoenvironmental framework for the Kastritsa rockshelter, northwest Greece. *Antiquity* 74, 349-355.

Galoukas, S.F., Hajek-Tadesse, V. & Danielopol, D. 1995. *Cytherissa laccustris* (Ostracoda, Cythereidae) in the Pleistocene of Lesvos Island, (Greece, Aegean Sea): its palaeoecological interest. *Neues Jahrbuch für Geologie und Paläontologie* Monatshefte 1995, 105-110

Gaston, K.J. & David, R. 1994. Hotspots across Europe. *Biodiversity Letters*, 2, 108-116.

Griffiths H.I. 1995. European Quaternary freshwater Ostracoda: a biostratigraphic and palaeobiogeographic primer. *Scopolia* 34, 1-168.

Griffiths, H.I., Reed, J.M., Leng, M.J., Ryan, S. & Petkovski, S. 2002. The recent palaeoecology and conservation status of Balkan Lake Dojran. *Biological Conservation* 104, 35-49.

Griffiths, S.J., Griffiths, H.I., Altınsaçlı, S. & Tzedakis, P.C. 2002. Interpreting the *Tyrrehenocythere* (Ostracoda) signal from Palaeolake Kopais, central Greece. *Boreas* 31, 250-259.

Hiltermann, H. & Lüttig, G. 1969. Biofazies und Paläolimnologie der pliozänen und pleistozänen Seen im Megalopolis-Becken (Peloponnes). *Mitteilungen der Internationalen Vereinigung für Theoretische und angewandte Limnologie* 17, 306-314.

Holmes, P.F. 1937. Ostracoda of Lake Ohrid. *Archiv für Hydrobiologie* 31, 485-500.

Karanović, I. & Petkovski, T.K. 1999. Two interesting ostracod species from Montenegro (SE Europe). *Annales de Limnologie* 35, 123-132.

Keraudren, B. 1975. Essai de stratigraphie et de paléogéographie du Plio-Pléstocène égèen. *Bulletin de la Société Géologique Française* 17, 1110-1120.

Klie, W. 1934. Zur Kenntnis der Ostracoden-gattung Limnocythere. *Zeitschrift für Wissenschaftliche Zoologie (Abteilung B)* 3, 534-544.

Klie, W. 1939a. Studien über Ostracoden aus dem Ohridsee: I. Candocyprinae. *Archiv für Hydrobiologie* 35, 28-45.

Klie, W. 1939b. Studien über Ostracoden aus dem Ohridsee: II. Limnocytherinae und Cytherinae. *Archiv für Hydrobiologie* 35, 631-646.

Klie, W. 1941. Studien über Ostracoden aus dem Ohridsee: III. Erster Nachtrag. *Archiv für Hydrobiologie* 38, 254-259.

Krstić, N. 1975. The Pliocene ostracods of Metohija, 3. Upper Pliocene of the locality Žabrat (Đakovica). *Bulletin du Muséum d'Histoire Naturelle, Belgrade, Séries A* 30, 191-219.

Krstić, N. 1985. New specia of the Quaternary ostracodes from Vojvodina. *Radovi Geoinstituta* 1981, 197-206.

Krstić, N. 1988. Some Quaternary ostracods of the Pannonian Basin with a review of a few neglectoida. In: *Evolutionary Biology of Ostracoda* (T. Hanai, N. Ikeya & K. Ishizaki, eds.), pp. 1063-1072. Kodansha, Tokyo.

Krstić, N. 1998. *Kaspibrakićni, endemićni I slatkovodni ostrakodi Pliocena u Metohiji*. 13. Kongres Geologa Jugoslavije (Herceg Novi). Knjiga II: Regionalna geologija, stratigrafija i paleontogija: pp. 283-297.

Komarek, Z. 1953. Herkunft der Süsswasser-Endemiten der dinarischen Gebirge. *Archiv für Hydrobiologie* 48, 269-349.

Lüttig, G. 1968. Die Ostrakoden des Megalopolis-Beckens (Peloponnes) und die Grenze Tertiär/Quartär. *Giornale di Geologia (Seria 2)* 35, 73-82.

Maness, T.R. & Kaesler, R.L. 1985. On *Tyrrhenocythere amnicola* (Sars). *Stereo-Atlas of Ostracod Shells* 12, 69-72.

Mikulić, F. 1961. Nove Candona vrste iz Ohridsekog jezera. *Bulletin du Muséum d'Histoire Naturelle de Belgrade* B17, 86-107.

Novevska, V., Naumovski, T., Gjorgonoska, D. & Mitić, V. (eds.) 1985. *Bibliography for Lake Ohrid*. Station Hydrobiologique, Ohrid.

Petkovski, T.K. 1958. Zwei neue Ostracoden aus dem Dojransee. *Biološki glasnik Societati Scientarium Naturalium Croatica* 11, 29-34.

Petkovski, T.K. 1960a. Süsswasserostracoden aus Jugoslawien VII. *Fragmenta Balcanica Musei Macedonici Scientiarum Naturalium* 3, 99-106.

Petkovski, T.K. 1960b. Zur Kenntnis der Crustaceen des Prespasees. *Fragmenta Balcanica Musei Macedonici Scientiarum Naturalium* 3, 117-131.

Petkovski, T.K. 1969a. Einige Neue und Bemwerkenswerte Candoninae aus dem Ohridsee und einigen anderen Fundorten in Europa (Crustacea - Ostracoda). *Acta Musei Macedonicic Scientiarum Naturalium* 11, 81-111.

Petkovski, T.K. 1969b. Zwei Neue *Limnocythere*-arten aus Mazedonien (Crustacea - Ostracoda). *Acta Musei Macedonicic Scientiarum Naturalium* 12, 1-18.

Petkovski, T.K. & Keyser D. 1992. *Leptocythere ostrovskensis* sp. n. (Crustacea, Ostracoda, Cytheracea) aus dem See Vegoritis (Ostrovsko Ezero) in NW Greichenland. *Mitteilungen aus den Hamburgischen Zoologischen Museum und Institut* 89, 227-237.

Polinski, V. 1929. Limnološka ispitivanja Balkanskog poluostrva I. Reliktna fauna gastropoda Ohridskog jezera. *Glas Srpske Kraljevske Akademije* 137, 129-182.

Radoman, P. 1985. *Hydrobioidea, a superfamily of Prosobranchia (Gastropoda) II. Origin, zoogeography, evolution in the Balkans and Asia Minor*. Monograph 1, Institute of Zoology, Beograd.

Ryan, S. & Griffiths, H.I. 2001. The decline and probable extinction of *Graecoanatolica macedonica* (Gastropoda: Orientallinidae) in Balkan Lake Dojran. *Journal of Conchology* 37, 261-265.

Serafimova-Hadžišče, J. (ed.) 1985. *Edition jubilaire consacrée à l'honneur de la 50ème Anniversaire de la Fondation de l'Institut Hydrobiologique - Ohrid, premier livre*. Station Hydrobiologique, Ohrid.

Sherbakov, D. 1999. Molecular phylogenetic studies on the origin of biodiversity in Lake Baikal. *Trends in Ecology and Evolution* 14, 92-95.

Sokač, A. 1976. Pleistocenska fauna Ostrakoda iz nekih busotina u istocnoj slavoniji (sjeverna Hrvatska). *Geološki Vjesnik* 29, 159-172.

Sokač, A. 1978. Pleistocene ostracode fauna of the Pannonian Basin in Croatia. *Palaeontologia Jugoslavica (Jugoslavenska Akademija Znanonosti i Umjetnosti, Academia Scientiarum et Artium Slavorum Meridionalium)* 20, 1-51.

Stanković, S. 1957. Ohridsko jezero i njegov živi svet. Kultura, Beograd.

Stanković, S. 1960. The Balkan Lake Ohrid and its living world. *Monographiae Biologicae* 9, 1-357.

Tzedakis, P.C. 1993. Long-term tree populations in northwest Greece through multiple Quaternary climatic cycles. *Nature* 364, 437-440.

Tzedakis, P.C. 1994. Vegetation change through glacial-interglacial cycles: a long pollen sequence perspective. *Philosophical Transactions of the Royal Society* B345, 403-432.

Tzedakis, P.C. 1999. The last climatic cycle at Kopais, central Greece. *Journal of the Geological Society of London* 156, 425-434.

Tzedakis, P.C., Lawson, I.T., Frogley, M.R., Hewitt, G.M. & Preece, R.C. 2002. Buffered tree population changes in a Quaternary refugium: evolutionary implications. *Science* 297, 2044-2047.

16. THE POTENTIAL OF DIATOMS AS BIODIVERSITY INDICATORS IN THE BALKANS

JANE M. REED*

INTRODUCTION

Patterns of biodiversity and endemism are better understood for higher organisms than for microscopic organisms. Microscopic organisms are extremely diverse, and discussion of biodiversity even at its most basic level (species richness) is hampered by the number of species which remain undescribed, or for which definition of species boundaries is poorly understood (World Conservation Monitoring Centre, 1992; Gaston, 1996). Diatoms (Bacillariophyceae) are unicellular, siliceous algae. Even at the genus level, the number of recognised diatom genera has risen from 70 in the 19th century, to >1000 (Round, 1996), as a result of new research, improved microscope technology, and major, ongoing changes in taxonomic nomenclature.

The term, biodiversity, encompasses genetic diversity, species diversity (in space and time) and ecosystem diversity (World Conservation Monitoring Centre, 1992). Diatom-based studies of genetic diversity (*e.g.* Medlin *et al.*, 1991) suggest that current diatom species classifications are too broad, and that some alleged single species may in truth comprise complex species groups. While these results are relevant to taxonomic issues (below), the research is in its infancy compared to some other groups (*e.g.* small mammals; Zima, this volume), and genetic data are not considered further here.

Within the Balkan Peninsula, spatial patterns of species richness have received the most attention. Extensive field research was carried out in the mid-20th century by the pioneering German diatomist, Friedrich Hustedt, who described the diatom flora of inland waters across much of Europe, and by the Croatian, Anto Jurilj, on the waters of the ancient graben, Lake Ohrid, in particular (Jurilj, 1954). Hustedt's Balkan research (Hustedt, 1945) included the ancient lakes of Ohrid, Prespa and Plitvicer and inland waters of Montenegro, Bosnia and Herzegovina and other regions. Of many new species described, some appeared to be cosmopolitan, whereas others appeared to be site-specific or regionally endemic. Their combined output highlighted the high biodiversity of the ancient lakes, resulting in a list of *ca.* 350 species and varieties for Lake Ohrid alone, many of which were new. It also underlined that the two checklists contained many taxa which had not been encountered by the other worker in their sampling, in spite of their use of shared taxonomic concepts (Jurilj, 1954).

Similar studies have been carried out on Lake Skutari (= Shkodra [Albania]; Skadar [Montenegro]), on the borders of Albania and Montenegro. An earlier series of research projects on the Montenegran side was similar to those on Lake Ohrid, resulting in a checklist of 134 species, with new species descriptions (Petković, 1981 and papers cited therein). Recent taxonomic lists of the phytoplankton compiled for its Albanian waters (Rakaj *et al.*, 2000) supported the high diatom biodiversity of the lake (claiming 242 species and varieties) but, in strong contrast to the earlier

H.I. Griffiths et al. (eds.), Balkan Biodiversity, 273–289.

research, argued that the flora was largely cosmopolitan rather than exhibiting high levels of endemism.

There have been few taxonomic publications over the last two decades relating to diatoms of the Balkans (*e.g.* Economou-Amilli, 1979, 1982), but the general revision of taxonomy and nomenclature in comprehensive texts (most notably, Krammer & Lange-Bertalot, 1986, 1988, 1991a, 1991b for species descriptions of diatoms of European inland waters, and Round *et al.*, 1990 for a major revision of the genera) and specialised taxonomic works (*e.g.* Lange-Bertalot, 1980; Lange-Bertalot & Simonsen, 1978) are relevant. Additional floristic data may be drawn from a number of ecological studies of individual lakes (*e.g.* Stojanov, 1975, 1983; Stojanovski, 1991; Stojanovski & Krstić, 1995, and Temponeras *et al.*, 2000 for Lake Dojran, on the border between the Former Yugoslav Republic of Macedonia [FYROM] and Greek Macedonia; Plenković-Moraj *et al.*, 2002 for the Plitvice lakes, Croatia; Moustaka-Gouni, 1993, Tryfon *et al.*, 1994; Danielidis *et al.*, 1996; Tafas & Economou-Amilli, 1997; Tryfon & Moustaka-Gouni, 1997; Economou-Amilli & Tafas, 2000 for lakes in Greece). Most of this research was aimed at elucidating phytoplankton dynamics in relation to eutrophication, in some cases with an emphasis on intra-annual species diversity (*e.g.* Moustaka-Gouni, 1993). The importance of lakes as centres of biodiversity and endemism has been acknowledged as a context in the cases of Lake Dojran (FYROM; Stojanovski & Krstić, 1995) and Mikri Prespa (Greece) (Crivelli & Catsadorakis, 1997 [preface to a special issue of which Tryfon & Moustaka-Gouni, 1997 forms a part]). In spite of its obvious botanical interest, there has been no recent floristic study of Lake Ohrid, although diatom monitoring is in progress (S. Krstić & S. Spaulding, personal comment).

A further important source of taxonomic (and ecological) data has come recently into the public domain, in the form of accessible data on species assemblage composition and associated water chemistry from the modern environment (surface sediment) of lakes throughout Europe and Africa, which are accessible through the European Diatom Database Initiative (EDDI) website (http://craticula.ncl.ac.uk:8000/Eddi/jsp/index.jsp) (Battarbee *et al.*, 2000). This EU project was set up to with the aim of harmonising the taxonomy of different research laboratories, in order to combine data-sets which had been collected for the derivation of regional, quantitative *transfer functions* for reconstruction of past changes in water quality in palaeolimnological studies of environmental change (see Battarbee *et al.*, 1999; Fritz *et al.*, 1999; Hall & Smol, 1999, respectively for recent reviews). The Balkan Peninsula stands in stark contrast to most of the rest of Europe, in that very little transfer function data have been derived from the region; only three small mountain lakes in Slovenia are included (Starolesnianske Pleso, Terianske Pleso and Zielowny Staw), from the UCL mountain lake pH data-set (N.G. Cameron *et al.*, unpublished). This is a consequence of the various research aims. pH data-sets (for studies of lake acidification) tend to be derived from oligotrophic, mountain lakes (as with the above Slovenian sites), salinity data-sets (for climate change) are from semi-arid regions with saline lakes (within Europe: Spain and European Turkey; Reed, 1998 and unpublished data), and total phosphorus data-sets (for eutrophication) have tended to focus on north-western Europe (*e.g.* Wunsam & Schmidt, 1995; Bennion *et al.*, 1996). A Turkish diatom (and ostracod) salinity transfer function, which includes some lakes of European Turkey and a limited number of fresh, Balkan lakes, is in progress (Reed, in preparation).

The other issues surrounding diatom biodiversity - those of temporal changes in species diversity, and ecosystem diversity, have received little attention in any study region (Williams, 1994; Kociolek & Spaulding, 2000). Most studies are biogeographic, focusing on the history of specific, unusual taxa over time (*e.g.* Williams, 1994; Kociolek & Stoermer, 1988 cited in Kociolek & Stoermer, 2001 for taxa of the Pacific Rim and the North American Great Lakes, respectively). Few (*e.g.* Sayer *et al.*, 1999 for a eutrophic UK pond) have considered temporal trends in the diversity of fossil assemblages as a whole.

To address these issues, the large body of fossil diatom data from lake sediment records which is accumulating through diatom-based palaeoenvironmental research offers potential for assessing temporal trends in species biodiversity. Since diatoms are an integral part of lacustrine ecosystems (Mason, 1998; Wetzel, 2001), and their ecological tolerances are now well understood for a wide range of environmental variables (*e.g.* Smol, 2002), this could also include temporal shifts in ecosystem diversity.

Compared to other parts of Europe, diatom-based palaeoecological research is scarce in the Balkans. The study of a *ca.* 30,000 year diatom record from Lake Ohrid (Roelofs & Kilham, 1983) stands out for two reasons, because it is the only long diatom record for the Balkans, and because it attempts to relate changes in diatom species assemblage composition to both climate and biodiversity. The study only presents preliminary results of diatom analysis, however, and the chronology is uncertain. Apart from a study of Holocene environmental change in Croatian coastal lagoons (Wunsam *et al.*, 1999), other diatom-based study sites have been undertaken on remote mountain lakes. A rigorous multi-proxy study of Late Glacial and Holocene climate, vegetation and palaeolimnological change in southwest Bulgaria (Stefanova *et al.* (2003) confirms the potential of diatoms for elucidating the temporal dynamics of aquatic ecosystems, but most papers focus only on short-term (last *ca.* one to two centuries) environmental change (Lake Mikri Prespa, Greece: Stevenson & Flower, 1991; Slovenian mountain lakes: Brancelj *et al.*, 2000, 2002; Lake Dojran, FYROM: Griffiths *et al.*, 2002).

From the foregoing, it is already clear that the degree of ongoing debate surrounding taxonomic nomenclature in the diatoms has an important bearing on the assessment of diatom species diversity and endemism in the Balkans, as does the difficulty of obtaining a representative sample from individual lakes. Rather than compile and analyse a species checklist from the available Balkan literature, the discussion of spatial patterns in species diversity (below) focuses instead on how close we are to being able to describe species diversity, by reference to issues surrounding taxonomic nomenclature and sampling strategy in studies of Lake Dojran (FYROM). The paper then discusses novel ways in which diatom data may contribute to biodiversity assessment, in studies of temporal trends in diversity at the level both of the species and the ecosystem. Since long-term palaeoecological data from the Balkans are scarce at present, the discussion also draws on examples from the author's research in the (approximately) neighbouring regions of Mediterranean Spain and the continental interior of south-central Turkey.

SPATIAL TRENDS IN SPECIES DIVERSITY: THE EXAMPLE OF LAKE DOJRAN

The study site

Lake Dojran (Dojransko ezero [Macedonia], Doïrani [Greece]) is a freshwater doline located in the karst region of the southern Balkans, straddling the borders of Macedonia (FYROM) and Greece (Figure 1). It has a surface area of *ca.* 40 km^2 and lies between 41° 10' - 41° 15'N, and 22° 12' - 22° 15'E. The region has an arid climate and the lake is a vital water resource for surrounding populations and agriculture (Griffiths *et al.* 2002). The early 1990s onwards saw renewed interest in research on the lake, with strong concerns over the impact on the important fishing and tourism industries of a major reduction in lake levels and possible associated eutrophication. There is debate as to the relative contribution of different factors. The primary cause is considered by the Macedonians as being the internationally-agreed water abstraction policy (for irrigation) in operation on the Greek side (summarised in Griffiths *et al.*, 2002). Greek limnologists regard this as contributing 'only a minor part' (Temperonas *et al.*, 2000) to lake-level reduction, and suggest instead that natural climate variability (decreased precipitation and increased evaporation) has been the major factor.

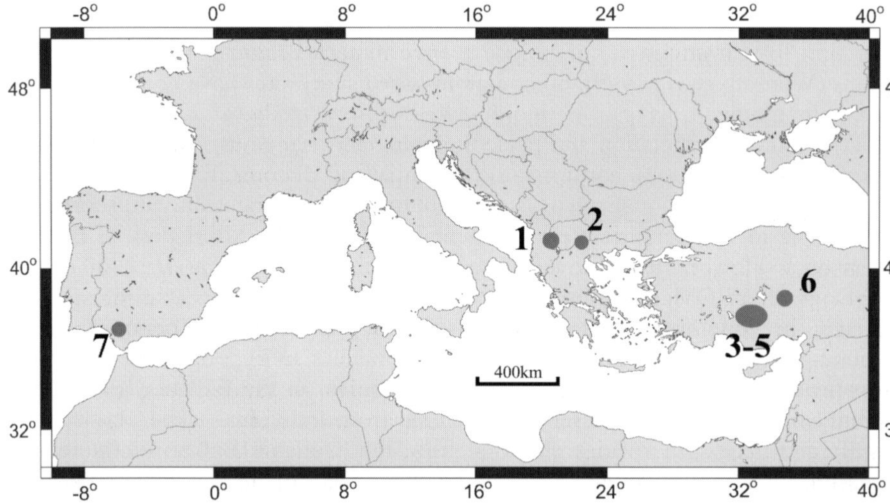

Figure 1. Map showing key sites referred to in the text. 1 = Lake Ohrid, Albania/Macedonia [FYROM]; 2 = Lake Dojran, Macedonia [FYROM]/Greece; 3-5 = Konya Basin, Turkey: Akgöl Düden, Süleymanhacı Gölü, Pinarbası; 6 = Eski Acigöl, Cappadocia, Turkey; 7 = Laguna de Medina, Cádiz, Spain.

In comparison with other lakes of the Balkans, the amount of recent diatom-based research on Lake Dojran has been fairly intensive as a result. In terms of modern limnology, there has been regular monitoring of the periphyton (shallow, littoral flora) and plankton on the Macedonian (FYROM) side, by P. Stojanovski (formerly Stojanov) and colleagues (Stojanov, 1975, 1983; Stojanovski, 1991; Stojanovski & Krstić, 1995), and a study of phytoplankton succession on the Greek side (Temperonas *et al.*, 2000). A palaeolimnological study of recent environmental impact (Griffiths *et al.*, 2002) used diatoms and other indicators as proxies for changes in water quality based on a short (47 cm) lake sediment core from waters to the south-west of the lake centre, in the FYROM.

The influence of taxonomic nomenclature on the assessment of species richness

In a comparison of monitoring results before and after environmental impact, Stojanovski & Krstić (1995) argued that 108 out of 257 algal taxa were thought to have disappeared from the lake since 1983, including many which were new to Macedonia, and all the new, undescribed species and varieties found in earlier research by Stojanovski. As a focus for discussion, these latter taxa are listed in Table 1.

Table 1. Names of endemic or new diatom species, varieties and forms thought to have disappeared from Lake Dojran (FYROM) during period 1983 - 1995 (data from Stojanovski & Krstić, 1995), which were new to the Balkans.

	Diatom species
New species/endemics	*Cyclotella petrowskae*
	Cymbella juriljii
	Navicula rara
New forms/varieties	*Melosira granulata*
	Synedra rumpens
	Synedra vaucheriae
	Pinnularia nobilis var. *macedonica*
Balkan endemics (Hustedt, 1945)	*Nitzschia macedonica*
	Nitzschia balcanica

Firstly, aside from the fact that it is logical to restrict enumeration of diatom diversity to the species level (Mann, 1999), it is difficult to accept that the varieties and forms in Table 1 are indisputably endemic. All the taxa are 'difficult', exhibiting wide morphological variation which can overlap that of other species. In contrast to the recognition of distinct varieties of *Synedra rumpens* and *S. vaucheriae* (now reassigned to the genus, *Fragilaria*) in Table 1, both of these species have been reassigned themselves to varieties of the species *F. capucina* by Krammer & Lange-Bertalot (1991a) although, as Mann (1999) points out, the authors were aware that this probably hid a complex array of different species. Equally, the wide morphological variation exhibited by the species, *Melosira granulata* (now reassigned to the genus, *Aulacoseira*) is illustrated in Figure 2 (variation in valve dimensions, height:width ratio and character of pores, or *punctae*). Krammer & Lange-Bertalot (1991a) recognised only one additional variety (*A. granulata* var. *angustissima* - see Figure 2) and a curved morphotype. Even at the species level, if morphological characteristics tend to overlap those of other species, their acceptance may only be justifiable on the demonstrable consistency of morphological variation, based on exhaustive taxonomic study. This was the case with *Cyclotella trichonidea* described from Lake Trichonis, Greece (Economou-Amilli, 1979); it falls within the range of variation of *Cyclotella ocellata* (which, again, may be a species complex; *cf.* discussions by Mann & Droop, 1996; Mann, 1999) but its status as a separate species is accepted by Krammer & Lange-Bertalot (1991a) on the basis of the consistent pattern of variation in the striae.

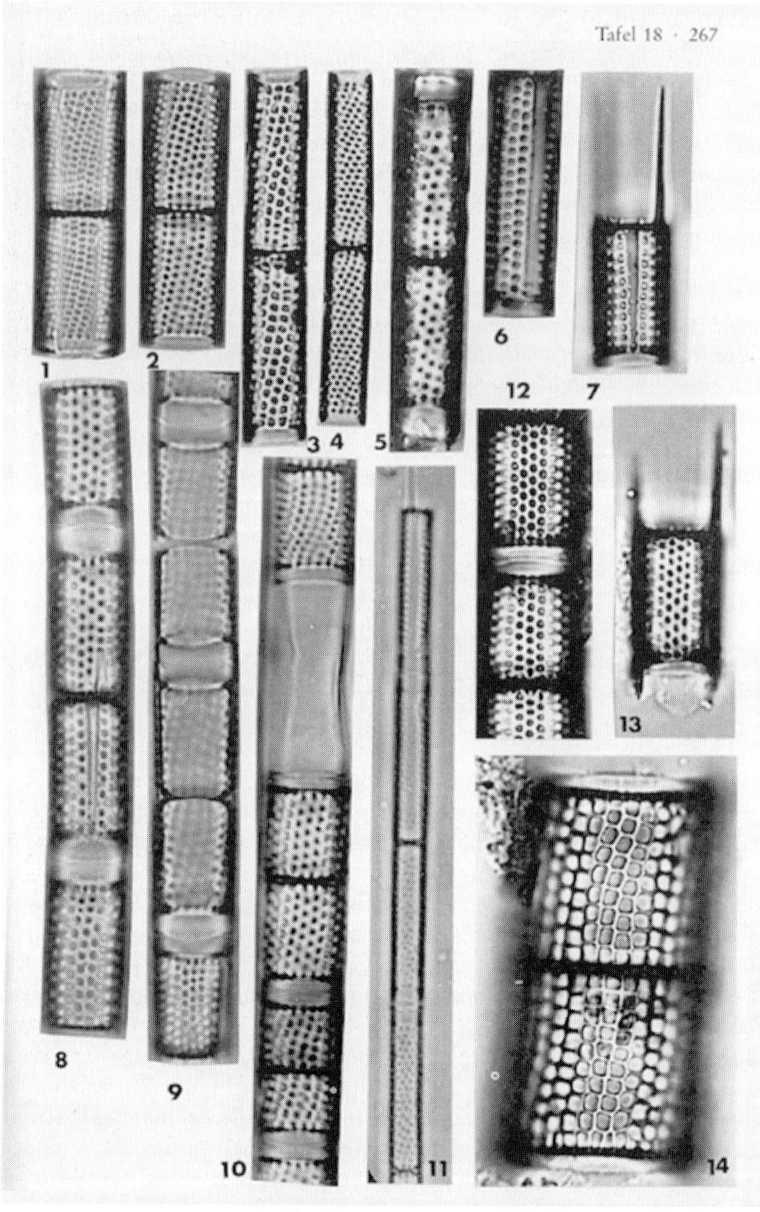

Figure 2. Aulacoseira granulata, *showing the variability in valve dimensions of a single species (reproduced from Krammer & Lange-Bertalot, 1991a, Tafel 18). Based on height:width ratio, the authors recognise only one additional variety - that of* A. granulata *var.* angustissima *O. Müller, Simonsen (depicted as no. 13). Another,* (?) A. granulata *var.* valida *Hustedt, Simonsen (no. 14) lacks a type description. (Reproduced by permission of Spektrum Akademischer Verlag).*

Secondly, the 'new species' listed in Table 1 are also not recognised in the Krammer & Lange-Bertalot (1986-1991) series of volumes, and original species descriptions could not be found by the author, so their status as endemics is again unclear. Photographs of *Cyclotella petrowskae* and *Navicula rara* are presented in Stojanov (1983) and Stojanovski & Krstić (1995), respectively (reproduced here as Figures 3a and 4). On the basis of the author's taxonomic experience, being the *Reed* in Griffiths *et al.* (2002), the depiction of *C. petrowskae* is actually that of the diatom, *Stephanodiscus* af. *medius*, with typically large areolae, striae which continue to the centre of the valve, and spines, clearly separating it from the genus *Cyclotella*. For comparison, *S. medius* is shown in Figure 3b. The depiction of this valve is probably a simple mistake. As corroborative evidence, however, a highly distinct *Cyclotella* taxon (unidentified) was indeed observed by Griffiths *et al.* (2002) as extremely rare (1-2 valves out of 500) in the sediment core. Alternately, Stojanovski & Krstić's (1995) taxon may have been that ascribed to the morphologically variable *Cyclotella ocellata* by Griffiths *et al.* (2002) (see discussion of *C. trichonidea* above) - a species which is still common in the lake. *Navicula rara* is more convincing as a valid species and possible endemic; small *Navicula* species can be notoriously difficult to separate, but the depiction of *N. rara* is clearly the same as a small, distinct taxon which was designated *Navicula* sp. 1 by Griffiths *et al.* (2002). Thus, while many difficult groups of species often appear to intergrade morphologically, making definition of species boundaries and assessment of endemism highly controversial, there may be some distinct taxa (and possible endemics) in Lake Dojran, but more rigorous taxonomic research is necessary.

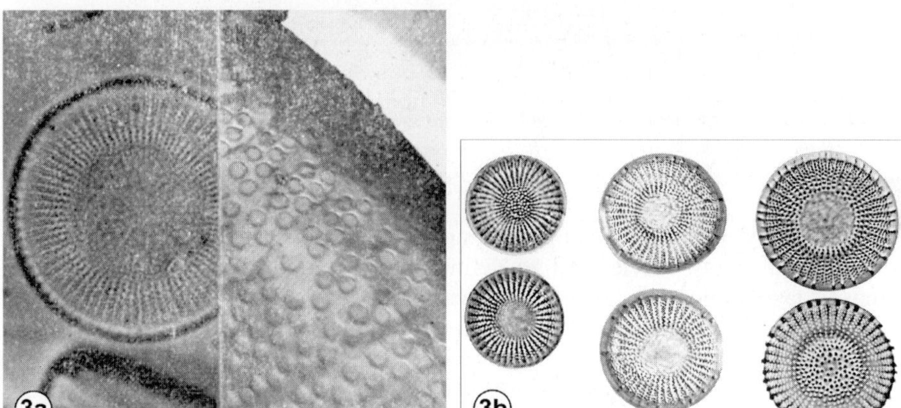

Figure 3. 3a: Cyclotella petrowskae *from Lake Dojran, FYROM (reproduced from Stojanov,1983, Plate 3, Fig. 4 - as* C. petrovskae *Jerković, Stojanov). (Reproduced by permission of the publishers of the Annuaire de la Faculté de Biologie de l'Université „Kiril et Metodij" - Skopje). 3b:* Stephanodiscus medius *Håkansson (reproduced from Krammer & Lange-Bertalot, 1991a, Tafel 75). (Reproduced by permission of Spektrum Akademischer Verlag).*

Finally, *Nitzschia macedonica* and *N. balcanica* were both described as regional endemics by Hustedt (1945). They have since been reassigned to the morphologically-variable species, *N. fonticola* and *N. capitellata*, respectively, by

Krammer & Lange-Bertalot (1988). If this is accepted, their status as regional endemics does not stand. It is interesting to note, however, that Mann & Droop (1996) focused on *N. macedonica* as an example of a species for which separate species status should possibly have been maintained. Another indication that this is the case is that the author, as a palaeoecologist, adheres mainly to the Krammer & Lange-Bertalot (1986-1991) series in order to be consistent with taxonomic guidelines set down by other workers worldwide (as do many diatom palaeoecologists; Kociolek & Stoermer, 2001). In this approach, there is a tendency towards 'lumping' rather than 'splitting'. It is pertinent, therefore, that in Lake Dojran, the author was content to include *N. balcanica* as *N. capitellata*, but separated *N. macedonica* as *Nitzschia* sp. 1 on the basis that it was sufficiently morphologically distinct from the widespread species designated *N. fonticola* in regional data-sets worked on from Spain (Reed, 1998), Turkey (Reed, in preparation) and Kazakhstan (Juggins & Reed, unpublished; see EDDI website), and also from the African concept of the species (Gasse, 1983).

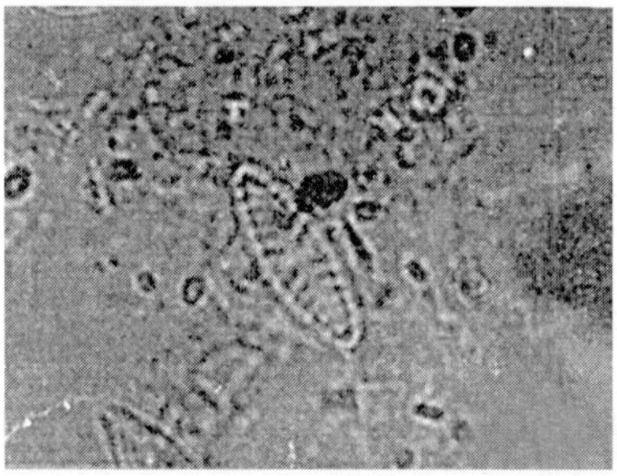

Figure 4. Navicula rara *from Lake Dojran, FYROM (reproduced from Stojanovski & Krstić, 1995, Table V, Fig. 1); in spite of poor photographic quality the valve morphology is clear. (Reproduced by permission of the publishers of the Annuaire de la Faculté de Biologie de l'Université „Kiril et Metodij" - Skopje.)*

In summary, it is clear that there is much taxonomic work to be done before the boundaries of many species can be defined with confidence. In a single lake the extensive taxonomic revision which has occurred over the last twenty years (towards lumping initially, but latterly with a reverse trend apparent; Mann, 1999) may not, from the foregoing, have a huge influence on the value of the total number of species. Its influence would be clearly significant in attempts to compare between different sites and regions, and in assessment of relative degrees of endemism, however.

The influence of sampling strategy on the assessment of species richness

The difficulties of obtaining a representative sample of diatoms from lakes were alluded to above in reference to work on Lake Ohrid and Lake Skutari, where additional research has tended to result in significant extension of the taxonomic list. Both planktonic and periphytic diatoms can exhibit marked seasonal variation, and assemblage composition varies between different habitats (open water, plants, mud, stone, sand etc); the diversity of periphytic (shallow water) algae may be particularly high in epiphytic (*on plants*) communities of lakes (Wetzel, 2001). Monitoring data therefore have the disadvantage that they might not reveal the full diversity of the flora. In Lake Ohrid, Jurilj (1954) notes that he found 130 taxa not encountered by Hustedt and, conversely, that Hustedt had listed 103 taxa not found by himself, in spite of their corresponding regularly and using shared taxonomic concepts. As a potential solution, it has often been noted that a lake-centre sediment-core surface sample (*ca.* 0.5 cm sediment from at the sediment-water interface) is more likely to integrate the flora of the lake as a whole (Battarbee, 1986).

In Lake Dojran, it is clear that the sediment core provides a more representative sample than spot-sample monitoring data. As noted, Stojanovski & Krstić (1995) reported the disappearance of major elements of the diatom flora since 1983. In the lake sediment core, however, all those identifiable taxa from Table 1 (discussed above) were still present in the uppermost core samples *ie.* had not disappeared. There was slight evidence for eutrophication based mainly on a relative increase in *Aulacoseira granulata,* but no evidence for a major change in diatom species assemblage composition over time (Griffiths *et al.*, 2002). Stojanovski & Krstić (1995) reported major habitat loss, with the development of black sapropel and a shift towards dominance by pollution-tolerant diatoms such as *Hantzschia amphioxys* and *Navicula pygmaea*. These latter taxa are rare in the sediment core. It may be surmised that, while major ecological change was clear at the sampling sites used for monitoring (and, from personal observation, along extensive stretches of the littoral zone), the overall loss of habitats across the lake as a whole was less extensive than it appeared to be from these studies. It should be noted, however, that Dojran is relatively shallow and wind-stressed; a lake-centre sample from a deep, plankton-dominated lake (such as Eski Acigöl, below) may show lower representativity of littoral assemblages. Establishing guidelines for consistent sampling of biodiversity in lakes would not, therefore, be a simple process.

An additional influence on the assessment of species richness lies in the type of sampling equipment used. In studies on Lake Dojran, open-water sampling was carried out using a variety of techniques, ranging from a plankton net (papers by Stojanovski and colleagues), preparation from water samples (Temperonas *et al.*, 2000), and a Glew (Glew, 1991) gravity sediment corer (Griffiths *et al.*, 2002). Various common species enumerated by Griffiths *et al.* (2002) were absent from the pre- and post-impact monitoring data of Stojanovski and colleagues. These mainly comprised small valves of *Stephanodiscus parvus*, *S. minutulus*, various small *Fragilaria* species and low abundance of tiny (<5 μm diameter) planktonic species. At the same time, large robust taxa such as *Caloneis* and *Pinnularia* spp. which were common in the monitoring data, were present only at low relative (%) abundance in the sediment diatom counts. Based on total phytoplankton analysis, Temperonas *et al.* (2000) noted that Dojran is dominated to *ca.* 85% by nanoplankton during the summer bloom, for which there is also no evidence in the other monitoring research. The bias towards absence of the smallest valves from the Macedonian monitoring data is presumably the consequence of using a plankton net with a relatively wide mesh as opposed to counting the entire size range of diatoms, as is possible with the other two techniques.

TEMPORAL TRENDS IN SPECIES DIVERSITY

The potential for diatoms to provide information on temporal trends in Pleistocene-Holocene species diversity is assessed from the results of palaeoenvironmental research on a variety of lakes in the Balkans, Turkey and Spain (Figure 1). These comprise the deep graben, Lake Ohrid (FYROM; Roelofs & Kilham, 1983), a shallow saline lake (Süleymanhacı Gölü), a small karstic spring (Pinarbası) and a sinkhole (Akgöl Düden) in the Konya Basin, southern central Turkey (Reed *et al.*, 1999 and unpublished data), a crater lake, Eski Acigöl, in Cappadocia, western central Turkey (Roberts *et al.*, 2001), and a shallow, saline solution basin, the Laguna de Medina, in Cádiz, south-west Spain (Reed *et al.*, 2001). Species diversity was estimated for all sites with the exception of Ohrid, by applying the Brillouin index (Pielou 1969, 1975 cited in Magurran, 1988) to diatom percentage species abundance data. Chronology was estimated from radiocarbon ^{14}C and other dating methods (see individual papers). Discussion of trends in diversity for Ohrid relies on qualitative descriptions by the authors. The full results of diatom analysis for the Turkish and Spanish sites are given in the relevant papers.

Trends in diversity are shown on a chronological timescale in Figure 5. The two lakes, Eski Acigöl (ESK) and Laguna de Medina (MDN) are less diverse than the other three lakes, Pinarbası (PN), Süleymanhacı (SH) and Akgöl (AGL), with mean values around 0.5 compared to *ca.* 1.0 - 1.4, respectively. PN and SH have the most complete Pleistocene records, which exhibit marked oscillation; there is slight evidence for an increase in diversity over time, but no clear trend. Both appear to show a major reduction in diversity at *ca.* 30 k yr. BP; this is the opposite of the trend observed by Roelofs & Kilham (1983) for Ohrid, however, where an increase at this time was interpreted as driven by middle Pleniglacial warming from *ca.* 30-20,000 yr. BP.

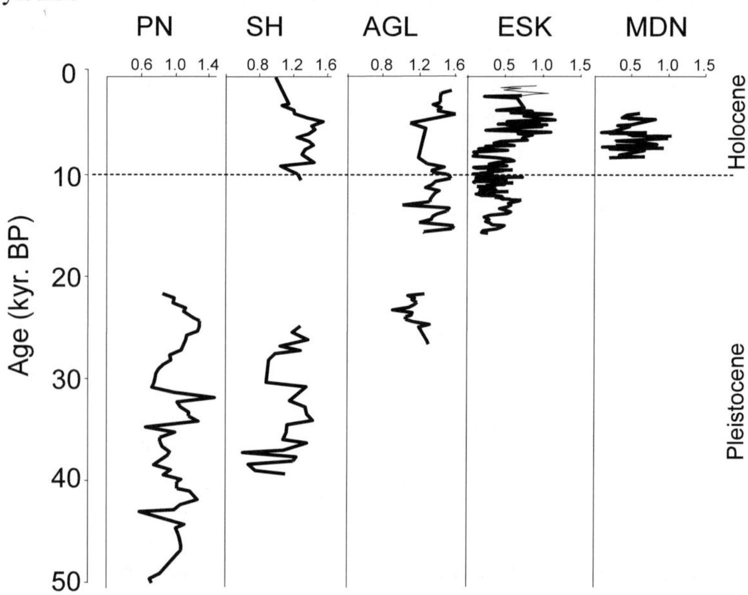

Figure 5. Temporal trends in diatom biodiversity reconstructed from the fossil data of five lake sediment records in Turkey and Spain using the Brillouin index (x axis values). Lake codes are as follows: PN (Pinarbası), SH (Süleymanhacı Gölü), AGL (Akgöl Düden), ESK (Eski Acigöl), and MDN (Laguna de Medina). See text for geographic location, lake characteristics and sources of data.

Comparing Pleistocene with Holocene diversity, Roelofs & Kilham (1983) noted a significant increase in diversity with the onset of the Holocene (marked by a shift from low-diversity plankton in the full glacial, to diverse planktonic assemblages in the Holocene), which they again related to climatic amelioration. Of the two sites which contain evidence for the Pleistocene-Holocene transition (AGL and ESK), there is no evidence for an increase in diversity. ESK does exhibit a marked increase in diversity during the Holocene, but this occurs later, in the mid-Holocene. SH, the other site with both Pleistocene and Holocene data, has a large gap in the Late Glacial; values for the Holocene are similar to those prior to 25 k yr. BP.

Thus, there are no clear long-term trends apparent from this selection of lakes. From the author's knowledge of the individual lakes, and changes in the diatom flora, it is clear that the absolute values for estimated diversity are mainly a function of two factors: habitat diversity at the coring site, and, in some cases, dissolution of diatom frustules under conditions of high salinity or turbidity. The low values for ESK compared to other sites, for example, are a consequence of its having been very deep during the Pleistocene and early-mid Holocene, such that fossil samples are dominated by extremely abundant but low diversity assemblages of planktonic taxa, with low representativity of shallow, littoral species; the mid-Holocene increase in diversity reflects a sharp increase in the relative abundance of periphytic taxa as the lake level fell, and does not necessarily indicate that the diversity of assemblages across the lake as a whole had increased. An apparent decrease in diversity at the core top relates more to poor diatom preservation as the lake began to dry out and became saline; assemblages are dominated by 1-2 robust taxa which are resistant to dissolution, and less fragile taxa are not preserved. The apparently low diversity of MDN is also an artefact of diatom dissolution in a shallow, saline lake. The other three sites, PN, SH and AGL, have always been relatively shallow, and are dominated by diverse assemblages from a variety of shallow-lake habitats, exhibiting higher apparent diversity as a consequence.

In summary, it cannot be assumed that extreme forcing functions such as climate change will have a predictable effect on diatom diversity in lakes, even with respect to major events such as the Pleistocene-Holocene transition within similar climate zones. It is possible to relate species shifts to changes in climate, but the relationship is indirect and the response varies between lakes depending on a number of environmental variables such as morphometry, water depth, open *vs.* closed hydrology, nutrient status, habitat availability, salinity etc.

TEMPORAL TRENDS IN ECOSYSTEM DIVERSITY

Assessing current threats to ecosystem biodiversity: human-induced loss of biodiversity

In the context of threats to the biodiversity of aquatic ecosystems as a whole, this paper would not be complete without a general comment concerning the value of diatoms (and palaeolimnological data in general) as indicators of changes in ecosystem status, or 'health'. As described in the Introduction, most Balkan diatom research has been restricted either to establishing taxonomic checklists or to elucidating seasonal phytoplankton dynamics. The potential to integrate diatom-based research in multi-proxy studies of ecosystem dynamics remains virtually untapped in the region. There is a whole literature on the strength of diatoms as indicators of past environmental change in aquatic ecosystems, and lakes in particular (*e.g.* Smol, 2002; papers in Stoermer & Smol, 1999). Owing to their sensitivity to pH, nutrients and salinity, they have been particularly valuable in lake acidification, eutrophication and climate change research which, as noted earlier, has been strengthened by the development of transfer functions to quantify past environmental change. Since species assemblage composition is so closely tied to water chemistry, and can give an additional insight into other ecosystem dynamics such as water depth, aquatic macrophyte cover, etc, there is clear potential for their use as indicators of changes in ecosystem status related to current environmental impacts such as eutrophication, which at their extreme pose a threat to the overall biodiversity of aquatic ecosystems. The idea of using diatom-based palaeolimnology to establish the natural 'baseline' of lakes prior to eutrophication, as a means of establishing restoration targets for the ecosystem as whole, has gained general acceptance, for example (Moss *et al.*, 1996).

The disappearance of ecosystems over time: natural loss of biodiversity

A second - and novel - potential use of diatoms in ecosystem research, which has become apparent to the author through the development of transfer functions, is the use of coupled modern and palaeolimnological data to identify the regional loss of specific ecosystems through time *ie.* loss of ecosystem biodiversity. The existence of large regional data-sets of modern diatom data (74 samples for Spain, *ca.* 100 samples for Turkey, for example) can provide an insight into the modern flora of a region which was previously lacking. While a taxonomic checklist may offer the potential, as has been described above, to argue that certain individual taxa may have disappeared from a single site (or even region), the modern data-sets are comprised of diatom counts of around 500 species from each lake, and provide comprehensive percentage relative abundance data on species in the diatom assemblage *as a whole*.

The example of the Holocene record of the Laguna de Medina, Spain (Reed *et al.*, 2001) was alluded to above. From *ca.* 7000 - 5900 yr. BP, assemblages were dominated in a shallow, permanent saline lake by planktonic *Cyclotella choctawhatcheeana*, together with benthic *Diploneis didyma* and, at lower abundance, a range of other benthic taxa including *Nitzschia granulata*, *Achnanthes fogedii*, *Mastogloia braunii* and *Campylodiscus echineis.* A range of these species is common in Holocene African lake sediments (*e.g.* Fontes & Gasse, 1991) and Holocene Dalmatian coastal lagoon sequences (Wunsam *et al.*, 1999), but they are absent both in the Spanish (Reed, 1998) and African (Gasse *et al.*, 1995) modern flora. In MDN, they occur with the ostracod, *Cyprideis torosa*, which is also now restricted to shallow coastal lagoons in Spain (Reed, unpublished data); the assemblage clearly indicates an ecosystem of permanent, shallow, chloride-dominated waters which has disappeared from inland waters of the region. Similar examples are apparent with the loss of the deep-water Pleistocene assemblages of Eski Acigöl, and other crater lakes in France and Italy, where modern analogues cannot be found in the immediate region (P. Rioual, personal comment; Ryves *et al.*, 1996).

With access to fossil and modern data, the comparison could be quantified using modern analogue matching techniques (such as MAT; Juggins, unpublished software) to assess similarity between fossil and modern assemblages. The ability to provide a rigorous means of establishing the degree to which former ecosystems have been lost or survive to the present day could furnish nature conservation managers with a strong argument for the conservation value of individual lake ecosystems.

CONCLUSIONS

Spatial patterns of species richness

From the outset, this paper has highlighted the difficulties in using current data on diatom distribution as an indicator of spatial patterns in species richness and endemism in individual lakes, the Balkan region, or in comparison with other regions globally. This view may not be universally acknowledged (*cf.* Rothfritz *et al.*, 1997, who argue that diatom systematics are well enough understood to use them as biodiversity indicators in river systems), but it certainly applies to any attempt to base conclusions as to regional biodiversity on published research, or on comparison between workers with different taxonomic concepts or sampling techniques.

Other attempts to assess regional species diversity have also highlighted the problems of establishing an accurate estimate of species richness. Edlund *et al.* (2001) noted that the apparent diversity of Mongolia rose from 516 to 547 diatom taxa simply as a result of having carried out additional research. They suggested that current statistics on regional diatom biodiversity for the regions of the Netherlands, the Great Lakes of North America and the African Rift Valleys clearly reflected the amount of research carried out rather than the relative levels of biodiversity one would predict intuitively.

Published taxonomic and ecological research in the Balkans has been rather piecemeal, and betrays a high degree of variation in taxonomic nomenclature between different workers. From analysis of the methods and results of various research projects on Lake Dojran, the problems are twofold. Firstly, there is still heated debate as to the definition of species boundaries, which may have a more significant influence on comparison between sites and regions, and assessment of endemism, than it does the assessment of species richness *per se*. There is a clear need in the Balkans for more extensive and rigorous taxonomic research to build on the earlier work by Hustedt and co-workers, and also - equally importantly - to develop a reliable data-set for a Balkan freshwater transfer function.

Secondly, the seasonality and diversity of diatoms is such that the results will vary according to sampling strategy (which has varied between different studies) and are unlikely to be representative of an entire lake flora. Since morphometry and habitat availability varies so much between lakes it would be difficult to establish a simple set of sampling guidelines to ensure consistent representativity between sites; a lake-centre sediment core sample appears to offer the most acceptable solution.

The definition of pattern and process in diatom species diversity in general will not be resolved until there is some consensus as to whether or not diatoms are cosmopolitan or exhibit high levels of endemism. At one extreme, it has been suggested that diatoms exhibit low global species diversity (but high local diversity) owing to their high dispersal ability and long evolutionary history, resulting in absent or weak biogeographic patterns (Hillebrand & Azovsky, 2001; Hillebrand & Blenckner, 2002); Cocquyt (2000) argued that the diatoms exhibit low endemism in the ancient East African Rift lake, Tanganyika, for example. At the other extreme are the protagonists of high global diversity and endemism, such as Mann & Droop

(1996), Mann (1999) or Kociolek & Stoermer (2001), who argue that the Krammer & Lange-Bertalot (1986-1991) species boundaries are too broad. They support their arguments with evidence from genetics, breeding experiments and rigorous taxonomic study which has sometimes produced strong evidence for the existence of endemics, both for regions (*e.g.* Kociolek & Spaulding, 1998) and individual sites such as Lake Baikal (Flower, 1993). In this paper, the difficulties of establishing whether alleged endemics exist in Lake Dojran were highlighted; some distinct species may be endemic, but others belong to 'difficult' groups where it is a matter of debate as to how to classify them. It was noted that the poor morphological correlation of many variable, 'widespread' taxa in Lake Dojran with the standard European floras is equally evident in other regions such as Kazakhstan, so this is not conclusive evidence in itself of peculiarly high regional variation.

Temporal trends in biodiversity

Diatom-based palaeoenvironmental research in the Balkans has been very limited to date, a situation which must be rectified. As a direct application to biodiversity studies, it was suggested that fossil diatom data could be translated into chronological data on changes in diatom species diversity. From the examples considered here, however, the trends are not predictable and do not correlate with external environmental change, but tend to vary instead with internal factors such habitat availability and diatom preservation potential. In a multi-proxy study of diversity changes in diatoms, aquatic pollen and chironomids with eutrophication, Sayer (1999) noted similarly that the response was complex, and varied due to differences in habitat requirements between diatoms and chironomids rather than simply according to nutrient enrichment. While such analyses may be of use in summarising fossil diatom data, they are unlikely to provide a major new source of data on pattern and process in diatom biodiversity.

In contrast, the potential value of diatoms is very clear in ecosystem biodiversity research which uses diatoms as environmental indicators rather than focusing on them as organisms. The range of potential applications, which remains to be exploited in the Balkans, was described briefly above. The most obvious are those which centre on elucidating the effects of pollutants on aquatic ecosystems, an issue which is central to the theme of biodiversity. A novel use of diatom palaeolimnological data which has been suggested here is to match fossil assemblages with modern regional diatom data from transfer functions, in order to establish whether former ecosystems have been lost to a region, or, conversely, to assess how 'rare' a modern ecosystem is, to provide an important tool for nature conservationists.

In summary, a lot of research remains to be done in the Balkans, both in terms of modern systematic research and in terms of limnology and palaeolimnology. This paper has explored how close we are to being able to explore pattern and process in diatom biodiversity in the region. Both in terms of diatoms as unicellular algae, and diatoms as environmental or palaeoenvironmental indicators, there is enormous potential for future research.

ACKNOWLEDGEMENTS

This paper is dedicated to the memory of Huw and his boundless enthusiasm. It was written while in receipt of a Leverhulme Special Research Fellowship. I would like to thank Sveto Krstić (Skopje, Macedonia) for his friendship and advice on Macedonian diatom research, and the referee, Anson Mackay (UCL London), for his helpful and positive comments on the manuscript.

REFERENCES

Battarbee, R.W. 1986. Diatom Analysis. In *Handbook of Holocene palaeoecology and palaeohydrology* (B.E. Berglund, ed.), pp. 527-570. John Wiley, Chichester.

Battarbee, R.W., Charles, D.F., Dixit, S.S. & Renberg, I. 1999. Diatoms as indicators of surface water acidity. In *The Diatoms: Applications for the Environmental and Earth Sciences* (E.F. Stoermer & J.P. Smol, eds.), pp. 85-127. Cambridge University Press, Cambridge.

Battarbee, R.W., Juggins, S., Gasse, F., Anderson, N.J., Bennion, H. & Cameron, N.G. 2000. European Diatom Database (EDDI). An Information System for Palaeoenvironmental Reconstruction. *European Climate Science Conference*, Vienna City Hall, Vienna, Austria, 19-23 October 1998: 1-10.

Bennion, H., Juggins, S. & Anderson, N.J. 1996. Predicting Epilimnetic Phosphorus Concentrations Using an Improved Diatom-Based Transfer Function and Its Application to Lake Eutrophication Management. *Environmental Science & Technology* 30, 2004-7.

Brancelj, A., Šiško, M., Lami, A., Appleby, P., Livingstone, D.M., Rejec-Brancelj, I. & Ogrin, D. 2000. Changes in the trophic level of an Alpine lake, Jezero v Ledvici (NW Slovenia), induced by earthquakes and climate change. *Journal of Limnology* 59 (Suppl. 1), 29-42.

Brancelj, A., Šiško, M., Muri, G., Appleby, P., Lami, A., Shilland, E., Rose, N.L., Kamenik, C., Brooks, S. & Dearing, J.A. 2002. Lake Jezero v Ledvici (NW Slovenia) - changes in sediment records over the last two centuries. *Journal of Paleolimnology* 28, 47-58.

Cocquyt, C. 2000. Biogeography and species diversity of diatoms in the Northern Basin of Lake Tanganyika. *Advances in Ecological Research* 31, 125-150.

Crivelli, A.J. & Catsadorakis, G. 1997. Preface. *Hydrobiologia* 351, ix-xi.

Danielidis, D.B., Spartinou, M. & Economou Amilli, A. 1996. Limnological survey of Lake Amvrakia, western Greece. *Hydrobiologia* 318, 207-218.

Economou-Amilli, A. 1979. Two new taxa of *Cyclotella* KÜTZING from Lake Trichonis, Greece. *Nova Hedwigia* 31, 467-477.

Economou-Amilli, A. 1982. SEM-Studies on *Cyclotella trichonidea* (Bacillariophyceae). *Archiv für Hydrobiologie. Suppl. 63, 1 Algological Studies* 30, 25-34.

Economou-Amilli, A. & Tafas, T. 2000. Distribution patterns of *Cyclotella trichonidea* EC.-AM. sensu lato in the plankton succession of the lakes Trichonis and Avrakia, Greece. *Archiv für Hydrobiologie* 147, 559-575.

Edlund, M.B., Soninkhishig, N., Williams, R.M. & Stoermer, E.F. 2001. Biodiversity of Mongolia: Checklist of diatoms, including new distributional reports of 31 taxa. *Nova Hedwigia* 72, 59-90.

Flower, R.J. 1993. A taxonomic re-evaluation of endemic *Cyclotella* taxa in Lake Baikal, Siberia. *Nova Hedwigia* 106, 203-220.

Fontes, J.C. & Gasse, F. 1991. Chronology of the major palaeohydrological events in NW Africa during the late Quaternary: PALHYDAF results. *Hydrobiologia* 214, 367-372.

Fritz, S.C., Cumming, B.F., Gasse, F. & Laird, K. 1999. Diatoms as indicators of hydrologic and climatic change in saline lakes. In *The Diatoms: Applications for the Environmental and Earth Sciences* (E.F. Stoermer & J.P. Smol, eds.), pp. 41-72. Cambridge University Press, Cambridge.

Gasse, F. 1986. *East African Diatoms. Taxonomy, ecological distribution.* Bibliotecha Diatomologica. Volume 11. J. Cramer, Stuttgart.

Gasse, F., Juggins, S. & Ben Khelifa, L. 1995. Diatom-based transfer functions for inferring past hydrochemical characteristics of African lakes. *Palaeogeography, Palaeoclimatology, Palaeoecology* 117, 31-54.

Gaston, K.J. 1996. Species richness: measure and measurement. In *Biodiversity. A Biology of Numbers and Difference*, (K.J. Gaston, ed.), pp. 77-113, Blackwell, Oxford.

Griffiths, H.I., Reed, J.M., Leng, M.J., Ryan, S. & Petkovski, S. 2002. The conservation status and recent palaeoecology of Balkan Lake Dojran. *Biological Conservation* 104, 35-49.

Hall, R.I. & Smol, J.P. 1999. Diatoms as indicators of lake eutrophication. In *The Diatoms: Applications for the Environmental and Earth Sciences* (E.F. Stoermer & J.P. Smol, eds.), pp. 169-182. Cambridge University Press, Cambridge.

Hillebrand, H. & Blenckner, T. 2002. Regional and local impact on species diversity - from pattern to process. *Oecologia* 132, 479-491.

Hillebrand, H. & Azovsky, A.I. 2001. Body size determines the strength of the latitudinal diversity gradient. *Ecography* 24, 251-256.

Hustedt, F. 1945. Diatomeen aus Seen und Quellgebieten der Balkan Halbinsel. *Archiv für Hydrobiologie* 40, 867-973.

Jurilj, A. 1954. [Flora and vegetation of diatoms from Ohrid Lake in Yugoslavia]. *Yugoslavian Academy of Science, Zagreb [JAZU]* 26, 99-190. (In Serbo-Croatian with English abstract).

Kociolek, J.P. & Spaulding, S.A. 1998. Valve morphology and systematic position of *Navicula walkeri* (Bacillariophyceae), a diatom endemic to Oregon and California (USA). *Nova Hedwigia* 67, 235-245.

Kociolek, J.P. & Spaulding, S.A. 2000. Freshwater diatom biogeography. *Nova Hedwigia* 71, 223-241.

Kociolek, J.P. & Stoermer, E.F. 2001. Taxonomy and ecology: a marriage of necessity. *Diatom Research* 16, 433-442.

Krammer, K. & Lange-Bertalot, H. 1986. Bacillariophyceae 1. Teil: Naviculaceae. In *Süsswasserflora von Mitteleuropa. Bacillariophyceae* (H. Ettl, G. Gärtner, J. Gerloff, H. Heynig & D Mollenhauer, eds.). Vol. 2/1. Gustav Fischer Verlag, Stuttgart.

Krammer, K. & Lange-Bertalot, H. 1988. Bacillariophyceae 2. Teil: Bacillariaceae, Epithemiaceae, Surirellaceae. In *Süsswasserflora von Mitteleuropa. Bacillariophyceae* (H. Ettl, G. Gärtner, J. Gerloff, H. Heynig & D Mollenhauer, eds.). Vol. 2/2. Gustav Fischer Verlag, Stuttgart.

Krammer, K. & Lange-Bertalot, H. 1991a. Bacillariophyceae 3. Teil: Centrales, Fragilariaceae, Eunotiaceae. In *Süsswasserflora von Mitteleuropa. Bacillariophyceae* (H. Ettl, G. Gärtner, J. Gerloff, H. Heynig & D Mollenhauer, eds.). Vol. 2/3. Gustav Fischer Verlag, Stuttgart.

Krammer, K. & Lange-Bertalot, H. 1991b. Bacillariophyceae 4. Teil: Achnanthaceae. In *Süsswasserflora von Mitteleuropa. Bacillariophyceae* (H. Ettl, G. Gärtner, J. Gerloff, H. Heynig & D Mollenhauer, eds.). Vol. 2/4. Gustav Fischer Verlag, Stuttgart.

Lange-Bertalot, H. 1980. Zur systematischen Bewertung der bandförmigen Kolonien bei *Navicula* und *Fragilaria. Nova Hedwigia* 33, 723-787.

Lange-Bertalot, H. & Simonsen, R. 1978. A taxonomic revision of the *Nitzschiae Lanceolatae* Grunow. 2. European and related extra-European fresh water and brackish water taxa. *Bacillaria* 1, 11-111.

Magurran, A.E. 1988. *Ecological Diversity and Its Measurement.* Chapman & Hall, London.

Mann, D.G. 1999. The species concept in diatoms. *Phycologia* 38, 437-495.

Mann, D.G. & Droop, S.J.M. 1996. 3. Biodiversity, biogeography and conservation of diatoms. *Hydrobiologia* 336, 19-32.

Medlin, L.K., Elwood, H.J., Stickel, S. & Sogin, M.L. 1991. Morphological and genetic variation within the diatom *Skeletonema costatum* (Bacillariophyta): evidence for a new species, *Skeletonema pseudocostatum. Journal of Phycology* 27, 514-524.

Mason, B. 1998. *Ecology of Fresh Waters. Man and Medium, Past to Future.* 3rd edition. Blackwell, Oxford.

Moss, B., Madgwick, J. & Phillips, G. 1996. *A guide to the restoration of nutrient-enriched shallow lakes.* The Broads Authority, Norwich.

Moustaka-Gouni, M. 1993. Phytoplankton succession and diversity in a warm monomictic, relatively shallow lake - Lake Volvi, Macedonia, Greece. *Hydrobiologia* 249, 33-42.

Petković, S. 1981. Phytoplankton. In *The biota and limnology of Lake Skadar* (G. Karam & A. Beeton, eds.). Titograd, pp. 163-189.

Plenković-Moraj, A., Horvatinčić, N. & Primc-Habdija, B. 2002. Periphyton and its role in tufa deposition in karstic waters (Plitvice Lakes, Croatia). *Biologia, Bratislava* 57, 423-431.

Rakaj, M., Hindak, F. & Hindakova, A. 2000. Phytoplankton species diversity of the Albanian part of Lake Shkodra in 1998-1999. *Biologia* 55, 329-342.

Reed, J.M. 1998. A diatom-conductivity transfer function for Spanish salt lakes. *Journal of Paleolimnology* 19, 399-416.

Reed, J.M., Roberts, N. & Leng, M.J. 1999. An evaluation of the diatom response to Late Quaternary environmental change in two lakes in the Konya Basin, Turkey, by comparison with stable isotope data. *Quaternary Science Reviews* 18, 631-646.

Reed, J.M., Stevenson, A.C. and Juggins, S. 2001. A multi-proxy record of Holocene climate change in southwest Spain: the Laguna de Medina, Cádiz. *The Holocene* 11, 705-717.

Roberts, N., Reed, J.M., Leng, M.J., Kuzucuoğlu, C., Fontugne, M., Bertaux, J., Woldring, H., Bottema, S., Black, S., Hunt, E. and Karabıyıkoğlu, M. 2001. The tempo of Holocene climatic change in the eastern

Mediterranean region: new high-resolution crater-lake sediment data from central Turkey. *The Holocene* 11, 721-736.

Roelofs, A.K. & Kilham, P. 1983. The diatom stratigraphy and paleoecology of Lake Ohrid, Yugoslavia. *Palaeogeography, Palaeoclimatology, Palaeoecology* 42, 225-245.

Rothfritz, H., Jüttner, I, Suren, A.M. & Ormerod, S.J. 1997. Epiphytic and epilithic diatom communities along environmental gradients in the Nepalese Himalaya; implications for the assessment of biodiversity and water quality. *Archiv für Hydrobiologie* 138, 465-482.

Round, F.E. 1996. What characters define diatom genera, species and infraspecific taxa? *Diatom Research* 11: 203-218.

Ryves, D.B., Jones, V.J., Guilizzoni, P., Lami, A., Marchetto, A., Battarbee, R.W., Bettinetti R. & Devoy, E.C. 1996. Late Pleistocene and Holocene environmental changes at Lake Albano and Lake Nemi (central Italy) as indicated by algal remains. *Memorie dell'Istituto Italiano di Idrobiologia* 55, 119-48.

Sayer, C.D., Roberts, N., Sadler, J., David, C. & Wade, P.M. 1999. Biodiversity changes in a shallow lake ecosystem: a multi-proxy palaeolimnological analysis. *Journal of Biogeography* 26, 97-114.

Smol, J.P. 2002. *Pollution of Lakes and Rivers. A Paleoenvironmental Perspective.* Arnold, London.

Stefanova, I., Ognjanova-Rumenova, N, Hofmann, W & Ammann, B. 2003. Late Glacial and Holocene environmental history of the Pirin Mountains (SW Bulgaria): a paleolimnological study of Lake Dalgoto (2310m). *Journal of Paleolimnology* 30, 95-111.

Stevenson, A.C. & Flower, R.J. 1991. A Palaeoecological Evaluation of Environmental Degradation in Lake Mikri Prespa, NW Greece. *Biological Conservation* 57, 89-109.

Stoermer, E.F. & Smol, J.P. 1999. *The Diatoms: Applications for the Environmental and Earth Sciences.* Cambridge University Press, Cambridge.

Stojanov, P. 1975. [Beitrag zur kenntnis des Phytoplanktons im Dojran-See.] *Annuaire de la Faculté de Biologie de l'Université „Kiril et Metodij" -Skopje* 27/28, 221-236 (in Macedonian with German abstract).

Stojanov, P. 1983. Algal flora of the periphyton of Lake Doiran. *Annuaire de la Faculté de Biologie de l'Université „Kiril et Metodij" - Skopje* 36, 95-109 [in Macedonian with English abstact].

Stojanovski, P. 1991. [Changes in Dojran Lake (Macedonia) microflora due to unreasonable use of the lake's waters.] In [*Sostojbite i perspectivite za zaštita na Dojranskoto ezero. Zbornik na trudovi od sovetvanjeto vo Star Dojran.*] (M. Gasevski, ed.), pp. 90-98. Dviženje na Ekologistite na Makedonija, Skopje (in Macedonian with English abstract).

Stojanovski, P. & Krstić, S. 1995. Lake Doiran's forced dystrophy as a direct consequence of the anthropogenic influence. *Godišen zbornik, Biologija* 48, 139-173 (in Macedonian with English abstract).

Temponeras, M., Kristiansen, J. & Moustaka-Gouni, M. 2000. Seasonal variation in phytoplankton composition and physical-chemical features of the shallow Lake Doïrani, Macedonia, Greece. *Hydrobiologia* 424, 109-122.

Tafas, T. & Economou-Amilli, A. 1997. Limnological survey of the warm monomictic lake Trichonis (central western Greece). II. Seasonal phytoplankton periodicity - a community approach. *Hydrobiologia* 344, 141-153.

Tryfon, E. & Moustaka-Gouni, M. 1997. Species composition and seasonal cycles of phytoplankton with special reference to the nanoplankton of Lake Mikri Prespa. *Hydrobiologia* 351, 61-75.

Tryfon, E., Moustaka-Gouni, M., Nikolaidis, G. & Tsekos, I. 1994. Phytoplankton and physical-chemical features of the shallow Lake Mikri-Prespa, Macedonia, Greece. *Archiv für Hydrobiologie* 131: 477-494.

Wetzel, R.G. 2001. *Limnology* 3rd edition. Academic Press, London.

Williams, D.M. 1995. Diatom biogeography: some preliminary considerations. In *Proceedings of the 13th International Diatom Symposium* (D. Marino & M. Montresor, eds.), pp. 311-319. Biopress Ltd., Bristol.

World Conservation Monitoring Centre, 1992. *Global Biodiversity. Status of the Earth's living resources.* Chapman and Hall, London.

Wunsam, S. & Schmidt, R. 1995. A diatom-phosphorus transfer function for Alpine and pre-alpine lakes. *Memorie dell'Istituto Italiano di Idrobiologica* 53, 85-99.

Wunsam, S., Schmidt, R. and Müller, J. 1999. Holocene lake development of two Dalmatian lagoons (Malo and Veliko Jezero, Isle of Mljet) in respect to changes in Adriatic sea level and climate. *Palaeogeography, Palaeoclimatology, Palaeoecology* 146, 251-281.

17. THE CURRENT STATUS OF ADRIATIC FISH BIODIVERSITY

LOVRENC LIPEJ* AND JAKOV DULČIĆ

INTRODUCTION

The western part of the Balkan Peninsula is bordered by the Adriatic Sea, a small semi-enclosed sea connected to the Eastern Mediterranean via the Strait of Otranto. It was named by ancient Greek geographers, referring to the city Adria, at the mouth of the Po River (Cushman-Roisin *et al.*, 2001). Apart from the Black Sea, the Adriatic is the northernmost part of the Mediterranean; this has a significant influence on its physical characteristics even in its southernmost waters. The Adriatic is 783 km long, with a mean width of 243 km. Its surface area, including the islands, amounts to 138,595 km^2 or about 4.6% of the total Mediterranean area. In the south, the Adriatic Sea is separated from the Ionian Sea by the 72 km wide Strait of Otranto. The eastern coast is steep and composed of limestone, with a rapidly deepening narrow shelf and a large number of both small and large islands, whilst the western coastal shelf is wider due to accumulation of river sediments, and has almost no islands. The Adriatic Sea is rather shallow, with the shelf making up about 74% of the sea bed. It is generally divided into three geographic regions: the Northern Adriatic, the Middle Adriatic and the Southern Adriatic (Figure 1) (Gačić *et al.*, 2001). The North Adriatic is very shallow and strongly influenced by the rivers of northern Italy, and the Po in particular. The Middle Adriatic is deeper, reaching 280 m in the Jabuka Pit. It is separated from the Southern Adriatic by the Palagruža Sill (180 m depth). The Southern Adriatic is much deeper (the South Adriatic Pit reaches 1,233 m). The Split-Gargano transect runs along the Sill in the Middle Adriatic; the thermohaline structure in this area shows strong temporal variability since it is exposed to influences from both the Northern and Southern Adriatic (Grbec *et al.*, 1998). The dynamics along the transect are also controlled by the topographic effect of the Palagruža Sill (Zore-Armanda & Bone, 1987).

The present-day flora and fauna of the Adriatic Sea result from the numerous geological, geographical, climatic and biological influences occurring during its formation (Jardas, 1996). The influence of geographical, geomorphological, climatic and other environmental factors is crucial even today; indeed the peculiarities of the Adriatic ichthyofauna actually depend on these factors. Although the Adriatic Sea is part of the Mediterranean, it is an independent biogeographical and ecological subunit, which is manifest in the composition and properties of its life communities (biocoenoses).

The first research into the Adriatic ichthyological fauna occurred in the 18th century and was reported in the book "*Ichthyologia massiliensis*" (Brünnich, 1768; cited in: Jardas, 1985). Lists of the Adriatic fish species were made by a number of authors during the 19th and 20th centuries (*e.g.* Naccari, Nardo, Martens, Lanza, Carrara, Plučar, Stossich, Kolombatović, Kosić, Bini, Tortonese, Šoljan, and Jardas; reviewed in Jardas, 1985). Some of these researchers focused on the occurrence of

H.I. Griffiths et al. (eds.), Balkan Biodiversity, 291–306.

new and rare species in the Adriatic Sea (reviewed in Jardas, 1985; Dulčić *et al.*, 1999a). A systematic study of the Adriatic ichthyofauna began in the first quarter of the 19th century, when several fish species were described for the first time, but there is still a lack of knowledge of several topics.

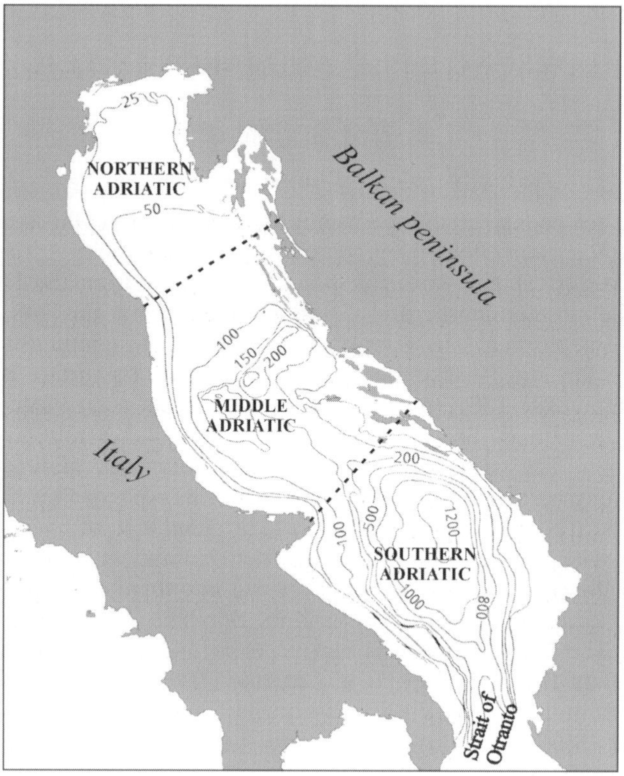

Figure 1. Map showing the major divisions of the Adriatic Sea.

The long-term monitoring of a number of abiotic and biotic factors in the Adriatic Sea, particularly in its central part, where systematic data collection has been carried out for almost thirty years, shows some significant changes in the ecosystem. During the 1973-1998 period, changes in the quantitative and qualitative composition of the Adriatic ichthyofauna were noted and related to oceanographic changes in the Adriatic Sea (Dulčić *et al.*, 1999a; Dulčić & Grbec, 2000).

The impact of climatic change may be playing an important role in long-term trends in hydrographic properties of the Adriatic Sea (Grbec, 1997). The variable impact of the Levantine Intermediate Water influences primary and secondary production (Buljan, 1953; Pucher-Petković & Zore-Armanda, 1973; Marasović *et al.*, 1996). During the study, an increase in catches of sardine (*Sardina pilchardus*), bogue (*Boops boops*), horse-mackerel (*Trachurus trachurus*), Spanish mackerel (*Scomber japonicus)* and some species from the family Thunnidae (Vučetić & Alegria-Hernandez, 1988) has been observed, accompanied by a total collapse of the European anchovy catch. Some short-lived events were also recorded, such as the sudden appearance of large quantities of gilt sardine (*Sardinella aurita*) in the central and northern parts of the Adriatic during the 1974-1976 period (Kačić, 1975; Regner, 1977). A significant correlation is found between inter-annual salinity variation in the intermediate layer in the Middle Adriatic and the small pelagic fish catch (Zore-Armanda *et al.*, 1999). The phenomenon of 'ingression' (see below) is probably the main control on the presence of some new fish species, as well as on the increase in the abundance of some relatively rare species (Jardas, 1996; Pallaoro, 1988; Dulčić *et al.*, 1999a).

The aim of this paper is to present the current state of knowledge of Adriatic fish biodiversity.

OCEANOGRAPHIC PROPERTIES OF THE ADRIATIC SEA

Even in the deepest water layers temperatures are almost always at or above 11-12°C. The summer temperature of the open sea surface is usually 22-25°C (Zore-Armanda *et al.*, 1999), although extremes vary between 3° and 29°C. This vertical temperature gradient is characteristic of 'anathermic' water bodies. The Southern Adriatic is 8-10°C warmer than its central and northern parts during the winter. In other seasons, the horizontal temperature distribution is more uniform. Generally, the open sea is warmer than the eastern coastal waters. Temperature generally decreases with depth, and the thermocline occurs at 10-30 m depth during the warmer season. The autumnal isothermal layer in the Southern Adriatic has been recorded at 18-19°C (Zore-Armanda *et al.*, 1999), and annual temperature variations are generally high. In the coastal areas, the monthly temperature maximum lags a month behind the maximal insolation, and occurs in August. The minimum generally occurs in February, two months after the insolation minimum, although in open sea areas the typical lag is one month (Zore-Armanda *et al.*, 1999; Gačić *et al.*, 2001). Interannual temperature variations are also high. Spectral analysis performed on the monthly mean surface temperature series measured at Cape Marjan in Split shows a periodicity of 2.8 years (Zore-Armanda *et al.*, 1999). An approximately 3-year period is also visible in the surface temperature data at the Split-Gargano transect (Zore-Armanda, 1969).

The Adriatic basin exhibits relatively high salinity. In the open southern part, salinity varies between 38.4 to 38.9 psu (Grbec *et al.*, 1998). Salinity is lower, and also more variable, in the northern part and in coastal zones, and the lowest salinity has been recorded close to the mouth of the Po River. Of the three clearly distinguished layers over the South Adriatic Pit, the intermediate layer has the highest salinity. Measurements in the coastal and open waters of the Middle and Southern Adriatic basins have shown a constant salinity increase for several decades (Grbec *et al.*, 1998).

Surface and bottom layer currents tend to run from the Adriatic into the Mediterranean, and Mediterranean water enters the Adriatic in the intermediate layer. In the winter, very cold dense water accumulates in the Northern Adriatic, which sinks to the deep layers of the Jabuka Pit and is then advected across the Palagruža Sill (Zore-Armanda, 1963). The transect area is also under the influence of more saline water advected from the Southern Adriatic. The most important feature of the Mediterranean waters advecting into the Adriatic in the intermediate layer is their high salinity (Buljan & Zore-Armanda, 1976). This high salinity is a property of the Levantine Basin, which has one of the highest world ocean salinities (>39 psu) (Tziperman & Mallanote-Rizzoli, 1991; Morcos, 1972).

Intensification of the inflow of the Mediterranean waters, termed 'ingression' (Buljan, 1953; Buljan & Zore-Armanda, 1976), results in a transient salinity increase in the Middle Adriatic. Since the temperature of the Levantine waters is higher than that of the Adriatic waters, ingressions affect the water temperature as well (Zore-Armanda, 1969). Different water types are observed in the Middle Adriatic in different years (Zore-Armanda, 1963), although Northern Adriatic Water (NAW) is observed only in the bottom layer, while Levantine Intermediate Water (LIW) appears in the intermediate layer of the Palagruža Sill.

In years with a relatively high air pressure gradient between the northern and southern parts of the Adriatic, higher intermediate water salinity is established. In years with a lower air pressure gradient, advection from the Mediterranean is weaker, and LIW may not be present at all. This suggests that atmosphere-ocean system feedback processes play an essential role in the thermohaline processes, and fluctuations in the atmosphere are closely related to thermohaline fluctuations in the sea (Grbec *et al.*, 1998). Atmospheric influence also appears to affect the ingressions in the Middle Adriatic, where higher salinity in the intermediate layer indicates a stronger advection of the Mediterranean water (Zore-Armanda *et al.*, 1999). The horizontal pressure gradient over the eastern Mediterranean turns out to be the most important factor enhancing water exchange between the two basins (Zore-Armanda, 1963, 1969). This pressure difference seems to be controlled by the location of the Icelandic cyclone and the Siberian anticyclone centres (Zore-Armanda, 1969, 1972), and such changes can be related to the conditions of a wide area of the North Atlantic and Europe (Zore-Armanda, 1974). The Adriatic shows considerable interannual variations in the oceanographic parameters, as evident by the amount of LIW present. Changes in the air pressure field over the North Atlantic, Europe and the Mediterranean dictate the activity of small moving cyclones over the Adriatic, and influence the amount of the bottom water formed in the region. This leads to variable exchange rates between the Adriatic and the Eastern Mediterranean - the variable air pressure gradient which, in turn, acts directly on the intensity of water exchange between the Adriatic and the Ionian Sea. The advection mechanism appears to be related to the distribution of large-scale low and high-pressure centres. It seems that a more north-eastern cyclone location increases the Ronal pressure gradient, which in turn increases water advection from the Mediterranean into the Adriatic (Grbec *et al.*, 1998). The variability of local atmospheric parameters is related to larger scale atmospheric variability (Zore-Armanda, 1972), and probably to the North Atlantic Oscillation Index (Rogers, 1990). Salinity change in the intermediate layer is highly correlated with the horizontal pressure gradient between the northern and southern parts of the Adriatic, thus the salinity change in the intermediate layer is a consequence of the wind induced advection.

The currents observed are generally low speed and have variable directions, and the current field shows a simple cyclonic circulation regime. Generally, streamlines follow the isobath lines along either coast, but wavelike patterns or meanders appear in response to the topography of the Middle Adriatic (Zore-Armanda *et al.*, 1999). Observations over the last decade show that the characteristics of the deep thermohaline circulation in the Eastern Mediterranean Sea have changed substantially. The source of the most dense waters is now the Aegean Sea rather than the Adriatic. This new source has proved to be very efficient since the estimated formation rate for the period 1989-95 was more than 1 Sv, about three times more than the dense water formation rate of the Adriatic Sea (Lascaratos *et al.*, 1999).

In addition, human impact on at least some of the oceanographic properties has become clear over the past three decades, and increasingly urgent problems of sea pollution have become apparent, which might influence toxic phytoplankton blooms as well as biodegradation of the environment and its biodiversity.

THE ADRIATIC ICHTHYOFAUNA

The Adriatic Sea is comparatively well studied, with a centennial tradition of biological research. However, new taxa of marine flora and fauna are still being recorded each year, including undescribed species. Jardas (1996) listed 407 fish species and subspecies in the Adriatic ichthyofauna, representing 117 families. There are 353 species and subspecies of Osteichthyes, 79% of the known species and subspecies of the Mediterranean (in total about 575 species and subspecies) (Jardas, 1996), and 54 are cartilaginous fishes. A Mediterranean fish biodiversity survey (Quignard & Tomasini, 2000) lists a total of 402 fish species for the Adriatic (347 Osteichthyes, 52 Chondrichthyes and 3 Agnatha). A comparison with ichthyofaunas of other Mediterranean regions reveals that the Adriatic is quite rich in terms of fish diversity, as only western areas close to the Gibraltar, such as the Catalonian Sea Strait, boast more fish species (Table 1).

Table 1. Diversity of fish (sensu lato) in the Adriatic Sea in comparison with other seas and regions. Data for the Adriatic Sea were obtained by new species added to the survey by Jardas (1996), whereas other data were obtained from Quignard & Tomasini (2000).

	Chondrichthyes	Osteichthyes	Agnatha	Total
Adriatic Sea	52	374	3	429
Gibraltar	88	570	3	661
Catalonia	64	395	2	463
Gulf of Lyon	61	289	2	352
North Africa	71	349	2	422
Aegean Sea	41	310	0	351
Levant	56	297	0	350
Black Sea	5	138	0	143
Red Sea	68	1180	0	1248
Mediterranean	86	575	3	664

Some authors suggest that endemism in the Mediterranean has decreased in the last few years, since species that had previously been found only locally, generally in very restricted areas, have recently been found in other areas throughout the Mediterranean and have thus lost their endemic status. According to our survey, only five species in the Adriatic are endemic, *Acipenser nacarii, Syngnathus taenionotus, Pomatoschistus canestrinii, Speleogobius trigloides* and the recently described *Gobius kolombatovici* (Kovačić & Miller, 2000). During the last decade, several papers have been published on the occurrence of new fish species in the Adriatic Sea, bringing up a number of the Adriatic fish species to 429, belonging to 118 families. Fish species not recorded in the checklist made by Jardas (1996) are listed in Table 2. The best represented families are now Gobiidae (50 species),

Sparidae and Labridae both with 18 species and Blenniidae (16 species), but almost 74% of the Adriatic fish families are represented by three species or fewer.

Table 2. New fish species in the Adriatic Sea, not mentioned in the list of the Adriatic ichthyofauna by Jardas (1996). Abbreviations: CS – changed taxonomic status, FR – first record, FCR – first confirmed record, NS – new species for science, GT – Gulf of Trieste (the northernmost part of the Northern Adriatic), NA – Northern Adriatic, MA – Middle Adriatic, SA – Southern Adriatic. Lessepsian migrants so far recorded in the Adriatic are denoted with asterisks.

Species	Family	Data	Area	Source
Apletodon incognitus	Gobiesocidae	FR	NA	Hofrichter & Patzner (2000)
Coelorinchus occa	Macrouridae	FR	SA	Ungaro *et al.* (2001)
Coelorinchus sp. n.	Macrouridae	NS	SA	Ungaro (pers. comm.)
Cataetyx alleni	Bythitidae	FR	SA	Ungaro *et al.* (2001)
Diaphus metopoclampus	Myctophidae	FR	SA	Fabiano & Fabiano (1977)
Didogobius splechtnai	Gobiidae	FR	NA	Herler & Patzner (2002)
*Epinephelus coioides**	Serranidae	FR	GT	Parenti & Bressi (20001)
Epinephelus aeneus	Serranidae	FR	SA	Glamuzina *et al.* (2000)
Gammogobius steinitzi	Gobiidae	FR	NA	Kovačić (1999)
Gobius couchi	Gobiidae	FR	NA	Kovačić (2001)
Gobius kolombatovici	Gobiidae	NS	NA	Kovačić & Miller (2000)
*Hemiramphus far**	Hemiramphidae	FR	SA	Collette & Parin (1986)
Lebetus guilleti	Gobiidae	FR	NA	Herler & Kovačić (2002)
*Leiognathus klunzingeri**	Leiognathidae	FR	SA	Dulčić & Pallaoro (2002)
Lepidion lepidion	Moridae	FR	SA	Ungaro *et al.* (2001)
Mycteroperca rubra	Serranidae	FR	SA	Glamuzina *et al.* (2002)
*Parexocetus mento**	Exocoetidae	FR	SA	Parin (1986)
Plectorhinchus mediterraneus	Haemulidae	FR	GT	Lipej *et al.* (1996)
Pomatoschistus norvegicus	Gobiidae	FR	NA	Stefanni (2000)
*Saurida undosquamis**	Synodontidae	FR	SA	Rakaj (1995)
*Sphyraena chrysotaenia**	Sphyraenidae	FR	SA	Pallaoro & Dulčić (2001)
Tylosurus acus imperialis	Belonidae	FR	SA	Bello (1995)
Valenciennelus tripunctulatus	Sternoptychidae	FCR	SA	Dulčić (2001)
Vanneaugobius dollfusi	Gobiidae	CS	MA	Pallaoro & Kovačić (2000)

Lack of knowledge

The list of 429 species and subspecies recorded in the Adriatic so far cannot be regarded as complete (Jardas, 1996). It is difficult to determine whether some fish species caught in the Adriatic in fact live there or only occasionally sojourn (*sensu* Jardas, 1996) (for example *Regalecus glasne* and *Lophotus lacepedei*), and there are several cases where a species is represented by a single find. Secondly, the greatest part of the Southern Adriatic basin, whose depth reaches as far as 1,330 m, had still not been sufficiently explored (especially at depths of >500 m) and knowledge of the deep demersal ichthyofauna is poor. Ungaro *et al.* (2001) list three species of deep demersal fish, *Lepidion lepidion, Coelorinchus occa* and *Cataetyx alleni*, and Dulčić (2001) reported another bathyal species *Valenciennelus tripunctulatus*, all recorded in the Adriatic Sea for the first time.

There are also some unsolved systematic (taxonomic) and status issues regarding certain fish species. Recently, Pallaoro & Kovačić (2000), found two specimens of a small gobiid *Vanneaugobius dollfusi* among ichthyological material of the Institute of Oceanography and Fisheries in Split, which were previously determined as *V. pruvoti*. Recently, a new species of a clingfish, *Apletodon incognitus*, was described by Hofrichter & Patzner (1997). Previously, this cryptic species has been considered as part of *Apletodon dentatus* or *Diplecogaster bimaculata*.

New techniques

The most important reason for the increase in perceived biodiversity of the Adriatic ichthyofauna is possibly the discovery of a large number of species outside their usual area of distribution. This may be due to an increase in traditional prospection activities, or to the use of more modern techniques (diving, underwater filming, use of narcotics, *etc.*) which allow the exploration of otherwise inaccessible habitats (Quignard & Tomasini, 2000). *In situ* observations allow one to acquire information on species that are hardly detectable with other sampling methods (Vacchi *et al.*, 1999). By such methods Miller (1992) recorded a new species of gobiid *Didogobius schlieweni* (Figure 2), and Kovačić & Miller (2000) recently discovered a new goby from the Kvarner Archipelago (Northern Adriatic). Most new records made using these newer techniques are of so-called *cryptobenthic* fishes, those that always live inside burrows (such as caves, cavities, holes, clefts) or below cover (stones, boulders, shells) and are therefore not visible from above (Miller 1979, 1996; Patzner, 1999). Recently adopted techniques, such as visual census, non-destructive diving (Harmelin, 1987; Harmelin-Vivien & Francour, 1992), and the use of narcotisers, have enabled some apparently "rare" benthic fish to be recorded in the Adriatic. However, few Adriatic institutes are currently using such techniques and the majority of ichthyologists are only concerned with economically important fish species.

Figure 2. Didogobius schlieweni *is a poorly known and rare gobiid species, which was discovered for the very first time in the northern Adriatic by Miller (1992) (Photo: M. Richter).*

Increasing research effort

The description of new species of European fishes is considered to be a rather rare event. According to Costello *et al.* (1996), 95% of all marine fish species occurring in Britain and Ireland were known by 1900. New techniques led to the discovery of some species new to science, and in the last thirty years a number of new gobiid species have been described for the very first time in the Adriatic Sea: *Pomatoschistus bathi* Miller, 1982, *Speleogobius trigloides* Zander & Jelinek, 1976, *Didogobius schlieweni*, Miller, 1992 (Figure 2); *Gobius kolombatovici* Kovačić & Miller, 2000. There is also some evidence of new finds of bigger fishes associated with use of new methodologies. The Sandbar shark *Carcharhinus plumbeus* was believed to be very rare (Nardo, 1827, cited in Brusina, 1888), but now should be considered a common species due to some new records in the northernmost area and better surveys of fish markets in different parts of the Adriatic. This shark has also often been misidentified, which seem to be a common problem with other shark species as well (*sensu* Lipej *et al.*, 2000a).

Temperature-related factors affecting fish diversity

Fish biodiversity in the Adriatic Sea increases from north to south, and there are a range of factors which may affect this pattern, but the main factor seems to be temperature. Several boreal species are found in the Northern Adriatic (*e.g. Sardina pilchardus, Sprattus sprattus phalericus, Merlangius merlangius euxinus, Platichthys flesus*) (Jardas, 1996), whilst the Middle Adriatic Jabuka Pit supports several mesopelagic and bathypelagic fishes (for example *Maurolicus muelleri*), and also thermophilous elements (*Sparisoma cretense, Xyrichthys novacula, Coryphaena hippurus, Thalassoma pavo, etc.*). The Southern Adriatic is characterised by bathyphilic species and numerous thermophilic elements, occurring only in this part of the Adriatic Sea (Jardas, 1996).

In the last few decades, several studies have confirmed the spreading of some warm water species towards the north. Changes in species distribution related to temperature fluctuation have been observed in different parts of the Mediterranean, the Ligurian Sea (Vacchi *et al.*, 1999), and in the Adriatic (Dulčić *et al.*, 1999a). In certain seasons, such as in summer 1992, the temperature of surface waters along the Eastern Adriatic coast was extremely high. At the same time, lower layers in the open sea were cooler than the long-term average and it was assumed that stable conditions reducing vertical mixing had become established due to the lack of wind (Zore-Armanda & Grbec, 1993). The extended dry period of 1988-93 and the exceptionally cold winters of 1987 and 1992-93 affected the entire Eastern Mediterranean, and changes in the circulation patterns in the LIW, themselves possibly linked to meteorological anomalies, appear to have played an important role in the redistribution of salinity (Lascaratos *et al.*, 1999). Based on the long-term mean for the 1961-1990 period (17.16°C), the 1985-1987 sea surface temperatures (SST) anomalies for Split, Middle Adriatic were *ca.* +0.15°C, and the 1990-1995 SST anomalies +0.30°C (Dulčić *et al.*, 1999a). These two periods also produced many records of rare or new fish species (Dulčić *et al.*, 1999a). Long-term measurements in the coastal and open waters of the Middle and Southern Adriatic show increases in salinity up to 1993-94. Interannual fluctuations of water exchange between the Northern and Southern Adriatic have influenced the long-term fluctuation of a variety of parameters like salinity, temperature, transparency and nutrient salts, the latter having a significant influence on fluctuations in measured chlorophyll a concentration (Zore-Armanda *et al.*, 1999). Human activity has also altered at least some of the oceanographic properties over the past three decades. Mediterranean water flowing into the Adriatic is relatively nutrient-rich, affecting primary and secondary production, and is also warmer. Numbers of thermophilous fish species have increased in the Adriatic over the last 25 years, and several species, formerly considered rare, are increasing in abundance.

Four categories of biological response to climatic change can be distinguished (Cushing & Dickson, 1976): the appearance of indicator species, the appearance of new populations, the increase or decrease in fish stocks based on year–class strength, and structural changes in the ecosystem. Dulčić & Grbec (2000) report that the first two responses are seen in the first occurrence of 15 subtropical and tropical species (*e.g. Ruvettus pretiosus*; Bettoso & Dulčić, 1999), and an increase in frequency of occurrence of other southern species (*Sardinella aurita* and *Balistes carolinensis*). A striking example of a new occurrence is *Plectorhinchus mediterraneus*, found in the Gulf of Trieste and Piran Bay in the Northern Adriatic (northernmost record – no haemulids have been recorded from the Adriatic before) (Lipej *et al.*, 1996; Dulčić & Lipej, 1997).

The northward spreading of southern species occurs in several phases (Francour *et al.*, 1994). In the first stage only single specimens appear, then more adult individuals are subsequently observed. In the second phase, larval stages occur in the area, and finally the southern species achieves the status of a new settler. Recently, Dulčić & Pallaoro (2000) reported a northernmost record of the cleaver wrasse (*Xyrichthys novacula*) and the northward spreading of the Mediterranean parrotfish (*Sparisoma cretense*) in the Adriatic. Guidetti & Boero (2001) observed several parrotfish juveniles close to the Strait of Otranto, but did not regard this as establishment of the second phase of Francour *et al.* (1994). The presence of juveniles in the area was attributed to eggs and larvae being carried by currents from other spawning areas, rather than local spawning.

In 1994, the Eastern Adriatic produced frequent first records of some larvae and juveniles of some thermophilic species, *e.g.* larva of the mesopelagic species *Trachipterus trachypterus* (Dulčić, 1996), juvenile of the grey triggerfish *Balistes carolinensis* (Dulčić *et al.*, 1997a), and juvenile of *Trachinotus ovatus* (Dulčić *et al.*, 1997b). The authors linked these occurrences with specific, unusual climatological and oceanographical conditions in 1994 and the input of intermediate waters (50-100 m) to the Middle Adriatic, which increased both salinity and temperature (Marasović *et al.*, 1996). Larval and juvenile stages of several other fish species were also recorded for the first time in the Middle and Southern Adriatic: larva of *Schedophilus medusophagus* (Dulčić, 1998), larvae of *Brama brama* and *Coryphaena hippurus* (Dulčić, 1999), larva and juvenile of *Luvarus imperialis* (Dulčić *et al.*, 1999b), and juveniles of *Balistes carolinensis* (Dulčić *et al.*, 1997a). All these records could be interpreted as the second phase of the spreading theory of Francour *et al.* (1994) in the Eastern Adriatic.

Other categories of biological response (Cushing & Dickson, 1976) are seen in the changes in migration of *Sprattus sprattus phalericus* and the drastic collapse of the European anchovy stock since 1985 (Dulčić & Cetinić, 1993). Mass mortalities of *Sardinella aurita* were recorded along the eastern Apulian and central Croatian coasts in January 2002, when an abrupt fall in seawater temperature occurred. This observation and other available information suggest that thermal shock is the most likely mechanism to explain this phenomenon (Guidetti *et al.*, 2002).

The evidence outlined above demonstrates coherent changes in the Adriatic ichthyofauna that may be related to long-term temperature change. Unusual occurrences in marine life may be used as indicators of the changing ocean conditions (Mearns, 1988), and Stephens *et al.* (1988) argue that changes in fish assemblages can reflect changes in oceanographic conditions, and may be the first indication of an environmental shift. According to these authors, temperature is the most important large-scale variable that could affect fish populations.

BIOLOGICAL INVASIONS

Species introductions into the Adriatic Sea have not been studied systematically as yet, although existing data imply the spreading of some algal species towards the north. Most data concern the spread of the allochthonous tropical green alga *Caulerpa taxifolia* in different parts of the Adriatic. According to some authors, the settling of this accidentally introduced species in the new environment caused a progressive decline in biodiversity, but at present there are no studies regarding the impact on fish diversity.

De Min & Vio (1997) reported the occurrence of at least 12 alien mollusc species in the Northern Adriatic, and believed that seven of these should be considered new settlers in the Gulf of Trieste.

After the construction of the waterway between the Eastern Mediterranean and the Gulf of Suez – the Suez Canal – in 1869, hundreds of Erythrean species traversed the channel and settled in the Mediterranean. This process is called Lessepsian migration, after Ferdinand Marie de Lesseps, the French diplomat and engineer who built the canal. Lessepsian migration is considered to be an important factor in the increase of Mediterranean fish diversity. According to Golani (1998), at least 54 Lessepsian fish (elements of the recent Indo-Pacific ichthyofauna of the Lessepsian biogeographic zone) species were recorded from the Eastern Mediterranean. Six Lessepsian species were recently recorded in the Adriatic Sea (Table 2): *Epinephelus coioides, Sphyraena chrysotaenia, Hemirhampus far, Leiognathus klunzingeri, Parexocoetus mento* and *Saurida undosquamis.* The occurrence of the orange-spotted grouper (*Epinephelus coioides*) in the Gulf of Trieste (Parenti & Bressi, 2001) is particularly interesting, since this Lessepsian migrant had been previously recorded only from the coast of Israel and was considered a rare and recent invader (Golani, 1998). The other species were amongst the first Erythrean invaders of the Eastern Mediterranean more than thirty years ago, and were then recorded as common or very common in the Aegean Sea and off the Anatolian coast (Ben-Tuvia, 1966).

Temperature again is the most important abiotic factor in determining the dispersal of Lessepsian fish (Ben-Tuvia & Golani, 1995; Pallaoro & Dulčić, 2001). Although the impact of the Lessepsian migrants on the Adriatic environment is not known, Golani (1993) shows that the impact of Lessepsian migration on the Levant basin ecosystem has been immense. However, some authors reported that the diet of Lessepsian predators, such as the brushtooth lizardfish *Saurida undosquamis,* consisted mainly of other Lessepsian fish species and Lessepsian crustaceans (Golani, 1993).

OTHER PHENOMENA

The coastal Eastern Adriatic has been heavily exploited. Changes in fish and edible invertebrate communities, manifested as changes in quantitative and qualitative catch structure, could occur in response to a range of causes, although intensive fishing is the most likely (Jardas *et al.*, 1998). Records from the last 30 years show that the quantity of fish and edible invertebrates in the experimental trammel bottom set catches from the coastal Eastern Adriatic have steadily declined. The same catch decline trends have been recorded in coastal areas of the Middle Adriatic, where the catch per unit effort has been reduced by 63-69%, and in the Southern Adriatic where the catch per unit effort fell by 24% between 1972-1987 (Jardas *et al.*, 1998). These trends are primarily due to intensive fishing activities, which reduce population density, and affect interspecific and intraspecific dynamics, as the vulnerability to fishing activity varies with species.

One of the most unusual events in recent years in the Adriatic Sea is the occurrence of the Basking shark (*Cetorhinus maximus*). This endangered shark species was considered rare (Brusina, 1888; Jardas, 1996; Barrull & Mate, 1999), but in the last decade the number of records increased markedly. Barrull & Mate (1999) survey all historical and recent records of this shark in the entire Mediterranean in the period from 1795 to 1998, and find that on average only three records per year were found in the whole region. In the last decade (and especially in the 1999-2001 period) more than 34 records have been reported from the Adriatic Sea alone (Lipej *et al.*, 2000b; Zuffa *et al.*, 2001) (Figure 3), mostly in the Northern Adriatic and the northernmost area of the Gulf of Trieste (Figure 4). Zuffa *et al.* (2001) present three possible reasons for the 'abnormal occurrence' of the Basking Shark. It could be related to climate change, to changes in zooplankton abundance, or to some unknown aspect of shark metabolism and/or behaviour. Longer term records may shed light on this phenomenon.

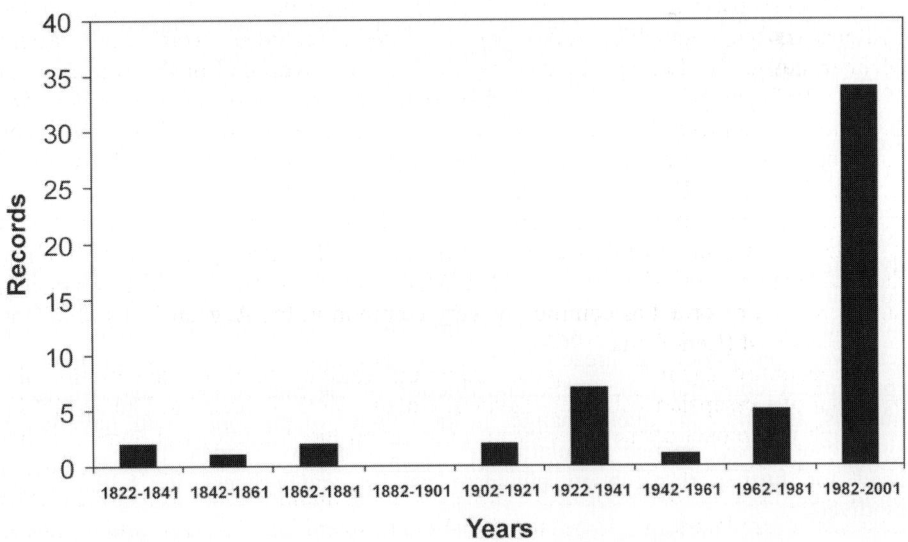

*Figure 3. Occurrence of the Basking Shark (*Cetorhinus maximus*) in the Adriatic Sea in the period from 1822-2001. Note the increase in number of records during the last decade.*

*Figure 4. The Basking shark (*Cetorhinus maximus*) grazing in Slovenian coastal waters (northern Adriatic) in May 2001. (Photo: C. Mlinar).*

CONCLUSION

Although the Adriatic Sea is considered to be a relatively well studied part of the Mediterranean Sea, this review has underlined the significant number of new species of marine flora and fauna which are still being recorded. Particular emphasis has been given to the rising number of new records of fish in the last thirty years in the Adriatic, even including some undescribed species. This has been attributed in part to increased prospection activity, coupled with changes in techniques which allow access to previously inaccessible habitats, and observation *in vivo*. It is also clear, however, that these changes are also attributable to real changes in population dynamics during this period. Most importantly, the increase in number of species correlates clearly with interannual shifts in climatic and oceanographic processes, as do events such as episodes of mass mortality. The impact of other potential factors is less well understood at present, but it is also likely that biological invasion (particularly consequent upon the construction of the Suez Canal), overfishing by humans, and the consequent changes in the balance of the food chain, have had a significant impact on fish populations over the last few decades. There is a clear need for further research to elucidate pattern and process in Adriatic fish biodiversity.

REFERENCES

Barrull, J & Mate, I. 1999. Registros de Tiburon peregrino (*Cetorhinus maximus*) en aguas del Mediterraneo. *Bolletin de la Asociación Española de Elasmobranquios* 2, 37-52.

Bello, G. 1995. *Tylosurus acus imperialis* (Osteichthyes: Belonidae), a fish new to the Adriatic Sea. *Cahiers de Biologie Marine* 36, 197-199.

Ben-Tuvia, A. 1966. Red Sea fishes recently found in the Mediterranean. *Copeia* 2, 254-275.

Ben-Tuvia, A. & Golani, D. 1995. Temperature as the main factor influencing the Lessepsian migration. *Proceedings of the International Colloqium Maison Environment, Montpelier. Montpelier,* 6-7 April 1995, pp. 159-162.

Bettoso, N. & Dulčič, J. 1999. First record of the oilfish *Ruvettus pretiosus* in the Northern Adriatic. *Journal of the Marine Biological Association of the United Kingdom* 79, 1145-1146.

Brusina, S. 1888. [Sharks of the Adriatic Sea and the Red Sea.] *Glasnik Hrvatskog Prirodoslovnog družtva* 3, 67-236. (In Croatian with English summary)

Buljan, M., 1953. Fluctuations of salinity in the Adriatic. Izvješća-Reports. The M.V. «Hvar» cruises-researches into fisheries biology 1948-49. *Izvješća-Reports* 2(3), 64 pp.

Buljan, M. & Zore-Armanda, M. 1976. Oceanographical properties of the Adriatic Sea. *Oceanography. Marine Biology Annual Review* 14, 11-98.

Collette, B.B. & Parin, N.V. 1986. Hemiramphidae. In *Fishes of the North-western Atlantic and the Mediterranean,* Vol. 2 (P.J.P. Whitehead, M.-L.-Bauchot, J. Nielsen, & E. Tortonese, eds.), pp. 620-622, UNESCO, Paris.

Costello, M. J., Emblow, C.S. & Picton, B.E. 1996. Long-term trends in the discovery of marine species new to science which occur in Britain and Ireland. *Journal of Marine Biological Association of the United Kingdom* 81, 717-718.

Cushing, D.H. & Dickson, R.R. 1976. The biological response in the sea to climatic changes. *Advances in Marine Biology* 14, 1-122.

Cushman-Roisin, B., Gačić, M., Poulain, P.-M. & Artegiani, A. 2001. *Physical oceanography of the Adriatic Sea*, Kluwer Academic Publishers, Dordrecht.

De Min, R. & A. Vio, A.. 1997. Molluschi esotici nell'Alto Adriatico. *Annals for Istrian and Mediterranean Studies* 5, 243-54.

Dulčić, J. 1996. First record of ribbon fish larva, *Trachipterus trachypterus,* from the eastern Adriatic. *Cybium* 20, 101-102.

Dulčić, J. 1998. First record of the cornich blackfish *Schedophilus medusophagus* (Pisces: Centrolophidae) larvae from the Adriatic Sea. *Journal of the Marine Biological Association of the United Kingdom* 78, 1935-1038.

Dulčić, J. 1999. First record of larval *Brama brama* (Pisces: Bramidae) and *Coryphaena hippurus* (Pisces: Coryphaenidae) in the Adriatic Sea. *Journal of Plankton Research* 6, 1171-1174.

Dulčić, J. 2001. On the occurrence of three meso- and bathypelagic fishes in the Adriatic Sea. *Cybium* 25, 285-288.

Dulčić, J. & Cetinić, P. 1993. Estimate of spawning biomass of anchovy *Engraulis encrasicolus* in the eastern part of Adriatic from 1989 to 1990 by means of egg surveys. *Acta Ichthyologica et Piscatoria* 2, 67-76.

Dulčić, J. & Grbec, B. 2000. Climate change and Adriatic ichthyofauna. *Fisheries Oceanography* 9, 187-191.

Dulčić, J. & Lipej, L. 1997. New records of marine fishes from the Slovenian coastal waters. *Falco* 12, 35-40.

Dulčić, J. & Pallaoro, A. 2000. Some new data on *Xyrichthys novacula* (Linnaeus, 1758) and *Sparisoma (Euscarus) cretense* (Linnaeus, 1758) from the Eastern Adriatic. *Annals for Istrian and Mediterranean Studies* 11, 35-40.

Dulčić, J. & Pallaoro, A. 2002. First record of the Lessepsian migrant *Leiognathus klunzingeri* (Pisces: Leiognathidae) from the Adriatic Sea. *Journal of the Marine Biological Association of the United Kingdom* 82, 523-524.

Dulčić, J., Kraljević, M. , Kršinić, F. & Pallaoro, A. 1997a. Occurrence of fingerlings of grey triggerfish *Balistes carolinensis* Gmelin, 1789 (Pisces: Balistidae) in the eastern Adriatic. Annales. *Annals for Istrian and Mediterranean Studies* 4, 271-276.

Dulčić, J., Pallaoro, A. & Kraljević, M. 1997b. First record of pompano fingerling *Trachinotus ovatus* (Linnaeus, 1758) (Pisces: Carangidae) in the eastern middle Adriatic. *Natura Croatica* 1, 61-65.

Dulčić, J., Grbec, B. & Lipej, L. 1999a. Information on the Adriatic ichthyofauna – effect of water warming. *Acta Adriatica* 40, 33-43.

Dulčić, J., Jardas, I. & Kačić; I. 1999b. New records of the early history stages of louvar *Luvarus imperalis* Rafinesque, 1810, from the eastern Adriatic. *Annals for Istrian and Mediterranean Studies* 9, 3-9.

Fabiano, M. & Fabiano, A.M. 1977. Pesci batifili raccolti nel corso delle campagne di pesca N/R »S. LoBianco«. Quade*rni del Laboratorio per la Tecnologia della Pesca* 3, 165-174.

Francour, P., Boudoresque, C.F., Harmelin, J.G., Harmelin-Vivien, M.L. & Quignard, J.P. 1994. Are the Mediterranean Waters Becoming Warmer? Information from Biological Indicators. *Marine Pollution Bulletin*, 28 523-526.

Gačić, M., Poulain, P.-M., Zore-Armanda, M. & Barale, V. 2001. Overview. In *Physical oceanography of the Adriatic Sea* (B. Cushman-Roisin *et al.*, eds.), pp. 1-44, Kluwer Academic Publishers, Dordrecht.

Glamuzina, B., Tutman P., Geffen A.J., Kožul V. & Škaramuca, B. 2000. First record of white grouper, *Epinephelus aeneus* (Serranidae) in the south eastern Adriatic. *Cybium* 24, 306-308.

Glamuzina, B., Tutman P., Kožul V. & Škaramuca, B. 2002. The first recorded occurrence of the Mottled Grouper, *Mycteroperca rubra* (Serranidae), in the southeastern Adriatic Sea. *Cybium* 26, 156-158.

Golani, D. 1993. Trophic adaptation of Red Sea fishes to the eastern Mediterranean environment – review and new data. *Israeli Journal of Zoology* 39, 391-402.

Golani, D. 1998. Distribution of Lessepsian migrant fish in the Mediterranean. *Italian Journal of Zoology* 65 (Suppl.), 95-98.

Grbec, B. 1997. Influence of climatic changes on oceanographic properties of the Adriatic sea. *Acta Adriatica* 38, 3-29.

Grbec, B., Morović, M., Zore-Armanda, M. 1998. Some new observations on the long-term salinity changes in the Adriatic Sea. *Acta Adriatica* 39, 3-12.

Guidetti, P.& Boero, F. 2001. Occurrence of the Mediterraean parrotfish *Sparisoma cretense* (Perciformes: Scaridae) in south-eastern Apulia (south-east Italy). *Journal of Marine Biological Association of the United Kingdom* 81, 717-718.

Guidetti, P., Boero, F. & Dulčić, J. 2002. Mass mortality of *Sardinella aurita* in the Adriatic and Ionian Sea. *Cybium* 26, 317-319.

Harmelin J.G., 1987. Structure et variabilité de l'ichtyofaune d'une zone rocheuse protégée en la Méditerranée (Parc national de Port-Cros, France). *Marine Ecology* (*Pubblicazioni della Stazione Zoologica di Napoli*) 8, 263-284.

Harmelin-Vivien, M. & Francour, P. 1992. Trawling or visual censuses? Methodological bias in the assessment of fish populations in seagrass beds. *Marine Ecology* (*Pubblicazioni della Stazione Zoologica di Napoli*) 13, 41-51.

Herler, J. & Kovačić, J. 2002. *Lebetus guilleti* (Teleostei: Gobiidae) in the northern Adriatic sea: first record and details on the species' morphology. *Annals for Istrian and Mediterranean Studies* 12, 177-188.

Herler J. & Patzner R. 2002. New records of *Didogobius splechtnai* Ahnelt & Patzner, 1995 (Pisces: Gobiidae) from the Tyrrhenian and northern Adriatic Sea. *Cybium* 26, 153-155.

Hofrichter, R. & Patzner. R. 1997. A new species of *Apletodon* from the Mediterranean Sea and eastern Atlantic with notes on the differentiation between *Apletodon* and *Diplecogaster* species. *Senckenbergiana Biologica* 77, 15-22.

Hofrichter, R.& Patzner, R.A. 2000. Habitat and Microhabitat of Mediterranean Clingfishes (Teleostei: Gobiesociformes: Gobiesocidae). *P.S.N.Z.I. Marine Ecology* 21, 41-53.

Jardas, I. 1985. [Check-list of the fishes (sensu lato) of the Adriatic Sea (Cyclostomata, Selachii, Osteichthyes) with respect of taxonomy and established number.] *Biosistematika* 11, 45-75 (In Croatian with English summary).

Jardas, I. 1996. [*The Adriatic ichthyofauna.*] Zagreb: Školska knjiga (in Croatian with English summary).

Jardas, I., Pallaoro, A. , Kraljević, M., Dulčić, J. & Cetinić, P. 1998. Long-term changes in biodiversity of the coastal area of the Eastern Adriatic: fish, crustacean and cephalopoda communities. *Periodicum Biologorum* 100, 19-28.

Kačić, I. 1975. [Observations on gilt sardine *Sardinella aurita* in the Adriatic.] *Morsko ribarstvo* 27, 11-13. (in Croatian).

Kovačić, M. 1999. *Gammogobius steinitzi* Bath, 1971, a fish new to the Adriatic Sea. *Natura Croatica* 8, 1-7.

Kovačić, M. 2001. The Kvarner population of *Gobius couchi* (Teleostei, Gobiidae), a fish new to the Adriatic Fauna. *Natura Croatica* 10, 1-10.

Kovačić, M. & Miller, P. J. 2000. A new species of *Gobius* (Teleostei: Gobiidae) from the northern Adriatic Sea. *Cybium* 24, 231-239.

Lascaratos, A., Roether, W., Nittis, K. & Klein, B. 1999. Recent changes in deep water formation and spreading in the Eastern Mediterranean Sea. *Progress in Oceanography* 44, 5-36.

Lipej, L., Spoto, M. & Dulčić, J. 1996. *Plectorchinchus mediterraneus* from off north east Italy and Slovenia – the first records of the family Haemulidae from the Adriatic Sea. *Journal of Fish Biology* 48, 805-806.

Lipej, L., Makovec, T., Soldo, A. & Žiža, V. 2000a. Records of the Sandbar Shark *Carcharhinus plumbeus* (Nardo, 1827) in the Gulf of Trieste (Northern Adriatic). *Annals for Istrian and Mediterranean Studies* 10, 199-206.

Lipej, L., Makovec, T., Orlando, M. & Žiža, V. 2000b. Occurrence of Basking Shark *Cetorhinus maximus* (Gunnerus, 1765) in the waters off Piran (Gulf of Trieste, Northern Adriatic). *Annals for Istrian and Mediterranean Studies* 10, 211-216.

Marasović, I., Grbec, B. & Morović, M. 1996. Long-term production changes in the Adriatic. *Netherlands Journal of Sea Research* 34, 267-273.

Mearns, A. J. 1988. The odd fish: unusual occurrences of marine life as indicators of changing ocean conditions In *Marine Organisms as Indicators* (D. F. Soule & G. S. Keppel, eds.), pp. 137-173, Springer, New York.

Miller, P. J., 1979. Adaptiveness and implications of small size in Teleosts. *Symposium of the Zoological Society of London* 44, 263-306.

Miller P. J., 1992: A new species of *Didogobius* (Teleostei: Gobiidae) from the Adriatic Sea. *Journal of Natural History* 26, 1413-1419.

Miller, P.J., 1996. The functional ecology of small fish: some opportunities and consequences. *Symposium of the Zoological Society of London* 69, 175-199.

Morcos, S., 1972. Sources of the Mediterranean Intermediate Water in the Levantine Sea. In *Studies in physical oceanography*, 2 (A.L. Gordon, ed.). Gordon & Breach, NewYork, pp. 185-206.

Pallaoro, A. 1988. [On the possibilities of the occurrence of certain fish species in the Middle Adriatic conected with the Adriatic ingression in years 1986/1987.] *Morsko ribarstvo* 3, 82-87. (In Croatian).

Pallaoro, A. & Dulčić, J. 2001. First record of the *Sphyraena chrysotaenia* (Klunziger, 1884) (Pisces, Sphyraenida) from the Adriatic Sea. *Journal of Fish Biology* 59, 179-182.

Pallaoro, A. & Kovačić, M. 2000. *Vanneaugobius dollfusi* Brownell, 1978 a rare fish new to the Adriatic Sea. *Journal of Fish Biology* 57, 255-257.

Parenti, P. & Bressi, N. 2001. First record of the orange-spotted grouper, *Epinephelus coioides* (Perciformes: Serranidae) in the northern Adriatic Sea. *Cybium* 25, 281-284.

Parin, N.V. 1986. Exocoetidae. In *Fishes of the North-western Atlantic and the Mediterranean*, Vol. 2 (P.J.P. Whitehead, M.-L.-Bauchot, J. Nielsen, & E. Tortonese, eds.), pp. 612-619, UNESCO, Paris.

Patzner, R.A. 1999. Habitat utilization and depth distribution of small cryptobenthic fishes (Blenniidae, Gobiesocidae, Gobiidae, Tripterygiidae) in Ibiza. *Environmental Biology of Fishes* 55, 207-214.

Pucher-Petković, T. & Zore-Armanda, M. 1973. Essai d'évaluation et prognostique de la production en function des facteurs du milieu dans l'Adriatique. *Acta Adriatica* 15, 1-39.

Quignard, J. P. & Tomasini, J.A. 2000. Mediterranean fish biodiversity. *Biologia Marina Mediterranea* 7, 1-66.

Rakaj, N. 1995. Albanian Ichthyofauna. Shtëpia Botuese "Libri Universitar", Tiranë, 700 pp. (In Albanian).

Regner, S. 1977. On the plankton stages of gilt sardine *Sardinella aurita* Valenciennes, in the Central Adriatic. *Acta Adriatica* 17, 1-12.

Rogers, J.C., 1990. The association between the North Atlantic Oscillation and the Southern Oscillation in the Northern Hemisphere. *Monthly Weather Review* 112, 1999-2015.

Stefanni, S. 2000. First record of the Norway goby in the Adriatic Sea. *Journal of Fish Biology* 57, 828-830.

Stephens, J. S., Hose, J. H. & Lowe, M. S. 1988. Fish assemblages as indicators of environmental change in nearshore environments. In *Marine Organisms as Indicators* (D. F. Soule & G. S. Keppel, eds.) , pp. 91-103, Springer, New York.

Tziperman, N. & Mallanote-Rizzoli, P. 1991. The climatological seasonal circulation of the Mediterranean Sea. *Journal of Marine Research* 49, 411-434.

Ungaro, N., Marano G. & Rivas, G. 2001. Notes on ichthyofauna of the deep basin of the Southern Adriatic Sea. *Sarsia* 86, 153-156.

Vacchi, M, Boyer, M., Bussotti, S., Guidetti, P. & La Mesa, G. 1999. Some interesting species in the coastal fish fauna of Ustica Island (Mediterranean Sea). *Cybium* 23, 323-331.

Vučetić, T. & Alegria-Hernandez, V. 1988. Trends of annual catches or stock densities of some pelagic fishes in recent "Pelagia years" in the Adriatic. In *Fifth technical consultation of stock assessment in the Adriatic* (J. F. Caddy & M. Savini, eds.). *GFCM*, Rome, pp. 133-136.

Zore-Armanda, M., 1963. Les masses d'eau de la mer Adriatique. *Acta Adriatica* 10, 5-88.

Zore-Armanda, M., 1969. Temperature relations in the Adriatic Sea. *Acta Adriatica*, 13, 1-50.
Zore-Armanda, M., 1972. Response of the Mediterranean to the oceanographic/meteorological conditions of the Northern Atlantic. *Rapport du Comité International pour l'Exploration Mer Méditerranée* 21, 203-205.
Zore-Armanda, M. 1974. Formation of the Eastern Mediterranean water in the Adriatic. *Coll. Internat. C. N. R. S.* 215, 127-133.
Zore-Armanda, M. & Bone, M. 1987. The effect of bottom topography on the current system of the open Adriatic Sea. *Bolletino di Oceanologia Teorica e Applicata* 5, 3-18.
Zore-Armanda, M. & Grbec, B. 1993. Unusual sea temperatures conditions in the Adriatic in summer 1992. *Notes of Institute of Oceanography and Fisheries, Split* 77, 1-8.
Zore-Armanda, M., Grbec, B. & Morović, M. 1999. Oceanographic properties of the Adriatic Sea-A point of view. *Acta Adriatica* 40 (Suppl.), 39-54
Zuffa, M., Soldo, A. & Storai, T. 2001. Preliminary observations on abnormal abundance of *Cetorhinus maximus* (Gunnerus, 1765) in the Central and Northern Adriatic Sea. *Annals for Istrian and Mediterranean Studies* 11, 185-192.

PART 3: SUBTERRANEAN

18. A CENSUS OF THE OBLIGATE SUBTERRANEAN FAUNA OF THE BALKAN PENINSULA

BORIS SKET*, KALOUST PARAGAMIAN & PETER TRONTELJ

INTRODUCTION

With the exception of an isolated reference to a *cave fish* in China in 1540 (Chen *et al*., 1994), the first phase of zoological discovery in caves really began in the 18th and early 19th century with the amphibian *Proteus anguinus* (mentioned by Valvasor, 1689, but described scientifically by Laurenti, 1768) and the beetle *Leptodirus hochenwartii* (Schmidt, 1832). These discoveries, and a great number of others, were made in the Slovenian region of Kras (= Carso, or Karst), the region which gave its name to the geomorphological phenomenon of karst. Kras is located in the northwestern Dinarides, or Dinaric Alps. This was followed by the earliest discoveries of cave animals in other parts of the world (in the Caucasus, the Appalachians, the Pyrenees, and New Zealand), in the middle of the 19th century. Thanks to this history (Belles, 1992; Sket, 1996), the "classical Karst" won its appellation as the "cradle of speleobiology" (biospeleology). The Slovenian karst is a part of the wider Dinaric karst region, which in turn is the main karstic region of the western Balkans. By chance, the earliest finds of interstitial aquatic animals, which represent another type of subterranean fauna, were also made in this part of the world. These latter discoveries were made during the research of S. Karaman in the 1930s in Macedonia (Former Yugoslav Republic of Macedonia, FYROM), on sites located mainly along the river Vardar. It became evident later that these early phases of discovery had both taken place in a region which is particularly rich in cave and interstitial aquatic fauna. Discovery of the subterranean fauna began much later in the southern and eastern Balkans (Greece and Bulgaria, respectively). It should be noted, however, that Bulgarian speleobiologists have since been particularly rigorous in their research and have published a series of unusually detailed faunal lists (see below).

At the end of the 18th century, Hamann (1896) published an extensive monograph with descriptions of animal species from European caves, with "particular consideration of the cave fauna of Carniola"; Carniola (Krain, or Kranjska) was a province of the Austro-Hungarian Empire which occupied the north-western part of the Dinaric karst, and now comprises the largest part of Slovenia. The next listing of the Balkan cave fauna was made in the great, although uncritical, *Catalogus* by Wolf (1938). Gueorguiev (1977) produced a modern census of the terrestrial, obligate (troglobiotic) fauna of the whole Balkan Peninsula which included an extensive biogeographical analysis. Two publications contain lists of the aquatic subterranean fauna: one is combined with the epigean aquatic fauna of Europe (Illies, 1978), and another with the stygobiotic fauna of the whole World (Botoşăneanu, 1986). Sket (1999c) employed these data in his comparative analysis of stygobiotic faunal diversity of the Dinaric (western Balkan) region; he also used them as the main basis for a taxonomic analysis of the stygobiotic fauna worldwide.

H.I. Griffiths et al. (eds.), Balkan Biodiversity, 309–322.

It became clear that there was an urgent need to produce a complete list of the Balkan subterranean fauna, for a number of reasons: for basic biogeographic studies, for application to environmental protection policy planning, and for conservation of biodiversity. Such a list would also be valuable for planning future field research. In the 1980s we therefore started to compile a comprehensive list of the animal species found in subterranean habitats of Yugoslavia. With some long periods of stagnation, this work has continued to the present day, which finds the region subdivided into a number of independent states. The list has already enabled us to compile an inventory of caves of high priority for environmental protection in Slovenia. In addition, in combination with other published work, it is now possible to generate relevant data concerning the biodiversity of subterranean habitats across the Peninsula as a whole. This paper focuses on the preparation and results of the Balkan census.

METHODS

Definition of the study region

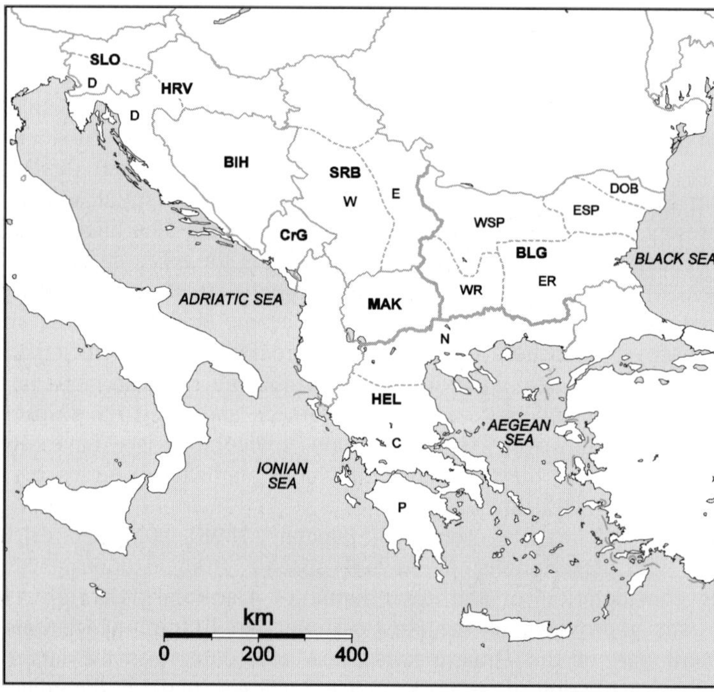

Figure 1. Map of the Balkan Peninsula, showing definition of the regions and subregions used in this study. SLO – Slovenija [Slovenia]; HRV – Hrvatska [Croatia], D – Dinaric territory; BiH – Bosna i Hercegovina [Bosnia and Hercegovina]; CrG – Crna Gora [Montenegro]; SRB – Srbija [Serbia], W and E – its western and eastern sectors, respectively; MAK – Makedonija [Macedonia, FYROM]; BLG – B'lgarija [Bulgaria], WSP and ESP – western and eastern Stara Planina, respectively, DOB – Dobrudža [Dobrugea], WR and ER – western and eastern Rodopi [Rhodopes]; HEL – Hellas [Greece], N and C – its northern and central subregion, respectively, P – Peloponnisos, Aegean Sea – Aegean islands.

The study concerns the south European Balkan Peninsula, and in most instances the study region is defined by the political borders of the main states (Figure 1). Thus, the northern border of Bulgaria excludes the Romanian part of Dobrugea, with its interesting and very distinctive fauna. The European parts of Turkey, and Albanian territory, are also excluded, since in both cases there is a serious paucity of published data. In the extreme west, in Croatia and Slovenia, the non-Dinaric regions are included and considered as separate, corresponding approximately to non-Balkan regions. More specifically, regions considered as non-Balkan comprise, firstly, the valley of the Sava River which runs through Croatia and Slovenia, with its rich interstitial fauna, together with all territory to the north of it and, secondly, in Slovenia, the whole karst of the Southern Calcareous Alps, and the so-called *isolated karst* between the Dinarides and the Alps in the west of the country. The small Italian portion of the Dinaric karst is also excluded; its fauna is mainly conspecific with the neighbouring region of Slovenia.

A clear biogeographical partitioning of the region is yet to be achieved. Since nature protection activities are mainly performed within the confines of political borders, however, there is a strong argument for analysing subterranean faunas mainly within existing administrative units. Most analyses have been carried out in three subregions: the western Balkans (countries of former Yugoslavia), Greece (including the islands), and Bulgaria.

Data collection and classification

The compilation of a faunal census for the western Balkans relied mainly on scattered published literature, together with unpublished data from our own collections. Unlike the interstitial fauna, the older primary data for caves could be regarded as superfluous due to the existence of lists in the secondary literature (Wolf, 1938). Much more recent and more high quality lists were available for the particularly rich coleopteran assemblages of Croatia and Montenegro, published by the late E. Pretner (1973, 1977) and Jalžić & Pretner (1977), together with an unpublished census for Slovenia which had been bequeathed by Pretner. For these states, therefore, only the most recent data had to be added. The taxonomic completeness of the lists was verified by reference to other compilations which lack site data: Illies (1978) and Botoşăneanu (1986) for the aquatic stygobiotic fauna, Gottstein Matočec *et al.* (2001) for Croatia, and the incomplete series '*Catalogus Faunae Jugoslaviae*' (edited by the Slovenian Academy of Sciences and Arts) for some particular taxonomic groups.

Following an initial period using card files, data were recorded using Excel. Each entry represents an individual taxon in a specific locality; an initial plan to include data on the position of the animal in the cave was later sacrificed for the sake of simplicity and versatility. A particular problem appeared to be the identification of localities since older, mostly foreign authors often published heavily corrupted topographical names or mentioned only the names of little-known caves and small villages which are difficult to locate on maps. Even the position of the locality in a state can be questionable. In ancient times, for example, the name "Dalmatien" (Dalmacija, or Dalmatia) also referred to the Hrvatsko Primorje (the Croatian Littoral), parts of Hercegovina and parts of Montenegro, in addition to Dalmacija proper. The toponym of "Istria" was similarly widely conceived.

Over 15,000 supposedly valid entries have accumulated to date in the census for the western Balkans alone (Table 1). Prior to editing, a number of additional entries from neighboring countries had been included erroneously. It had been noted during data analysis that a high degree of redundancy existed, after which 1,172 superfluous entries were identified for Slovenia. Their removal reduced the number of unique taxon-site records from 7,793 to 6,621. It is assumed that a much lower degree of redundancy occurs in the lists of the other countries.

Table 1. Number of entries (species/location) recorded within different administrative regions of the western Balkan Peninsula at the time of writing

State	
Slovenia	6,620
Croatia	3,200
Bosnia & Hercegovina	2,910
Montenegro	1,460
Serbia	720
Macedonia	555
Total number of records	15,465

The sources of published data are still far from being completely exhausted, new data are being published and some of our rich collections are still awaiting taxonomic evaluation. There has been a significant reduction in the rate of generation of new data, however, which suggests that the census is ready for preliminary analyses.

The ecological classification of taxa in the subterranean fauna can also present a problem. With the current state of knowledge, none of the current classification systems, however simple, allows one to classify the majority of species reliably. In order to minimise sources of error, it was decided to simplify the classification further, to consideration only of obligate subterranean animals (*i.e.* terrestrial troglobionts and aquatic stygobionts) *versus* non-specialised guests. For a very high proportion of taxa even this classification cannot be made with any degree of certainty, however. One has to take into account that troglomorphism is not really limited to troglobionts and stygobionts, and that not all of them are troglomorphic. On the other hand, some species are only known from caves simply because the surface fauna in the proximity of the cave has not been investigated. It is also possible that some surface animals retreat fully underground in the hot, dry summer of the Dinaric karst, where the caves are visited only rarely by explorers in wetter months. In spite of this, for the sake of simplicity and practicality, for the classification as troglobiont (or stygobiont) the combination of troglomorphism and absence of surface data for a taxon was often used. Only supposedly specialised species or troglomorphic (*i.e.* specialised) populations of troglophilous species were considered.

For Bulgaria, lists were compiled from the published data by Gueorguiev & Beron, (1962), Beron & Gueorguiev (1967), Beron (1994), and Pandourski (1998). The classification of the authors was accepted, and non-obligate species were excluded.

For Greece, the progress of speleobiological research has been summarised (Gueorguiev, 1977; Beron 1987; Casale & Giachino, 1994), but few complete lists have been published. Pesce & Maggi (1983) provide the most comprehensive list of the stygobiotic fauna, while the most recent list for troglobiotic species (Gueorguiev, 1977) is now out of date. We compiled faunal lists for Greece based mainly on scattered data from around 420 published papers and, to a much lesser extent, from our own collections. We included only those species which were either classified by the authors or exhibited specialised, troglomorphic characters.

Taxonomy

For the purposes of this study, identification was made to the level of species rather than subspecies. It was assumed that the species level had been defined in the sense of the "biological species concept", since in most groups the authors were evidently trying to follow this. An important exception might be the case of the false scorpions, Pseudoscorpiones, judging from the low number of nominal subspecies. The potential influence on the results of using a different species concept is discussed below. For some groups, specialists kindly checked taxonomic names and the ecological classification of species.

RESULTS AND DISCUSSION

Species richness and species distribution

The supposedly obligate subterranean fauna of the Balkan Peninsula (Tables 2-3), with the inclusion of the non-Balkan regions of Slovenia and Croatia, is composed of *ca.* 975 terrestrial and >650 aquatic species. Exclusion of the non-Balkan parts of Slovenia and Croatia would reduce these numbers by 60 in the case of troglobionts, and 70 in the case of stygobionts.

At present, the only possible comparison is with species richness statistics for the subterranean fauna of North America. Although the recent list by Peck (1998) is already out of date, and some groups have definitely been seriously undersampled in America, the results are nevertheless similar to those of the Balkans. The total species numbers for both troglobionts and stygobionts follow the same pattern, but are considerably higher in the Balkans – for stygobionts by as much as 50%.

Almost 80% of both troglobiotic and stygobiotic species of the Balkan Peninsula are restricted in their distribution to the western Balkans. It has been suggested previously (Sket, 1996, 1999a) that the stygobiotic fauna of the western Balkans is by far the richest in the world in comparison with other regions of sub-continental size. The second richest (Botoşăneanu, 1986) is the eastern Balkans (including Bulgaria, together with large sectors of Serbia, Macedonia [FYROM; Former Yugoslav Republic of Macedonia], and Romania), and the Pyrenean-Aquitanian Province. In the present calculation, the number of listed stygobionts was reduced from 396 taxa (species plus subspecies) to 330 species, but this still excluded a large part of Slovenia. The entire country of Slovenia, with 169 stygobiotic taxa, clearly exhibited the highest species richness within the Dinaric (or western Balkan) region, and only slightly less rich than other regions ten times its surface area in size. (At the time of writing, the totals have reached 155 species for Dinaric Slovenia and an additional 50 for its non-Dinaric parts.)

The situation is different for the terrestrial cave fauna. While the Dinaric karst region is again the richest within the Peninsula, its maximum species richness is attained in the southeast. Following administrative boundaries, the highest species number belongs to Bosnia and Herzegovina, followed by Croatia; in fact, the "hot spot" seems to comprise southeastern Hercegovina, southeastern Dalmacija (= Dalmatia, in Croatia) and the southwestern sector of Montenegro.

Approximately 200 obligate subterranean species were found in Greece, half of which occurred in the Aegean islands. A lack of stygobionts in the northern district is responsible for a remarkable difference in numbers between terrestrial and aquatic species; in other parts of Greece, the numbers are more or less equal. The situation is similar in Bulgaria, within which the northwestern district, the western Stara Planina, is host to *ca.* 60% of all known obligate subterranean fauna.

Endemism

The only territory where some species break the endemism of the three subregions (western Balkans, Greece and Bulgaria), remarkably, is eastern Serbia, which both orogenically and biogeographically has more in common with the east, and shares several species with Bulgaria. The cave oligochaete *Delaya bureschi* (Michaelsen) is distributed between Slovenia and Bulgaria. In the Greek cave fauna, only two troglobiotic species, both of which are Trichoniscidae woodlice, are distributed beyond the Greek political borders. *Trichoniscus rhodopiensis* Vandel, which is common in caves of eastern Greek Macedonia and Thraki, is also known from caves of Bulgaria (Beron, 1987), while *Trichonethes kosswigi* Strouhal is found in Rhodos and southwestern Turkey (Argano & Manicastri, 1988). Apart from these exceptions, which relate more to the artificiality of boundary definition than anything else, there are hardly any species in common between the three subregions.

*Table 2. Number of species, genera and families of obligate subterranean animals in the three regions of the Balkan Peninsula. The western Balkans (WB) comprises Slovenia, Croatia, Bosnia and Herzegovina, Montenegro, Serbia, and Macedonia (FYROM); the non-Balkan (*i.e. *non-Dinaric) territory of Slovenia and Croatia is included. Coleoptera are shown separately, partitioned into families. Ann – Annelida, Ara – Arachnida, Cru – Crustacea, Ins – Insecta, Mol – Mollusca, Myr – Myriapoda, Tur – Turbellaria.*

			Western Balkans			**Greece**			**Bulgaria**	**Balkan Peninsula**
AQUATIC FAUNA	families	genera	**species**	families	genera	**species**	families	genera	**species**	**species**
Porifera	1	1	**1**							**1**
Tur: Temnocephalida	1	5	**13**							**13**
Tur: Tricladida	3	6	**12**							**12**
Cnidaria	1	1	**1**							**1**
Mol: Gastropoda	6	41	**148**	1	4	**8**	1	11	**13**	**169**
Mol: Bivalvia	1	1	**1**							**1**
Nemertini	1	1	**1**							**1**

Ann: Polychaeta	2	3	**3**							**3**
Ann: Oligochaeta	4	14	**22**	2	3	**3**	1	1	**1**	**26**
Ann: Hirudinea	2	3	**4**				1	1	**1**	**5**
Cru: Cladocera	1	1	**4**							**4**
Cru: Copepoda	8	30	**113**	8	23	**37**	2	9	**25**	**175**
Cru: Ostracoda	2	15	**36**	1	6	**8**	2	2	**3**	**47**
Cru: Bathynellacea	2	2	**2**				2	3	**7**	**9**
Cru: Decapoda	1	2	**4**							**4**
Cru: Thermosbaenacea	2	2	**2**	1	1	**1**				**2**
Cru: Mysidacea	1	1	**1**							**1**
Cru: Isopoda	7	10	**41**	4	5	**13**	5	6	**15**	**69**
Cru: Amphipoda	10	16	**107**	5	6	**18**	3	3	**16**	**141**
Ver: Amphibia	1	1	**1**							**1**
			WB			**GR**			**BUL**	
Total stygobionts	55	154	**517**	21	47	**88**	17	37	**81**	**686**
TERRESTRIAL FAUNA	families	genera	**species**	families	genera	**species**	families	genera	**species**	**species**
Tur: Tricladida	1	1	**2**							**2**
Mol: Gastropoda	11	16	**36**	5	6	**13**				**49**
Ara: Araneae	9	27	**97**	5	10	**10**	1	2	**2**	**109**
Ara: Pseudoscorpiones	3	11	**100**	2	4	**23**	2	4	**9**	**132**
Ara: Opiliones	4	6	**16**	1	1	**1**	3	4	**4**	**21**
Ara: Palpigradi	1	1	**3**							**3**
Cru: Isopoda: Oniscida	6	24	**70**	5	10	**26**	2	10	**20**	**116**
Myr: Chilopoda	3	7	**28**				1	2	**4**	**32**
Myr: Diplopoda	14	29	**71**	6	10	**11**	5	10	**17**	**99**
Ins: Diplura	1	1	**4**	2	2	**4**	1	1	**3**	**11**
Ins: Collembola	8	12	**27**				2	3	**7**	**34**
Ins: Thysanura	1	1	**1**							**1**
Ins: Coleoptera	5	83	**326**	3	13	**25**	3	14	**33**	**384**
Ins: Diptera	1	1	**2**							**2**
			WB			**GR**			**BUL**	
Total troglobionts	68	220	**783**	29	56	**113**	20	30	**99**	**995**
Coleoptera:										
Carabidae		15	**110**		1	**8**		3	**17**	
Cholevidae		58	**196**		8	**13**		10	**14**	
Pselaphidae		8	**20**		3	**4**		0	**0**	

*Table 3. Numbers of obligate subterranean animal species in the states of the western Balkans, including their non-Balkan and non-Dinaric territories (†). *The division of Serbia into western (tending to the Dinarides) and eastern (margins of the Carpathians) sectors is also shown. Abbreviations: see Table 2.*

	Slovenia-Dinaric	Sovenia-non-Dinaric†	Croatia-Dinaric	Croatia-Non-inaric†	Bosnia & Herzegovina	Montenegro	Serbia	Macedonia	Serbia*
AQUATIC FAUNA									W+E
Porifera	0	0	1	0	0	0	0	0	0
Cnidaria	1	0	1	0	1	0	0	0	0
Tur: Temnocephalida	9	0	5	0	4	1	0	0	0
Tur: Tricladida	5	3	4	0	1	1	0	0	0
Mol: Gastropoda	33	16	54	3	38	22	9	7	8+1
Mol: Bivalvia	1	0	1	0	1	0	0	0	0
Ann: Hirudinea	2	0	2	0	1	1	0	0	0
Ann: Oligochaeta	19	3	3	1	0	0	0	0	0
Ann: Polychaeta	1	0	3	1	1	0	0	0	0
Cru: Amphipoda	18	11	33	10	16	21	13	26	5+8
Cru: Bathynellacea	1	2	0	2	0	0	1	1	1+0
Cru: Cladocera	2	0	0	0	1	0	0	1	0
Cru: Copepoda	34	5	14	1	9	21	21	29	4+17
Cru: Decapoda	3	0	3	0	3	2	0	0	0
Cru: Isopoda	14	7	12	4	3	5	8	9	4+4
Cru: Mysidacea	0	0	0	0	1	0	0	0	0
Cru: Ostracoda	14	2	3	2	1	6	0	8	0
Cru: Thermosbaenacea	0	0	1	0	1	0	0	0	0
Ver: Amphibia	1	0	1	0	1	1	0	0	0
TERRESTRIAL FAUNA									
Tur: Tricladida	1	0	0	0	1	0	0	0	0
Mol: Gastropoda	6	3	25	0	13	5	0	0	0
Ara: Araneae	12	5	44	0	32	29	4	2	3+1
Ara: Pseudoscorpiones	13	2	47	0	27	9	10	11	2+8
Ara: Opiliones	0	0	7	0	7	3	2	1	1+1
Ara: Palpigradi	3	0	1	0	0	0	0	0	0
Cru: Isopoda	7	1	32	0	19	6	12	8	5+7
Myr: Chilopoda	3	1	13	0	19	2	2	0	1+1
Myr: Diplopoda	14	12	29	0	23	4	8	0	1+6
Ins: Coleoptera	45	38	70	0	152	70	8	10	5+3

High endemism also occurs within subregions, when considering separately the individual administrative (state) borders. The western Balkans was divided into the territories of Slovenia, Croatia, Bosnia and Herzegovina, Montenegro, Serbia, and Macedonia (FYROM). Similarly, Greece was divided into the northern mainland (Greek Macedonia + Thraki), central Greece (Sterrea Ellada + Thessalia + Ipeiros + Ionian Islands), the Peloponnisos, and the Aegean Islands (including Kriti). Bulgaria was divided into five zones, generally following Gueorguiev (1977), but approximating the zones to administrative districts. Whenever one of these territorial divisions reaches a number of 20 obligate subterranean species or more, be it at the level of one taxonomic group (*e.g.* order) or the complete fauna, 40-60% of species are effectively endemic. To avoid overstatement, cases with lower species numbers were not considered.

There were nevertheless exceptions. Macedonia's comparatively poor terrestrial cave fauna is endemic to a level of almost 90%. The western Stara Planina karst of Bulgaria exhibits almost 85% endemism, probably owing to the poverty (or poor knowledge) of faunas in surrounding districts. Over 95% endemism occurs in the terrestrial cave fauna of the Aegean Islands, but it is interesting to note that the aquatic fauna here does not exhibit marked endemicity. Very high levels of endemism also occur in Greek continental terrestrial faunas, although at lower absolute numbers (13-34 species). The 75% endemism of Trechinae beetles in Dinaric Slovenia is also noteworthy, a figure which would be much higher if Alpine species were included; this is probably mainly a function of the high speciation rate within the genus *Anophthalmus*. Similarly, it is evident that the artificiality of the borders is also obscuring the particularly high endemism of leptodirine beetles in the southeastern Dinaric karst (in the neighbouring districts of Hercegovina, Dalmacija and Montenegro). The relatively high endemism of more than 65% in Pseudoscorpiones between the neighboring Croatia and Bosnia and Hercegovina is equally remarkable.

A trial with the "phylogenetic species concept"

The biological species concept (BSC, which conceives species as a potential reproductive community) is often criticised. The phylogenetic concept (PSC) is preferred instead by many (*cf.* Kottelat, 1997). Since most state legislature does not consider subspecies as worthy of legal protection, the PSC may have some advantages in its application. Therefore, we have tried to calculate the possible influence of the phylogenetic species concept.

According to one of the PSC definitions, the species is “the smallest (evolutionary) unit appropriate for analysis” (Mayden, 1997). The potential effect of the PSC was simulated by assigning all subspecific taxa, *i.e.* all identifiable entities, the status of species. This caused a highly variable degree of increase in the total number of entities within different group. *Niphargus* (Amphipoda) and Pseudoscorpiones both increased to *ca.* 150%, compared to 200-400% for *Monolistra* (Isopoda) and some beetle genera (*Antroherpon, Anophthalmus, Neotrechus*), and up to 900% for the beetle genus *Typhlotrechus*. In this latter genus, only three groups had been recognised as species using the biological species approach.

The application of the PSC resulted in a similar effect on the estimated degree of endemism. The comparatively low endemism of *Niphargus* species in Slovenia or Croatia (35%) increased to 70 % or more using the PSC, and in Macedonia from 70 % to 100%; the genus *Anophthalmus* increased from 75% to 95% within Dinaric Slovenia.

Taxonomic peculiarities of the obligate subterranean fauna in the Balkan Peninsula

In almost all regions, the Coleoptera are the most important contributors to terrestrial cave biodiversity. In the western Balkans they represent 45% of all troglobiotic species, in Bulgaria 33%, while in Greece, at only 22%, they equal the woodlice (Isopoda: Oniscidea). However, a remarkable number of new coleopteran species from Greece is in the course of being described at the time of writing (P.M. Giachino, *pers. comm.*). In all three regions the group is richly represented by Cholevidae: Leptodirinae and by Carabidae: Trechinae, and much less so by Pselaphidae (troglobiotic pselaphids are not yet reported from Bulgaria). Other systematic groups are more or less rare. Small curculionids (genus *Troglorhynchus*) are often considered as troglobionts but we tend to consider them edaphobionts, depending on the presence of tree roots. Highly troglomorphic carabid beetles, resembling *Aphaenopsis* in the Dinarides and *Pheggomisetes* in the Stara Planina, are fairly common and are distributed beyond the Balkans (*Aphaenopidius* in the Slovenian Alps, *Aphaenops* in SW Europe, *Sinaphaenops* in the Far East, *etc*; *cf.* Casale *et al.*, 1998). The high diversity of the Trechinae in the Balkans mainly derives from the high speciation rate of the less troglomorphic genera of *Anophthalmus* in the northwest, and *Duvalius* and *Neotrechus* in the southeast of the Dinarides. *Duvalius* is also species-rich in Bulgaria and Greece. The cholevids are much more diverse. Since the edaphobiont *Bathyscia*-type original is almost egg-shaped, troglomorphism may be particularly pronounced in them. Beyond the Balkan Peninsula, only the so-called pholeuonoid type (spindle-shaped trunk with long appendages) may be found (Jeannel, 1924) which is represented *e.g.* by *Pholeuon* in Romania or *Antrocharis* in France. The exceptional Canadian *Glacicavicola* Westcott (Peck, 1974) is morphologically a step further, resembling a number of Dinaric genera such as *Astagobius*, the Greek *Elladoherpon*, or the Bulgarian *Genestiellina*. However, only in the Dinaric karst is the more extreme leptodiroid type developed, represented by the unique species of *Leptodirus* in the northwest and by almost 30 species of *Antroherpon* (Pretner, 1968), a genus nearly endemic to Hercegovina and Montenegro.

Very high diversity is also attained by Pseudoscorpiones, Araneae, Diplopoda, and Isopoda Oniscidea. The diversity of spiders in the Dinaric karst is also exceptional, owing mainly to a high number of small *Troglohyphantes* species (Linyphiidae), but also to comparatively large numbers of representatives of seven genera and 25 species of Dysderidae. Woodlice are mainly represented by Trichoniscidae, some of which have attained very large bodies as cave-dwellers, and may exhibit amphibious habits; both traits are true of the Dinaric *Titanethes* and the Karpathian *Bureschia*.

As in other stygobiotic faunas, Crustacea are by far the dominant group (Sket, 1999b); Copepoda with 175 and Amphipoda with 140 species are the leading groups, both with over a hundred species in the western Balkans alone. By far the most diverse genus is the amphipod *Niphargus* with an enormous morphological span. The high number of gastropods is also exceptional (170 species, with 150 in the western Balkans and its surroundings), mostly comprising Hydrobiidae. This number is close to one half of the known stygobiotic gastropod fauna occurring worldwide.

Some typical representatives of the Dinaric cave fauna are depicted in Figure 2. From taxonomic and ecological perspectives, the most intriguing members of the Balkan cave fauna, or more precisely of the Dinaric cave fauna, are the unique stygobiotic (exceptionally some troglobiotic) members of some higher taxa, as follows:

Eunapius subterraneus Sket & Velikonja (Porifera: Spongillidae) is the only stygobiotic freshwater sponge;

Velkovrhia enigmatica Matjašič & Sket (Hydrozoa: Bougainvilliidae) is the only known freshwater stygobiotic cnidarian;

Congeria kusceri Bole (Mollusca: Bivalvia: Dreissenidae) is the only reliably stygobiotic clam;

Marifugia cavatica Absolon & Hrabě (Polychaeta: Serpulidae) is the only stygobiotic tube-worm;

Proteus anguinus Laurenti (Amphibia: Proteidae) is the only non-American stygobiotic amphibian; a troglomorphic population of *Rhynchodemus henrici* Bendl and an undescribed species from Slovenia seem to be the only troglobiotic representatives of terricolous planarians (Turbellaria: Tricladida Terricola).

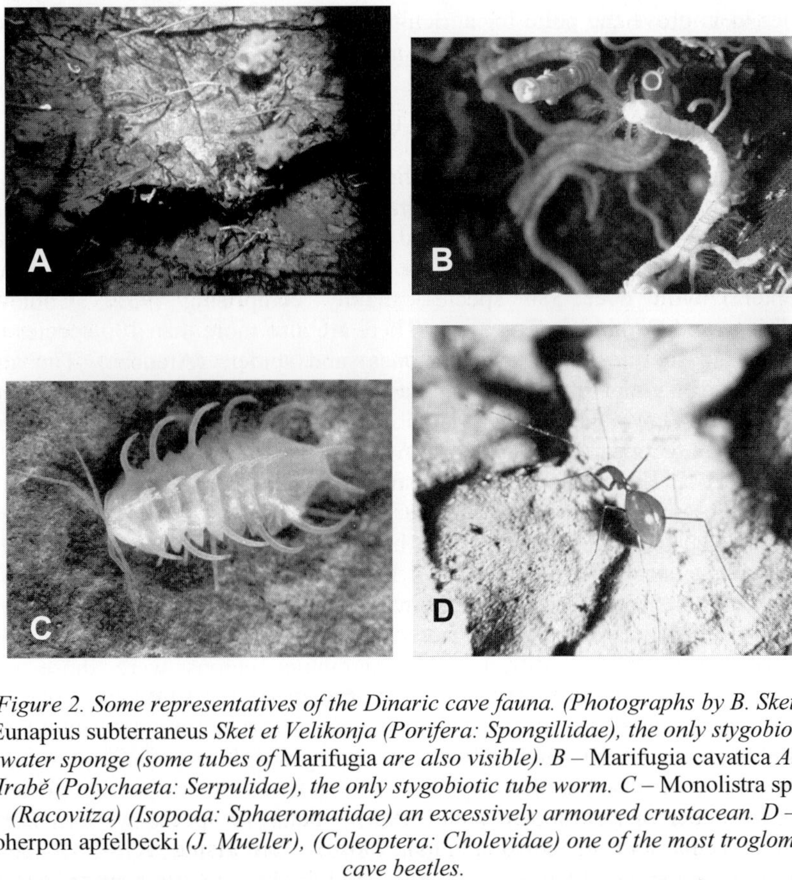

Figure 2. Some representatives of the Dinaric cave fauna. (Photographs by B. Sket.) A – Eunapius subterraneus *Sket et Velikonja (Porifera: Spongillidae), the only stygobiotic freshwater sponge (some tubes of* Marifugia *are also visible). B –* Marifugia cavatica *Absolon et Hrabě (Polychaeta: Serpulidae), the only stygobiotic tube worm. C –* Monolistra spinosa *(Racovitza) (Isopoda: Sphaeromatidae) an excessively armoured crustacean. D –* Antroherpon apfelbecki *(J. Mueller), (Coleoptera: Cholevidae) one of the most troglomorphic cave beetles.*

Considering the poverty of their groups in hypogean freshwater habitats, the four species of stygobiotic Copepoda: Calanoida and three species of Cladocera are also interesting statistics.

At least three interesting ecological assemblages of the Dinaric karst are noteworthy (Sket, 1999a). An epizoic protozoan faunula of no less than seven

Ciliata: Peritricha and two Suctoria was found on a sole specimen of the isopod *Monolistra spinosissima* (Racovitza) (Hadži, 1940).

A habitat resembling *hygropetricum* (Sket, 2001) in the Vjetrenica cave was found to be inhabited by an originally aquatic amphibian animal, the amphipod *Typhlogammarus mrazeki* Schaeferna and by an originally terrestrial amphibian, the beetle *Hadesia vasiceki* J. Mueller. The presence of similar ecological assemblages has been noted in other parts of the Dinarides, but is very rare outside this region. It is interesting to note that the marker species are morphologically similar, but mutually not closely-related leptodirine beetles.

Under very specific environmental conditions in the sink-cave Crnulja in Popovo polje, very dense and extensive colonies of *Marifugia* and therefore thick layers of its tubes have formed a kind of hypogean tufa. This was previously inhabited by an interstitial fauna of oligochaetes, acarines, and some crustaceans, but human activities to improve the polje for agriculture have now prevented regular flooding of the cave, thus killing *Marifugia* and its associates.

CONCLUSIONS

The comprehensive census of the obligate subterranean fauna from the still expanding list for the Balkan Peninsula (including non-Balkan parts of Slovenia and Croatia) revealed *ca.* 1,000 terrestrial and 680 aquatic species, with a high number of subspecies not considered here. By far the richest terrestrial group is the beetles (Coleoptera) with over 380 species, mainly comprising (93%) Cholevidae: Leptodirinae and Carabidae: Trechinae. There are also more than 100 species of the arachnid false scorpions (Pseudoscorpiones) and spiders (Araneae). The aquatic fauna (from caves and interstitial habitats) is clearly dominated by three groups, comprising 140-175 species each: amphipod shrimps (Amphipoda), tiny snails (Gastropoda) and copepods (Copepoda).

In spite of having used administrative boundaries rather than natural biogeographic provinces to subdivide the Balkan Peninsula, very few species were found to be distributed across boundaries between the three main subregions defined in the study: the western Balkans or Dinaric subregion, the eastern Balkans (equating with Bulgaria) and the southern Balkans (Greece). The only cases where the use of administrative boundaries was suspected as causing a serious reduction of faunal contrasts between different biogeographic regions were those of the transitional regions of (eastern) Serbia and Macedonia (FYROM). Comparatively high endemism was also exhibited within different districts (in the western Balkans the districts are the recent states) of the three subregions; for most animal groups this amounted to 40 to 60%, but often higher.

The by far richest subregion in the Balkan Peninsula is the western Balkans, consisting in this study of Slovenia, Croatia, Bosnia and Herzegovina, Montenegro, Serbia and Macedonia (FYROM). In addition to the highest numbers of "common" subterranean biota, the most complete assortment of animal orders is also represented here. Included are some groups which are not represented in caves outside the Balkans at all: freshwater sponges, cnidarians, clams, and tubeworms. The only non-American stygobiont amphibian inhabits the Dinaric karst and only here do the cholevid beetles attain the extremely troglomorphic "leptodiroid" body shape.

For some groups, the differences in numbers are definitely a result of biased investigation attempts; such is the case with Oligochaeta. Nematoda have been deleted from this census for that reason. This is definitely not the case with the other groups, such as Gastropoda. The numbers presented will change in subsequent publications. Additional data will be discovered in the literature, views on the ecology of some species will change, the taxonomic status of some known taxa will appear higher or lower, and, finally, new species are continually being discovered and described even in the best-investigated groups and regions, such as Coleoptera in Slovenia (compare Giachino & Etonti, 1996; Polak, 2002).

As a final point, it has been stated clearly that the census was compiled according to administrative entities (states). As such, it has high potential as the starting point for the definition of nature protection policies in individual states.

ACKNOWLEDGEMENTS

We would like to thank a number of students who carried out paperwork for the lists. S. Polak (Postojna) entered the data from E. Pretner's manuscript list of Slovenian cave Coleoptera. Some specialists kindly checked (or are still checking) the lists for possible synonymy: K. Christiansen (Grinell) for Collembola, D. Danielopol (Mondsee) for Ostracoda, C. Deltshev (Sofija) for Araneae, J.-J. Geoffroy (Paris) for Diplopoda, I. Kos (Ljubljana) for Chilopoda, T. Novak (Maribor) for Opiliones, B. Sambugar (Verona) for Oligochaeta, H. Schmalfuss (Stuttgart) for Isopoda Oniscidea, I. Sivec (Ljubljana) for Diptera, and K. Tarman (Ljubljana) for Acarina. A. Brancelj (Ljubljana) offered his unpublished list of Copepoda in Slovenia. This work was co-funded by the contract n° EVK2-CT-2001-00121-PASCALIS of the Fifth Research and Technological Development Framework Program supported by the European Community.

REFERENCES

Argano, R. & Manicastri, C. 1988. Il Genere Trichonethes Strouhal, 1953 di Turchia (Isopoda, Oniscoidea, Trichoniscidae). *Fragmenta Entomologica* 21, 9-16.

Belles, X. 1992. From dragons to allozymes. A brief account of the history of biospeleology. In: *The natural history of biospeleology* (A.I. Camacho, ed.), pp. 3-24. Museo Nacional de Ciencias Naturales, Madrid.

Beron, P. & Gueorguiev, V. 1967. Essai sur la faune cavernicole de Bulgarie. II. Resultats des recherches biospeologiques de 1961 à 1965. *Izvestija na zoologičeskija institut s muzej BAN* 24, 151-212.

Beron, P. 1987. Results of the studies of the cave fauna of Greece. *Biologica Gallo-Hellenica* 12, 125-131.

Beron, P. 1994. Resultats des recherches biospeleologiques en Bulgarie de 1971 à 1994 et liste des animaux cavernicoles bulgares. *Serie Tranteeva* 1, 1-137.

Botoşăneanu, L.1986. *Stygofauna Mundi*. Brill and Backhuys, Leiden.

Casale A. & Giachino P. M. 1994. Greece. In: *Encyclopedia biospeologica*, T. I (C. Juberthie & V. Decu, eds.), pp. 711-717. Société de Biospeleologie, Moulis, Bucharest.

Casale, A., Vigna Taglianti, A. & Juberthie, C. 1998. Coleoptera Carabidae. In: *Encyclopedia biospeologica,* T. II (C. Juberthie & V. Decu, eds.), pp. 711-717. Société de Biospeleologie, Moulis, Bucharest.

Chen, Y.R., Yang, J.X. & Zhu, Z.G. 1994. [A new fish of the genus Sinocyclocheilus from Yunnan with comments on its characteristic adaptation (Cypriniformes: Cyprinidae)]. *Acta zootaxonomica sinica* 19, 246-253. (in Chinese with English abstract).

Giachino, P.M. & Etonti, M. 1996. *Prospelaeobates* gen.nov e due sp.n. di Leptodirinae delle isole del Quarnero e dell'Istria (Coleoptera: Cholevidae). *Acta entomologica slovenica* 4, 63-71.

Gottstein Matočec, S., Bakran-Petricioli, T., Bedek, J., Bukovec, D., Buzjak, S., Franičević, M., Jalžić, B., Kerovec, M., Kletečki, E., Kralj, J., Kružić, P., Kučinić, M., Kuhta, M., Matočec, N., Ozimec, R.,

Rađa, T., Štamol, V., Ternjej, I. & Tvrtković, N. 2001. Croatia. In: *Encyclopedia biospeologica* T. III (C. Juberthie & V. Decu, eds.), pp. 2237-2287. Société de Biospeleologie, Moulis, Bucharest.

Gueorguiev, V. & Beron, P. 1962. Essai sur la faune cavernicole de Bulgarie. *Annales de speleologie* 17, 285-436.

Gueorguiev, V.B. 1977. *La faune troglobie terrestre de la peninsule Balkanique. Origine, formation et zoogéographie*. Izdatelstvo na B'lgarska Akademija na Naukite, Sofija.

Hadži, J. 1940. Favnula epizoičnih infuzorijev na jamski mokrici (*Microlistra spinosissima* Rac.). *Razprave matematično-prirodoslovnega razreda Akademije znanosti in umetnosti v Ljubljani* 1, 121-148.

Hamann, O. 1896. *Europaeische Hoehlenfauna. Eine Darstellung der in den Hoehlen Europas lebenden Tierwelt mit besonderer Beruecksichtigung der Hoehlenfauna Krains*. Costenoble, Jena.

Illies, J. 1978. *Limnofauna europaea*. G. Fischer, Stuttgart.

Jalžić, B. & Pretner, F. 1977. [Beitrag zur kenntnis der Koleopterenfauna aus den Höhlen und karstschächten Kroatiens]. *Krš Jugoslavije* 9, 239-274 (in Croatian with German abstract).

Jeannel, R. 1924. Monographie des Bathysciinae. Biospeologica V. *Archives de Zoologie Experimentale et generale* 63, 1-436, pl. 1-498.

Kottelat, M. 1997. European freshwater fishes. *Biologia (Bratislava)* 52 (Supplement 5), 1-271.

Laurenti, J. L. 1768. *Specimen medicum exhibens synopsis reptilium emendatam*. Viennae.

Mayden, R.L. 1997. A hierarchy of species concepts: the denouement in the saga of the species problem. In: *Species: the units of biodiversity* (M.F. Claridge, H.A. Dawah & M.R. Wilson, eds.), pp. 381-424. Chapman & Hall, London.

Pandourski, I.S. 1998. Catalogue of stygobiontic higher crustaceans (Crustacea, Malacostraca) from Bulgaria. *Acta zoologica bulgarica* 50, 23-33.

Peck, S.B. 1974. Biology of the Idaho lava tube beetle, *Glacicavicola*. *NSS Bulletin* 36, 1-3.

Peck, S.B. 1998. A summary of diversity and distribution of the obligate cave-inhabiting faunas of the United States and Canada. *Journal of Cave and Karst Studies* 60, 18-26.

Pesce, G. L., Maggi, D. 1983. Ricerche faunistiche in acque sotterranee freatiche della Grecia meridionale ed insulare e stato attuale della conoscenza sulla stigofauna di Grecia. *Natura* 74, 15-73.

Polak, S. 2002. *Spelaeodromus sneznikensis* sp. n. from Slovenia (Coleoptera: Cholevidae: Leptodirinae). *Acta entomologica slovenica* 10, 5-12.

Pretner, E. 1968. Coleoptera. Fam. Catopidae. Subfam. Bathysciinae. *Catalogus faunae Jugoslaviae* 3, 1-59.

Pretner, E. 1973. [Fauna Coleopterologica subterranea Croatiae]. *Krš Jugoslavije* 8, 101-239 (in Croatian with German abstract).

Pretner, E. 1977. [Übersicht der unterirdischen Koleopterenfauna Montenegros]. *Glasnik Odjelenja prirodnih nauka CANU* 2, 91-186 (in Serbo-Croatian with German abstract).

Schmidt, F. 1832. *Leptodirus Hochenwartii* n.g. n.sp. *Illyrisches Blatt*, 21, 9.

Sket, B. 1996. Biotioc diversity of hypogean habitats in Slovenia and its cultural importance. *Biodiversity - International Biodiversity Seminar, Proceedings*, pp. 59-74, UNESCO (Gozd Martuljek)

Sket, B. 1999a. Diversity and singularity of hypogean fauna in the Dinaric region. *Abstracts, 14th international Symposium of Biospeleology* (Makarska), pp. 21-25.

Sket, B. 1999b. The nature of biodiversity in hypogean waters and how it is endangered. *Biodiversity & Conservation* 8, 1319-1338.

Sket, B. 1999c. High biodiversity in hypogean waters and its endangerment – the situation in Slovenia, Dinaric karst, and Europe. *Crustaceana* 72, 767-779.

Sket, B. 2001. The hygropetric habitat in caves and its inhabitants. *Abstracts of the XVth International Symposium of Biospeleology* (Intervales), pp. 42-43.

Valvasor, J.W. 1689. *Die Ehre des Herzogthums Crain*. Endtner, Nurenberg.

Wolf, B. 1934-1938. *Animalium cavernarum catalogus*. Junk' s-Gravenhage.

19. DIVERSITY OF COPEPODA (CRUSTACEA) IN THE UNSATURATED ZONE OF KARSTIC CAVES OF SLOVENIA

ANTON BRANCELJ* AND TANJA PIPAN

INTRODUCTION

The subclass Copepoda is divided into ten orders, within which around 12,000 taxa are known at present. Some species are external parasites or live in symbiosis with other organisms, and others are free-living (Huys & Boxshall, 1991). Six orders are composed entirely of free-living representatives, occupying a diversity of habitats from the deep-sea floor to high-mountain glaciers. Representatives of four orders (Calanoida, Harpacticoida, Gelyelloida and Cyclopoida) occupy subterranean freshwaters, with a high number of stygobitic (*i.e.* living exclusively in caves and groundwater) and endemic taxa. The order Gelyelloida is represented by only two groundwater species of European karstic systems, which are endemic to France and Switzerland (Rouch, 1991), and by one species collected recently from interstitial stream habitats in South Carolina, USA (Dole-Olivier *et al.*, 2000). The other three orders have representatives distributed throughout the World but most of them are known from Europe.

The earliest research on subterranean copepods dates back to the beginning of the 20th century, with Graeter's studies of cave-dwelling Copepoda (Chappuis, 1927). Chappuis (*ibid.*) compiled a list of 77 Copepoda occurring in subterranean environments. Among these taxa (some of which were later synonymised), it was noted that most of them were trogloxenes (not native to the groundwater) or troglophiles (found regularly in the caves near the entrances or in shallow gravel beds) but that few were stygobionts, a finding which is still valid today. Most of the stygobionts were discovered in caves of Romania and Serbia. At about the same time, Kiefer began his study of subterranean Copepoda in the Balkans and North Italy, and described several additional new species (Kiefer, 1930, 1931, 1932, 1933). In spite of this, there has been very little research until recently on the cave copepods of the Balkan Peninsula, the available data having been derived entirely from some samples collected from jets of percolating water or their associated puddles. Other than the work of Chappuis and Kiefer, Petkovski (1959, 1983, 1984) and Karanović & Bobić (1998) have contributed most to these data. More recently, data which are emerging from some Dalmatian caves indicate that the fauna is rich, with a clear need for intensive study (Brancelj, in preparation).

Kiefer and Chappuis studied material from caves (in Slovene *jama*, pl. *jame*) within the borders of modern Slovenia (Podpeška jama, Postojnska jama and Škocjanske jame: for details see Brancelj, 2001) in the late 1930s. Twelve stygobitic taxa were reported (seven taxa of Harpacticoida and five of Cyclopoida), of which two were later synonymised. With ten taxa, Slovenia was at the time the region with the highest reported number of stygobitic Copepoda. There were virtually no data on the associated fauna in these early studies.

H.I. Griffiths et al. (eds.), Balkan Biodiversity, 323–332.

The second period of research on Copepoda began in the early 1960s, with the work of the Macedonian, Trajan Petkovski. Several papers on new taxa of subterranean Copepoda from Slovenia, and their associated fauna, were published between 1954 and 1985 (for details see Brancelj, *ibid.*). This increased the total number of reported Copepoda from cave waters to about 20, 15 of which were endemic. Systematic surveys were still lacking, apart from studies of the fauna from three jets of dripping water from Planinska jama and a spring near Žužemberk (Petkovski, 1983, 1984).

The situation changed after 1985, when intensive investigations of different types of cave water were carried out by Brancelj (Postojna - Planina Cave System, Škocjanske jame, Osapska jama, Kompoljska jama) (Petkovski & Brancelj, 1985; Brancelj 2001). Several new species were discovered, most of which were endemic, and a species checklist of subterranean species was prepared with details of their distribution.

The fourth research period began in 2000, when detailed studies were initiated into the microdistribution and seasonal dynamics of Copepoda in dripping water and puddles of percolating water in several cave systems. This included caves which had been sampled previously (Postojnska jama, Škocjanske jame). The main aims were: A) to establish the total number of taxa in jets of water in the unsaturated karstic zone; B) to detect differences in copepod fauna between different water bodies in individual caves, and C) to detect differences in copepod fauna between caves in neighbouring geographic areas.

Because of the scarcity of data on the copepods of percolating water from other regions of the Balkan Peninsula, the focus of this paper is on the diversity of Copepoda in seven caves from Slovenia. On the assumption that the same basic principles apply to the wider region, generalisations can be made from the results of this study, concerning the Dinarides (Dinaric region) or even the Balkans as a whole.

MATERIAL AND METHODS

The seven cave study sites are located within a rectangle of dimensions of *ca.* 30 x 60 km in southern and south-west Slovenia (Figure 1). Cave length varies from 75 m (Velika Pasjica) to the >14 km-long labyrinth of Postojnska jama. Two types of data set are used for this study. The first set is based on published literature. The second data set was collected during the period 2000-2001. The latter is restricted to studies of jets of percolating water and their associated puddles. Individual water jets and puddles were treated as separate sampling sites, with the aim of detecting differences between jets and puddles, and to establish the seasonal dynamics of Copepoda assemblages in specific water bodies.

Water from jets or drip flows was collected using a funnel and filtered through a 60- μm net. A known volume of water from puddles was filtered with the same type of net. Samples were stored in 4% formaldehyde solution.

In some caves (Postojnska jama, Pivka jama and Črna jama), jets were sampled once a week for one year (50 samples per location). Otherwise, samples were collected once a month (10 - 12 samples per location). Puddles were sampled four times during the year (four samples per location). The total number of sampling sites within each cave was around 10, with a maximum of 17.

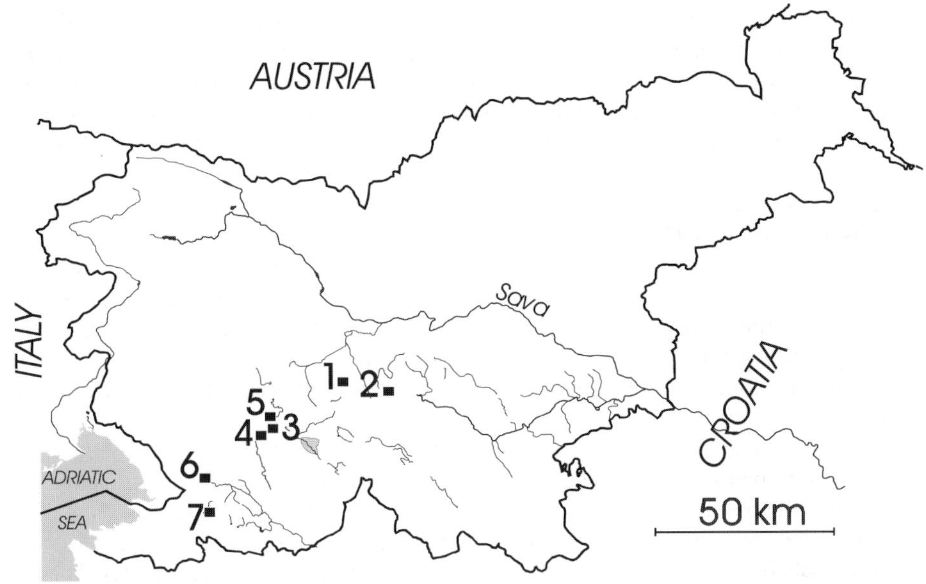

Figure 1: Map showing the location of the seven study sites in south-western Slovenia (1: Velika Pasjica; 2: Županova jama; 3: Pivka jama, 4: Postojnska jama, 5: Črna jama; 6: Škocjanske jame, 7: Dimnice). (jama =cave, in Slovene)

The Sørenson index of similarity (S) (Odum, 1971) was applied to evaluate similarity between different caves based on species presence in both jets and puddles.

$$S = 2C / A + B$$

Where: C = number of species common to both samples
A = number of species in sample A
B = number of species in sample B

The clustering of cave sites based on species assemblage composition was investigated further using TWINSPAN (Hill *et al.*, 1975). Multivariate ordination techniques (ter Braak & Verdonschot, 1995, ter Braak & Šmilauer, 2002) were used to assess patterns of species variation in relation to four measured environmental variables (annual precipitation, average temperature, ceiling thickness and volume of puddles). The data-set is small (seven samples and four environmental variables, with species expressed as dummy variables), so a linear rather than unimodal assumption was assumed. The influence of the four measured environmental variables on species assemblage composition was investigated initially using the direct gradient analytical technique, redundancy analysis, but after forward selection the Monte Carlo test of significance of the environmental parameters gave non-significant results. Patterns of variation of the species data were therefore investigated instead simply using the indirect gradient analytical technique, principal components analysis (PCA), using CANOCO, version 4.5 (ter Braak & Šmilauer, 2002).

RESULTS

Thirty-seven taxa were collected from the seven caves, in which sampling sites were restricted to *juvenile* water, that is, associated with water percolating from the cave ceiling (Table 1). Eleven taxa belonged to a group of troglophilic or ubiquitous taxa (*i.e.* widely distributed in different water bodies) which are also common in subterranean environments. The remainder (26 taxa), belong to the stygobionts. Four of these (genera *Parastenocaris* and *Nitocra*) have uncertain taxonomic status and need further study. Five taxa are recognised as new species (scientific descriptions in preparation), and three others (*Morariopsis dumonti* Brancelj, 2000, *Parastenocaris andreji* Brancelj, 2000 and *Bryocamptus borus* Karanović & Bobić, 1998) have been described recently.

Table 1. Copepoda taxa reported from jets of percol0ating water and associated puddles from seven caves in central and south-western Slovenia (O = jets; X = puddles).

	Velika Pasjica	Županova jama	Pivka jama	Dimnice	Črna jama	Škocjanske jame	Postojnska jama	O + X; SUM OF CAVES
Acanthocyclops kieferi (Chappuis, 1925)							X	0+1; 1
Attheyella crassa (G.O. Sars, 1863)						X		0+1; 1
Bryocamptus balcanicus (Kiefer, 1933)	X	OX	O	X	OX		OX	4+5; 6
Bryocamptus cf. *borus* Karanovic & Bobic, 1998				X				0+1; 1
Bryocamptus dacicus (Chappuis, 1923)			OX		X		X	1+3; 3
Bryocamptus n.sp.			OX			X		1+2; 2
Bryocamptus pygmaeus (Sars, 1863)				X				0+1; 1
Bryocamptus pyrenaicus (Chappuis, 1923)	X	OX	X	X				1+4; 4
Bryocamptus typhlops (Mrazek, 1893)	X	X	X			X	X	0+5; 5
Bryocamptus zschokkei (Schmeil, 1893)						X		0+1; 1
Diacyclops langouidoides (Lilljeborg, 1901)		OX					X	1+2; 2
Diacyclops languidus (Sars, 1863)						X		0+1; 1
Elaphoidella cvetkae Petkovski, 1983	X	O	OX		OX	O		4+3; 5
Elaphoidella n. sp. A	X	OX						1+2; 2
Elaphoidella n sp.B	X		O					1+1; 2
Elaphoidella kieferi Petkovski & Brancelj, 1985						OX		1+1; 1
Elaphoidella stammeri Chappuis, 1936		O						1+0; 1
Epactophanes richardi Mrazek, 1893			X					0+1; 1
Maraenobiotus brucei (Richard, 1898)			OX					1+1; 1
Megacyclops viridis						X		0+1; 1

(Jurine, 1820)								
Morariopsis dumonti Brancelj, 2000	X	OX						1+2; 2
Morariopsis scotenophila (Kiefer, 1930)				OX	X	X	X	1+4; 4
Moraria sp. B		O						1+0; 1
Moraria sp. A					O	O		2+0; 2
Moraria stankovichi Chappuis, 1924		O						1+0; 1
Moraria varica (Graeter, 1910)	X		OX				X	1+3; 3
Moraria poppei (Mrazek, 1893)	X		OX		X	OX	OX	3+5; 5
Nitocra sp.				O			O	2+0; 2
Parastenocaris cf. *andreji* Brancelj, 2000		X		O				1+1, 2
Paracyclops fimbriatus (Fischer, 1853)						X	X	0+2, 2
Parastenocaris nolli alpina Kiefer, 1969	X	O	O	OX	OX	OX		5+4; 6
Parastenocaris type 1				O	O			2+0; 2
Parastenocaris type 2		O	O	O	O	OX	OX	6+2; 6
Parastenocaris type 3		O				O		2+0; 2
Phyllognathopus vigueri (Maupas, 1892)	X	OX					X	1+3; 3
Speocyclops infernus (Kiefer, 1930)	X	OX	OX	OX	OX	OX	OX	6+7; 7
cf. *Stygepactophanes* sp.				O	O	O		3+0; 3
TOTAL (O + X) SUM:	**12**	**14+9**	**11+10**	**8+7**	**8+7**	**9+13**	**5+12**	
TAXA PER CAVE	**12**	**16**	**14**	**12**	**11**	**17**	**13**	

The basic characteristics of the study sites (Table 2) indicate variability in size, water temperature and thickness of the ceiling. After intensive sampling, a minimum of 11 and maximum of 17 taxa were found in each cave. The number of taxa does not appear to correlate with any of the environmental parameters (including lengths of the caves). A single species (*Speocyclops infernus* Kiefer, 1930) occurred in all seven caves, and usually in both jets and puddles. Six taxa were also common, being

*Table 2: Summary of environmental data for the seven study sites, for the period 1999 - 2000. ppt. = precipitation; temp. = temperature; *rainfall data for 10 months only.*

	Cave length (m) (dry part only)	Annual ppt. (mm)	Average ceiling thickness (m)	Average temp. °C (puddles)
Velika pasjica	75	1,235*	5	8.8
Županova jama	682	2,019	25	9.1
Pivka jama	794	1,548	55	6.9
Dimnice	2,020	1,827	50	6.0
Črna jama	3,294	1,827	50	4.8
Škocjanske jame	4,300	1,577	80	10.0
Postojnska jama	13,500	1,827	80	9.1

collected in six and five caves, respectively. The majority of taxa were found in only one or two caves (13 and 11 taxa, respectively) (Table 3). Eight taxa, (20 % of taxa) were found only in jets or drip flows of percolating water. The distribution of ten taxa was restricted to puddles filled by percolating water (which are occasionally contaminated by other types of water).

In terms of geographic location, there is no clear correlation between distance between caves and similarity in species assemblage composition (expressed as the Sørenson index) (Figure 2); the caves are located in three distinct groups (Figure 1), yet these are not grouped even at the first level split. One taxon (cf. *Stygepactophanes*) which is present in caves Dimnice, Črna jama and Škocjanske jame drives the first level split. The same clustering of the caves was confirmed by TWINSPAN analysis, wherein an additional ten preferential taxa were connected to those caves.

Table 3. The number of taxa observed to be in common between different numbers of caves (total number of taxa = 37).

7 caves = 1 taxon
6 caves = 3 taxa
5 caves = 3 taxa
4 caves = 2 taxa
3 caves = 4 taxa
2 caves = 11 taxa
1 cave = 13 taxa

The sum of taxa present in pairs of both caves, which was used for the additional calculation of the similarity between each pair of caves (21 calculations in total) varied between 16 (Dimnice and Črna jama) and 27 (Županova jama and Škocjanske jame). The median value of the sum of taxa was 20. The highest similarity was observed between the pairs of caves, Velika Pasjica and Pivka jama (sum of taxa = 17, taxa in common = 9), Velika Pasjica and Županova jama (sum of taxa = 19, taxa in common = 9) and Dimnice and Črna jama (sum of taxa = 16, taxa in common = 7). The component caves for both these pairs are located relatively far apart (25, 15 and 33 km, respectively). Similarity was found to be low in more closely adjacent caves < 3 km apart (Pivka jama, Postojnska jama and Črna jama).

The results of PCA (Figure 3) indicated that 50.6 % of the cumulative variance in the species data was contained within the first two axes. The sites and species are widely separated in the direction of both Axis 1 and Axis 2, underlining the distinctness of the fauna of individual caves, although the proximity of Škocjanske jame, Črna jama and Dimnice in the direction of Axis 1 is in accord with the above results which indicated the clustering of these sites in terms of species assemblage composition. There is no obvious clustering of sites from which to infer the clear influence of any of the measured environmental variables on species assemblage composition. Postojnska jama, which plots in the centre of Axis 1 and Axis 2, is the only site which appears to show any great degree of overlap with others. It is inhabited by the common species (*Speocyclops infernus, Parastenocaris* type 2, *Moraria poppei*, *Bryocamptus typhlops* and *B. balcanicus*) which tend also to occur in other sites (Table 1). Four species, plotted in the centre of Axis 1 and Axis 2 (*Acanthocyclops kieferi*, *Speocyclops infernus*, *Parastenocaris nolli alpina* and *Bryocamptus dacicus*), seem to be the most ecologically tolerant, but the data-set is insufficient at present to identify precise ecological relationships with environmental

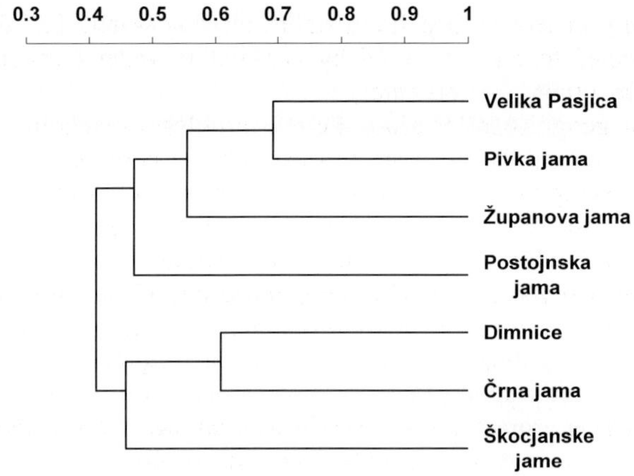

Figure 2. Cluster diagram constructed from cumulative species data for the presence of copepod taxa in the seven caves of south-western Slovenia, based on the Sørenson index of similarity (see text for explanation).

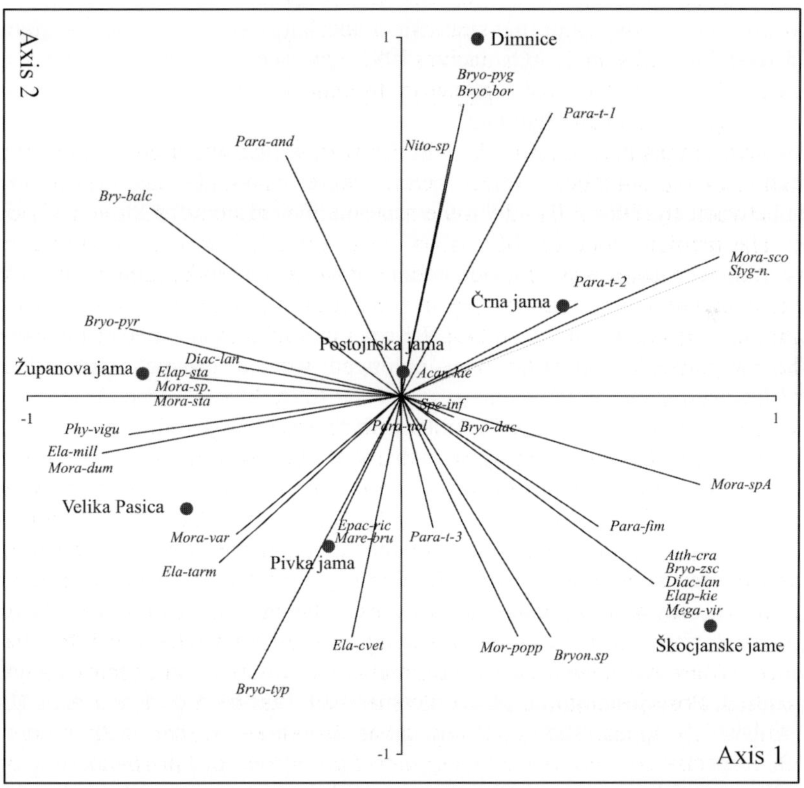

Figure 3. Diagram showing results of principal component analysis (PCA) of a data set of 37 species collected in 1999 and 2000 from seven locations in Slovenia. (Taxon codes are derived from the first three or four letters of the genus name and three or four letters of the species name - see Table 1 for full species names.)

variables. It is also interesting to note the clear separation of Škocjanske jame, which is dominated by most of the taxa which are classed as epigean or ubiquitous (*i.e. Attheyella crassa, Paracyclops fimbriatus, Bryocamptus zschokkei, Diacyclops languidoides, Megacyclops viridis*), possibly indicating the influence on species assemblage composition of more intense human activity in the cave (see *Conclusions*), and equally that other sites are relatively well sheltered at present from outside influences.

CONCLUSIONS

In a compilation of published data on Copepoda from European caves, Kiefer (in Illies, 1978) recorded a total of *ca.* 250 taxa, around 150 of which were endemic to particular parts of Europe. Approximately 90 taxa were recorded from caves in the Balkans, 25 of which were endemic. The main Balkan study regions comprised Macedonia (Former Yugoslav Republic of Macedonia; FYROM), Serbia and the Dinarides - including Slovenia (Figure 4). In Slovenia at that time, around 65 species had been recorded from caves, 16 of which were considered endemic. Thus, from Slovenia and the Balkans as a whole, around one quarter of Copepoda from caves were considered endemic. Most had been collected from the saturated karstic zone or from interstitial sampling sites. Since the saturated zone in both systems can extend over large distances (Camacho, 1992) one would predict low numbers of endemics. This was the first indication that the opposite was true, with high biodiversity in groundwater habitats.

Research on Balkan Copepoda was limited during the 1990s apart from in Slovenia, where a number of new species were recorded, most of which were endemic (Brancelj, 2001). By 2000, the total number of taxa for Slovenia had risen to 110 taxa, with 20 endemics. More than 50 % (about 70 taxa) also occurred in cave waters, and 32 of them were considered as stygobitic taxa. The density of endemics could be estimated as *ca.* one species per 1000 km^2 across Slovenia as a whole or, based more specifically on the area of the cave region, *ca.* one species per 300 km^2. All but one of the endemic taxa belonged to the groups Cyclopoida and Harpacticoida.

These statistics changed significantly with the results of systematic sampling of percolation waters in the epi-karst, which is an unsaturated or vadose zone (Camacho, 1992). These contained an unusually rich fauna, with several new species (Brancelj, 2002). The fauna has two unique characteristics: restricted distribution of species, resulting in endemism, and low abundance of specimens, resulting in low frequency of occurrence.

As noted, only four species were common, being present in six or all of the seven-cave data set. In contrast, a high proportion of taxa were relatively rare (11 taxa occurring in two caves, and 13 taxa in only one cave). Eight of these 24 are new to science. Comparable studies for other parts of Europe are very scarce (Delay, 1968; Gibert, 1986) and show a somewhat lower number of species per cave (five and 12, respectively, compared to an average of 13.6 for Slovenia). It is worth noting that several species in both jet and puddle samples of the Slovenian data set were represented by a single specimen. This supports the hypothesis of Brancelj (2002) that jets and puddles in the galleries are simply sink populations (Brancelj & Culver, 2003) and that the occurrence of specimens (species) there is random or accidental. This is also supported by species-area curves reaching their asymptotic values after several samplings (Brancelj, 2002).

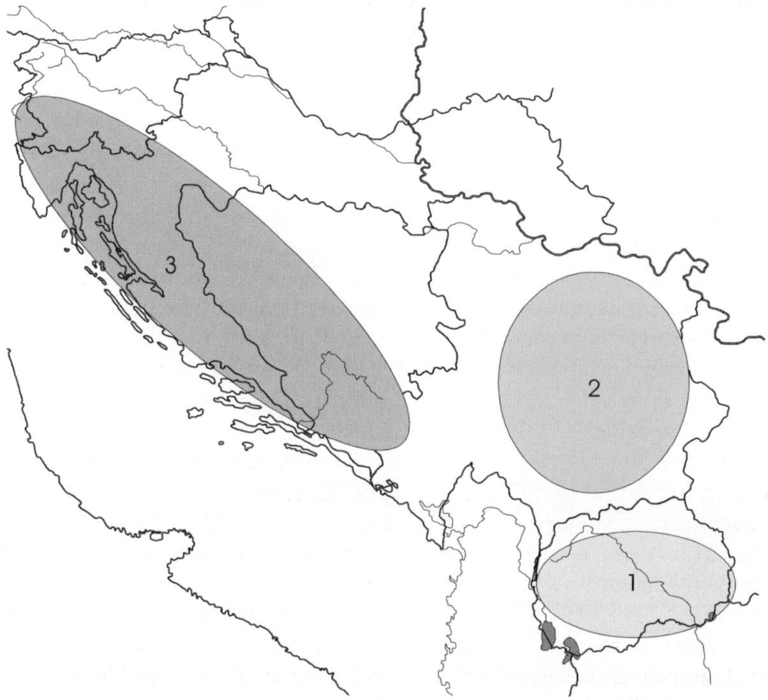

Figure 4. Map showing the three regions of the Balkan Peninsula (1- Macedonia [FYROM], 2- Serbia; 3- Dinarides) within which extensive Copepoda research had been carried out prior to the late 1970s.

Judging from the situation in Slovenia, where intensive studies on copepods from the unsaturated zone have been carried out, we can expect an extremely rich endemic fauna for the rest of the Balkans, particularly in the Dinarides. To extrapolate the results of the present study, the epi-karstic zone would be expected to contain the highest number of endemics. The main mechanism driving the evolution of endemism amongst copepods in these environments is probably the very localised and isolated hydrological system in a zone of vertical percolation, where there is very limited horizontal communication with neighbouring jets. Each jet, with its associated pool on the floor of the gallery, functions as an isolated water pool in which the stygobitic fauna is trapped. Only a few species, which are probably more mobile and resistant to desiccation, can move from one local hydrological system to another.

In the case of Škocjanske jame, where an unusually high concentration of epigean/ubiquitous taxa were found in puddles filled with percolating water, the potentially harmful impact of human activity on the stygobitic fauna could be recognised. In particular, the presence of large-body specimens of *Megacyclops viridis* in the puddles suggests the transport of some copepods consequent upon a variety of human activities needing a water supply (*e.g.* construction or cleaning of paths). This would pose a significant threat to the local stygobitic fauna from the epi-karstic zone. A similar observation was recorded in Lepe jame, the part of Postojnska jama which is open to tourists (Brancelj, 1987).

Finally, this study has shown the potential for elucidating the ecological relationships of cave faunas with a suite of environmental variables, and there is a need for extensive future research to increase the size of the current data-set.

REFERENCES:

Brancelj, A. 1987. Cyclopoida and Calanoida (Crustacea, Copepoda) from the Postojna-Planina Cave System. *Biološki vestnik* 35, 1-16.

Brancelj, A. 2001. Male of *Moraria radovnae* Brancelj, 1988 (Copepoda: Crustacea), and notes on endemic and rare copepod species from Slovenia and neighbouring countries. *Hydrobiologia* 453/454, 513-524.

Brancelj, A. 2002. Microdistribution and high diversity of Copepoda (Crustacea) in a small cave in central Slovenia. *Hydrobiologia* 477, 59-72.

Brancelj, A & Culver, D. 2003. Epicarst Communities. In *Encyclopedia of Caves* (D. Culver & W. White, eds.), Academic Press, San Diego.

Camacho, A.I. (ed.). 1992. *The natural history of biospeleology*. Monografias. Museo nacional de ciencias naturals, Madrid.

Chappuis, P.A. 1927: Die Tierwelt der Unterirdiscen Gewässer (Band III). In *Die Binnengewässer. E. Schweizerbart'sche Verlagsbuchhandlung* (A. Thienneman, ed.), Gustav Fischer Verlag, Stuttgart.

Delay, B. 1968. Recherches sur les eaux souterraines - 2- Données sur le peuplement de la zone de percolation temporaire. *Annales de Spéléologie* 23,705-716.

Dole-Olivier, M.J., Galassi, D.M.P, Marmonier, P. & Creuzé des Châtelliers, M. 2000. The biology and ecology of lotic microcrustaceans. *Freshwater Biology* 44, 63-91.

Gibert, J. 1986. Ecologie d'un systeme karstique Jurassien. Hydrogéologie, dérive animale, transits de matières, dynamique de la population de *Niphargus* (Crustacée). *Mémoires de Biospéologie* 12, 1-376.

Hill, M.O., Bunce, R.G.H. & Shaw, M.W. 1975. Indicator species analysis, a divisive polythetic method of classification, and its application to a survey of native pinewoods in Scotland. *Journal of Ecology* 63, 597-613.

Huys, R. & Boxshall, G. 1991. *Copepod evolution*. The Ray Society, London.

Illies, J. (ed.). 1978. *Limnofauna Europaea*. 2nd edition. Gustav Fischer Verlag, Stuttgart. pp. 532.

Karanović, T. & Bobić, M. 1998. Two new species of Copepoda Harpacticoida from East Serbia (Balkan peninsula): *Parastenocaris serbica* sp.n. and *Bryocamptus (R.) borus* sp.n.. *Crustaceana* 71, 171-184.

Kiefer, F. 1930. Neue höhlenbewohnende Ruderfusskrebse. *Zoologische Anzeiger*, 9/10, 222-228.

Kiefer, F. 1931. Wenig bekannte und neue Süsswasser-Copepoden aus Italien. *Zoologische Jahrbücher (Systematik)* 5/6, 697-712.

Kiefer, F. 1932. Neue Süsswassercopepoden aus Jugoslawien. *Zoologische Anzeiger* 101, 49- 60.

Kiefer, F. 1933. Neue Süsswassercopepoden aus Jugoslawien. *Zoologische Anzeiger* 101, 309-318.

Odum, E.P. 1971. *Fundamentals of Ecology*. 3rd edition. W.B.Saunders Company, Philadelphia.

Petkovski, T.K. 1959. *Morariopsis kieferi* n.sp. ein neuer Höhlenharpacticide (Crustacea Copepoda) aus Dalmatien. *Fragmenta balcanica Musei Macedonici Scientiarum naturalium* 2, 197-200.

Petkovski, T.K. 1983. Neue höhlenbewohnende Harpacticoida (Crustacea, Copepoda) aus Slovenien. *Acta Musei Macedonici Scientiarum naturalium* 16, 177-205.

Petkovski, T.K. 1984. Neue und seltene copepoden (Crustacea) aus Jugoslawien. *Acta Musei Macedonici Scientiarum naturalium* 17, 135-164.

Petkovski, T.K. & Brancelj, A. 1985. Zur Copepodenfauna (Crustacea) der Grotten Škocjanske Jame in Slovenien. *Acta Musei Macedonici Scientiarum naturalium* 18, 1-25.

Rouch, R. 1991. Copepoda. In: *Encyclopaedia biospeleologica* (C. Juberthie & V. Decu (eds.), pp. 105-111.

ter Braak, C. J. F. & Šmilauer, P. 2002. *CANOCO Reference Manual and CanoDraw for Windows User's Guide. Software for Canonical Community Ordination (version 4.5).* Biometris, Wageningen.

ter Braak, C.J.F. & Verdonschot, P.F.M. 1995. Canonical correspondence analysis and related multivariate methods in aquatic ecology. *Aquatic Sciences* 57, 255-289.

SUBJECT INDEX

A

Abies, 25, 32, 38, 40, 52, 54-55, 57, 62, 72, 74-76
Abieti-Fagetum, 74
Ablepharus kitaibelii, 178-179, 185, 190
Acanthocyclops kieferi, 326, 328
Acantholingua, 210, 212
Accubogammarus, 210
Acer, 25, 32, 38, 52, 56, 72, 74-76
 A. monspesulanum, 76
Acheulean industry, 149, 157-158
Achnanthes fogedii, 284
Acipenser nacarii, 295
Acomys
 A. cilicicus, 97
 A. minous, 92, 97, 102
 A. nesiotes, 97
Acrocentric chromosomes, 113, 115, 124, 126
Acroloxus, 212
Aculepeira talishia, 197
Adaptation, 60, 147, 152, 159-161, 176, 180, 304, 321
Adaptive radiation, 97, 259
Adjusted species richness, 87, 92, 101
Adriatic, The 1-2, 4, 13-14, 17, 19-20, 50, 52-55, 67-68, 70, 81, 83, 89-90, 105, 113, 160, 167-171, 174, 177, 181, 187, 189, 193, 203-215, 221, 255, 261-262, 266-267, 289, 291-306
Adriatic coast, 9, 20, 81, 89, 105, 168, 298
Adriatic plain, 160
Adriatic Sea, 1-2, 17, 52-53, 59, 68, 160, 215, 291-292, 295-296, 298, 300-306
Adriatic Triangle, 169-171, 174, 177, 181, 187
Adriohydrobia, 210
Adrioinsulana, 210
Aegean, The 5, 9, 14, 19-21, 59, 102, 135, 140, 142-145, 158, 162, 175, 193, 204, 208, 214, 221-222, 237, 239, 260-262, 265, 267, 269, 271, 294-295, 300, 310, 314, 316-317
Aegean island, 135, 140, 142-144
Aegean Sea, 140, 142, 208, 233, 260, 269, 271, 294-295, 300, 310
Aegeopotamos, River, 216
Aesculapian snake, 170
Africa, 6, 16, 25, 45, 88-89, 101, 106, 143, 149, 151, 154-156, 162, 164, 175, 177, 183, 213-214, 220, 239, 259-260, 274, 287, 295
African chameleon, 170
African Rift lakes, 5, 17, 219, 235, 285
Agnatha, 295
Agriculture, 38, 75, 161, 276, 320
Aiduronisi, island of, 193
Akgöl marsh, 35, 37, 42, 276, 282
Alaska, 121
Albania, 1, 8-11, 14-15, 17, 31, 35, 38, 46, 53, 55, 66, 81, 96, 100, 105, 109, 157, 162, 167, 184, 194, 220-221, 223, 226, 230, 238-239, 252-253, 255-256, 259, 262, 273, 276
Alburnoides, 210-211, 215
 A. bipunctatus, 215
Alburnus alburnus alborella, 210
Algae, 3, 273, 280, 286
Algal bloom, 246
Algeria, 87-88, 100, 103, 106
Allactaga
 A. euphratica, 137-139
 A. major, 137-138
Alleröd interval, 56
Allocricetulus eversmanni, 137-138
Allopatric, 92, 176, 180, 196
Allozyme, 130, 132, 187, 221
Alnus, 26, 32, 52, 54, 56-60, 66, 70
 A. glutinosa, 58, 66
Alpine crested newt, 172
Alpine lakes, 221, 287
Alpine newt, 171, 175, 179, 182, 186
Alpine orogeny, 11, 13-16, 21, 254, 268
Alpine shrew, 113, 130
Alps, 4, 9, 11, 15, 17, 19-20, 50, 60, 69, 71, 74, 77, 81, 83, 89, 97, 103, 110, 113, 115-116, 118, 129, 167-168, 179, 195, 215, 268, 309, 311, 318
Altai Mountains, 180
Altitudinal 18, 76, 89, 169, 186
 - range, 76, 169, 186
Amesoda, 222
Amnicolidae, 204
Amorgos, island of, 193
Amphibia, 171, 182-186, 315-316, 319

Amphibian, 3-5, 167-169, 170-183, 186-187, 309, 319-320
Amphipod, 208, 217, 253, 315, 318, 320
Amphipoda, 204, 217, 316-318, 320
Amphisbaenia, 171
Amphisbaenian, 169
Ampullariidae, 244
Amygdalus, 27
Anagastina, 210, 255
Anatolia, 5, 15, 26, 28, 35, 41-42, 45, 47-48, 118, 135, 137-140, 144, 175, 186, 204, 208, 211-215, 239, 265-266
Anchovy, 292, 299, 303
Ancient DNA, 49, 153. *See also* DNA
Ancient lake, 2, 5, 17, 219-220, 232, 235-237, 239, 243-244, 250, 252-257, 259, 271, 273
Ancona, 70
Ancylidae, 244
Ancylus, 212, 247
  A. fluviatilis, 248
Andros, island of, 193
Angola, 89, 102, 106
Anguis, 175, 189
Annelida, 314
Anodonta
  A. anatina, 221, 230, 248
  A. cynaea, 221, 232, 249
  A. c. deplanata, 221, 230
  A. c. gravida, 221
Anophthalmus, 317-318
Antalya, 139, 145
Anthropogenic impact. *See* Human impact
Antrocharis, 318
Antroherpon, 317-318
  A. apfelbecki, 319
Antrohyphantes, 196
Anura, 171, 185-186
Aoos, River, 212
Apennines, 14, 127, 167-168, 174-175
Aphaenopidius, 318
Aphaenops, 318
Aphaenopsis, 318
Apidima Cave, 151, 154-155, 162-164
Apletodon
  A. dentatus, 297
  A. incognitus, 296
Apodemus
  A. agrarius, 106, 112, 130
  A. ciscaucasicus, 122
  A. epimelas, 137, 140-142
  A. flavicollis, 106, 111-112, 128-129, 132-133, 137
  A. fulvipectus, 139
  A. hermonensis, 137, 139
  A. microps, 122, 131
  A. mosquensis, 122
  A. mystacinus, 89, 92-93, 104, 106, 112, 118, 127, 131, 137, 140-142, 145
  A. sylvaticus, 89, 106, 112, 122, 130, 132
  A. uralensis, 106
  A. vohlynensis, 122
Appalachian Mountains, 309
Apulian, 13, 299
Aquatic, 3, 5-6, 16, 201, 203-216, 243, 248-249, 254, 258, 261, 268, 275, 283, 286, 309, 311-314, 317, 320, 332
Arabia, 88, 100-102, 106, 139
Arachnida, 200, 314
Arago, 149
Aral Sea, 205, 213, 265, 270
Aralo-Caspo-Pontic distribution, 205
Araneae, 193, 199-200, 315-316, 318, 320-321
Arboreal pollen. *See* pollen, arboreal
Archaeological site, 29, 54, 57, 60, 161, 259
Archeolacerta, 171, 172
Area-adjusted species richness, 87-88, 92, 101
Areopolis, 151
Aridity, 35, 40, 42, 62, 64, 250
Armenia, 106
Arnissa, 135-137, 144
Arrhenius equation, 87, 106
Artemisia, 27, 31-32, 35, 40, 72
Arvicola
  A. mosbachensis / terrestris, 137
  A. terrestris, 106, 114, 130-133, 140
Arvicoline, 96
Asia, 2, 4, 16, 25-29, 31-32, 35, 38-42, 45, 47, 50, 73, 86, 88, 92-93, 97, 99-102, 106, 116, 118, 126, 128, 130, 139-141, 143, 151, 175, 183-184, 195, 213-214, 217, 220-221, 225, 239-240, 272
Asia Minor, 2, 4, 16, 86, 88, 92-93, 97, 99-102, 106, 116, 118-119, 126, 128, 130, 139-141, 143, 175, 184, 195, 217, 221, 240, 272
Asprochaliko, 158, 162
Astacus torrentiu, 209
Astagobius, 318
Asteraceae/Compositae, 27
Astipalea, island of, 193
Atapuerca, 149, 151-152

Athens, 102, 135-137, 145, 162-164, 182, 200, 217, 259
Athos Peninsula, 28
Atmospheric circulation, 45, 50
Atmospheric CO_2, 62
Attheyella crassa, 326, 330
Atticus cytotype. *See Microtus thomasi*
Attiko-Beotia region, 203, 213-214, 261
Atyid shrimp, 210
Aulacoseira granulata, 277-278, 281
Aulopyge, 210
Aurignacian, 154, 156
Autochthonous, 144, 171, 173-174, 181, 189
Autocorrelation coefficient, 90-95, 108
Autosomal fusion, 115
Autosome. *See* chromosome
Axis axis, 110
Azerbaijan, 106

B

Bacho Kiro cave, 154-156, 163
Baicallidae, 244, 269
Baikal, Lake, 5, 17, 115, 212, 219, 232, 235-237, 240, 269, 272, 286-287
Balistes carolinensis, 299, 303
Balkan Mountains, 3, 11, 310, 314, 317-318
Balkan Peninsula, 1-5, 9-10, 14, 28, 31, 50, 79, 80, 85-88, 91-92, 95, 99-101, 105, 107, 109-110, 121-122, 126-127, 135, 140, 147-148, 159, 167-169, 172-184, 193-200, 203-217, 221, 225-226, 234, 255, 271, 273-274, 291, 309-314, 317-318, 320, 323-324, 331
Balkan refuge, 97. *See also* Glacial refuge
Balkan snow vole, 140
Balkanides. *See* Balkan Mountains
Balkanostenasellus, 205, 209
  B. skopljensis, 209
Balkhash, Lake, 138
Baran, River, 157
Barbastella barbastellus, 105
Barbus
  B. albanicus, 213-214
  B. barbus, 209
  B. cyclolepis, 208-209, 213
  B. euboicus, 213
  B. graecus, 211, 213-214
  B. macedonicus, 209
  B. peloponensis, 212
  B. plebejus, 210
  B. prespensis, 213
  B. tauricus bergi, 208
  B. thessalus, 209
Barrier, 16-17, 19-20, 42, 50, 141, 151, 161, 168, 175, 176, 215
Barusia, 196
Basking shark, 301-302
Bathymetry, 244, 256
Bathynellacea, 315-316
Bathypelagic fish, 298, 303
Bathyscia, 318
Bátorliget marsh, 53, 57, 68
Batrachofauna, 167, 174-176, 181
Bat, 5, 81, 83-85, 87, 90-92, 101, 104-107, 115, 132-133
Bavelian, 140
Bay of Trieste, 16, 19-20, 168, 193, 296, 299-301, 305
Beetle, 309, 317-318, 319-320, 322
Belgrandia, 204
Belgrandiella haesitans, 248
Belonidae, 296, 303
Benedictiidae, 244
Bern Convention, 237-238
Betfia, 140, 145
Betula, 32, 38-39, 52, 54-55, 57, 59-60, 67, 70, 72, 74
  B. pendula, 60
Beyşehir, Lake, 41, 46
Biarmed chromosome, 113, 124
Bicanestrinia, 208, 211, 214
Biharian, 140, 145
Bilzingsleben, 149, 157
Biochronology, 102, 135, 144
Biocoenosis, 291
Biodiversity, 2-9, 15-16, 19, 21, 28, 47, 65-66, 69, 77, 87, 102, 109, 135, 143, 147, 160-161, 167, 179, 181, 183, 219, 243, 258, 261, 270-271, 272-275, 281-289, 291, 293, 295, 297-298, 300, 302, 304-305, 310, 317, 322, 330
Biogenetic reserve, 169
Biogeographic island, 86
Biogeographic origins, 5
Biogeographic pattern, 167, 184, 285
Biogeographic province, 320
Biogeography, 46, 65-66, 77, 80, 102-103, 131, 167, 169, 172, 175, 177, 181, 183-184, 186, 214, 219, 220, 233, 234, 261, 262, 265, 267, 270, 271, 287, 289, 291, 309, 311
Biological invasion, 302
Biological species concept, 313, 317
Biome, 49, 169
Biospeleology, 5

Bison bonasus, 110
Bistra Planina, 120
Bithynia
B. cf. *meridionalis*, 247
B.graeca, 247-248
B.tentaculata, 248
Bithyniidae, 244
Bivalve, 5, 219-220, 225, 230, 232-240, 243, 247-248, 254
Bivalvia, 238, 239-240, 314, 316, 319
Biwa, Lake, 219, 236, 239
Black Drin River, 253
Black rat, 128, 132, 140
Black Sea, 9, 14, 18-20, 53, 56, 67, 83, 139-140, 168, 193, 203-204, 207-208, 214, 216, 233, 265, 268, 291, 295
Blanus strauchi, 181, 189
Bled, Lake, 74
Blenniidae, 296, 305
Blind mole, 92
Bogue, 292
Bolivia, 89, 101, 106
Bombina
B. bombina, 172, 176, 186-187
B. variegata, 172, 176, 186-187
Bone artefact, 149
Boops boops, 292
Bordul Mare, 159
Boreal forest. *See* Coniferous forest
Boreal sector, 121
Boreal species, 138
Borneo, 89, 104, 106
Bosna i Hercegovina. *See* Bosnia and Herzegovina
Bosnia. *See* Bosnia and Herzegovina
Bosnia and Herzegovina, 1, 10, 14, 16, 105, 109, 196, 233, 273, 312, 314, 316, 318, 320
Bosphorus land bridge, 118, 175
Bosphorus Strait, 9, 14, 116, 118, 127
Bosphorus-Dardanelles, 135, 139, 193
Bottleneck, 66, 176, 183
Bougainvilliidae, 319
Boxgrove, 149
Brama brama, 299, 303
Brown frog, 172, 175
Brunhes/Matuyama boundary, 151
Brushtooth lizardfish, 300
Bryocamptus
B. balcanicus, 326, 328
B. borus, 326
B. cf. *borus*, 326
B. dacicus, 326, 328
B. pygmaeus, 326
B. pyrenaicus, 326
B. typhlops, 326, 328
B. zschokkei, 326, 330
Bukovina, 19
Bulgaria, 1, 2, 6, 9-10, 14-16, 20, 32, 38-39, 48, 53, 55, 58-59, 65, 83, 99, 102, 105, 109, 113, 116, 119, 128-131, 135, 137, 144, 154, 156, 159, 163, 169, 180, 184, 193-194, 196, 199-200, 205-206, 208, 220, 222-223, 230, 238, 275, 289, 309-318, 320, 322
Bulgarocerberus, 205, 209
Bureschia, 318
Buxus sempervirens, 74
Bythinella, 204, 207-208, 210
B. charpentieri, 248
B. dacica, 207-208
B. schmidti, 204
Bythynellidae, 204, 210, 212-213

C

Caenozoic, 177, 225
Calanoida, 319, 323, 332
Calcareous Alps, 311
Caloneis, 281
Campylodiscus echineis, 284
Candona, 263, 266, 268, 272
C. angulata meridionalis, 265-267
Candonid ostracod, 263
Candoninae, 263, 272
Cantabrian Mountains, 180
Cape Marjan, 293
Cappadocia, 276, 282
Capra ibex, 110
Capreolus
C. pygargus, 122
C. capreolus, 112, 122
Carabidae, 315, 318, 320-321
Caras, 208
Carboniferous, 12, 15
Carcharhinus plumbeus, 298, 305
Cardiid bivalve, 254
Carinodreissena, 226, 233, 247, 249
Carinurella, 210
Carniola, 309
Carnivore, 6, 109, 121
Carpathian Mountains, 9, 11, 14-15, 19-20, 56, 81, 83, 85, 96, 110, 132, 167, 186, 254, 268, 316
Carpinus, 25, 28, 32, 38-40, 52, 55-57, 60, 62, 70, 72, 74-75
C. betulus, 40, 52, 57, 70, 72, 75
C. orientalis, 25, 39, 52, 55, 57, 74-75

Caspian Sea region, 17, 28, 139, 213, 225, 233, 265, 268
Castanea, 28, 39-40, 76
Castor fiber, 110, 137
Cataetyx alleni, 296
Catalonia, 295
Caucasus Mountains, 5, 11, 88-89, 100, 106, 126, 139, 152-153, 162-163, 174, 208, 268, 309
Caudata, 171, 173, 183-184
Caulerpa taxifolia, 300
Cave, 2, 3, 5-6, 21, 29, 55, 59, 74, 103, 135, 148-157, 159, 162-163, 195-196, 199-200, 309, 311-314, 317-328, 330, 332
Cave spider, 196
Cavernisa, 207
C-banded karyotype, 111
Cedar, 28
Cedrus, 41
Central Europe, 2, 9, 16, 20, 25, 28, 31, 50, 58, 60, 67, 69, 77, 79, 88, 99-101, 106, 111, 113, 116, 121, 127-128, 138-139, 164, 173, 175-176, 185, 205, 208-209, 214, 220, 225, 233, 239, 259
Centre of speciation, 175, 197
Centre of origin, 181
Centric fusion, 112, 118
Centromere, 115
Ceratonia, 27
Cerna, 207, 208
Cetina, River, 210
Cetorhinus maximus, 301-303, 305-306
Chalcalburnus belvica, 212, 217
Chalkidiki, 149
Chamaeleo, 170, 181, 189
  C. africanus, 170, 189
Charcoal, 29, 32, 38, 45, 48, 54-60
Chenopodiaceae, 31, 33, 35, 40
Chenopod, 72
C-heterochromatin, 119, 132
Chilopyrgulinae, 212
China, 138, 149, 309
Chionomys nivalis, 95, 97, 102-103, 106
Chios, island of, 135-137, 139-140, 143-144
Chiroptera, 81, 103, 105, 115, 128, 132, 133
Chlorophyll, 299
Cholevidae, 315, 318-322
Chondrichthyes, 295
Chondrostoma, 208-210, 213
  C. nasus, 208, 213
  C. prespense, 213
  C. vardarense, 208, 209
Chromosomal diversification, 94
Chromosomal polymorphism, 110, 130-132
Chromosomal polytypy, 113
Chromosomal variation, 109-110, 116, 121-122, 126-129
Chromosome number, 120, 124
Chronology, 7, 47, 145, 150, 157, 164, 249, 264, 275
Ciliata, 320
Cingulipisidium crassum, 233
Circum-Mediterranean, 25, 167, 173, 177, 205, 210, 213
Clactonian industry, 158
Cladocera, 315-316, 319
Classical Karst, 309
Cleaver wrasse, 299
Clethrionomys glareolus, 106, 112, 114, 131, 133, 137
Clethrionomys-Microtus clade, 96
Climate, 3, 9, 11, 17, 19-21, 25-26, 28-29, 32, 35, 39-40, 45-48, 50-51, 61-62, 64-67, 73-76, 152, 157, 159, 162, 169, 244, 259, 271, 274-276, 283-284, 287-289, 301. *See also* aridity, humidity, moisture, precipitation, seasonality, Siberian anticyclone, temperature, wind
Climate change, 38-39, 41, 47-48, 71, 74, 160, 177, 287-288, 292, 299, 303-304
Climate, continental, 18, 62
Climate, Mediterranean, 17-18, 62
Climatic oscillation, 49, 51, 58, 61-62, 65, 67, 76
Climatic reversal, 35, 39, 61
Climatic zone, 18, 283
Clinal increase, 111
Clinal variation, 90
Clingfish, 296
Cluj, 20
Cnidaria, 314, 316
Cnidarian, 319
Cobitis, 205, 208-214
  C. punctilineata, 208
  C. strumicae, 208
  C. elongata, 206, 208
  C. meridionalis, 187, 190-191, 212-213
  C. taenia, 209
  C. vardarensis, 209
Coefficient of Community, 99-101, 107
Coelorinchus occa, 296
Coevolution, 250, 260

Coggyria, 74
Coleoptera, 314-322
Collembola, 315, 321
Colonisation, 4, 16, 58, 60-61, 67, 72-74, 86, 97, 127-128, 140, 147, 176, 179, 181, 196, 233, 237
Commensal house mouse, 86, 105
Common hamster, 109, 131
Common shrew, 92, 115, 127-128, 130, 133
Como, Lake, 221
Congeria, 210, 225, 234, 237, 240
Congeria kusceri, 210, 225, 234, 237, 240, 319
Coniferous forest, 25-26, 32, 35, 40, 52, 54-56, 59, 62, 70, 76, 89-90
Conservation, 3, 6, 29, 47, 49-50, 64, 183, 220, 237-238, 271, 285, 287-288, 310
Conspecific population, 126, 176, 180
Constanta, 19
Contact zone, 81, 178, 185
Continental plate tectonics, 11-12, 21.
Conventional staining, 110, 117
Convergent origin, 236
Copepod, 332
Copepoda, 315-332
Corbiculidae, 219-220, 235, 238-239
Corfu, island of, 157, 199
Cornus mas, 74
Corylus, 32, 38-39, 41, 52, 54-55, 57-62, 67, 70, 74-77
  C. avellana, 59, 67
Coryphaena hippurus, 298-299, 303
Cosmopolitan distribution, 69, 195, 222, 251, 273-274, 285
Cotinus, 74
Crater lake, 17, 31, 282, 284
Crayfish, 206, 209, 213, 216
Crenobiotism, 215
Cres, island of, 53, 56, 67, 90
Crested newt, 176, 183-186
Cretaceous, 11, 13, 16, 222
Crete, island of, 3, 12, 15, 19-20, 45, 92, 97, 102, 104-105, 142-144, 183, 186, 193-194, 196, 199, 269, 213
Cricetulus
  C. griseus, 112, 129
  C. migratorius, 106, 132, 137, 140
Cricetus cricetus, 106, 112, 114, 129, 131, 137-138
Crimea, 208
Crkvice, 19
Crna Gora, 310
Črna jama, 324-328
Crnulja cave, 6, 320
Croatia, 2, 9-11, 14, 17, 38, 53, 55-56, 65-67, 105, 109, 140, 151-152, 154, 156-159, 163-164, 169-170, 172, 174, 180, 187, 194, 196, 221, 263, 272, 274, 288, 310-317, 320, 322
Croatian coast, 16, 275, 299, 311
Crocidura
  C. leucodon, 89, 105, 114, 127-128
  C. suaveolens, 89, 105, 111-112
  C. zimmermanni, 92, 97, 104
Crustacea, 217, 270, 272, 314, 318, 322, 332
Crustacean, 209, 213, 217, 253, 261, 300, 320, 322
Cryphoecina, 196
Cryptic taxa, 32, 59-60, 121, 172, 181, 297
Ctenophthalmus nifetodes, 96-97
Curata, 159
Cyclopoida, 323, 330, 332
Cyclotella, 277, 279, 287
  C. choctawhatcheeana, 284
  C. ocellata, 277, 279
  C. petrowskae, 277, 279
  C. trichonidea, 277, 287
Cymbella juriljii, 277
Cyprideis torosa, 284
Cyprinidae, 204, 216-217, 271, 321
Cyprus, 86, 97, 103
Cyrtodactylus kotschyi, 171, 189
Cytheroid ostracod, 263
Cytheroidea, 263
Cytogenetic markers, 119
Czech Republic, 2, 106

D

Dabriana, 204, 206
Dalmacija. *See* Dalmatia
Dalmatia, 11, 14, 16, 38, 56, 59, 74, 89, 100, 167, 203-204, 206, 209-212, 214-216, 233, 261, 266-267, 284, 289, 311, 314, 323
Dalmatien. *See* Dalmatia
Dama dama, 110
Dansgaard-Oeschger (D-O) event, 59, 61
Danube, 9, 14-15, 19, 83, 97, 122, 132, 168, 185, 193, 196, 203-210, 212, 214-216, 226, 230, 233-234
Danubian, 194, 208-209, 212, 214-216, 261
Dardanelles. *See* Boshporus-Dardanelles
Dead Sea-Jordan rift, 40

Decapoda, 204, 315-316
Deciduous forest, 25, 27-28, 32, 38-40, 46, 52, 54-56, 58-60, 62, 65, 71, 89-90
Deep demersal fish, 296
Deer, 97, 105, 164
Delaya bureschi, 314
Deme, 176
Dendrocoelum, 212
Desert, 26, 151
Diacyclops
  D. languidoides, 326, 330
  D. languidus, 326
Diaphus metopoclampus, 296
Diatom, 5, 35, 273-289
Didogobius
  D. schlieweni, 297-298
  D. splechtnai, 296, 304
Dimnice, cave, 325-328
Dinaric Alps. *See* Dinarides
Dinaric karst, 89, 174, 225, 234, 240, 309, 311-313, 317-320, 322
Dinaric region, 194, 196, 233, 310, 322, 324
Dinaric snow vole, 125-126
Dinarides, 3, 9, 11, 14, 16-17, 19-20, 81, 83-84, 89-90, 93, 97-99, 101, 113, 168, 174, 215, 268, 309, 311, 316, 318, 320, 324, 330-331
Dinaromys bogdanovi, 89, 93, 95-96, 102-104, 106, 125-126, 137, 140
Diplecogaster bimaculata, 297
Diploid number, 113, 119
Diplopoda, 315-316, 318, 321
Diplura, 315
Diptera, 315, 321
Disjunct distribution, 59
Disjunct range, 208, 221
Dispersal, 32, 39, 42, 60, 76, 84, 128, 135, 139-143, 167, 176, 214-217, 243-244, 261, 270, 285, 300
Divergence, 64, 66, 77, 97, 119, 126, 176, 179, 183-184, 244, 271
Djerdap Canyon, 83
DNA, 49, 57, 66-67, 109, 152-154, 162-163, 176, 180-182, 185-186, 234, 270-271.
Dniester, River, 212
D-O event. *See* Dansgaard-Oeschger event
Dobrudja. *See* Dobrugea
Dobrugea, 9, 11, 15, 20, 135, 138, 169-170, 174, 181, 187, 194, 196
Dojran, Lake, 11, 17, 226-227, 252, 256-257, 260, 262, 265, 271-272, 274-277, 279-281, 285-287, 289
Dojransko jezero. *See* Dojran
Doline, 15, 276
Doïrani. *See* Dojran
Drava, River, 10, 203, 205-206
Dreissena, 5, 225-230, 233-235, 237, 239-240, 249
  D. blanci, 226-227, 233-234, 258
  D. bugensis, 226-227, 229, 233-234
  D. cristellata, 233
  D. filifera, 233
  D. polymorpha, 225-227, 229-230, 234, 239-240
  D. rostriformis, 233
  D. stankovici, 133, 226-227, 229-230, 233-234
Dreissenidae, 219-220, 225, 230, 235, 240, 319
Dreissenid, 234
Drifting, 141
Dryomys nitedula, 106, 137
Duvalius, 318
Dysdera, 195
Dysderidae, 195, 199, 318

E

Earthworm, 69
East African Rift lake. *See* African rift lake
Eastern Mediterranean, *See* Mediterranean, Eastern
Eastern spadefoot toad, 172, 186
Ecology, 8, 13, 21, 39, 46-48, 65-68, 77, 97, 102-103, 129-131, 183-186, 198, 239-240, 259-260, 272, 287-288, 304-305, 321, 332
Ecosystem, 3-7, 61, 84, 109, 161, 165, 273, 275, 283-286, 289, 292, 299-300
Ecosystem degradation, 84
Ecotone, 69
Edaphobiont, 318
EDDI, *See* European Diatom Database Initiative
Edessa, 38
Egypt, 88, 100, 104, 106
Elaphe longissima romana, 170
Elaphoidella
  E. cvetkae, 326
  E. kieferi, 326
  E. stammeri, 326
Elburz Mountains, 268

Eliomys quercinus, 89, 106, 114, 116, 127, 130, 133
Elladoherpon, 318
Ellobius talpinus, 137-138
Elster glacial, 139
Emericiidea, 204
Emirkaya, 139, 144
Emmericia, 210
Emmericiidae, 210
Emys orbicularis, 176, 185, 188
Endemism, 2-6, 28, 64, 69, 80, 87, 92-98, 101, 105, 115, 124-128, 140, 142, 144, 167-169, 173-174, 177, 181-182, 187, 193, 195-200, 204-213, 215-216, 219, 223, 225-226, 232-239, 243-244, 247-248, 250, 252-257, 261, 263, 266, 271, 273-275, 277, 279-280, 285-288, 295, 314, 316-318, 320, 323-324, 330-332
England, 139, 195
Environmental heterogeneity, 86
Eocene, 16, 210, 213
Eolagurus luteus, 137-138
Epactophanes richardi, 326
Epetesicus serotinus, 105
Epigean habit, 215, 309, 330-331
Epigravettian industry, 156
Epinephelus
  E. aeneus, 296, 304
  E. coioides, 296, 300, 305
Epirus, 68, 162, 164-165, 245, 250, 260
Eptesicus nilssonii, 105
Ereğli marshes, 35, 37
Ericacea, 76
Erinaceus
  E. concolor, 89, 105, 113, 121, 129
  E. europaeus, 110, 113, 121
Eski Acigöl, 276, 281-282, 284
Estuary, 225, 233, 261
Euboea, island of, 213, 271
Eucommia, 70
Euglesa, 222
Eunapius subterraneus, 210, 319
Eurasia, 13, 66, 132, 144, 149, 154, 177, 217, 220, 233, 236
Eurasian, 12-13, 28, 138, 140, 204
European Diatom Database Initiative, 274, 280, 287
European souslik, 139
Euro-Siberian, 74, 213, 221, 232, 261
Euryhalinity, 204, 210, 265
Eurytopic organism, 5, 245, 257
Eutrophication, 274, 276, 281, 284, 286-287
evergreen forest. *See* Coniferous forest
Eversmann´s hamster, 138
Evolution, 2, 4-6, 9, 11-13, 21, 52, 64-68, 96-97, 103-104, 129, 131-133, 144, 147-149, 151, 153, 155-156, 162, 165, 172, 175, 177-186, 214-215, 217, 219, 225, 232, 234-235, 237, 239-241, 243, 257, 259-262, 268, 270-272, 285, 317, 331-332
Evros, River, 169-170, 174, 187, 208
Exocoetidae, 296, 305
Exotic rodent, 86
Extinction, 16, 28, 31, 50, 64, 76, 110, 135, 138, 142, 147, 156, 160-161, 165, 176, 192, 234, 251, 256, 260, 269, 272

F

Fageiella, 196, 200
Fagus, 25, 28, 32, 38-40, 52, 54-60, 66, 70, 72, 74-76
False scorpion, 313, 320
Faunal complex, 126, 197
Faunal exchange, 4, 139, 175, 215
Faunal heterogeneity, 79-80, 98-99
Faunal homogeneity, 98
Faunal innovation, 140
Faunal remains, 135, 149, 152
Faunal renewal, 135
Faunal similarity, 99
Feldhofer cave, 152, 154
Feral, 86
Ferrissia, 248, 258
  F. clessiniana, 258
Field vole, 138
Finland, 106
Fire-bellied toad, 172, 186
Fish, 4, 29, 59, 203-216, 220, 246, 253, 261-262, 269, 291-305, 309, 321
Flea, 96-97
Folkia, 196, 199
Fontogammarus, 210
Foraminifera, 51, 63
Former USSR, 11, 102-103, 173, 185, 220, 225, 239
Former Yugoslav Republic of Macedonia, 11, 81, 100, 105, 110-111, 115, 118-119, 208, 220, 222, 225-226, 230, 234, 274-277, 279-280, 282, 309-310, 313-314, 316, 320, 330-331
Fossil, 4-6, 29, 46, 49, 58-59, 67, 96-97, 135-144, 149, 151-152, 155-156, 158, 164, 220, 233, 237, 240, 243, 247, 249-250, 252, 255-259, 261, 263-269, 271, 275, 282-284, 286

Fossil evidence, 4, 58, 67, 261
Fossil record, 49, 58, 97, 139-141, 149, 158, 250, 266, 270
Fossil site, 135-136, 138, 140, 269
Fossorial, 169
Fragilaria
F. capucina, 277
F. vaucheriae, 277, 288
France, 115, 139-140, 164-165, 180, 195, 200, 284, 304, 318, 323
Fraxinus, 25, 38, 56, 74-76
F. ornus, 25, 74, 76
Freshwater, 4-5, 167, 204-205, 208, 213-221, 225, 233, 235-240, 244, 255, 259-267, 270-271, 276, 285, 319-320, 322
Frog, 77, 179, 184-186
Fuente Nueva, 151
Full glacial. *See* Pleniglacial
FYROM. *See* Former Yugoslav Republic of Macedonia

G

Gajtan cave, 157
Gammaridae, 210, 269
Gammogobius steinitzi, 296, 304
Garda, Lake, 221
Garden dormouse, 116, 127, 132-133
Gastropod, 5, 219, 232, 235, 238, 241, 243-245, 247-248, 250-260, 318
Gastropoda, 217, 236, 238-240, 259-260, 272, 314-316, 320-321
Gavdos, island of, 193
G-banding, 120, 125, 128-130
Gelyelloida, 323
Gene flow, 121, 155, 176, 186, 215
Gene structure, 77
Genestiellina, 318
Genetic differentiation, 172
Genetic exchange, 176
Genetic isolation, 64, 176
Genetic markers, 121, 180
Genetic signature, 58.
Genome, 109, 130, 152, 155, 176, 185
Genomic structure, 58-59
Genotypic variation, 172
Geographic barrier, 176
Geographic origin, 142
Geographic pattern, 84, 87, 93
Geographic range, 115, 181
Geography, 6, 9, 11, 29, 49-50, 64, 77, 86, 92, 103, 118-119, 124, 164, 169, 175-176, 178, 181, 194, 256, 271, 291
Geology, 11, 15-16, 21-22, 66, 68, 177, 186, 245, 259, 272
Georgia, 106
Germany, 2, 106, 152, 157-158, 195, 231, 260
Ghab, 40
Gilt sardine, 292, 304-305
Glacial, 2, 4, 6, 17, 28, 31-32, 35, 38-42, 49-53, 59-64, 71-72, 94, 135-136, 139-140, 143-144, 157, 160, 171, 174, 177, 179-181, 251, 264-265, 267, 276, 283. *See also* Last Glacial, Last Glacial Maximum, Pleniglacial.
Glacial climate, 35
Glacial period, 50, 67, 135
Glacial refuge, 2, 4, 32, 49, 59-60, 71, 94, 140, 174, 176, 180-182
Glacial relict, 180
Glacial, penultimate, 135-136, 139, 143
Glacial-interglacial cycle, 31, 48, 68
Glacicavicola, 318, 322
Glis glis, 89, 97, 106, 129, 137
Glochidia, 220
Gobiesocidae, 296, 304-305
Gobiidae, 295-296, 304-305
Gobio, 209-211
G. couchi, 296, 304
G. gobio, 215
G. kessleri banarescui, 209
G. uranoscopus elimeius, 209
Gobius kolombatovici, 295-296, 298
Gölhisar, Lake, 41-42, 44
Gondwanaland, 13, 177
Gorizia, 193
Graben, 17, 273, 282
Graecanatolica, 213
G. macedonica, 256, 260, 272
Gramineae, 27, 35, 39-41
Gramineae/Poaceae, 27, 35, 39-41
Gramousti, Lake, 38, 53, 55, 68
Gran Dolina, 151
Gray hamster, 140
Graziana, 204, 207
G. lacheineri, 207
Great jerboa, 138-139
Greece, 1-4, 6, 9-12, 15, 17, 21, 28, 32-33, 38-39, 45-48, 50, 52-53, 55, 58-59, 62-68, 72, 74, 83, 96, 100, 103, 105, 109, 113, 120, 129, 131-132, 135, 137, 139-140, 144-145, 149, 151, 154, 156-165, 170, 179, 181, 186, 193-194, 196, 199-200, 204, 208, 211-213, 215, 217, 220-226, 233-234, 238, 243, 245-246,

250, 256, 258-260, 262-265, 267, 270-289, 309-312, 314, 316, 318, 320-321
Greek islands, 14, 101, 174
Grey triggerfish, 299, 303
Grossuana serbica, 208, 209
Groundwater, 14, 255, 262, 271, 323, 330
Guenther's vole, 119
Gulf of Corinth, 86
Gulf of Lyon, 295
Gulf of Suez, 300
Gura Dobrogei, cave, 135-136
Gymnocephalus schraetser, 205
Gyraulus, 212, 236, 240, 247, 250, 259-260
G. cf. *piscinarum*, 248
G. crista, 248
G. janinensis, 247-248, 250

H

Habitat, 29, 50, 59, 64, 84, 89-90, 97, 101, 128, 140, 159-160, 169, 171, 177, 184, 219, 221, 235, 237, 244-245, 251, 257, 261, 269, 280-281, 283, 285-286, 297, 302, 310, 319-320, 322-323, 330
Habitat change, 251
Habitat heterogeneity, 177
Habitat requirements, 97, 251, 286
Hadesia vasiceki, 320
Hadzia, 205, 209
H. gjorgjevici, 209
Haemulidae, 296, 305
Hamster, 94, 109, 131-132, 138-140
Hantzschia amphioxys, 281
Haplotype, 58, 176, 180
Harpacticoida, 323, 330, 332
Harvest mouse, 122, 132
Haua Fteah, 151
Hauffenia, 204
Hedgehog, 109, 129
Heidelberg, 149, 164, 240
Heinrich events, 61-63
Hellas, 310
Hemiramphidae, 296, 303
Hemirhampus far, 296, 300
Herpestes auropunctata, 110
Herpetofauna, 4, 167, 174-176, 181, 183-185
Herzegovina. *See* Bosnia and Herzegovina
Heterochromatin, 111-114, 120-122, 126, 128, 130
Heterochronic change, 179
Heterochronic phenomena, 167, 182
Heterochrony, 179, 184
Heterozygosity, 176
High Karst Zone, 16
Hippeutis complanatus, 248
Hirudinea, 315-316
Holandriana, 208
Holarctic distribution, 4, 28, 69, 93, 121, 195, 204, 212-213, 215, 232-233, 237
Holocene, 6, 8, 25, 31-32, 35, 38-42, 45-48, 52, 55, 57, 61, 65-66, 68-76, 96-97, 102, 104, 127, 139, 144, 156, 161, 179, 249, 250-251, 259, 264-266, 275, 281, 283-284, 287-289. *See also* Postglacial
Holoeuropean distribution, 195, 197
Holomediterranean distribution, 195
Hominid, 147-164
Homo
H. (sapiens) taenarius, 151
H. erectus, 149, 151, 152, 154, 156
H. ergaster, 154
H. habilis, 156
H. heidelbergensis, 148-149, 151-152, 156, 158-160, 164
H. neanderthalensis, 148, 151, 153, 156, 160
H. sapiens, 147-149, 151, 154-156, 160
Homozygosity, 58, 176
Horatia
H. epirana, 248
H. macedonica, 209
Hornbeam, 25, 66
Horse-mackerel, 292
Hot spot, 2-3, 6, 8, 64, 79, 81-85, 98, 107-108, 169, 219, 243, 314
Hrvatska, 272, 310
Hrvatsko Primorje, 311
Hrvatsko Zagorje, 151, 153, 163
Hula, Lake, 35, 40, 46
Huleh, Lake. *See* Hula
Human evolution, 147
Human impact, 6, 28, 39, 40-41, 48, 87, 105, 128, 295, 299
Humidity, 19, 42
Hungary, 14, 17, 19, 32, 53, 57, 60, 68, 70, 72, 105, 138, 157, 205, 268
Hušnjak, 152
Hybrid zone, 172, 178-179, 182, 186
Hybridisation, 167, 178-179, 182, 184
Hybridogenesis, 167, 179, 182
Hydrobia janinensis, 247
Hydrobiidae, 204, 210, 213, 244, 318
Hydrobiid, 244, 250-252, 255, 257

Hydrobioidea, 204, 217, 240, 272
Hydrobioid, 204, 210, 213, 215
Hydrology, 14, 47, 268, 283
Hydrozoa, 319
Hypogean habit, 3-5, 204, 208-210, 215, 319-320, 322

I

Iberia, 101, 151. *See also* Spain
Iberian Peninsula, 4, 59, 127, 151, 167-168, 172, 174-175, 180, 214
Icariella, 196
Ice Age, 46, 67, 147
Ice volume, 51, 61, 71
Icelandic cyclone, 294
Ichthyofauna, 291-292, 295-297, 300, 303-305
Iezerul Calimani, 53, 56
Illinois, 106
Immigration rate, 177
Imros, island of, 193
Index of Faunistic Change, 95, 99, 107
India, 89, 106, 225
Indiana, 106, 163
Indicator species, 299
Ingolfiella, 209
Ingression, 293-294, 305
Insecta, 314
Insectivora, 81, 103-105, 128-133, 144
Insectivore, 79-81, 83-85, 87, 89-92, 101, 105, 107, 109
Insolation, 293
Interglacial, 4, 28, 31, 39, 49-50, 64, 135, 139, 157, 176, 182, 249, 260, 267, 272
Interstadial, 56-57, 59-60, 62
Interstitial, 5, 309, 311, 320, 323, 330
Inter-Tropical Convergence Zone, 45
Intraspecific differentiation, 172, 180
Intraspecific karyotype variation, 110
Introduced species, 110, 189, 300
Ioannina, Lake, 4, 17, 31-33, 35, 46, 50, 52-53, 55, 62-64, 72, 203, 211, 243, 245-246, 250-251, 258, 260, 271
Ionian. *See* South Adriatic-Ionian
Iowa, 106
Ipeiros, 316
Iran, 5, 28, 33, 36, 46, 48, 88, 100, 103, 106, 208, 214, 225
Isernia, 157
Islamia, 213-214
Island arcs, 15, 140-142
Island effect, 88
Island endemic, 87, 92
Isonzo, River, 210
Isopod, 209-210, 253, 320
Isopoda, 204, 212, 315-319, 321
Israel, 35, 45, 129, 139, 186, 212, 225, 300
Isthmus of Corinth, 86
Istranca Mountains, 11, 118
Istranca race, 115
Istria, 1-2, 10, 53, 55, 59, 65, 169, 209, 311, 321
Italian Peninsula, 88, 172
Italy, 1, 17, 31-32, 34-35, 46-48, 55, 66, 96, 102, 104, 106, 140, 157, 183, 204, 210-215, 225, 265, 284, 289, 291, 304-305, 323
IUCN, 237

J

Jabuka Pit, 291, 293, 298
Jaccard coefficient. *See* Coefficient of Community
Japan, 121, 239
Java, 149
Jird, 140
Jordan, 40, 46, 225
Juglans, 32, 39-40, 76
Jugogammarus, 210
Julian Alps, 9, 19-20
Juniperus, 25, 27, 32, 38, 52, 54, 57, 59-60, 62-63, 70, 72, 76
Jura Mountains, 15
Jurassic, 13, 220, 222

K

Kalamakia, 159
Kalymnos, island of, 140, 141, 144
Kapela Mountains, 169
Karain, 139
Karamık Batalığı, Lake, 41
Karpathos, island of, 142-143
Karst, 5-6, 14, 16-17, 55-56, 67, 69, 72, 144, 149, 151, 174, 252, 256, 276, 282, 288, 309, 311, 317, 323-324, 330-332
Karyotype, 4, 102, 109-133
Karyotypic race, 113, 115-116, 118
Karyotypic sibling species, 121
Kasos, island of, 143
Kastoriás, Lake, 17, 221
Kastro, 158
Katsika, Mount, 149
Kazakhstan, 138-139, 280, 286
Kis-Mohos Tó, 53, 57

Kleisoura cave, 156
Klithi, 68, 154, 160, 162, 165
Kokkinopilos, 157, 159, 164
Kompoljska jama, 324
Konya basin, 35, 37, 47, 276, 282, 288
Kopais, Lake, 38, 52-54, 62-63, 65-68, 264-265, 270-272
Koper, 2, 7, 19, 65, 74, 103, 238
Korea, 73, 77
Korrission lagoon, 157
Kos, island of, 140, 252, 255, 260, 321
Kosovo, 89, 179, 267
Krapina cave, 151-153, 156, 158-160, 162-165
Kras, 14, 16, 21, 309
Kriti. *See* Crete, island of
Kritimys, 137, 139, 142-143
  K. aff. *kiridus*, 143
  K. sp., 137, 139, 143
Krka, River, 168, 210
Krovili, 158
Kufonisi, island of, 193
Kupa, River, 168
Kvarner, 169, 297, 304

L

La Adam cave, 135-137
La Sima, 152
Labridae, 296
Lacertilia, 171
Lacustrina, 222
Lacustrine environment, 4, 21, 29, 31, 47-48, 66, 233, 234-236, 246, 249, 259-262, 266, 268-269, 271, 275
Lagoon, 157, 208, 261, 275, 284, 289
Lagurus lagurus, 137-139
Lake, 6, 17, 21, 29, 35, 38, 41, 46-48, 54, 56, 66-67, 183, 203, 212, 215-216, 219, 222, 230, 233, 235, 237, 243-246, 249-258, 261, 263, 266-267, 269, 271, 273-277, 279-289. *See also* Ancient lake
Lake level, 29, 35, 47, 54, 66-67, 251-252, 255, 257, 271, 276, 283. *See also* Water depth
Land corridor, 139, 143
Larix, 57, 59, 60, 70, 72
Lasconia, 196
Last glacial, 31-32, 46-52, 61-67, 72, 76, 264-265, 267
Last Glacial Maximum, 47, 50-52, 54-55, 57-61, 66, 160, 251
Last interglacial, 46, 61, 249, 267
Late Glacial, 35, 38, 40-41, 46
Late Glacial-Holocene transition, 38, 41
Latitudinal belt, 79, 91
Latitudinal gradient, 3, 175
Latitudinal pattern, 86, 91
Latomi, 135, 136
Laudakia stellio, 181, 189
Laurasian, 13, 177
Lebanon, 28
Lebetus guilleti, 296, 304
Leiognathidae, 296, 303
Leiognathus
  L. klunzingeri, 296, 300, 303
  L. klunzingeri, 296
Lepidion lepidion, 296
Lepthyphantes, 195, 200
  L. centromeroides, 195
  L. spelaeorum, 195
Leptocythere, 263, 267, 272
  L. fluviatilis, 266
  L. karamani, 266
  L. picturata, 266
  L. pseudoproboscidea, 266
  L. saljanica, 266
Leptodirinae, 318, 320-322
Leptonetella, 196
Lesbos, island of, 142
Lesotho, 106
Lessepsian migrant, 296, 300
Lessepsian migration, 300, 303
Lesser horseshoe bat, 115, 116
Lesser mole-rat, 121
Lesser white-toothed shrew, 111
Leuciscus, 210, 217, 271
Levantine Basin, 293
Levantine Intermediate Water, 292, 294
LGM, *See* Last Glacial Maximum
Libyan Sea, 143
Ligurian Sea, 298
Limnos, island of, 193
Limnosbaena, 205, 210
Linyphiidae, 195, 199-200, 318
Lithic industry, 156-157, 159, 162
Lithoglyphidae, 204, 213
Lithoglyphulidae, 204, 206, 213
Lithoglyphulus, 204, 210
Lithoglyphus, 204, 207
  L. naticoides, 207
Lithogyphulidae, 210
Littoral, 5, 72, 244-245, 251, 256-257, 276, 281, 283
Lizard, 167, 169-170, 172-176, 180
Ljubljana Moor, 52-53, 56
Loach, 208-209

Lophotus lacepedei, 296
Louros, River, 157
Luciobarbus, 211, 214
Lugano, Lake, 221
Lutra lutra, 106
Luvarus imperialis, 299
Lymnaea stagnalis, 248
Lymnaeidae, 244

M

Macedonia, *See* Former Yugoslav Republic of Macedonia
Macedonia-Thessaly region, 203, 208-209, 214
Macedoniella, 196
Mäcin Mountains, 15
Macrouridae, 296
Macrovipera, 174, 181, 191-192
  M. schweizeri, 174, 192
Maghreb, 143
Malacostracean, 204, 210
Malaprespia, 212
Malawi, Lake, 17
Maliq, Lake, 35, 38, 46, 53, 55, 66
Malo Jezero, Lake, 38
Mammal, 3-5, 28, 59, 67, 77, 79-80, 82-85, 91-92, 107, 110, 121, 124, 127-133, 144-145, 184, 273
Maraenobiotus brucei, 326
Marginal population, 97
Marica, River, 169
Marifugia, 6, 319-320
  M. cavatica, 319
Marine core, 29, 52, 54-55, 63
Marine Isotope Stage, 57, 60, 63
Marine Oxygen Isotope Record, 157
Marissa, River, 14
Maritsa, River, 222
Maritza, River, 205, 208
Marmara, Sea of, 14, 193
Marmota marmota, 110
Marsh, 38, 53, 57, 244, 246
Martes
  M. foina, 106
  M. martes, 106
Martino's vole, 95-97
Massif Central, 115
Mastogloia braunii, 284
Mauer, 149, 164
Maurolicus muelleri, 298
Mediterranean, 2-4, 6, 8, 11, 13, 14, 16-21, 25-31, 35, 40, 42, 45-48, 59, 61-62, 65, 67, 69, 72, 76, 84, 86-87, 89-90, 103-104, 116, 127-128, 140, 144, 159, 167, 170-171, 177, 182-186, 195-197, 214-215, 217, 220-221, 225, 232, 234, 237, 240, 269-270, 275, 288, 291, 293-295, 298-306
Mediterranean climate, 17-18, 62
Mediterranean, Eastern, 5, 9, 11, 20-22, 45-46, 65, 102, 184, 186, 291, 294, 298, 300, 305-306
Mediterranean parrotfish, 299
Mediterranean, Western, 61, 182, 184
Megacyclops viridis, 326, 330-331
Megalakkos, 160
Megalópolis Basin, 263, 266
Melania
  M. curvicosta, 247
  M. rhodensis, 247
  M. tourneri, 250
Melanopsis
  M. costata, 247
  M. costatus, 250
  M. gorceixi, 255
Meles meles, 106
Melosira granulata, 277
Merck's rhino, 153
Meriones tristrami, 140
Merlangius merlangius euxinus, 298
Mesocricetus, 92-94, 106, 122, 126, 132, 137, 139
  M. auratus, 122
  M. newtoni, 93, 106, 122, 126, 132, 137, 139
Mesolithic, 75
Mesopelagic fish, 298-299
Mesozoic, 11-16, 177, 245
Messinian Salinity Crisis, 13, 16, 214-215
Mesta, River, 208
Metacentric chromosomes, 113, 115
Metamorphism, 15
Metohia, 210
Metohija, 89, 179, 263, 267-268, 271
Mezmaiskaya cave, 152-153
Michigan, 106
Miconos, island of, 193
Microcerberus stygius, 209
Microcondylaea
  M. bonellii, 220-221, 230, 232-233, 237-238
  M.compressa, 220, 237
Micromammal, 135, 139, 142, 144
Micromys
  M. danubialis, 122
  M. minutus, 106, 132
Micropyrgula stankovici, 212, 215

Micropyrgulidae, 204, 213
Microtus
M. agrestis, 106, 112, 133, 137-138
M. arvalis, 106, 112, 114, 122, 124, 127-128, 130, 132-133, 137
M. duodecimcostatus, 124
M. epiroticus, 124, 130
M. felteni, 93, 106, 122, 124, 126
M. gerbei, 124
M. gregalis, 137-138
M. guentheri, 92-93, 106, 112, 114, 119, 127-128, 133, 137, 139
M. lusitanicus, 124
M. majori, 117-118, 122, 126-127, 130
M. multiplex, 93, 106, 112, 114, 122, 124
M. nivalis, 103-104, 137, 145
M. obscurus, 124
M. oeconomus, 106, 137-138
M. rossiaemeridionalis, 106, 112-113, 122-124, 127, 133, 137
M. savii, 124
M. subarvalis, 124, 133
M. subterraneus, 106, 112-114, 117-118, 122, 127, 130, 137
M. tatricus, 93, 110, 122, 124, 132
M. thomasi, 90, 93, 106, 114, 120, 124, 126
Micrpyrgulidae, 212
Middle East, 26, 46, 48, 87, 101, 115-116, 126-127, 139, 195, 239
Migration, 16, 28, 49, 50-51, 58, 60, 62, 72, 139-140, 143-144, 175, 185, 268, 299-300
Mikri Prespa, Lake, 256, 260, 262, 274-275, 289
Miniopterus schreibersii, 106
Minotauria, 196, 199
Miocene, 11, 13, 16, 139, 171, 180, 196, 213, 215, 217, 254, 259-260, 269
Mirabad, Lake, 35, 40, 42
MIS. *See* Marine Isotope Stage
Modern human, 151-164
Moisture availability, 41-42, 59, 62, 64, 257
Moldova, 105
Mole, 93, 95, 126
Mole lemming, 138
Mole rat, 120, 129-131, 139
Molecular, 52, 58-60, 64, 97, 109-110, 155, 219, 220, 223, 233-234, 237, 270
Molecular data, 58, 97, 219, 237
Molecular marker, 51, 60
Mollusc, 69, 102, 219-221, 225, 230, 233-234, 237-243, 245, 247-249, 252, 258-260, 264, 269, 271, 300
Mollusca, 238, 239-240, 314, 319
Momonisia, 209
Monachus monachus, 110
Monfalcone, 193
Mongolia, 89, 106, 138, 285, 287
Monk seal, 110
Monolistra, 210, 317
M. spinosa, 319
M. spinosissima, 320
Monosbaena, 210
Monospecific assemblage, 92, 96, 207, 209-210, 212, 215
Monte Carlo test, 325
Montenegro, 1-2, 10-12, 14, 19, 21, 81, 110, 132, 180, 183, 193-194, 196, 211, 221, 253, 255, 271, 273, 310-312, 314, 316-318, 320
Monticchio, Lago di, 32, 34, 48
Moraca, River, 255
Moran's autocorrelation, 90-91, 93, 95
Moraria
M. poppei, 327, 328
M. stankovichi, 327
M. varica, 327
Morariopsis dumonti, 326-327
Moridae, 296
Morocco, 88, 100, 106
Morovitsa cave, 135-137, 144
Morphological characteristics, 151, 179, 191, 233, 277
Morphological variation, 147, 154, 156, 277
Mouse, 88, 105, 111, 113, 118, 122, 138-142
Mouse-tailed dormouse, 138-139
Mousterian, 103, 152-153, 156, 158-159, 162
mtDNA, 152, 154, 176, 180. *See also* DNA
Mujina pećina cave, 159
Multi-proxy data, 29, 35, 41, 45, 270, 275, 283, 286, 288-289
Multivariate statistics, 179, 325
Murid, 139, 142-143
Muridae, 102-103, 129-130, 132, 144
Mus, 81, 86, 105-106, 112-113, 122, 129, 132, 137, 142, 183
M. musculus, 86, 112, 122
M. domesticus, 86, 112-113, 132
M. macedonicus, 106
M. spicilegus, 106, 122

Muscardinus avellanarius, 106
Musculium, 222, 232, 235-236
M. hartmanni, 236
M. lacustre, 223, 249
M. nyanzae, 235-236
M. stuhlmanni, 235-236
M. titicacense, 232, 235-236
Mussel, 206, 210, 213, 254
Mustela
M. erminea, 106, 110, 112
M. eversmanii, 106
M. lutreola, 106
M. nivalis, 106, 112, 114, 121, 127-128, 132
M. vison, 110
M. putorius, 106
Mustelidae, 81, 104, 106, 129, 132
Mustelid, 81, 83, 85, 87, 90-91, 101, 105, 107
Mutation, 118
Mycenaean, 170
Mycteroperca rubra, 296, 304
Myocastor coypus, 110
Myomimus, 92-94, 106, 137-139
M. roachi, 106, 138
Myotis
M. alcathoe, 121-122, 129
M. bechsteini, 105
M. blythi, 105
M. brandtii, 105
M. capaccinii, 105
M. dasycneme, 105, 110
M. daubentonii, 105
M. emarginatus, 105
M. myotis, 105
M. mystacinus, 105, 114, 121-122
M. nattereri, 105
Myriapoda, 314
Mysidacea, 204, 315-316

N

Najades, 220, 221, 237, 240
Nannospalax
N. leucodon, 92-94, 106, 114, 120-122, 127-128, 130, 137, 139
N. nehringi, 92
N. nehringi, 137
Narrow-skulled vole, 138
Natrix tessellata, 176, 184, 191
Navicula
N. pygmaea, 281
N. rara, 277, 279-280
Neanderthal, 151-164
Near East, 32, 46-48, 93-94, 141, 156, 158, 215
Near Eastern, 78, 157
Nei's genetic distance, 97
Nemertini, 314
Neogene, 5, 11, 183, 186, 225, 259, 263, 269
Neolithic period, 28, 38, 42, 47, 75, 161
Neomys
N. fodiens, 105
N. anomalus, 105
Neopisidium, 222, 232
Neotethys. *See* Tethys
Neotrechus, 317-318
Nepal, 224-225
Neretva, River, 81, 210
Nesticus, 196
Nestos, 208
Neumayria janinensis, 247
New Mexico, 88-89, 102, 106
New Zealand, 309
Niche, 261
Niphargus, 317-318, 332
Nistru, River, *See* Dneister
Nitocra, 326-327
Nitzschia
N. capitellata, 279
N. fonticola, 279, 280
N. balcanica, 179, 184, 186, 188, 277, 279-280, 332
N. granulata, 284
N. macedonica, 277, 279
Non-arboreal pollen. *See* Pollen, non-arboreal
NORs distribution, 114
North America, 50, 89, 106, 164, 204, 213, 217, 220, 225, 285, 288, 313, 323
North American, 74, 102, 105, 204, 213, 239, 275
North Atlantic, 25, 51, 61, 63-65, 67, 294, 305
North Atlantic Oscillation Index, 294
Norway, 106, 305
Nyctalus
N. lasipterus, 105
N. leisleri, 105, 111-112
N. noctula, 105
Nyctereutes procyonoides, 110
Nyssa, 70

O

Oak, 25, 28

Obligate subterranean species, 5, 309, 312-320
Ochridaspongia, 212, 216
Odocoileus virginianus, 110
Ohio, 106
Ohrid, Lake, 4-5, 17, 203-204, 209-216, 219-223, 226-227, 230-238, 243, 250, 252-263, 267, 269, 271-276, 280-282, 288
Ohridopyrguinae, 212
Oldowan industry, 156-157
Olea, 27, 35, 39-40, 59, 76
O. europea, 27
Oligocene, 11, 196, 213, 215
Oligochaeta, 315-316, 321
Oligochaete, 320, 253, 314
Olm, 179
Olt, River, 157
Oltet, River, 157
Olympus, Mount, 28
Ondatra zibethica, 110
Oniscida, 315
Oniscidea, 318, 321
Opiliones, 200, 315-316, 321
Orange-spotted grouper, 300, 305
Orce, 157
Orientalina cursa kicavica, 212
Orientalinidae, 207, 212-213
Orientalininae, 212
Orientallinidae, 204, 210, 260, 272
Orogeny, 11-13, 21, 62, 252, 259, 269. *See also* Alpine orogeny
Orography, 169
Orthrias
O. barbatulus, 209
O. brandti, 208-209, 217
O. pindus, 211
Oryctolagus cuniculus, 110
Osapska jama, 324
Osteichthyes, 217, 295, 303-304
Ostracod, 46, 102, 253, 259, 261-263, 265, 271-272
Ostracoda, 259, 261, 270-272, 315-316, 321
Ostrya, 25, 38, 39, 52, 55, 57, 74, 76
O. carpinifolia, 25, 74, 76
Oviparity, 180, 182-183, 245
Ovis musimon, 110

P

Pachychilon
P. macedonicum, 209
P. pictum, 212
Paedogenesis, 179-180, 182
Paedomorphosis, 179, 182, 183
Paladilhiopsis (*Paladilhiopsis*) *janinensis*, 248
Palaearctic, 4, 87-88, 93-94, 100-102, 104, 122, 138, 140, 175, 179, 195, 213, 215, 220, 225, 232, 239, 261
Palaeobiogeographical, 139, 269
Palaeolimnology, 29, 283-286
Palaeolithic, 6, 67-68, 75, 147, 153, 156-165
Palaeosol, 54, 57
Palaeotethys. *See* Tethys Sea
Palaeotropic, 195
Palaeozoic, 12-13, 15-16
Palaiokastro, 157
Palpigradi, 315-316
Pamvotis, Lake, 46, 243, 245-267, 271
Pangaea, 12-13
Pannon, Lake, 17, 252, 254-255, 257, 259-260
Pannonia, 69
Pannonian origin, 233
Pannonian Plain, 14, 17, 19-20, 169
Parabythinella, 212
Paracyclops fimbriatus, 327, 330
Paralimnocythere, 263
Parallel evolution, 234, 237
Parapatric, 92
Parastalita, 196
Parastenocaris, 326-328, 332
P. andreji, 326-327
P. nolli alpina, 327-328
Paratethyan fauna, 215
Paratethys. *See* Tethys Sea
Pardosa incerta, 197
Parexocetus mento, 296, 300
Pelister race, 115
Pelister, Mount, 111, 115, 118
Pelobates, 172, 176, 183, 186, 188
P. syriacus, 172, 186, 188
Peloponnese. *See* Peloponnesos
Peloponnesos, 12, 15-16, 20, 81, 83, 85-87, 93, 101, 113, 142, 158, 169-171, 174, 187, 211, 213, 310, 316
Peneios, River, 157
Peninsula effect, 85, 101
Pennsylvania, 89, 103, 106
Peonic, Lake, 256
Peracarid, 209, 213
Pericentric inversion, 112, 114
Peripannonian region, 169-170, 174, 181, 187
Periphyton, 276, 289

Peritricha, 320
Permafrost, 176
Permian, 15
Petralona, 148-151, 156, 159-164
Petralona skull, 148, 151
Phagocata, 212
Pheggomisetes, 318
Phenotypic variation, 109, 172
Pholeuon, 318
Phoxinellus, 210-211, 217
Phoxinus phoxinus, 215
Phyllognathopus vigueri, 327
Phylogenesis, 5, 109, 118, 121, 126-128, 142, 148, 151, 158, 169, 172, 175, 181, 185, 219, 233, 237, 261, 272, 317
Physella acuta, 248
Physidae, 244
Physiology, 244
Phytoplankton, 246, 273-274, 276, 281, 283, 289, 295
Picea, 26, 32, 38-39, 52, 54-57, 59-60, 70, 72, 74
Pinarbası, Lake, 276, 282
Pindhos Mountains. *See* Pindus Mountains
Pindus Mountains, 3, 11, 14-15, 19, 62, 64, 81, 83-84, 93, 97, 99, 101, 194, 196-197, 245
Pine, 28, 41, 74, 130-131
Pine vole, 113, 117-118, 120, 124, 127, 131
Pinnularia, 277, 281
P. nobilis var. *macedonica*, 277
Pinus, 25, 27, 32, 38, 41, 52, 54-55, 57, 59-60, 62-63, 70, 72, 74, 76
P. cembra, 72
P. halepensis, 25, 27, 76
P. mugo, 74, 76
P. sylvestris, 70, 72
Pipistrellus
P. kuhlii, 105
P. nathusii, 105
P. pipistrellus, 105
P. savii, 105
Piran Bay, 299, 305
Pirin Mountains, 48, 65, 197, 200, 289
Pisidium, 5, 222, 224-225, 230-240
P. amnicum, 223, 232-233, 237
P. annandalei, 224-225, 240
P. bulgaricum, 222
P. casertanum, 222-223, 232-233, 236, 239, 249
P. conventus, 236
P. crassum, 231, 233, 236
P. dilatatum, 236
P. edlaueri, 223, 231-233, 236-237
P. fistulosum, 235-236
P. henslowanum, 223, 230, 232-234, 237
P. kawamurai, 236
P. lauricochae, 235, 236
P. maasseni, 223, 233, 236-237
P. milium, 223, 233, 249
P. moitessierianum, 223, 230, 232, 234, 237
P. nitidum, 223, 231-234, 236-237, 247, 249
P. obtusale, 223, 249
P. parenzani, 222
P. personatum, 222-223, 232-233
P. pirothi, 235-236
P. ponderosum, 236
P. pseudosphaerium, 223, 230
P. raddei, 232, 236
P. sogdianum, 223-225, 232, 234, 237
P. subtruncatum, 223, 232-233, 236-237
P. supinum, 223, 230
P. tenuilineatum, 223, 232, 233
P. vincentianum, 225
Pistacia, 25, 27, 35, 38-40, 52, 54-55, 59, 74
P. lentiscus, 27
P. terebinthus, 25, 76
Pitymys, 125, 129-133
Pivka jama, 324-328
Planinska jama, 324
Planorbarius corneus, 248
Planorbidae, 236, 240, 244, 259
Planorbis, 212, 247
P. carinatus, 248
P. macedonicus, 212
P. planorbis, 247-248
Plant diversity, 3
Plant macrofossil, 29, 45, 51, 56-57, 60
Plantago, 27, 39-40, 75
P. lanceolata, 75
Platanus, 39-40, 76
Plate tectonics, *See* Continental plate tectonics
Platichthys flesus, 298
Plecotus
P. auritus, 106
P. austriacus, 106
Plectorhinchus mediterraneus, 296, 299
Pleistocene, 7, 32, 35, 40-42, 47, 50, 66-68, 70, 77, 93, 96-97, 102, 104-105, 135, 138-149, 154, 156-164, 171, 176-

177, 180, 182-183, 215, 233, 245, 249-256, 260, 264-266, 269, 271-272, 281-284, 289
Pleniglacial, 32, 35, 39, 48, 52, 57, 78, 282-283
Plesiomorphic, 115, 118, 126
Pleven, 20
Pliocene, 16, 28, 70, 72, 77-78, 96-97, 102, 139, 143, 171, 197, 199, 210, 215, 233, 249, 252, 254, 258, 266-267, 269, 271
Pliomyi, 96
Plio-Pleistocene, 246-247, 249-256, 264-265
Plitvice, Lakes, 17, 274, 288
Po, River, 59, 221, 291, 293
Poaceae/Gramineae, *See* Gramineae/Poaceae
Podarcis, 171, 190
P. erhardii, 171
P. melisellensis, 171
P. sicula, 171
Podpeška jama, 323
Polje, 6, 157, 245, 255-256, 320
Pollen data, 35, 40-42, 45-46, 62, 71, 75
Pollen record, 6, 40-41, 46, 52, 55, 57, 67, 72
Pollen, arboreal, 41, 52, 54, 61-63
Pollen, non-arboreal, 42
Pollution, 111, 237, 246, 259, 281, 295
Polychaeta, 315, 316, 319
Polymorphic karyotype, 4, 110, 112-113
Polymorphism, 110-113, 127, 129-130, 132, 176
Polytypic karyotype, 4, 110, 113, 120, 127, 208
Polytypic species, 113, 127, 208
Pomatoschistus
P. bathi, 298
P. canestrinii, 295
P. norvegicus, 296
Pond terrapin, 176
Pontian, 268
Pontic, 175, 195
Pontides Mountains, 268
Ponto-Aegean, 261
Pontobelgrandiella, 207
Pontodreissena, 226, 233
Pontoniphargus racovitzai, 208, 217
Popovo polje, 6, 320
Population crash, 50, 61-62
Population genetics, 162, 178-179
Populus, 57
Porifera, 314, 316, 319
Portugal, 61, 63, 106
Posavina, 264
Poso, Lake, 219, 238
Possostrea, 219
Postglacial, 4, 38, 45, 49, 58-60, 65-68, 90, 127-128, 176, 181. *See also* Holocene
Postojna, 16, 321, 324, 332
Postojna-Planina caves, 332
Postojnska jama, 16, 323-328, 332
Potomida littoralis, 220-221, 232, 234
Praomys pomeli, 143
Precambrian, 15
Precipitation, 18-20, 26-27, 35, 40, 45, 61-62, 74, 276, 325, 327
Preglacial, 174, 177, 182, 205
Prehistory, 78, 147, 149, 159, 161
Pre-Holocene, 25, 57
Preluca Tiganului, Lake, 32, 53, 56
Prespa, Lake, 4-5, 17, 203-204, 211-212, 216-217, 219-221, 223, 226, 230, 232-237, 240, 252-253, 256-257, 262, 270, 273, 289
Prespiana, 212
Prespolitorea, 212
Prespopyrgula, 212
Prevlaka, 266
Proasellus, 212
Prokletije Mountains, 169, 181
Prosobranchiate, 204, 212
Prosobranch, 244- 245, 250-251, 253, 255-257
Proteidae, 186, 319
Protelsonia, 204
Proteus, 174, 179, 186-187, 309
P. anguinus, 187, 309, 319
P. anquinus, 179
Protoleptoneta, 196
Provincialism, 98
Pselaphidae, 315, 318
Pseudanodonta complanata, 221
Pseudohoratiine, 212
Pseudophoxinus, 211-214
P. beoticus, 213
P. marathonicus, 213
P. minutus, 212
Pseudoscorpiones, 313, 315-318, 320
Pterocarya, 70
Pula maar, 70, 72
Pulmonate, 212, 244-245, 247, 250-251, 253, 255-257
Punikve, 158
Pupicina cave, 56
Pygeretmus pumilio, 137

Pyrenean Peninsula, 88
Pyrenean-Aquitanian Province, 313
Pyrenees, 4, 17, 50, 180, 204, 213, 215, 309
Pyrgula, 210-211, 247, 250
P. acicula, 258
P. annulata, 210-211
P. brusinai, 247, 250
P. eugeniae, 247
P. falkneri, 258
P. pambotis, 258
Pyrgulidae, 204, 210, 213

Q

Quantitative data, 79, 167, 269, 274, 292, 301
Quaternary, 4, 6, 12, 14, 16, 45-51, 61, 64-69, 77-78, 144-145, 157, 162, 165, 177, 180, 183, 186, 197, 205, 251, 259-261, 265, 271-272, 287-288
Quercetum-mixtum, 74
Quercus, 25, 27-28, 32, 35, 38-41, 46, 52, 54-60, 62, 65, 70, 72, 74-76
Q. coccifera, 27, 76
Q. ilex, 27, 74-76
Q. pubescens, 74

R

Radiation, adaptive. *See* Adaptive radiation
Radix
R. auricularia, 248
R. peregra, 248
Rafting, 97
RAMSAR, 28-29
Rana, 169, 171, 175-176, 179, 184-186, 188
R. catesbeiana, 169, 188
R. epeirotica, 179
R. esculenta, 179, 188
R. ridibunda, 171, 179, 184, 186, 188
R. shqiperica, 179
Ranae fuscae, 172, 175
Range fragmentation, 176
Rapoport rescue hypothesis, 175
Rat, 86, 92, 105, 128, 132, 140-141
Rattus, 81, 86, 105, 110, 112-114, 128, 137, 139, 143
R. casimcensis, 139
R. dobrogicus, 137, 139
R. norvegicus, 110, 112
R. rattus, 110, 112-114, 137, 140
Rb-subalpine cytotype, 120
Recolonisation, *See* Colonisation
Recombination, 179
Red Sea, 295, 303-304
Red-toothed shrew, 94
Refuge. *See* Glacial refuge
Regalecus glasne, 296
Relic, 142
Reproductive bimodality, 180
Reproductive capacity, 60
Reproductive mode, 168, 180, 182, 184
Reproductive relationship, 179
Reproductive strategy, 244
Reptile, 3-4, 77, 167-168, 169, 170-187
Reptilia, 171
Rezina marsh, 38, 48
Rheophilic fish, 208-209, 216
Rhinolophus
R. blasii, 105
R. euryale, 105
R. ferrumequinum,, 105
R. hipposideros, 105, 114-116, 127-128
R. mehelyi, 105
Rhodera, 196
Rhodes, island of, 15, 142-143, 269, 314
Rhododendron, 28
Rhodope province, 158
Rhodope-Moesia block, 13
Rhodopes Mountains, 3, 11, 14-15, 19, 65, 83, 101, 310
Rhodos. *See* Rhodes, island of
Rhynchodemus henrici, 319
Ribnica, 74
Rijeka, 168
Rila-Rhodopes Mountains, 3, 14, 55, 65, 83-84, 101, 197
River, 1, 4, 9-10, 14, 20, 27, 59, 72, 81, 97, 168-169, 175, 193, 203-217, 220-221, 225, 227, 233, 236, 240, 253, 255, 261, 289, 291, 293, 311
Robertsonian change, 112-114
Robertsonian fusion, 113-114
Rock mouse, 140-142
Rocky mouse, 118
Rodent, 82, 135
Rodentia, 81, 102-106, 128-145
Rodent, 79-93, 101-109, 128-129, 135, 141-142
Rodopi, *See* Rhodopes Mountains
Roman period, 45, 47, 70, 170
Romania, 1-2, 9-10, 14-15, 19-20, 31-32, 38-39, 46, 48, 53, 56, 68, 105, 110,

135, 139-140, 144-145, 157, 159, 169, 186, 194, 206, 208, 225, 313, 318, 323
Ronal pressure gradient, 294
Root vole, 138
Ruscinian, 143
Russia, 87, 103, 105-106, 116, 124. *See also* former USSR
Rutilus, 210, 212-213
R. basak, 212
R. ohridanus, 212, 213
R. prespensis, 213
R. ylikiensis, 213
Ruvettus pretiosus, 299, 303

S

Sabanejewia, 210
Sadleriana, 207
S. fluminenesis, 207
S. sadleriana, 207
Sahara, 151
Saint Petersburg, 116, 225
Salamandra, 171, 175, 180, 182, 184, 187
S. salamandra, 180
Saline lake, 274, 283, 287
Salinity, 244, 246, 274, 283-284, 293-294, 298-299, 303-304
Salix, 32, 57, 59-60, 70, 72, 74
Salmonidae, 204
Salmothymus, 210, 212
S. obtusurostris, 210
Sand viper, 179
Sandalja cave, 55, 65, 157
Sandbar shark, 298
Sarajana, 207
S. apfelbecki, 207
Sarajevo, 16, 20, 102, 163, 217
Šara-Pindhos Mountains. *See* Pindus Mountains
Sardina pilchardus, 292, 298
Sardine, 292
Sardinella aurita, 292, 299, 304-305
Saudi Arabia, 139
Saurida undosquamis, 296, 300
Sava, River, 10, 20, 97, 168, 193, 204-205, 311
Scandinavia, 121, 172, 179, 214
Scardinius graecus, 213
Schedophilus medusophagus, 299, 303
Sciadopitys, 70
Sciurus
S. anomalus, 122
S. vulgaris, 106, 122
Scomber japonicus, 292
Scutari, Lake. *See* Skutari, Lake
Scutari-Peć Line, 14
Sea of Marmara. *See* Marmara, Sea of
Sea surface temperature, 61, 63
Sea-level, 52, 54, 59, 66, 245
Seasonality, 42, 48, 75, 285
Sedmo Rilsko, 53, 65
Segmentina nitida, 248
Semi-desert, 139
Semisalsa steindachneri, 248
Sequoia, 70
Serbia, 1-2, 10, 15, 110, 133, 170, 180, 183, 188, 191, 193-194, 206, 263, 310, 312-316, 320, 323, 330-332
Serbo-Macedonian Massif, 11, 15
Serpentes, 171
Serpulidae, 319
Serranidae, 296, 304-305
Sexual strategy, 244
Shark, 298, 301
Shkoder, 15, 20
Shkodra, Lake. *See* Skutari, Lake
Siberia, 115, 121, 139, 205, 213-214, 287
Siberian anticyclone, 35, 294
Sibling species, 96, 110, 121-124, 127, 133, 171, 200
Sibling vole, 123-124
Sicista
S. betulina, 106, 110
S. severtzovi, 122
S. subtilis, 106, 114, 122, 137-138
Silurian, 15-16
Silurus aristotelis, 211
Sima de los Huesos, 152
Sinai, 104, 106
Sinanodonta woodiana, 220
Sinaphaenops, 318
Sirina, island of, 193
Sister species. *See* Sibling species
Skadar, Lake. *See* Skutari, Lake
Skadarsko jezero. *See* Skutari, Lake
Skiros, island of, 193
Škocjanske jame, 16, 323-328, 331
Skutari, Lake, 11, 16-17, 169, 203, 209-212, 221, 226, 230, 252, 255, 257, 273, 280
Slavonia, 264
Slovakia, 60, 67
Slovenia, 2, 5, 7, 10, 12, 14, 16-17, 19-21, 32, 38-39, 45, 52-53, 56, 65-66, 69-74, 77, 96, 103, 105, 110, 135, 159, 180, 183, 194, 206, 222, 233, 238, 253, 258, 274, 287, 305, 309-332
Slovenija. *See* Slovenia

Snail, 4, 59, 203, 206-216, 259, 260, 320
Snake, 170, 174, 178-179, 184-185, 192
Snake-eyed skink, 178-179, 185
Snow vole, 95, 97
Soča, River, 20, 168, 193, 210
Social organisation, 159
Söğüt, Lake, 41, 43
Soil, 1, 14, 16, 25-26, 38-39, 42, 68-69, 74-75, 250
Sørenson index of similarity, 325, 329
Sorex
S. alpinus, 93, 105, 113, 129-130
S. araneus, 92, 105, 112-113, 115, 121, 127-133
S. minutus, 105, 121, 127
S. satunini, 121, 127
S. volnuchini, 121, 127
Souslik. *See* European souslik
South Adriatic-Ionian, 9, 14-15, 19, 50, 59, 62, 157, 163, 193, 203, 210-212, 214, 259, 261-262, 266-267, 291, 294, 304, 316
South Africa. *See* Africa
Southern birch mouse, 138
Southwest Asia. *See* Asia
South-west Asian jerboa, 138-139
Spain, 3, 11, 66, 106, 140, 152-153, 157, 180, 204, 274-276, 280, 282, 284, 288
Spalax graecus, 93, 106, 122
Spanish mackerel, 292
Sparidae, 296
Sparisoma cretense, 298-299, 304
Spatial autocorrelation, 93
Spatial pattern, 81, 84, 174, 273, 275, 285
Specialisation, 244
Speciation, 5, 28, 64, 66, 77, 94, 139, 141, 147, 151, 154, 156, 177, 181-182, 184, 199, 219, 235, 243-244, 260, 269, 317-318
Species area effect, 87, 92
Species complex, 125, 273, 277
Species density, 89, 91, 101
Species diversity, 170, 181, 240, 273-276, 281, 285-288
Species flock, 235, 240, 263
Species introduction. *See* Introduced species
Species pool, 81-82, 86, 93
Species richness, 3, 5-6, 28, 79, 81-83, 86-94, 98, 101, 106-107, 122, 169-170, 172, 174, 193, 197, 199, 273, 277, 280-281, 285, 313
Species-area curve, 331
Spelaeocaris, 204, 210
Speleogobius trigloides, 295, 298
Speocyclops infernus, 327-328
Spermophilus, 92-93, 106, 111, 114, 122, 137, 139
S. citellus, 93, 106, 111, 114, 122, 128, 131, 133, 137, 139
S. xanthoprymnus, 122
Sphaeriid, 222, 230, 232-236
Sphaeriidae, 219-221, 223, 230, 235-236, 238-240
Sphaerium, 222-223, 236, 238-239
S. baicalense, 236
S. corneum, 222-223, 236, 249
S. nucleus, 109, 222-223, 238
S. rivicola, 222-223
Sphaeromatidae, 204, 319
Sphyraena chrysotaenia, 296, 300, 305
Sphyraenidae, 296
Spider, 4, 193-200, 318, 320
Split, 70, 293, 296, 298
Sponge, 5, 210, 212, 319-320
Spongilla stankovici, 212
Spongillidae, 319
Sprattus sprattus phalericus, 298-299
Squalius, 210
Srbija. *See* Serbia
Sri Lanka, 88-89, 106
Stability, ecological, 64
Stability, environmental, 4-6
Stable isotopes, 29, 35-36, 42, 45-46, 164, 288
Stagnicola palustris, 248
Stalagtia, 196, 199
Stalita, 196
Stara Planina Mountains. *See* Balkan Mountains
Starolesnianske Pleso, Lake, 274
Steinheim, 149, 250, 259-260
Stenotopic organism, 245, 251, 257, 261
Stephanodiscus
S. medius, 279
S. minutulus, 281
S. parvus, 281
Steppe, 26-27, 32-33, 35, 40-42, 49, 72, 138-140, 175
Steppe lemming, 138-139
Steregoiu, 32, 46
Sternoptychidae, 296
Sterrea Ellada, 316
Stone Age, 161-165, 259
Stratigraphy, 15, 31, 47, 52, 54, 57, 102, 149, 152, 154, 158, 163, 249, 252, 264, 288
Struma, River, 208, 209, 214

Stygepactophanes, 327, 328
Stygobiont, 5, 312-315, 320, 323, 326
Stygobiotic fauna, 309, 311, 313
Stygobitic, 323, 330-331
Subalpine cytotype, 120
Subsidence, 13, 15, 17, 21, 31, 245
Subspecies, 93, 96, 126, 131, 169, 171-180, 185, 187-188, 206, 208-209, 212-213, 217, 221-222, 225-226, 233-234, 237, 295-296, 313, 317, 320
Subterranean environment, 5, 89, 103, 196, 217, 234, 240, 307-313, 320, 323-326
Subtropical, 17, 70, 72, 77, 299
Sucho Ezero, 53
Suctoria, 320
Sulawesi, 219, 235, 239
Sulcia, 196
Süleymanhacı Gölü, 276, 282
Suncus etruscus, 105
Supernumerary chromosome, 110, 111, 132
Superspecies, 121, 171, 175-179, 183, 186
Sus scrofa, 112, 114, 130, 133
Swanscombe, 149
Swaziland, 106
Sweden, 60, 66, 106, 145
Switzerland, 113, 195, 323
Sympatric speciation, 235
Synanthropic rodent, 105
Synedra
S. rumpens, 277
S. vaucheriae, 277
Syngnathus taenionotus, 295
Synklepton, 179
Synodontidae, 296
Syntopic population, 180
Syria, 211-212, 221, 225
Systematics, 129, 144, 217, 240

T

Tadarida teniotis, 106
Tailed amphibian, 169
Tailless amphibian, 169, 173
Talpa
T. caeca, 90, 92-93, 105, 114, 121
T. europaea, 105, 121-122
T. levantis, 93, 105, 121
T. stankovici, 93, 105, 114, 122, 126
Tandem fusion, 120
Tanganyika, Lake, 17, 219, 235-237, 240, 259-260, 285, 287
Taphonomy, 153
Taul Zanogutii, 53
Taxodium, 70
Taxon-area cladogram, 175, 181
Taxonomy, 3, 5-6, 80-82, 92, 94, 97, 101-105, 107, 110, 119, 121-122, 127, 129-133, 135, 169, 171-172, 174, 179, 181-187, 200, 220-222, 234, 239, 260-261, 270, 273-275, 277, 279-280, 283-288, 296, 304, 309, 311-313, 317-318, 321, 326
Tectonic basin, 17, 254, 268
Tectonism, 12-17, 21, 31, 177, 196, 245, 252, 254-255, 257, 268-269.
Tegenaria, 195, 200
Telestes, 205, 210, 217
T. polylepis, 205
Temnata cave, 155-156
Temnocephalida, 314, 316
Temperate zone, 5, 17, 21, 38-39, 51, 61-63, 70, 73, 84, 167, 244, 293-294, 298-300, 325, 327
Temperature, 18-20, 35, 39, 51, 61-63, 70, 73, 84, 167, 244, 293-294, 298-300, 325, 327
Tenaghi Philippon, 52-53, 62-63, 68
Terianske Pleso, Lake, 274
Terranigra, 206, 209, 216
Terricola, 113, 124-125, 129, 319. *See also Microtus*
Tertiary, 11, 13-14, 16, 50, 140, 142, 144, 171, 177, 197, 233, 236, 254-255, 268
Testudines, 171, 173
Tethyan origin, 203-204, 209
Tethys Sea, 11-13, 17, 21, 184-185, 215, 269
Thailand, 89, 103, 106, 224
Thalassoma pavo, 298
Theodoxus
T. doricus, 255
T. licherdopoli, 247
T. micans, 247
T. quadrifasciatus, 247
T. saulcyi, 213
T. subrelictus, 210
T. varius varius, 258
Theopetra, 154, 159, 163-164
Thermocline, 255, 293
Thermosbaenacea, 315-316
Thessalia, 316
Thessaloniki, 19
Thessaly, 65, 157, 164, 203, 207-216
Thiarid gastropod, 241, 244, 250-253, 255-257

Thiaridae, 240, 244
Thomas' pine vole, 120
Thrace, 45, 81, 105, 133, 162, 182, 188, 194, 203, 205, 208
Thracian Plain, 14
Thraki, 11, 16, 81, 93, 100, 158, 196, 203, 205, 207-214, 216, 234, 314, 316
Thunnidae, 292
Thysanura, 315
Tiberias, Lake, 212
Tilia, 25, 32, 38-39, 56, 74
Timişoara, 19
Titanethes, 318
Titicaca, Lake, 219, 232, 235-236
Topography, 4-5, 9-10, 20-21, 64, 86, 261, 291, 294, 306
Trachemys scripta, 169, 189
Trachinotus ovatus, 299, 303
Trachipterus trachypterus, 299, 303
Trachurus trachurus, 292
Trakya. *See* Thraki
Trans-Aegean dispersal, 140
Trans-Caucasia, 174
Transfer function, 274, 285, 288-289
Translocation, 121
Transmediterranean, 175
Transylvanian Alps. *See* Carpathian Mountains
Trechinae, 317-318, 320
Tree population, 4, 48-64, 68, 272
Triassic, 11-13, 15-16
Trichonethes kosswigi, 314
Trichonia, 212
  T. trichonica, 248
Trichoniscidae, 314, 318, 321
Trichoniscus rhodopiensis, 314
Tricladida, 314-316, 319
Tripotamia, 158
Triturus, 171-172, 175-187
  T. alpestris, 171, 175, 182-183, 186-187
  T. carnifex, 172, 176, 180, 187
  T. cristatus, 171-172, 175-177, 179, 182-187
  T. karelinii, 176
  T. vulgaris, 180
Troglobiont, 174, 225, 312
Troglobiotic, 309, 313-314, 318-319
Troglocaris, 210
Troglodyte, 234, 240
Troglohyphantes, 195-196, 199, 318
Troglomorphism, 312, 318
Troglomysis, 204, 210
Troglophile, 323
Troglorhynchus, 318
Trogloxene, 169, 323
Tropical zone, 3, 8, 17, 25, 28, 64, 69, 71, 88-89, 225, 239, 299-300
Tropidophoxinellus, 211
Tseravinas, Lake, 53, 55
Tsuga, 70
Tubeworm, 5-6, 320
Turbelarians, 210, 212
Turbellaria, 314, 319
Turkey, 3, 6, 9-17, 20, 28, 31-32, 35, 37, 41, 43-48, 97, 103, 105-106, 110, 113, 115, 118, 121, 129, 131, 133, 144-145, 159, 162, 193-194, 225, 233, 269, 274-276, 280, 282, 284, 288, 311, 314
Turnover, 89, 144
Turtle, 169, 185
Tuscany, 140
TWINSPAN, 325, 328
Tylosurus acus imperialis, 296, 303
Typhlogammaridae, 204, 210
Typhlogammarus, 210
  T. mrazeki, 320
Typhlotrechus, 317
Tyrrhenocythere
  T. amnicola, 265, 272
  T. donetziensis, 265

U

Ubiquitous distribution, 90, 261, 326, 330-331
Ukraine, 2, 87, 96, 132, 138-140, 222, 226, 238
Ulmus, 25, 32, 38-39, 52, 54-57, 60, 62, 70, 72, 74
Umbelliferae/Apiaceae, 40
Uniarmed chromosome, 113
Unio
  U. crassus, 221, 230, 232, 237
  U. crassus bruguierianus, 221
  U. crassus carneus, 221
  U. elongatulus, 221, 237
  U. mancus, 221, 232, 237
  U. mancus kruperi, 221
  U. pictorum, 220-221
  U. pictorum gaudioni, 221
  U. tumidus, 220
Unionacean mussel, 206
Unionidae, 219-220, 230, 232, 235, 238, 240
Urals, the, 60
Urmia, Lake, 40, 46
USA. *See* North America

USSR. *See* Former USSR

V

Valenciennelus tripunctulatus, 296
Valvata, 212, 247
- *V. cristata*, 247-248
- *V. piscinalis*, 247-248

Valvatid snail, 213
Valvatidae, 244-245
Van, Lake, 40, 42, 48
Vanneaugobius dollfusi, 296, 305
Vardar tectonic zone, 196
Vardar, River, 204-209, 212, 216-217, 234, 269, 309
Varkiza, 135, 136
Vegetation, 3-4, 6, 17, 19, 25-32, 35, 38, 40-41, 45-48, 61-70, 75-78, 89, 102, 161, 275, 288
Vegetational change, 38, 84
Vegoritis, Lake, 240, 262, 272
Vela Draga, 53
Velebit Mountains, 89, 169, 185, 191
Velika Pasjica cave, 324-326, 328
Velkovrhia enigmatica, 319
Vesanska cave, 56
Vespertilio murinus, 105
Vicariance, 175, 211, 214, 269
Victoria, Lake, 235-236
Villanyian, 104, 140
Vimba melanops, 208-209
Vindija, 151-158, 163-164
Vipava, River, 20, 168
Vipera ammodytes, 179, 186, 191
Vistula, 208, 212
Vitosha Mountains, 65, 128, 222
Viviparidae, 244
Viviparity, 180-182, 186
Viviparous lizard, 176, 180, 183
Viviparus
- *V. brevis*, 255
- *V. brusinai*, 247
- *V. clathrata*, 247, 250
- *V. hellenicus*, 258
- *V. mamillatus janinensis*, 248
- *V. megarensis*, 247

Vjetrenica cave, 320
Voidomatis, River, 160
Vojvodina, 264, 272
Volant mammal, 84, 91-92, 107
Volcanism, 14-15, 17, 31-32, 46, 66, 255
Vole, 92, 94-97, 113, 117-120, 123-127, 138-140
Vormela peregusna, 106
Vrana, Lake, 53, 56
Vulpes vulpes, 112

W

Wales, 195
Walnut, 74
Water depth, 53, 236, 283-284. *See also* Lake level
Water frog, 175-176, 179, 184-185
Water vole, 130-131, 140
Weasel, 121, 130, 132
Western Mediterranean. *See* Mediterranean, western
Whiskered bat, 121, 129
Widely-distributed species, 89, 174, 209, 212, 219, 233, 235-237
Widespread phenomenon, 38, 74-75, 79, 89-90, 94, 97, 121, 156, 195, 197, 269, 280, 286
Wind, 17, 20
Wisconsin, 106
Wood mouse, 111, 122, 130-131, 139
Woodlouse, 314, 318

X

Xeric, 27
Xinias, Lake, 35, 52-54, 66, 264, 271
Xyrichthys novacula, 298-299, 303

Y

Y chromosome. *See* Chromosome
Yarimburgaz cave, 157, 162
Yellow-necked mouse, 111, 129, 132
Younger Dryas, 35, 39-40, 46, 56-58, 61
Yugoslavia, 1-2, 9-10, 21, 31, 102-105, 129, 131-133, 163-164, 183-186, 193, 199, 240, 288, 310-311

Z

Zafarraya cave, 153
Zaghreb, 20
Zagros Mountains, 28, 33, 35, 40
Zajsan, Lake, 138
Zeribar, Lake, 33, 36, 40, 42, 48
Zielowny Staw, Lake, 274
Zingel
- *Z. streber balcanicus*, 209
- *Z. streber nerensis*, 206, 208

Zoogeography, 79, 96, 104, 135, 140-142, 167, 169, 175, 177, 181-183, 193-197, 199, 205-206, 213, 217, 267
Zootoca, 172, 176, 180, 183-185, 190
Z. vivipara, 172, 176, 180, 183, 185, 190
Zrmanja, River, 204, 210
Županova jama, 325-328
Žužemberk, 324